THE BIOPHYSICAL FOUNDATIONS OF HUMAN MOVEMENT

SECOND EDITION

Bruce Abernethy
*The University of Hong Kong
and The University of Queensland*

Stephanie J. Hanrahan
The University of Queensland

Vaughan Kippers
The University of Queensland

Laurel T. Mackinnon
The University of Queensland

Marcus G. Pandy
*The University of Melbourne
and The University of Texas at Austin*

HUMAN KINETICS

Library of Congress Cataloging-in-Publication Data

The biophysical foundations of human movement / Bruce Abernethy . . . [et al.].--2nd ed.
    p. ; cm.
  Includes bibliographical references and index.
  ISBN 0-7360-4276-8 (soft cover)
  1. Human mechanics. 2. Biophysics. 3. Kinesiology.
  [DNLM: 1. Movement--physiology. 2. Biomechanics. 3. Biophysics. 4. Exercise--
physiology. 5. Sports--psychology.   WE 103 B6159 2005] I. Abernethy, Bruce, 1958-
  QP303.B586 2005
  612.7'6--dc22   2004008595

ISBN: 0-7360-4276-8

The Web addresses cited in this text were current as of May 6, 2004, unless otherwise noted.

Acquisitions Editor: Michael S. Bahrke, PhD; **Developmental Editor:** Elaine H. Mustain; **Assistant Editors:** Maggie Schwarzentraub and Sandra Merz Bott; **Copyeditor:** Joyce Sexton; **Proofreader:** Eric Cler; **Indexer:** Robert Howerton; **Permission Manager:** Dalene Reeder; **Graphic Designer:** Andrew Tietz; **Graphic Artist:** Yvonne Griffith; **Photo Manager:** Kareema McLendon; **Cover Designer:** Keith Blomberg; **Photographer (cover):** © PhotoDisc; **Photographers (interior):** © Human Kinetics; figures 3.3b and 10.7 © EyeWire, Inc.; figure 3.3c © PhotoDisc; **Art Manager:** Kelly Hendren; **Illustrators:** Argosy, Louise Blood, Scott McLean, unless otherwise noted; **Printer:** Edwards Brothers

Printed in the United States of America   10 9 8 7 6 5 4 3 2 1

Human Kinetics
Web site: www.HumanKinetics.com

United States: Human Kinetics; P.O. Box 5076; Champaign, IL 61825-5076
800-747-4457
e-mail: humank@hkusa.com

Canada: Human Kinetics; 475 Devonshire Road Unit 100; Windsor, ON N8Y 2L5
800-465-7301 (in Canada only)
e-mail: orders@hkcanada.com

Europe: Human Kinetics; 107 Bradford Road; Stanningley; Leeds LS28 6AT, United Kingdom
+44 (0) 113 255 5665
e-mail: hk@hkeurope.com

Australia: Human Kinetics; 57A Price Avenue; Lower Mitcham, South Australia 5062
08 8277 1555
e-mail: liaw@hkaustralia.com

New Zealand: Human Kinetics; Division of Sports Distributors NZ Ltd.; P.O. Box 300 226 Albany; North Shore City; Auckland
0064 9 448 1207
e-mail: blairc@hknewz.com

To all those scholars, past and present, who have contributed to the academic credibility of the study of human movement.

CONTENTS

vi   Contents

PREFACE

This second edition of *The Biophysical Foundations of Human Movement,* like the first, has been written with three main purposes in mind.

The first purpose is to provide an introduction to key concepts concerning the anatomical, mechanical, physiological, neural, and psychological bases of human movement. Fulfilling this purpose involves considering the biophysical dimensions of the field of study known variously, in different parts of the world, as human movement studies, human movement science, kinesiology, or sport and exercise science and also examining the discipline–profession links in this field. Gaining an overview of the field provides students of kinesiology with an entree to more detailed study within any of the subdisciplines of human movement studies and helps lay the foundations for integrative, multi- and cross-disciplinary studies.

The second purpose is to provide an overview of the multidimensional changes in movement and movement potential that occur throughout the life span with the processes of growth, development, maturation, and aging. The third purpose is to provide a comparable overview of the changes that occur in movement and movement potential as an adaptation to training, practice, and other lifestyle factors. Fulfilling the second and third purposes is important as a means of exposing readers to fundamental issues in biology and in positioning the study of movement as a major topic within human biology. For these reasons we have deliberately selected *life span changes* and *adaptation* as key organizing themes for this book.

Gaining knowledge about the processes of growth, development, maturation, and aging aids understanding of key changes in movement potential throughout the life span that, because of their inevitability, directly affect all of us. While the processes of maturation and aging are inevitable, adaptation (through training, practice, and lifestyle decisions) offers humans some degree of control of their own destiny and capabilities. We hope that readers will allow this clear message about the important role of physical activity in the maintenance of health to affect their personal lifestyle decisions and will also communicate this message to others.

The text features three sections that reflect its three main purposes.

• The Introduction (chapter 1) examines the disciplinary and professional structure of contemporary human movement studies.

• The main text of the book provides an introduction to basic concepts, life span changes, and adaptations arising in response to training, within each of the five major biophysical subdisciplines of human movement. These subdisciplines correspond to the five parts that compose the main text:

- Part I: Functional anatomy
- Part II: Biomechanics
- Part III: Exercise physiology
- Part IV: Motor control
- Part V: Sport and exercise psychology

To explain these subdisciplines, one can make a crude analogy to automotive engineering. To understand and optimise the performance of a motor vehicle, the automotive engineer needs specific knowledge about the vehicle's material structure (its *anatomical basis*); its mechanical design characteristics (its *mechanical basis*); its motor, capacity, and fuel consumption (its *physiological basis*); its electrical wiring and steering mechanisms (its *neural basis*); and the characteristics and capabilities of its driver (its *psychological basis*). In the case of the human "vehicle", this information is provided, respectively, by the subdisciplines of functional anatomy, biomechanics, exercise physiology, motor control, and sport and exercise psychology and the interactions between them.

The structure of each of the five parts of the book is broadly the same. Each part begins with a brief introduction that defines the subdiscipline and provides information on its historical development, the typical issues and problems it addresses, the level(s) of analysis it uses, and relevant professional training and organizations. This introduction is followed by chapters that overview the basic concepts within the subdiscipline and then by chapters on life span changes and adaptation. The exercise physiology section also contains a chapter devoted to applications of fundamental principles of exercise physiology to health.

• The Afterword (chapter 22), devoted to multidisciplinary and cross-disciplinary approaches to human movement, provides some examples of contemporary issues in which the application and integration of knowledge from a number of the biophysical subdisciplines are fundamental to understanding. This part of the text demonstrates how integrating information from different subdisciplines provides the essential strength and prospective direction for human movement studies and for practice in the professions grounded in that academic discipline.

Throughout the book, special elements called "In Focus" highlight key organizations, individuals, and studies that have contributed to our understanding of human movement. While the body of the text does not, as a general rule, focus on specific research studies or methods, the special elements provide the reader with some feel for the research methods in the field as well as more detailed exposition of some of the pivotal studies that underpin current knowledge. In this way these special sections present a picture (albeit a limited one) of the kinds of research methods and information students can expect to encounter in more detailed investigations within each subdiscipline of human movement studies. Such investigations are typically undertaken in the second and subsequent years of formal courses in kinesiology or human movement studies. Research examples in the special elements come from around the world, highlighting the importance of positioning research and knowledge in a relevant social and cultural context and illustrating the genuinely international nature of research activity in this field.

A glossary at the end of the text defines key terms introduced throughout the book and is presented as an aid to student comprehension and review. The glossary presents definitions of terms that appear in bold in the text.

Because this is an introductory text, it provides a broad and illustrative rather than a detailed and exhaustive treatment of the topics included. Students with an interest in further study may refer to the many excellent specialized texts now available for each of the subdisciplines, a number of which are listed in the Further Reading and References section of each chapter.

This second edition includes both new information and formatting and presentation style changes. The section on the mechanical bases of human movement has been completely rewritten, and the chapter structure of this part of the text has been reorganized for greater compatibility with the other subdisciplinary sections. The final chapter (the Afterword) on integrative perspectives has been significantly expanded to provide more examples of the interdisciplinary research that is increasingly at the forefront of new understanding about human movement. Further, in keeping with the emerging development of Web-based resources, this second edition contains an expanded listing of supporting Web materials. Clear learning objectives have been added to the beginning of each chapter, and chapters now end with brief summaries. We trust that these many changes will enhance both the readability and pedagogical value of the text.

The study and understanding of human movement present an exciting challenge for students, scientists, and practitioners alike. Given how central an understanding of human movement and its enhancement is to a wide range of human endeavours, it is our hope that this second edition of *The Biophysical Foundations of Human Movement* will serve as a readable introduction for both students and professionals involved in the many disciplines grounded in an understanding of human movement/kinesiology—sport and exercise science, physical education, ergonomics, music and performing arts, physiotherapy, occupational therapy, nursing, medicine, health education and health promotion, and other rehabilitation and health sciences, to name but a few. We trust that the text will help convey to our readers some of the fascination that this subject matter holds for us.

ACKNOWLEDGEMENTS

In addition to those people acknowledged for their contribution to the first edition of this book, we wish to also express our sincere appreciation to:

- Those instructors who adopted the first edition for their courses and those students who read the text as part of their courses and provided valuable feedback on the first edition
- The reviewers of the first edition, including the two anonymous reviewers recruited by our publisher
- Kath Thornhill for cheerfully and skillfully undertaking the painstaking task of compiling and standardising the electronic versions of the manuscripts provided by the different authors, chasing up copyright permissions, maintaining the (many) electronic versions of the text, and generally helping keep the whole project and the author team on track
- Rainer Martens, Mike Bahrke, and the staff at Human Kinetics for their support, patience, and contributions
- Our families, friends, and colleagues for their support, understanding, and tolerance while this book was being written (again!).

Without the collective encouragement of each and every one of these people, there would not have been a second edition.

The introductory chapter, chapter 1, plus chapters 14-17 were written by Bruce Abernethy; Vaughan Kippers wrote chapters 2-5; Marcus Pandy, chapters 6-9; Laurel Mackinnon, chapters 10-13; and Stephanie Hanrahan, chapters 18-21. Chapter 22, befitting its title, was a collaborative, integrative effort.

INTRODUCTION
Human Movement Studies
As a Discipline and a Profession

Understanding how knowledge about human movement is organized and how human movement studies relates to other disciplines and professions is an important stepping stone to scholarly and professional development in this field.

> The major learning concepts in this chapter relate to
> - the focus and importance of human movement studies,
> - the characteristics of disciplines and professions,
> - the disciplinary and professional elements of human movement studies,
> - naming the discipline of human movement studies, and
> - the relationship between human movement studies and its key professions.

Human Movement Studies: Definition and Importance

Human movement studies is the systematic study of human movement. It concerns understanding *how* and *why* people move and the factors that limit and enhance our capacity to move.

Two key elements need to be emphasized in this simple definition of human movement studies. The first is the focus of the field, and the second is its methodology.

Focus of the Field

The unique focus of the field is human movement. This is true whether the movement involves, for example, a fundamental daily skill (such as walking, speaking, or reaching and grasping), executing a highly practiced sport or musical skill, exercising for health, or regaining the function of an injured limb. The study of human movement is important because movement is a central biological and social phenomenon.

The study of movement is central to the understanding of human biology because movement is a fundamental property, indeed indicator, of life (and biology is literally the study of life). Human movement, as we have noted in the preface, offers a valuable medium for the study of biological phenomena fundamental to developmental changes across the life span (changes that occur with aging as a consequence of internal body processes), to **adaptation** (changes that occur as an accommodation or adjustment to environmental processes), and to the interactions of genetic and environmental factors (nature and nurture) that dictate human phenotypic expression.

Human movement, especially that occurring in collective settings such as organized sport, exercise and rehabilitation settings, and health and physical education classes, also clearly has an essential social and cultural component. Understanding individual and group motives, as well as opportunities for and barriers to involvement in different types of human movement, for instance, provides an important window into the nature of human society, just as understanding the mechanisms of individual human movement provides an important window into the understanding of human biology.

Movement, in short, plays a fundamental role in human existence and therefore warrants our very best efforts to understand it. Moreover, knowledge about human movement is fundamental to optimising health and performance and preventing injury and illness.

Research As Fundamental to Understanding

A second key point that our definition of human movement studies must highlight is the importance of a systematic, research-based approach to the generation of knowledge. Because many aspects of the current practice of human movement are based as much on fads, folklore, tradition, and intuition as on sound, logical theory substantiated by systematically collected, reproducible data, it is imperative that the knowledge base for human movement studies rely on research conducted with a methodological rigor equal to that of other established biological, physical, and social sciences. Only through such an approach can fact be separated from fiction and a solid foundation for best practice in the professions, based on the knowledge made available through human movement studies, be established.

In the tradition of the scientific method, human movement studies aims not only at describing key phenomena but also at moving beyond *description* to *understanding* through explanation and prediction. Human movement studies therefore carry the twin goals of all fields of science:

1. Generating knowledge through understanding of basic phenomena
2. Applying knowledge for the benefit of society

The understanding of human movement that the field strives for affects all the areas and professions dealing with the enhancement of our capacity to move. Obvious examples are in sport and sport science, exercise, health promotion, physical education, ergonomics, medicine, and physical rehabilitation.

How is this vital knowledge about human movement organized and how is it translated into the practices of relevant professions? To answer this question we must consider the discipline and professions of human movement studies.

Disciplines and Professions

The principal function of a discipline is to develop a coherent body of knowledge that describes, explains, and predicts key phenomena from the domain of interest (or subject matter).

In contrast, professions, as a general rule, attempt to improve the conditions of society by providing a regulated service in which practices and educational or training programs are developed in accordance with knowledge available from one or more relevant disciplines. Practice in the profession of engineering, for example, is based on application of knowledge from disciplines such as physics, mathematics, chemistry, and computer science, while practice in the medical profession is based on knowledge from disciplines such as anatomy, physiology, pharmacology, biochemistry, and psychology. Established professions share a number of characteristics, including

1. an identified set of jobs or service tasks over which they have jurisdiction, monopoly, or both;
2. organization under the framework of a publicly recognized association;
3. identified educational competencies and formalized training and education criteria (this generally includes the mastery of complex skills and the presence of an evidence base for practice);
4. political recognition, usually through acts of government legislation (including, in some cases, establishment of licensing/registration boards); and
5. a code of ethics defining minimal standards of acceptable practice.

Disciplines therefore seek to understand subject matter, and professions seek to implement change based on this understanding. The emphases within disciplines and professions are often characterized as theory/research versus application/practice, but such a distinction is overly simplistic and potentially misleading. Applied research (including research on aspects of professional practice) is now an accepted part of the business of the discipline, just as the profession may frequently be the site for original research.

Is Human Movement Studies a Discipline?

The establishment of university departments of human movement studies, kinesiology, and the like, independent of traditional professions such as physical education teacher training, is predicated on the assumption that human movement studies possesses an organized body of knowledge much as "traditional" disciplines such as physics, chemistry, and psychology do, and that human movement studies is more than simply a loose collection of the applications of knowledge from other fields. Therefore an important practical and philosophical question for our field is whether there is, in fact, a unique, organized body of knowledge on how and why people move (to satisfy the criteria for a discipline of human movement studies) or whether human movement studies is simply the application of knowledge from other disciplines such as anatomy and physiology (and therefore, by definition, more a profession than a discipline). This question has generated much debate both within and beyond the field for at least the last 50 years, with the extent of the uniqueness and collective coherence of the knowledge within the field being the source of most contention.

The forebears of the modern field of human movement studies were primarily physical educators. A number of these, most particularly the physiologist/psychologist Franklin Henry and the motor developmentalist Lawrence Rarick, created strong arguments in the 1960s for both the importance and the existence of a discipline of human movement studies, claiming that such a field (or its precursor, physical education) asked questions that would not have arisen from **cognate disciplines.**

Rarick (1967), in a paper in the journal *Quest,* argued that

> "most certainly human movement is a legitimate field of study and research. We have only just begun to explore it. There is need for a well-organised body of knowledge about how and why the human body moves, how simple and complex motor skills are acquired and executed, and how the effects (physical, psychological and emotional) of physical activity may be immediate or lasting." (p. 51)

In a similar vein, Henry, in a much-quoted 1964 paper titled "Physical Education: An Academic Discipline," claimed the preexistence of a discipline base for the study of human movement. Henry wrote:

> "There is indeed a scholarly field of knowledge basic to physical education. It is constituted of certain portions of such diverse fields as anatomy, physics and physiology, cultural anthropology, history and sociology, as well as psychology. The focus of attention is on the study of man as an individual,

engaging in the motor performances required by his daily life and in other motor performances yielding aesthetic values or serving as expressions of his physical and competitive nature, accepting challenges of his capability in putting himself against a hostile environment, and participating in the leisure time activities that have become of increasing importance in our culture." (p. 32)

Henry's definition of the discipline base of our field has stood the test of time well except for some relatively minor changes, most obviously

- substitution of "human movement studies" for "physical education" (the latter now typically defined in a narrower professional sense),
- extension of the focus of the field beyond the person as an individual to the person as an element of a social system, and
- alteration of the language to acknowledge the equal involvement of females and males.

Structuring a Discipline of Human Movement Studies

Because movement potential and performance are known to be influenced by many things including biological factors (such as maturation, aging, training, and lifestyle), health factors (such as disease, disuse, and injury), and social factors (such as motivation, incentive, and opportunity), it is clear that a human movement studies discipline must draw heavily on the methods, theories, and knowledge of a wide range of other disciplines and provide them with an integrative focus on human movement. Relevant information for a discipline of human movement studies may be gleaned from biological science disciplines such as anatomy, physiology, and biochemistry; physical science disciplines such as physics, chemistry, mathematics, and computer science; social science disciplines such as psychology, sociology, and education; and disciplines in the humanities such as history and philosophy.

Figure 1.1 presents one possible way of conceptualising the organization of knowledge within a discipline of human movement studies. In this conceptualisation the discipline of human movement studies consists of the collective knowledge contained within and between each of the subdisciplines of functional anatomy, exercise physiology, biomechanics, motor control, and sport and exercise psychology and the pedagogy, sociology, history, and philosophy of sport and physical activity (as illustrated by the shaded box). The subdisciplines of functional anatomy, exercise physiology, biomechanics, motor control, and sport and exercise psychology (insofar as it focuses on the individual rather than group or societal behaviour) constitute the biophysical foundations of human movement and are covered in this text. The social psychology of sport and exercise, along with the pedagogical, sociological, historical, philosophical, political, and cultural aspects of physical activity and sport, constitutes the sociocultural foundations of human movement. Figure 1.1 clearly indicates that each subdiscipline draws theories, methods, and knowledge from one or more cognate disciplines. Note, however, that each specialist subdiscipline draws on only a subset of the knowledge in the cognate discipline(s) and that the subdisciplines can, and frequently do, generate theories, methods, and approaches of their own not acquired from the cognate discipline(s).

Figure 1.1 also suggests a number of other points:

- The clustering of disciplines into discipline groups and the selection of the subdisciplinary groups are necessarily subjective; and no two conceptualisations of the field, its interrelations, and the naming of its component parts are likely to be identical.

- The interrelationships of cognate disciplines and human movement studies subdisciplines are frequently reflected in the structuring of many tertiary programs in human movement studies; exposure to the cognate disciplines generally precedes exposure to the subdisciplines of human movement studies.

- The broad general education in the sciences and in the arts that students of human movement studies/kinesiology must first undertake, in order to understand the foundations of their own discipline, positions them well not only for careers in professions linked to human movement studies but also for further study in many cognate fields.

- The disciplines and subdisciplines are organized to present a generally progressive shift from left to right, from a focus on the micro phenomena to the more macro phenomena of human movement.

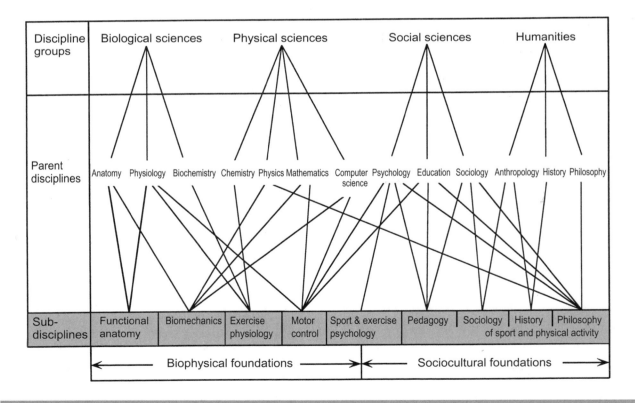

Figure 1.1 One possible conceptualisation of the structure of knowledge about human movement. The discipline of human movement studies is represented by the shaded area.

The subdisciplines are presented in figure 1.1 as insular components, having as much (or more) contact with the cognate discipline as with the other human movement studies subdisciplines. Because this representation is a reasonably accurate reflection of the current state of the field, there is increasing concern in many quarters that the growing differentiation and specialization within human movement studies may produce fragmentation and a loss of integrity within the discipline base. This fragmentation is reinforced by the dominant tendency in tertiary programs to teach each of the subdisciplinary fields in a separate, independent course. To this end the field of human movement studies currently is probably most accurately depicted as **multidisciplinary,** whereas the desirable direction is to make it more **cross-disciplinary** and ultimately **interdisciplinary** or **transdisciplinary** (figure 1.2). A good indicator of the maturation of the discipline will be the extent to which it becomes more interdisciplinary and advances knowledge by crossing the traditional (but arbitrary) boundaries between the subdisciplines and by synthesizing material from the subdisciplines rather than importing ideas from the "mainstream" disciplines.

Naming the Discipline of Human Movement Studies

While it is clear from the previous sections that we favour the use of the term **human movement studies** to describe the discipline, that term is far from universally accepted. The fact that finding a widely accepted name for the discipline remains a persis-tent source of debate perhaps indicates the relative immaturity and diversity of our field of study.

The term *human movement studies* was coined by the British physical educator and psychologist H.T.A. (John) Whiting in the early 1970s and has been adopted in parts of Europe, the United States, and Australia; it is also the title for one of the field's journals. Although the term is useful because it encapsulates the unique and unifying theme within the body of knowledge in our field, some consider the term too cumbersome, especially in a public marketing context. The alternative term **human movement science** has had some following, especially in Europe, and is probably appropriate for the biophysical aspects of human movement. This term becomes problematic in relation to aspects of the sociocultural foundations of human movement, which involve knowledge acquisition through methods different than those of traditional science.

In North America the term **kinesiology,** literally meaning the study of movement (from the Latin *kinein,* "to move," and the Greek *logos,* a branch of learning), is widely used. The American Academy of Kinesiology and Physical Education, which has championed the cause for universal usage of the term in degree titles and department names, defines kinesiology as both the "knowing about (theory) and the knowing how (practice) of movement in context." Knowledge drawn from experience of movement, as well as knowledge drawn from scholarly research, is seen as an integral part of the study of kinesiology. The term *kinesiology* has not as yet been widely adopted internationally, presumably because in many countries it either is poorly understood in general usage, is in general use already

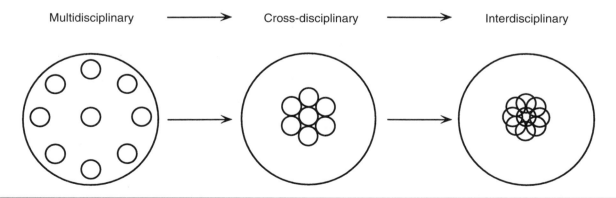

Multidisciplinary   →   Cross-disciplinary   →   Interdisciplinary

Figure 1.2 A desirable evolutionary progression for a discipline of human movement studies. The components shown by circles represent the subdisciplines identified in figure 1.1.
Adapted from Zeigler 1990.

in relation to some forms of alternative therapies practices that by and large are not evidence based, or is used in a scholarly context to refer exclusively to the mechanics of human movement.

In contrast to both human movement studies and kinesiology, the terms **exercise and sport science** and **physical education** are well understood by the general public but are much narrower in focus. In the case of physical education, its current usage is more closely tied to notions of a profession than to the concept of an all-encompassing discipline.

Professions Based on Human Movement Studies

As we noted earlier, professions, unlike disciplines, have a specific interest in the application of knowledge as a means of solving specific problems, enhancing the quality of life, and providing a service to society. Standards of practice in specific professions are typically controlled by professional bodies that impose minimal training/education requirements, set membership and accreditation criteria, and establish codes of professional ethics and the like. A wide range of professions draw on knowledge from the discipline of human movement studies, and each of these is represented by one or more professional bodies. Like other disciplines, human movement studies informs multiple professions, but in a nonexclusive way.

For example, human movement studies informs professions such as sports medicine, physical therapy, and health promotion; but these professions are also all informed by knowledge generated in disciplines like anatomical sciences, psychology, and pharmacology.

Traditionally the profession most linked to the knowledge base of human movement studies has been physical education, which is the principal historical forebear of human movement studies. Over time new professions, such as those related to exercise prescription, sport and exercise science, exercise rehabilitation and physical therapy, ergonomics, sports medicine, athletic training, coaching, sport management, and sport and exercise psychology, have emerged; all of these draw, to some degree, on the discipline of human movement studies.

Currently a number of national and international organizations represent key professions whose practices are grounded in human movement studies. To illustrate the plethora of organizations that exist, table 1.1 lists some of the major ones (along with their acronyms) in sports medicine; exercise science; and physical education, recreation, and dance. The introductions to parts I through V of this book provide information about the professional groups representing the specific interests of each biophysical subdiscipline of human movement studies. The following sections present additional details on the functions and goals of three of the major professional organizations.

International Council of Sport Science and Physical Education

The International Council of Sport Science and Physical Education (ICSSPE) was founded in Paris in 1958, originally under the name International Council of Sport and Physical Education. ICSSPE serves as an international umbrella organization that promotes and disseminates results and findings in the field of sport science and their practical application in cultural and educational contexts. Its aims are to contribute to the awareness of human values inherent in sport, to improve health and physical efficiency, and to develop physical education and sport to a high level in all countries. As a world organization, ICSSPE endeavours to bridge the gap between developed and developing countries and to promote cooperation among scientists and organizations from countries with different political systems.

ICSSPE has five stated fundamental objectives. These are to

- encourage international cooperation in the field of sport science;
- promote, stimulate, and coordinate scientific research in physical education and sport throughout the world and to support the application of its results in the many practical areas of sport;
- make scientific knowledge of sport and practical experiences available to all interested national and international organizations and institutions, especially to those in developing countries;
- facilitate differentiation in sport science while promoting the integration of the various branches; and
- support and implement initiatives with aims similar to its own and originated or developed by any other appropriate agency or organization in the field.

Table 1.1

Examples of Some Major Professional Organizations Relevant to the Discipline of Human Movement Studies

Region	Organization name	Acronym	Web site*
International	Association Internationale des Ecoles Superieures d'Education Physique	AISEP	www.aiesep.com
	Fédération Internationale de Médecine du Sportive	FIMS	www.fims.org
	International Council for Health, Physical Education, Recreation, Sport, and Dance	ICHPER·SD	www.ichpersd.org
	International Council of Sport Science and Physical Education	ICSSPE	www.icsspe.org
North American	American Academy of Kinesiology and Physical Education	AAKPE	www.aakpe.org
	American Alliance for Health, Physical Education, Recreation and Dance	AAHPERD	www.aahperd.org
	American College of Sports Medicine	ACSM	www.acsm.org
	Canadian Academy of Sport Medicine	CASM	www.casm-acms.org
	Canadian Association for Health, Physical Education, Recreation and Dance	CAHPERD	www.cahperd.ca/e/
European	British Association of Sport and Exercise Medicine	BASEM	www.basem.co.uk/index.php
	British Association of Sport and Exercise Sciences		www.bases.org.uk/newsite/home.asp
	Deutsche Vereinigung für Sportwissenschaft	DVS (Germany)	www.tu-darmstadt.de/dvs/
	European College of Sport Science	ECSS	www.dshs-koeln.de/ecss/HTML/
	European Federation of Sports Medicine Associations	EFSMA	www.efsma.org
	L'Association des Chercheurs en Activités Physiques et Sportives	ACAPS (France)	www.u-bourgogne.fr/ACAPS/

(continued)

Table 1.1

(continued)

Region	Organization name	Acronym	Web site*
European (continued)	Netherlands Association of Sports Medicine		www.sportgeneeskunde.com
	Physical Education Association of the United Kingdom	PEAUK	www.pea.uk.com/
Asian and South Pacific	Asian Federation of Sports Medicine	AFSM	www.afsm-asia.com/
	Australian Association for Exercise and Sports Science	AAESS	www.aaess.com.au/
	Australian Council for Health, Physical Education and Recreation	ACHPER	www.achper.org.au/
	Chinese Association of Sports Medicine		
	Hong Kong Association of Sports Medicine and Sports Science		www.fmshk.com.hk/hkasmss/home.htm
	Japanese Federation of Physical Fitness and Sports Medicine		
	Sports Medicine Australia	SMA	www.sma.org.au/
	Sports Medicine New Zealand	SMNZ	http://sportsmedicine.co.nz
Africa	Biokinetics Association of South Africa		www.biokinetics.org.za
	South African Sports Medicine Association		
South America	Confederación Sudamericana de Medicina del Deporte		
	Sociedade Brasileira de Medicina do Esporte		

This listing is by no means exhaustive. The introduction to each part of the book includes information on more specialized bodies.

*Web site information is provided where available.

ICSSPE conducts its scientific work in three main areas: sport sciences; physical activity, physical education, and sport; and scientific services (information dissemination). It disseminates information through a variety of publications, of which *Sport Science Review,* an international journal dedicated to thematic overviews of research in sport science, is the best known. ICSSPE serves as a permanent advisory body to the United Nations Educational, Scientific and Cultural Organization (UNESCO) and regularly conducts research projects on behalf of UNESCO and the International Olympic Committee (IOC). It has eight regional bureaus throughout the world plus close links to a number of other international organizations of physical education and sport science. Membership in ICSSPE is open to organizations and institutions rather than individual subscribers, and currently some 200 organizations and institutions are affiliated with ICSSPE. More information about ICSSPE can be obtained from www.icsspe.org.

International Federation of Sports Medicine

The Fédération Internationale de Médecine du Sportive (FIMS) is an organization made up of the national sports medicine associations of over 100 countries. It was founded in 1928 during a meeting in St. Moritz, Switzerland, of Olympic medical doctors with the principal purpose of promoting the study and development of sports medicine throughout the world.

The specific objectives of FIMS are to

- promote the study and development of sports medicine throughout the world;

- preserve and improve the health of humankind through physical fitness and sport participation;

- study scientifically the natural and pathological implications of physical training and sport participation;

- organize or sponsor international scientific meetings, courses, congresses, and exhibits in the field of sports medicine;

- cooperate with national and international organizations in sports medicine and related fields; and

- publish scientific information in the field of sports medicine and in related subjects.

FIMS hosts a major international conference in sports medicine every three years. In addition, FIMS produces position statements on aspects of health, physical activity, and sports medicine. The most influential of these position statements is probably the 1995 joint statement of FIMS and the World Health Organization (WHO) titled *Physical Activity for Health*. FIMS now also publishes a bimonthly electronic (e-)journal, *International SportMed Journal*. In addition to its member national associations, FIMS makes individual membership available to doctoral degree holders in medicine and related sciences who are affiliated with one of the member national bodies in sports medi-

cine. More information is available about FIMS at www.fims.org.

American College of Sports Medicine

The mission of the American College of Sports Medicine (ACSM) is to promote and integrate scientific research, education, and practical applications of sports medicine and exercise science to maintain and enhance physical performance, fitness, health, and quality of life. ACSM was founded in 1954 by a group of 11 physicians, physiologists, and educators. The organization has grown rapidly since its formation; with over 18,000 members from North America and more than 75 other countries, ACSM is easily the largest sports medicine and exercise science organization in the world.

ACSM's research and educational programs are broad ranging. Its annual four-day meeting, held in May of each year, is one of the major international conferences for the presentation and discussion of new sports medicine and exercise science research. Similarly the College's official journal, *Medicine and Science in Sport and Exercise* (first published in 1969), is one of the principal international journals in the field for the publication of original research; its other publications, such as *Exercise and Sport Science Reviews, Health and Fitness Journal,* and *Sports Medicine Bulletin,* are a valuable source of state-of-the-art reviews on key sports medicine and exercise science topics. In addition to its role in advancing the discipline through disseminating basic and applied scientific research on physical activity, ACSM is the peak professional body in North America for the certification of people seeking to work in clinical sports medicine, exercise science, and the health and fitness industry. This role has been facilitated by the development of key guidelines and policy documents, such as *ACSM's Guidelines for Exercise Testing and Prescription,* which have become the "gold standard" for professional practice internationally.

ACSM offers five types of memberships (Professional, Professional-in-Training, Graduate Student, Undergraduate Student, and Associate), details of which can be found on the College's Web site (www.acsm.org).

Relationships Between the Discipline and the Professions

Thus far, our discussion of discipline–profession relations may have created the impression that the flow of information between the two is unidirectional, with the role of the discipline that of generating knowledge that the professions can use as a basis for practice. While this information flow is important, it is essential to recognize that the ideal relationship between discipline and profession should be one of mutual benefit, with information flowing as much from the professions to the discipline as in the reverse direction. In particular, the professions are well positioned to provide questions, problems, observations, and issues that can function as a valuable guide and focus for the knowledge-seeking, research-based activities of the discipline. Observations made in practice frequently form the initial basis of hypothesis testing within the discipline.

Summary

Human movement studies is a discipline (an organized body of knowledge) concerned with understanding how and why people move and the factors that limit and enhance that capacity to move. The discipline, known most frequently in North America as kinesiology, goes by different names throughout the world. Human movement studies is important because movement is a fundamental biological and social phenomenon with profound implications for human health and existence. Knowledge about human movement comes from a range of subdisciplines spanning the focus from micro-level biophysical aspects of movement through to macro-level social and cultural phenomena related to organized forms of physical activity. As the discipline of human movement studies matures, one can expect that these subdisciplines will become less discrete and cross-disciplinary and that interdisciplinary studies will become more prevalent. Knowledge about human movement forms the basis for the activities of an increasingly wide range of professions.

Further Reading and References

Bouchard, C., B.D. McPherson, and A.W. Taylor, eds. 1992. *Physical activity sciences.* Champaign, IL: Human Kinetics.

Brooke, J.D., and H.T.A. Whiting, eds. 1973. *Human movement: A field of study.* London: Henry Kimpton.

Brooks, G.A., ed. 1981. *Perspectives on the academic discipline of physical education.* Champaign, IL: Human Kinetics.

Curtis, J.E., and S.J. Russell, eds. 1997. *Physical activity in human experience: Interdisciplinary perspectives.* Champaign, IL: Human Kinetics.

Henry, F.M. 1964. Physical education: An academic discipline. *Journal of Health, Physical Education, and Recreation* 35(7): 32-33, 69. (Reprinted in Brooks 1981, pp. 10-15.)

Hoffman, S.J., and J.C. Harris, eds. 2000. *Introduction to kinesiology: Studying physical activity.* Champaign, IL: Human Kinetics.

Newell, K.M. 1990. Kinesiology: The label for the study of physical activity in higher education. *Quest* 42: 269-278.

Rarick, G.L. 1967. The domain of physical education. *Quest* 9: 49-52. (Reprinted in Brooks 1981, pp. 16-19.)

Some Relevant Web Sites

American Academy of Kinesiology and Physical Education. 1998. Kinesiology: The field of study. [www.aakpe.org/aakpe2.htm#academy]

American College of Sports Medicine [www.acsm.org]

International Federation of Sports Medicine [www.fims.org]

International Council of Sport Science and Physical Education [www.icsspe.org]

PART I

ANATOMICAL BASES OF HUMAN MOVEMENT

The Subdiscipline of Functional Anatomy

*H*uman anatomy is the study of the structure of the human body at a number of different levels of organization, from the subcellular level (structures that can be seen with an electron microscope) through tissues (structures that can be seen with a light microscope) to organs (which can be seen with the unaided eye). At each level of organization, there is an assumed relationship between structure and function. The relationships between structure and function of bones, joints, and muscles are emphasized in chapter 2. When applied to the musculoskeletal system, functional anatomy is the study of movement and the effects of physical activity on the organs and tissues of the system. Functional anatomy therefore is dynamic anatomy, which considers both the short- and long-term effects of activity on the musculoskeletal system. It is obvious that functional anatomy overlaps with both physiology, because of its functional approach, and biomechanics, because functional anatomists consider the musculoskeletal system as mainly a mechanical system.

Typical Questions Posed and Problems Addressed

The subdiscipline of functional anatomy concerns itself with answering a range of questions related to physical activity and the musculoskeletal system. Some typical questions include these:

- What functions do bones perform?
- How strong are bones?
- How do muscles produce movement?
- What prevents dislocation of joints during movement?
- How can the size and shape of a person be described?
- Are children merely scaled-down adults?
- Why do older women in particular experience more fractures than other groups?
- What adaptations occur when a person begins a regular exercise program?
- Is there an optimal type and level of exercise for maintaining the integrity of the musculoskeletal system?

Levels of Analysis

Anatomy is a very visual science. In gross or **macroscopic anatomy** classes, the unaided eyes are used to study the structure of the human body, whether directly from a **cadaver,** a model, or a chart or indirectly through anatomical atlases, which include many illustrations or photos of different parts of the body. In the study of human movement, the observation of surface features is also important. Therefore, familiarity with *surface anatomy,* which requires the skills of observation (using the sense of vision) and **palpation** (using the senses of touch and pressure, particularly in the fingertips), is an important precursor to the analysis of human movement.

Light microscopes are used to aid visualization of structures, tissues, and the cells that form tissues, while **electron microscopes** are used to aid visualization of cells and structures within cells. Microscopes are used in functional anatomy to define the responses of tissues and cells to physical activity.

The discipline of anatomy includes many areas of study, only a few of which are covered in this text. The chapters that follow take both a systematic and a functional approach to anatomy by addressing the function and properties of structures that form the musculoskeletal system. The three main areas of study discussed are the skeletal system **(osteology)**, the joint system **(arthrology)**, and the muscular system **(myology)**.

Historical Perspectives

Anatomy is generally recognized as the oldest of the biomedical sciences. Although an understanding of history is not a prerequisite for the study of anatomy, an overview of the development of anatomy is valuable as a means of gaining a better appreciation of the discipline. Most people think that everything there is to know about anatomy must already be known, but this is not so. Anatomy research is still being published.

Key Developments

Manuscripts on the anatomy of the pig produced in Salerno during the period 1000 to 1050 A.D. argued that the Greek term *anatomy,* which derives from *ana* (meaning "apart") and *tome* (meaning "cutting"), was actually meant to convey more than the straight derivation implies. According to these documents, the Greeks meant that in anatomy, the "cuts" must be performed according to set rules.

Anatomy is still learned by many students who dissect human cadavers as part of their courses, following instructions written in dissection manuals. For many other students, the material has already been dissected by others and is ready for inspection of the important structures. Even if you never see a cadaver, you will see a model or a chart; and these are derived from the investigation of human bodies. Possibly the first anatomical illustrations were drawn in 1522 A.D. by Berengario da Carpi.

The well-known medical historian Charles

Singer dubbed Herophilus from Alexandria, in Egypt, the "Father of Anatomy" because Herophilus was probably the first to dissect both human and animal bodies for the purposes of anatomical instruction. At that time, however, anatomy was really only an investigative technique used by physiologists and surgeons. After a brief period of prominence in the Alexandrian school (300-250 B.C.) during the Hellenic period, there followed a period of Roman rule during which anatomy was almost exclusively based on animal dissection. The most famous anatomist of this period is Galen (129-200 A.D.), a Greek who was one of the greatest physicians of all time.

Galen wrote prolifically, on the basis of both dissection and experiment. It was Galen who demonstrated that the **arteries** (meaning "air carriers") actually carried blood. Many terms first used by Galen in relation to bones, joints, and muscles have survived as part of modern nomenclature. In fact, most anatomical terms are derived from Latin or Greek but are now often anglicised for ease of use. After this glorious period from 50 to 200 A.D., anatomy, and medical science in general, waned with the dominance of the "practical outlook" required by the Roman Empire during its famous decline. This demise is a classic illustration of how science depends on hypothesis- and curiosity-driven research, both of which may disappear when excessive demands for studies with immediately applicable results are made.

After the end of the Dark Ages, there was a great intellectual awakening in medicine during the 14th century A.D., led by scientists at the University of Bologna. A professor from Bologna, Mondino (1270-1326 A.D.), restored the techniques of anatomy, including anatomical dissection, and gave anatomy the theoretical foundation that would allow it to be considered a scientific discipline, distinct from surgery and physiology. From 1300 to 1325, Mondino personally dissected cadavers as part of the medical curriculum.

It was not only scientific anatomists who were interested in the structure of the human body. Artists also used scalpels to gain a better appreciation of the human form. Results of this understanding include the anatomical studies of Leonardo da Vinci (1452-1519 A.D.) and the anatomically correct paintings and sculptures of Renaissance artists such as Michelangelo (1475-1564 A.D.). The medical historian, Charles Singer, has stated that Leonardo da Vinci was not only "one of the greatest biological investigators of all time" but also pos-

Figure I.1 A drawing by Leonardo da Vinci titled *The Proportions of the Human Figure.*

sibly the greatest genius in human history. He was probably the first to question Galen's authoritative teachings. Today, his drawing illustrating the proportions of the human body (figure I.1) is perhaps the most popular and widely recognized symbol of the human body, representing the human as a physical and spiritual entity.

Galen's prolific writings survived to the Middle Ages, and his teachings influenced anatomical thinking to such an extent that dissections were performed merely to demonstrate his ideas. If the cadaver did not confirm Galen's teachings, then the cadaver was thought to be wrong; but of course it was not long before people questioned this approach.

It was left to Andreas Vesalius of Brussels (1514-1564 A.D.) to reform the science of anatomy. He did this as a professor of anatomy at the University of Padua, when he published *On the Fabric of the Human Body,* a work of seven books, in 1543, when he was only 28 years old! To Vesalius, "fabric" meant "workings," and the first two books on bones and muscles (reprints of which can still be purchased today) indicate that he regarded the human body as a machine designed to perform work. Vesalius did not completely reject the teachings of Galen, but he thought that dissection provided the opportunity to observe the human body and make comparisons with textbook descriptions. In this way previous knowledge could be confirmed or questioned. It is interesting to note that this major advance in the biological sciences was paralleled by a major publication in

the physical sciences. It was in the same year that Nicholas Copernicus published *On the Revolutions of the Celestial Spheres,* in which he argued that the earth was not the centre of the universe.

In the history of anatomy, teaching has had equal prominence with discovering new structures or techniques. Possibly two of the greatest teachers of human anatomy in the English language were brothers William Hunter (1718-1783 A.D.), an obstetrician, and John Hunter (1728-1793 A.D.), a surgeon. The Hunterian Museum and Art Gallery at Glasgow University houses books, pictures, and specimens donated by William, and the Hunterian Museum at the Royal College of Surgeons in London still includes some of the 13,000 specimens of human and animal material donated by John. Possibly the best-known modern anatomy textbook in the English language is *Gray's Anatomy,* which is now in its 38th British edition. Henry Gray (1827-1861 A.D.), who was a lecturer in anatomy and curator of the anatomy museum at St. George's Hospital, London, wrote the first two editions starting in 1858.

Key People

Most histories of anatomy concentrate on gross anatomy and later, **microscopic anatomy,** and tend to overlook many of the key innovators in the field of functional anatomy. Girolamo Mercuriale published a book in 1577 A.D. titled *The Practice of Gymnastics* in which he outlined the medical, athletic, and military aspects of exercise. The book provided instructions on correct exercise technique and claimed that exercise could improve health. Mercuriale discussed the effects of physical activity on healthy and diseased individuals and stressed that exaggerated exercise can cause damage, especially in competitive athletics. All these issues are still topical today and are addressed in the chapters that follow. Vesalius, too, had a dynamic approach to the musculoskeletal system, and Galileo noted in 1638 that there was a direct relationship between body mass, physical activity, and bone size.

This idea of the relationship between form and function was extended by a number of 19th-century German scientists, culminating in the publication in 1892 of a book titled *The Law of Bone Transformation,* written by a German engineer and anatomist, Julius Wolff. This book included what is now known as Wolff's law, which was based on engineering analyses of the small bony rods (trabeculae) that form spongy bone. Wolff summarized the work of a number of researchers by concluding that there was a strong correlation between the direction of the trabeculae and the lines of forces acting through the bone. The law related to the architecture of different bones, but others realized that there was also an adaptive relationship that is still of major interest. Research continues today on the mechanisms responsible for the biological adaptation of bone to various levels of exercise (see chapter 5).

Most actions of muscles had been determined from knowledge of muscle attachment sites and from lines of pull of the muscle related to the joint axes of rotation. Guillaume Duchenne (1806-1875 A.D.), a French physician, attempted to confirm these hypotheses, and his experiments and observations over many years culminated in the publication in 1865 of his influential text *Physiology of Motion: Demonstrated by Means of Electrical Stimulation and Clinical Observation and Applied to the Study of Paralysis and Deformities*. Of major interest to modern functional anatomists is the recording of the actions of almost all of the skeletal muscles in the human body.

Particularly since World War II, the converse of Duchenne's technique has been used. When a muscle is stimulated by its nerve, it responds in two ways—by producing an electrical signal and by producing a mechanical force. Nobody has yet directly measured the force within the muscle tissue of a living human, although tendon forces have been measured. It is possible to detect the electrical signal produced by muscle tissue using a variety of electrodes. This technique, known as electromyography, is used to describe muscle activity during human movement. The best-known name in this field for the last four decades has been John Basmajian (1921-) from North America, who has held professorships in both Canada and the United States. He is a former president of the American Association of Anatomists. Not only has Basmajian contributed greatly to the elucidation of dynamic human muscle function in a variety of situations, he has also been very involved in the teaching of gross and functional anatomy through his authorship of many books, including the well-known *Muscles Alive: Their Functions Revealed by Electromyography*.

Although the anatomy of adult humans is of major interest to anatomists, it is not their sole interest. The structural changes that occur during childhood, and later, have also been studied extensively by auxologists. The field of **auxology** is defined as the science of biological growth, and key concepts from this field are examined in some detail in chapter 4. Many of the recent researchers in this field have been trained at the Institute of Child Health at the University of London, in a program headed by the noted auxologist, Professor Jim Tanner. Among many books published by Tanner is one on the history of human growth studies, listed as a further reading source to supplement material in chapter 4.

Professional Training and Organizations

Anatomy is a foundation course for many professions including medicine and all the allied health professions. Many qualified professionals often find it useful to refresh their knowledge of anatomy. For example, people interested in sports medicine may attend a session in an anatomy laboratory as part of their continuing education. Likewise, surgeons, nurses, sport coaches, ergonomists, and others also find it valuable to continually improve their knowledge of anatomy.

Most countries or regions have an association, composed of practicing anatomists, that holds regular conferences to present the latest research of members and to discuss new directions in the teaching of anatomy. The International Federation of Associations of Anatomy (FIAA) holds a conference every five years. Members of the International Society for the Advancement of Kinathropometry (ISAK) have a particular interest in the relationships between body dimensions and human performance (see chapter 3). Functional anatomists frequently find more interest in the meetings of societies in biomechanics than in the meetings of general anatomy societies.

Further Reading and References

Agur, A.M.R., and M.J. Lee. 1999. *Grant's atlas of anatomy*. 10th ed. Philadelphia: Lippincott, Williams & Wilkins.

Backhouse, K.M., and R.T. Hutchings. 1998. *Clinical surface anatomy*. 2nd ed. London: Mosby-Wolfe.

Singer, C. 1957. *A short history of anatomy and physiology from the Greeks to Harvey (the evolution of anatomy)*. 2nd ed. New York: Dover.

Williams, P.L., L.H. Bannister, M.M. Berry, P. Collins, M. Dyson, J.E. Dussek, and M.W.J. Ferguson, eds. 1995. *Gray's anatomy*. 38th ed. Edinburgh, New York: Churchill Livingstone.

Some Relevant Web Sites

Anatomy on the Internet
[www.meddean.luc.edu/lumen/MedEd/GrossAnatomy/anatomy.htm]

Arthritis Foundation
[www.arthritis.org/]

Australasian Menopause Society
[www.menopause.org.au/]

The Bone and Joint Decade: Joint Motion 2000-2010
[www.bonejointdecade.org/]

Boys and Puberty, Girls and Puberty
[www.avert.org/yngindx.htm]

Centre for Physical Activity in Ageing
[www.cpaa.sa.gov.au/home.html]

Galen: A Biographical Sketch
[www.ea.pvt.k12.pa.us/medant/galbio.htm]

Growth charts and tables
[www.auxologia.com/tabelle_eng.html]

Health at Every Age
[www.healthandage.com/]

Hippocrates: The "Greek Miracle" in Medicine
[www.ea.pvt.k12.pa.us/medant/hippint.htm]

The Hunterian Museum and Art Galery, University of Glasgow, Scotland
[www.hunterian.gla.ac.uk/]

The Hunterian Museum of the Royal College of Surgeons of England
[www.rcseng.ac.uk/services/museums/history/hunterian_html]

Information about flexibility exercises from the American Academy of Orthopaedic Surgeons
[http://orthoinfo.aaos.org/fact/thr_report.cfm?Thread_ID=4&topcategory=Wellness]

International Society for the Advancement of Kinanthropometry
[www.isakonline.com]

International Society for the Advancement of Kinanthropometry (ISAK)—home page
[http://somatotype.net/ISAK/]

Internet resources for anthropometry
[www.library.unisa.edu.au/resources/subject/anthropo.asp]

National Osteoporosis Foundation
[www.nof.org/osteoporosis/]

Orthopedic Anatomy from the Southern California Orthopedic Institute
[www.scoi.com/anat.htm]

Osteoarthritis: Arthritis Foundation
[www.arthritis.org/conditions/DiseaseCenter/oa.asp]

Somatotype—Calculation and Analysis
[www.sweattechnologies.com/~somatotype/]

Strength training information from Georgia State University
[www.gsu.edu/~wwwfit/strength.html]

Vesalius the Humanist
[http://hsc.virginia.edu/hs-library/historical/antiqua/texth.htm]

The Visible Embryo
[www.visembryo.com/]

The Visible Human Project
[www.nlm.nih.gov/research/visible/visible_human.html]

CHAPTER 2

Basic Concepts of the Musculoskeletal System

*T*he purpose of this chapter is to introduce key concepts related to the structure and function of the skeletal system, the system of joints (the articular system), and the muscular system. We begin by describing the tools for measurement of these systems.

Tools for Measurement

Language is very important for basic **gross anatomy** because of the descriptions required to explain the position, relative size, and relationships of each anatomical feature. Learning anatomy from a textbook can be quite difficult but becomes easier with the aid of atlases that have artistic illustrations or photographs of the various structures.

The concepts introduced in this chapter are presented as general descriptions, but the reader should realize that many of these concepts have been verified experimentally using quantitative methods. Bone density in living humans can be measured using **radiological** and other more direct techniques. Bone structure can be visualized under a microscope; but because bone is a hard tissue, special preparation techniques are required. Chemical analyses can be performed to determine the composition of bone. Movement relies on the integrity of joints, and **goniometers** are examples of instruments used to measure ranges of joint motion. Muscles produce forces that have an external effect, and these effects can be measured

The major learning concepts in this chapter relate to
- equipment used in anatomy learning and research;
- mechanical and physiological functions of the skeletal system;
- components of bone and the different types of bone cells;
- structure of long bones and the mechanical properties of bone tissue;
- the functions, shapes, and organization of whole bones;
- classifying joints according to their structure or function;
- characteristic features of synovial joints;
- protection, lubrication, and wear of synovial joints;
- the joint as the functional unit of the musculoskeletal system;
- structural features of muscles and their main actions;
- distinguishing properties of skeletal muscle;
- the biological and mechanical bases of muscular contraction;
- the mechanical and electrical responses of muscle to stimulation; and
- the major factors that determine strength and range of joint motion.

by different types of **dynamometers** that basically determine muscle "strength." The electrical signal generated just prior to a muscle contraction can also be detected, recorded, and analysed using electronic equipment and computers.

The Skeletal System

The structure of the skeletal system is exquisitely designed to fulfil both its mechanical and its physiological functions. We first examine those functions and then show how both the composition

SOME EXCELLENT ANATOMICAL ATLASES

- Agur, M.R., and M.J. Lee. 1999. *Grant's atlas of anatomy.* 10th ed. Philadelphia: Lippincott, Williams & Wilkins.
- Netter, F.H., and J.T. Hansen. 2002. *Atlas of human anatomy.* 3rd ed. Teterboro, NJ: ICON Learning Systems.
- Rohen, J.W., C. Yokochi, and E. Lütjen-Drecoll. 2002. *Color atlas of anatomy: A photographic study of the human body.* 5th ed. Philadelphia: Lippincott, Williams & Wilkins.

and architectural structures of the skeletal system support them.

Functions of the Skeletal System

Humans have an endoskeleton and are **vertebrate** animals. Attached to our "backbone," or vertebral column, are other bones that form the framework of the human body, much like the beams and studs of a house or the chassis of a car. These analogies are imperfect, however, as they imply that the skeletal system has only mechanical functions when, as we shall see, the physiological functions of bone are equally important.

Mechanical Functions of the Skeletal System

The most obvious mechanical function of the skeletal system is support for weight bearing. The skeletal system also provides protection of internal organs—for example, of the brain by some bones of the skull and of the heart and lungs by the ribs. The major bones of the limbs provide rigid links between joints, and these bones also provide sites for muscle attachment. These functions of the skeleton, in providing linkages and sites for muscle attachment, facilitate human movement and form the basis of the mechanical models of the human body.

Physiological Functions of the Skeletal System

One must remember that bone is a living, dynamic tissue. The framework for a house is often made of timber that was once a living tree. Similarly, the bones in an anatomy laboratory are changed from their condition in the living body. Thus, in addition to its mechanical functions, the living skeletal system has important physiological functions. When subjected to large forces over a long period, the framework of the house may start to fail, just as the metal in a car may start to fatigue or rust. But living bone has the advantage that it may heal when broken and even carry out maintenance to prevent failure. Some bone researchers believe that the stimuli to bone adaptation include microcracks that form during increased levels of physical activity. (Adaptation in response to physical activity is explored further in chapter 5.)

Bone tissue is also involved in the storage of essential minerals such as **calcium** and **phosphorus.** Additionally, bone **marrow** produces blood cells and is part of the body's immune system.

The Composition of Bone

In a car, the metal in the engine tends to be thick and rigid because it must resist deformation, but the body panels are more cosmetic and also help to absorb the energy of impact during a collision. As with the metal in the car, the composition of bone in the skeleton is not uniform. This makes sense because each of the constituents of bone must accommodate different physiological and mechanical functions.

Mechanical Properties Provided by Components of Bone

It has been argued that both **stiffness** and flexibility are important properties of bone. In a car the axles must be stiff to resist deformation as they transfer the forces from the engine to the wheels, but springs are designed to deform while the shock absorbers dampen the motion. Engineers can use different materials to suit particular purposes; they choose steel, aluminium, plastic, or other alternatives depending on the purpose of the part. To produce the optimal mechanical properties, metallurgists develop alloys that consist of different types of metal bonded together.

Is bone like a homogeneous metal, such as copper or iron, or does it consist of various components? The answer is the latter; consequently, bone is described as a composite material. One useful analogy for bone is fibreglass, which consists of glass fibres cured in an epoxy resin. The final product, fibreglass, has mechanical properties that are superior to those of its individual components.

With regard to the mechanical properties of bone, comparison with other materials is useful. Bone is similar to wood from the oak tree in its properties of strength and stiffness. Bone is about as flexible as fibreglass, although weaker, but is stronger and more flexible than ordinary glass. Ceramics in general are about one-third to one-half as strong in **compression** as bone. The blocks of older car engines were manufactured from cast iron, which is similar to bone in its **tensile** strength; but bone is three times lighter (that is, less dense) and much more flexible. Bone is clearly a remarkable physical material.

About one-quarter of the mass of bone in the living body is water. Adult bone, after removal from the body and drying, is about two-thirds inorganic crystals consisting mainly of calcium and phosphorus. The main mechanical components of bone are **collagen** (most of the organic component) and calcium salts (most of the inorganic component). The collagen provides toughness and flexibility, so it contributes to the tensile strength; bone's hardness and rigidity are due mainly to its calcium salts, which also contribute principally to its **compressive strength**. According to one calculation, the optimal mineralization of bone, in terms of strength and flexibility properties, is two-thirds mineral to one-third organic material. This is the proportion of calcium salts to collagen in healthy adult bone. Almost all the body's calcium is stored in bone and is released from there as required by other tissues, such as skeletal muscle, which requires calcium to contract.

As already noted, organic collagen fibres compose most of the organic one-third of dry bone. Bone cells are another organic component of bone. In the matrix of bone, the basic cells are called **osteocytes.** Also associated with bone are bone-forming cells, called **osteoblasts.** Depending on the local environmental conditions, osteocytes can become osteoblasts and vice versa so that adaptive responses can occur in the presence of mechanical stimuli. There are also bone-eroding cells, called **osteoclasts.** Remodelling of bone, which is a continual process in adults, involves organized erosion and deposition of bone tissue by the various cells. A full cycle of remodelling takes about three months.

Types of Bone
Metallurgists and engineers have at their disposal a variety of building materials, whereas the two types of bone within the human skeleton are made of the same material. **Compact** and **spongy bone** differ mainly in their porosity, although compact bone is also more organized than spongy bone.

Spongy bone (also termed **cancellous** or **trabecular bone**) is a lattice meshwork of bony rods **(trabeculae)** (figure 2.1a). This meshwork arrangement makes spongy bone much more "springy" than compact bone. In spongy bone, each osteocyte is close to a nutrient supply because the bony tissue is surrounded by blood vessels and associated material.

Compact bone (also called **cortical** or **dense bone**) is much more solid than spongy bone. When bone is remodelled to become compact, bone cells may be too far from their nutrient supply because now the blood vessels are surrounded by bony tissue. If compact bone were not organized in a specific way, this lack of nutritive supply would result in the death of each cell. Compact bone contains a basic unit **(Haversian system** or **osteon)** that is repeated many times. Haversian canals, longitudinally arranged in the shafts of long bones, carry blood vessels and are surrounded by layers of bone **(lamellae).** In each lamella, osteocytes are contained within **lacunae.** There is a limit to the number of lamellae so that bone cells are not too far from their supply of nutrition. Figure 2.1b illustrates the microscopic organization of this compact bone in contrast with the organization of spongy bone (figure 2.1a).

Most of the calcium within bone is in the compact bone, but the calcium in the spongy bone is more easily released into the blood when required.

The Architecture of Bone
In engineering, an important concept is the relationship between strength and mass. Elite athletes have a good "**power**-to-weight" relationship, which is the same idea. The analysis of bone architecture allows us to conclude that the bones in our body are very **efficient** in that they are able to withstand large forces and at the same time are relatively light.

Bone Shape and Organization
Where different functional requirements exist, there are variations in the arrangements of the two types of bone, as well as differently shaped bones. Bone shape and predominant function have a specific structure–function relationship. The long bones of the limbs have many attached muscles and act as rigid links between major joints (figure 2.2a). The flat bones of the skull help protect the brain. Their structure, consisting of two layers of compact bone with spongy

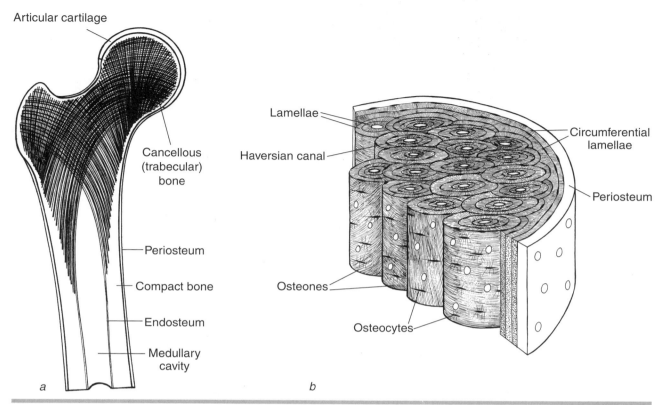

Figure 2.1 *(a)* The trabeculae of spongy bone and *(b)* the architecture of compact bone in the shaft of a long bone, shown in longitudinal and transverse sections.
Adapted from Watkins 1999.

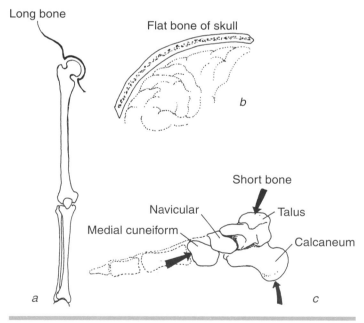

Figure 2.2 Examples of bones with different shapes that serve different functions. Muscles are attached to most bones to facilitate movement. *(a)* Long bones of the limbs, *(b)* flat bones of the skull, *(c)* short bones of the hindfoot.

bone in between (figure 2.2*b*), is similar to that of a bicycle helmet. The ribs protect vital organs in the chest such as the heart. The short bones of the hindfoot have a thin layer of compact bone around the outside and are filled with spongy bone (figure 2.2*c*). These weight-bearing bones resist the large compressive forces generated during running and help cushion **ground reaction forces.** All the bones illustrated in figure 2.2 consist of both compact and spongy bone, in differing proportions and structural arrangements.

Note the arrangements of compact and spongy bone illustrated in the different parts of a **vertebra** (figure 2.3). A vertebra is classified as an irregular bone, but it is more like a composite bone consisting of three main components. The vertebral body is the main weight-bearing part of the vertebra and is constructed like a short bone; in section, one can see the thin layer of compact bone surrounding a mass of spongy bone. The **lamina** of the vertebral, or neural, arch

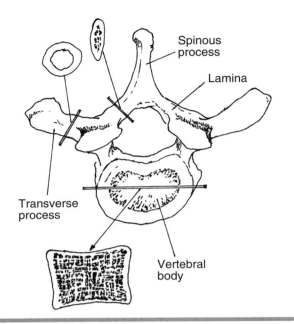

Figure 2.3 A vertebra is classified as an irregular bone, with three main components: the weight-bearing vertebral body, the "flat" lamina, and the long processes.

protects the spinal cord and is constructed like a flat bone; in section it looks like one of the ribs that help protect the heart and lungs. The transverse process has areas of attachment for ligaments and muscles, providing leverage for these structures, and is constructed like a long bone.

Amount, **density,** and distribution of bone material have major effects on the mechanical properties of a whole bone. Compare the resistance of a long plank of wood to bending in the vertical direction when it is supported at both ends and placed either on its narrow edge where it does not deform or on its wide side where it bends easily like a springboard. Also compare the deformation of high- and low-density foam mattresses under the weight of a person.

Architecture of Long Bones

The bones of the lower limb are often under compression during standing, while the bones of the upper limb may be subjected to **tension** while one is holding a load in the hand or performing a giant swing on the high bar (see figure 2.4). During forward bending of the trunk, there are **shear** forces between adjacent vertebrae. During walking, the long bones of the lower limb may be acted on by **bending** and torsional (twisting) forces. The compressive forces on the ends of the femur in the thigh are not aligned in the same straight line, causing bending, which also occurs because of the natural bend of the shaft of the femur. Twisting about the long axis of the foot immediately after heel strike in jogging, commonly called hindfoot pronation, produces **torsion** of the main leg bone, the tibia. These patterns of loading require optimal arrangement of the bone material, but what is the most efficient architecture? The shafts of

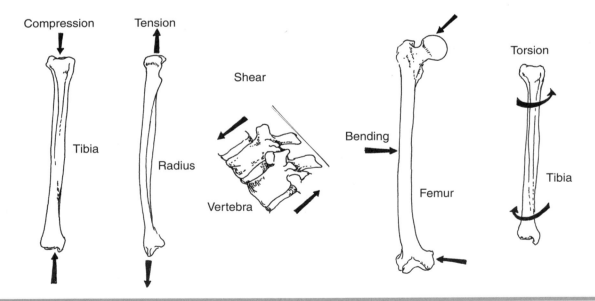

Figure 2.4 Examples of the different types of forces that may act within and between bones. Major forces that the long bones of the limbs, particularly the lower limbs, must withstand are compression along their long axes, bending, and twisting. The bones of the upper limb may be subjected to tension.

long bones are hollow, giving them mechanical advantages over a solid rod of the same mass. If a pipe and a solid rod contain the same amount of material, the pipe is stronger in bending than the rod, so a hollow shaft is more efficient in terms of the relationship between strength of the whole bone and its weight. A hollow shaft also resists twisting better than a solid rod. It appears from mechanical analyses that the shapes of our bones are close to optimal for the activities that most humans perform.

During movement, especially during landing from a jump, forces are transferred from one bone to the next via the joints. A large contact area between the bones results in less pressure on the ends of the bones, so expanded ends are advantageous. Much of the material forming the expanded ends is spongy bone that absorbs energy during impact. Everybody knows the difference between jumping on a trampoline and jumping off the trampoline onto a concrete floor. Compact bone is like concrete because it does not deform much but is very strong, whereas spongy bone is more like a trampoline that cushions the force of landing as it deforms. In fact, spongy bone is 10 to 15 times more flexible than compact bone. At the end of the long bone, a thin outer layer of compact bone protects the overlying cartilage during impact and transfers the forces to the underlying spongy bone (figure 2.5). The deformable **articular cartilage** acts like an air-filled bicycle tyre, which is supported by a light but solid rim, like the thin layer of compact **subchondral bone**. The trabeculae of the expanded ends line up along the major lines of force to perform a function similar to that of the spokes of a wheel, which help to support the rim and maintain its shape while also acting to cushion the forces from bumps in the road. Recall that the composition of compact and spongy bone is the same; just the organization is different.

The Articular System

The system of **joints** between the bones is called the articular system because the union of bones is an **articulation.** Joints are important because they allow movement, but they must remain stable during movement. The study of joints is called **arthrology.**

Classification of Joints

We all know that joints allow cars and bicycles to move. These are relatively simple joints com-

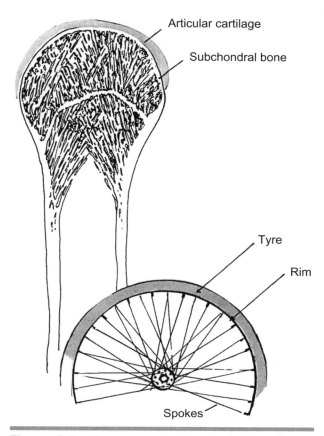

Figure 2.5 The expanded end of a long bone is compared with a bicycle wheel consisting of spokes, rim, and tyre.

pared with the synovial joints in the human body; in fact, engineers are still attempting to explain the functions of biological joints. The design of artificial joints for replacement of human joints such as hips and knees is progressing rapidly, but replacement joints still do not work as well as the natural joints.

To give some idea of the complexity of natural joints we can contrast them with the joints in a car. The bearing surfaces in the artificial joints of a car are usually smooth hard metal, whereas the articular cartilage is less smooth but quite deformable. Artificial joint surfaces are very regular, usually consisting of part of a circle or sphere, whereas human joint surfaces are ovoid (egg shaped), resulting in much more complex movement patterns. To remove debris in the joints of a car, we take the car to the local garage for an "oil change." Any effects of wear result in rapid deterioration of artificial joints; but within a synovial joint, cell debris can be removed continuously.

The materials between the bones forming the joints may differ, providing the basis for the structural classification of joints. All anatomical joints may be described as **fibrous, cartilaginous, or synovial.** Most of the major joints we are familiar with, such as the shoulder, elbow, knee, and ankle, are examples of synovial joints, so this section focuses exclusively on these joints.

The major function of joints is to allow movement but at the same time remain stable. Synovial joints allow a relatively large amount of movement; consider, for example, the ranges of motion that are possible at both the shoulder and elbow joints.

Characteristic Features of Synovial Joints

A number of characteristic features are associated with a typical synovial joint (figure 2.6). On the ends of the bones forming the joint is articular cartilage, which consists of collagen fibres in a liquid matrix. Articular cartilage has a water component of about 80% and has been likened to a sponge from which water can be squeezed. When not being deformed it absorbs water because of the dissolved chemicals that it contains. Articular cartilage forms a relatively

smooth bearing surface and also acts, through its capacity to deform, to cushion forces.

A **joint capsule** forms part of the boundary of the joint. The joint capsule contains a high proportion of collagen fibres and provides some intrinsic stability to the joint as well as some resistance to motion. Forming the inner layer of the joint capsule is the **synovial membrane,** which has a number of important functions. It produces the fluid within the joint and also removes the cell debris that results from "wear and tear" within the joint.

A major characteristic of a synovial joint is the cavity bounded by the articular cartilage and the synovial membrane of the joint capsule. In this joint cavity is a small amount of **synovial fluid** that contains constituents of blood, substances secreted by the synovial membrane, and some products resulting from abrasive wear within the joint.

The synovial fluid has three important functions: lubrication, protection, and nutrition. The **viscosity** of the fluid can change according to local environmental conditions, so it may be thick and protective, like grease, or thin and lubricating, like oil. Because articular cartilage does not receive an adequate blood supply, nutrients are supplied via the synovial fluid in contact with the cartilage. Pressure during movement helps squeeze fluid out from the cartilage. When the pressure is removed, the liquid can seep back into the cartilage. Physical activity thus promotes the nutritional function of synovial fluid.

To help maintain the integrity of the synovial joint, associated **ligaments** attach from bone to bone and cross the joint. Ligaments consist of collagen fibres, also a constituent of bone. In ligaments, the collagen forms about 90% of the structure, and the fibres tend to run parallel to each other. The ligaments help the joint capsule provide stability and also function to guide the joint's movements. In doing so, they provide some resistance to joint motion. Ligaments are basically passive structures that resist tensile forces, which tend to separate the bones forming the joint.

Classification of Synovial Joints

One can use the classical anatomical view to classify synovial joints on a descriptive basis or use the view of engineers, who are interested in movement between the joint surfaces in contact, to produce a system to explain joint lubrication and wear. It is not our aim in this section to provide details of the alternative classification systems; instead we summarize the structural and functional criteria on which each system is based.

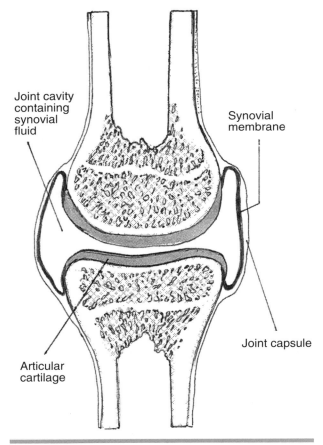

Joint cavity
containing
synovial
fluid

Synovial
membrane

Joint capsule

Articular
cartilage

Figure 2.6   Representation of a typical synovial joint indicating its characteristic features.

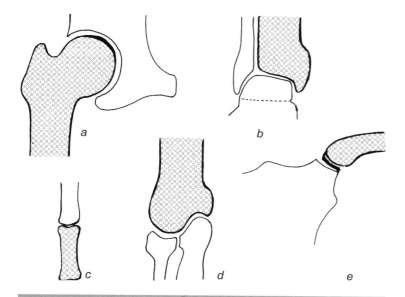

Figure 2.7 Representations of some types of synovial joints: (a) spheroidal (ball-and-socket) joint, (b) hinge-like joint, (c) simple joint, (d) compound joint, and (e) complex joint containing an intra-articular disk.

Range of Movements Allowed by Synovial Joints

Movements that occur at synovial joints have traditionally been described in terms of the three major planes of the body (figure 2.8). The **sagittal plane** is a vertical plane dividing the body into left and right parts; the **coronal (frontal) plane** is a vertical plane dividing the body into front (anterior) and back (posterior) parts. The **transverse plane** is horizontal. Anatomically, movements that occur in the sagittal plane have been termed **flexion** (when the angle between the limb segments decreases) and **extension** (when the angle increases) (figure 2.8). These gross descriptions of directions of movements of body segments can be contrasted with the more mechanical approach in which the terminology relates to the movement that occurs between the articular surfaces in contact.

Anatomically, each synovial joint can be classified on the basis of the approximate geometric form of its articulating surfaces. For example, the hip looks like a ball in a socket (figure 2.7a), much like the joint between a car and a trailer, so it is called a ball-and-socket joint. Synovial joints can also be classified according to the gross movements permitted. For example, because the ankle joint allows movement basically in only one plane, it is called a hinge joint (figure 2.7b). The ankle could also be classified as a uniaxial joint because it allows movement about only one principal axis. The joints forming the knuckles of the fingers are classified as biaxial because they allow motion about two principal axes (figure 2.7c). Note that the fingers can be moved in two directions at right angles—backward and forward and from side to side.

The complexity of organization of the joint structures is another criterion for classification. If a joint consists of two bones, with one pair of articulating surfaces, it can be classified as a **simple joint;** an example is the knuckles of the fingers (figure 2.7c). If there is more than one pair of articulating surfaces, such as within the elbow joint capsule (where there are three pairs), the joint is classified as **compound** (figure 2.7d). Sometimes synovial joints contain intra-articular structures such as a cartilaginous disk or **meniscus,** and these joints can be classified as **complex** (figure 2.7e).

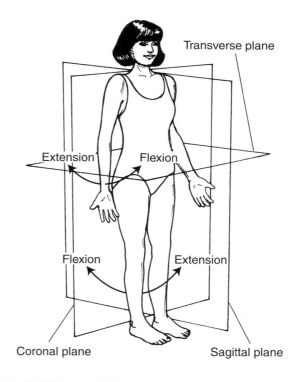

Figure 2.8 The major planes of the human body with respect to the anatomical position. The movements of flexion and extension of the elbow and knee are illustrated.

Adapted from Whiting and Zernicke 1998.

Using the engineering approach, we can describe the relative motion between articular surfaces as spin, slide, or roll (figure 2.9). The movements can be likened to the spinning of a top, the skidding of a tyre on a road, and the normal rolling of a tyre on the road, respectively. These movements have an effect on the frictional resistance. For example, resistance is much less when the wheels of a car are rolling freely than when the brakes are applied and the tyres slide along the road. Often, combinations of all movements occur during joint motion, such as during flexion and extension at the knee.

Joint Protection, Lubrication, and Wear

The structures in contact within a synovial joint are the articular cartilages on the ends of each bone. The **cartilage** is smooth and deformable, so it can cushion forces applied to its surface. The subchondral bone, which is the thin layer of compact bone under the cartilage, provides a solid base and helps protect the cartilage from damage. This thin layer of compact bone sits on a

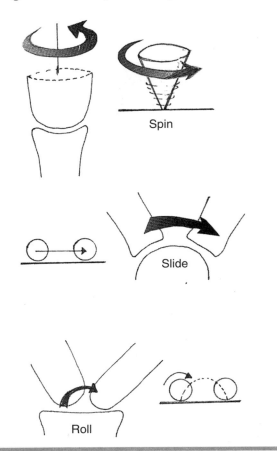

Figure 2.9 The motions of spin, slide, and roll between articular surfaces in a synovial joint.

Spin

Slide

Roll

more deformable network of bony rods forming the spongy bone (figure 2.5). This organizational structure is similar to that of a bicycle wheel with spokes, with the spokes maintaining the shape of the rim but also helping cushion forces from the ground. (A solid bicycle wheel would be too stiff for general purposes.)

The synovial fluid acts as a lubricant so that friction between articular cartilages in a synovial joint is less than between two blocks of smooth ice. Synovial joints have a number of important characteristics differentiating them from artificial joints. Movement at a synovial joint can best be described as oscillatory ("backward and forward"), but in most load-bearing joints in machines the angular motion is in only one direction ("round and round"). The load-bearing surface in a synovial joint is deformable cartilage, which is both elastic and porous; and the synovial fluid has particular chemical properties that allow it to act as a lubricant. This arrangement is much more complicated than that of most joints used in cars, in which the bearing surfaces are very hard.

The Joint As the Functional Unit of the Musculoskeletal System

Human movement studies and **functional anatomy** emphasize movement, and therefore the joint is the focus of functional musculoskeletal anatomy.

The characteristic features of a synovial joint have been shown (figure 2.6), but each joint has associated with it a number of other structures. These include the following:

- On either side of the joint are bones that act as levers and aid force cushioning as explained earlier (figure 2.5).
- **Skeletal muscles** have a role in movement because they cross joints and thus initiate and control movement. The forces they produce across the joint also stabilize it. Thus muscles are secondary stabilizers, in addition to the associated ligaments. If you contract all the muscle groups about a joint simultaneously so that no movement occurs, the joint becomes much more stable. The joint is also stiffer, as you would realize if somebody else attempted to move it.
- Muscle forces are transmitted to bony attachments via tendons; thus it is the muscle–tendon unit that is of major significance.

- Nerves are also associated with joints. As we will see in chapter 14, motor nerves within the **central nervous system** provide some control over the muscles producing actions at a joint; sensory nerves provide feedback about joint position and movement from a variety of sensors located in the joint capsule and ligaments, as well as in the muscles and tendons.

Injury to any intrinsic joint structure, or to the associated structures, will result in functional impairment of the entire joint and adjacent body segments. Functional impairment of one joint often results in a "chain reaction" affecting adjacent joints. For example, an ankle or knee injury will affect the whole lower limb and therefore adversely affect performance during walking or running. Because of altered function of the injured joint, and possibly pain associated with the injury, the other joints must attempt to compensate, usually resulting in a limping **gait**.

The Muscular System

The muscular system comprises the main effector structures for human movement. It is an important part of the **musculoskeletal system** because it produces joint motion (see part II on biomechanics); but it is also an important part of the neuromuscular system in that it is controlled by the central nervous system (see part IV on motor control).

The Structure of the Muscular System

We can identify major muscles that lie just beneath the skin, such as the "pecs," "lats," and deltoid and biceps muscles. An understanding of the mechanisms of action of these muscles relies on knowledge of muscular tissue at a number of levels of organization, including its microscopic structure.

Association of Muscles With Other Structures

As with bones and joints, the structure of muscles is related to their function. Muscles cross joints, and the major skeletal muscles of the trunk and limbs are attached to bones at both ends. The muscle–tendon unit therefore consists of a chain of structures, typically bone–tendon–muscle–tendon–bone.

The attachment sites of muscle and **tendon** are important because they determine the action of each muscle. Whenever a muscle shortens, it tends to pull the two attachment sites closer together; so the relationship between direction of pull of

the muscle and the axis of rotation of the joint determines the resulting joint action. This type of analysis is the basis for defining the actions listed in tables in all basic gross anatomy books. Functional anatomists, starting with Duchenne in the late 19th century, have used other techniques to attempt to unravel the complexity of muscle contraction when different muscles act simultaneously, or when joint position is changed from the "normal" anatomical position illustrated in figure 2.8.

Structural Features of Muscles

So far in this discussion, we have used the term *muscle* in relation to only one type of muscle, skeletal muscle. There are, however, other types of muscle tissue as discussed in part III (physiology). **Smooth muscle** is found in the walls of the digestive system and certain blood vessels. **Cardiac muscle** forms the major part of the walls of the heart. In this and subsequent functional anatomy chapters we restrict our discussion to skeletal muscle.

Not all muscles look the same. The "typical" muscle has a belly and tapers toward the tendon that attaches it to bone at each end, but there are other shapes. Muscles attach to bone either directly or via a tendon. Examples of muscles with different architectural arrangements are shown in figure 2.10. Muscle fibres are oriented in the direction of pull of the whole muscle (e.g., spindle) or at an angle to this direction (e.g., unipennate). The

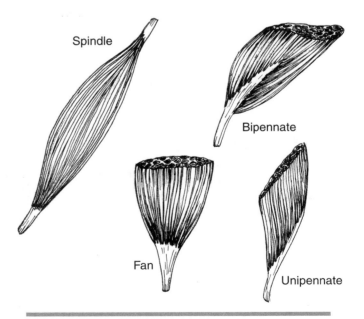

Figure 2.10 Examples of muscles with different shapes and different arrangements of fibres.

hip and thigh region contains examples of these shapes. *Semitendinosus* (one of the **hamstring muscles** in the back of the thigh) has a spindle shape, while another hamstring muscle *(semimembranosus)* is a unipennate muscle. *Rectus femoris* (one of the four muscles forming the quadriceps muscle group at the front of the thigh) is a bipennate muscle. Underneath the large *gluteus maximus* in the buttock is the fan-shaped *gluteus medius.*

Skeletal muscle cells are elongated and contain many nuclei; hence muscle cells are called fibres. The alternating thick and thin filaments produce characteristic striations that one can clearly see when viewing muscle tissue through an electron microscope. This appearance is produced by tens of thousands of repeating units in series that form each fibre. These repeating units, called **sarcomeres,** are visible in a photograph taken by an electron microscope. The sarcomere is the structural and functional unit of muscle. You can gain an idea of its structure by looking at figure 2.11. The function of the components of

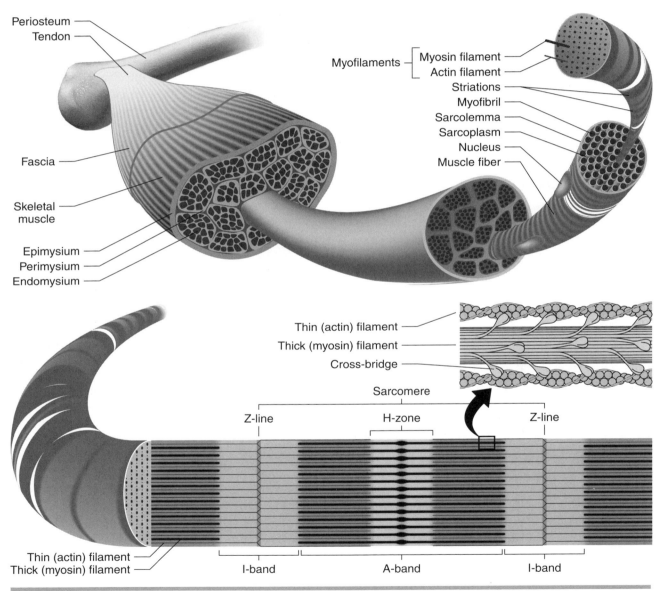

Figure 2.11   A typical muscle comprises bundles (fascicles) of muscle fibres. Each muscle fibre, in turn, contains many myofibrils that are made of repeating series of sarcomeres. In the M region at the middle of each sarcomere are myosin filaments. Overlapping slightly with each end of every myosin filament are actin filaments. The actin filaments stretch from their attachment to the Z disk at either end of the sarcomere toward the M region in the middle of the sarcomere.

Reprinted from Behnke 2001.

the sarcomere will be discussed under "Muscle Contractions." The connective tissue element of the muscle consists of thin sheets surrounding each fibre, each bundle of fibres, and the whole muscle itself.

The structural and functional unit of the neuromuscular system is a **motor unit** consisting of a nerve and the **muscle fibres** that it controls (figure 2.12). Muscles over which we have fine control, such as the small muscles in the hand, have relatively few muscle fibres per motor unit, whereas the large muscles of the lower limb, over which we have only relatively coarse control, have many more muscle fibres per motor unit.

Distinguishing Properties of Muscles

The macro- and the microstructure of muscle give it five properties that are crucial to its function.

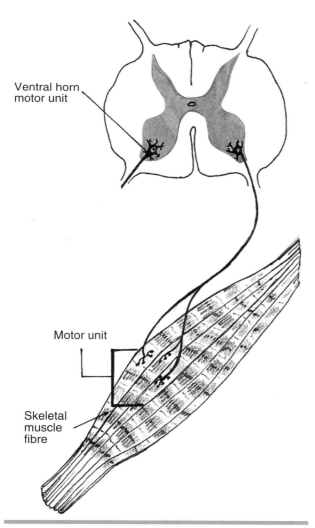

Figure 2.12 Illustration of a motor unit in a limb muscle where the nerve fibre originates in the spinal cord.

Muscle as a tissue has three main properties:

1. Its excitability in response to nerve stimulation
2. Its contractility in response to the stimulation
3. Its conductivity, which allows the electrical signal produced by the muscle fibres in response to neural stimulation to travel along those fibres

A whole muscle has two additional properties that arise primarily from its structure and the mechanical characteristics of the connective tissue within it:

1. Extensibility
2. Elasticity

Because muscle tissue responds to neural activation by producing a force, it is regarded as an active tissue, in contrast to the passive connective tissue component, which can only resist applied forces.

Muscle Contractions

Muscle fibres operate by producing a force that tends to shorten the muscle. Knowledge of the microstructure of muscle tissue is necessary for understanding this process.

In the lengthened position, there is little overlap between the longitudinal thick and thin protein filaments contained in each sarcomere. The muscle shortens (contracts) by increasing that overlap. Thus researchers proposed the **sliding filament hypothesis** of muscle contraction, which states that the shortening of the sarcomere results from the thick and thin filaments sliding toward one another. Further research led to the **cross-bridge hypothesis,** which states that muscles shorten when cross-bridges are formed as the thick myosin filaments attach themselves to the thin **actin** filaments connected to the **Z disk** on either end of the sarcomere. These "bridges" are the means by which the **myosin** filaments pull the actin filaments toward themselves, shortening the muscles. Both hypotheses, which are generally accepted, indicate that there are both upper and lower limits to sarcomere (and hence muscle) length.

Activation via a nerve supplying a muscle prompts both a mechanical response and an electrical response. The electrical response spreads over the surface of each muscle fibre, causing a series of steps to occur. Here we highlight only some of the major steps in a process known as **excitation–contraction coupling:**

1. In the neuromuscular (nerve–muscle) junction, a chemical is released from the end of the nerve fibre, which causes a rapid change in voltage in the muscle.

2. The electrical signal travels over the surface of, and along, the muscle fibres.

3. The electrical signal causes the release of calcium ions into the cytoplasm (intracellular fluid) of the muscle fibre. This is an important part of the process because it is the link between excitation and contraction.

4. Calcium ions expose active sites on the actin (thin) myofilaments, to which the myosin filaments immediately attach.

5. By means of these attachments (cross-bridges), each myosin filament pulls the actin filaments that overlap with it at either end toward its centre, producing cross-bridge cycling, which is the mechanical response to the electrical signal (figure 2.13).

The whole process takes a couple of milliseconds. Although almost all the calcium in the body is stored in the bones, the 1% that is unbound in skeletal muscle is essential for muscle contraction.

Cross-bridge cycling involves four steps (figure 2.13): (1) attachment of the myosin to the active site of the actin filament (when the active site is exposed in the presence of calcium ions); (2) pivoting of the myosin head to pull the actin toward it (power stroke); (3) release of the myosin head from the actin; (4) reactivation of the myosin (return to "straight head" configuration of myosin).

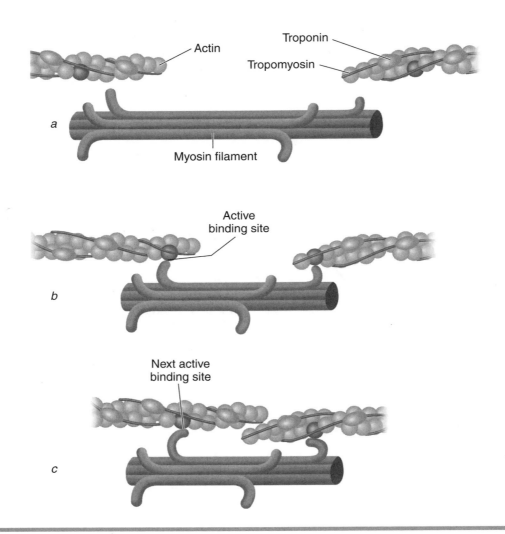

Figure 2.13   The thick filament binds to the active site on the thin filament, pulls the thin filament toward it, then detaches ready to repeat the process.

Reprinted from Wilmore and Costill 2004.

Electrical activity of a muscle can be detected using electrodes, in a technique called **electromyography (EMG).** Electromyography involves detecting the small electrical signal produced by muscle slightly before it contracts, and then using appropriate hardware and software to record and analyse the signal. A basic tool in functional anatomy, EMG has been used to determine muscle roles in movement situations, to provide biological feedback to a person attempting to improve motor performance, and to investigate the effects of strength training.

Mechanics of Muscular Action

The force produced by muscle activation places tension on the attached tendon or bone to produce joint motion. As we shall see, muscles do not always shorten when they produce force.

Types of Contractions

When muscles contract, they can perform three main actions. It is the muscular system that provides the power for the human body to perform work. Muscles cross joints so that their contractions produce movement, and whenever a muscle contracts it tends to shorten and pull the two bony attachments closer together. This action of a muscle is termed **concentric,** in contrast to an **eccentric** action that occurs when a muscle is activated but is lengthening. In this situation other forces (such as external loads) prevent the muscle from shortening. During eccentric actions, muscles are controlling the movement produced by the other forces. (For example, while lowering a weight, the muscle is controlling the movement produced by gravity working on the weight.) During an **isometric** action, the muscle is activated but the overall length of the muscle–tendon unit does not change, so this type of action is important for the stabilization of joints. Muscles may therefore act concentrically to produce movement, eccentrically to control movement, or isometrically to maintain posture and enhance joint stability. Joint stability is also a by-product of the dynamic types of action. In figure 2.14, the elbow flexor muscles act concentrically to move the load up against the resistance of gravity and then act eccentrically to control the downward movement. The joint action descriptions that often appear in tables in anatomy books are based on the assumption that the joint movement is produced by a concentric muscle action.

Explaining Joint Actions

The joint actions caused by muscles would be relatively simple to predict and explain if each

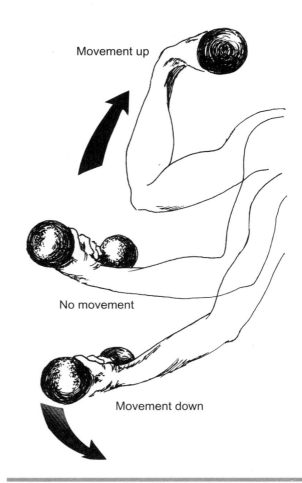

Figure 2.14 The three major types of skeletal muscle action are isometric, concentric, and eccentric. Isometric actions occur in static situations (no movement) whereas concentric (muscle shortening) and eccentric (muscle lengthening) actions occur during dynamic tasks.

muscle crossed only one joint, that is, if all muscles were **monoarticular.** However, in addition to monoarticular muscles, there are also muscles that cross two joints **(biarticular),** such as the hamstring muscle group at the back of the thigh, and others that cross more than two joints **(polyarticular),** such as the muscles in the forearm that bend the fingers. A basic rule for predicting joint movement is that when a muscle is activated it shortens and tends to produce all the joint actions of which it is capable. If these actions do not occur, the reason must be that external forces are present or that other muscles are preventing some of the actions. In these situations, even a purely mechanical analysis of the muscular system is quite complex.

THE JOINT-STABILIZING ROLE OF MUSCLE–TENDON UNITS

Bioengineers are fascinated with the human body, and many are involved in the design of replacement parts such as artificial hips and knees. A human synovial joint is analogous to a tent that relies on the interaction between the following structures for its stability: The bones with cartilaginous ends are like the poles of a tent that resist compressive forces, so the mechanical properties of bone and cartilage have been measured using specially designed equipment. A tent that relied solely on the poles would be highly unstable; it would simply blow over in windy conditions. For extra stability, guy ropes are attached to the poles, and these ropes must be flexible (or attached to springs) to perform their functions optimally.

Ligaments are flexible structures that stabilize joints. Their tension varies through the range of joint motion because of the complex shapes of joint surfaces, so joint stability is dependent on joint position. Joints must not be completely stable because they must also allow movement. Stabilizing features of joints, which tents do not share, are the muscle–tendon units that are under neural control and that produce the movements at the joints, control the movements, or both.

What is the primary mechanism that helps prevent our ligaments from tearing and our joints from dislocating? It is the stabilizing influence of the contraction of surrounding musculature. It is unfortunate in terms of injury prevention that the time required to damage a ligament is less than the time of a simple muscular reflex response to muscle stretch. This means that the stabilizing muscles must be activated prior to the imposition of external deforming forces or they cannot prevent joint dislocation and ligament tears. When you step on the edge of a hole that you have not seen, you are likely to sustain a sprained ankle because you have relied on the ligaments for stability and, unaided by surrounding musculature, the ligaments are usually not up to the job in that situation. If you see the same hole ahead of time, however, the surrounding muscles are activated before you step; the muscle–tendon units add stability and you are much less likely to sprain your ankle joint. The activation patterns of muscles during human movement is a major research area.

In a whole muscle, the shortest length is determined by how much the actin and myosin filaments in each sarcomere of the muscle tissue can overlap, but the upper limit is determined mainly by the extensibility of the muscle's connective tissue component. When joints are moved through their full range, most muscles in the human body normally operate well within their available length range. In some cases, however, the limits are reached. Biarticular and polyarticular muscles may be limited, for example, because their lengths are determined by motion at a series of joints.

Try these movements to demonstrate both the upper and lower length limits of muscles crossing more than one joint. Stand on one foot and bend the unsupported limb as far back as possible at the hip (extend the hip). Now try to bend (flex) the knee as far as possible (figure 2.15a). You will notice that the range of knee flexion is less than normal when the hip is flexed. (You can verify this by trying the same knee movement with the hip flexed.) Now lie on your back and bend (flex) one hip so that the thigh is vertical, then attempt to straighten the knee (figure 2.15b). Most people are unable to do this. If you can straighten your knee in this position, flex the hip more and try again. These two examples principally indicate the lower and upper length limits of the hamstring muscle group that crosses both the hip and knee joints.

To demonstrate the lower limit of the muscles that bend the fingers, first bend (flex) the wrist so the palm of the hand comes as close as possible to your forearm and then try to make a fist. Notice that this is a very weak grip compared with your strongest power grip. Notice the position of the wrist when the grip is strongest. The muscles that provide most of the strength for bending the fingers have their bellies in the forearm and tendons that cross a number of joints, including

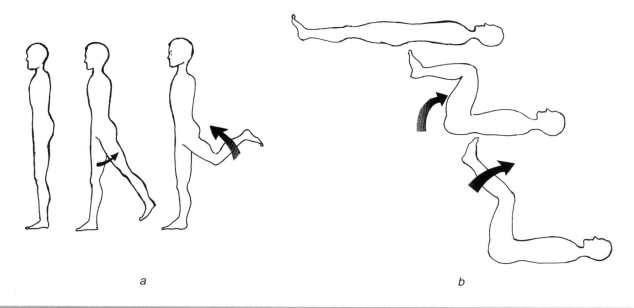

a   *b*

Figure 2.15   Exercises used to demonstrate the *(a)* shortened and *(b)* lengthened positions of the hamstring muscle group.

the wrist and others within the hand. When the wrist is flexed, the finger-flexing muscles are already shortened; they have almost run out of the capacity to shorten any further to produce finger flexion.

Limitations to Range of Joint Motion

When a joint reaches the end of its range of motion, this limitation has several possible causes in addition to the intrinsic features of the joint illustrated in figure 2.6. When joint motion occurs, the body segments bend such that the tissues on one side are compressed while those on the opposite side are stretched. The most obvious limiting factor is therefore the tension in the joint capsule and its associated ligaments on the side of the joint where stretching is occurring. In addition, and most commonly, stretch of the associated muscles and tendons also restricts the range of joint motion. Some of the resistance to joint motion through its range is also provided by stretching of the skin. Sometimes apposition of soft tissues on the compressive side of the joint restricts movement; and, rarely, apposition of bony parts forming the joint will restrict movement. This bony contact would occur only at the extreme end of the range and is potentially injurious in dynamic situations.

Determinants of Strength

What is **muscular strength?** Mechanically, strength is determined by both the force generated by muscular contraction and the leverage of the muscle at the joint. This concept, which is termed **moment of force,** is examined in detail in chapter 6. Muscle force is proportional to the cross-sectional area of the muscle.

Summary

The musculoskeletal system consists of bones, joints, and muscles. Bone tissue is both hard and tough. Compact bone is particularly strong, while spongy bone is better for shock absorption. Individual bones have a variety of shapes, with each shape being particularly suited for performing one of the functions of bone.

Joint mobility and stability tend to be competing requirements, so the structure of each joint is a compromise. The many types of joints are classified on the basis of their structure, the gross movements they allow, or the motions that occur between the articular surfaces in contact.

Skeletal muscle tissue looks striated in a longitudinal section under a microscope. A muscle fibre consists of up to tens of thousands of sarcomeres in series (joined end to end). Connective tissue is an important component of whole muscles and provides some of the properties of the muscle–tendon unit. The sliding filament hypothesis of muscle contraction describes what happens when muscle changes length and is based on electron micrographs of muscle at different lengths. According to the cross-bridge hypothesis, cross-bridges on the thick myofilaments pull the thin myofilaments toward them after excitation from the nerve that supplies the muscle. Length changes resulting from neural activation are dependent on the net effect of a number of forces. The muscle–tendon unit may act isometrically to stabilize joints, concentrically to produce movement, and eccentrically to control movement. The electrical response of muscle to neural stimulation can be detected and used as a tool in the study of normal muscle contraction. Joint motion may be limited by the length of muscle–tendon units, which may be increased by flexibility exercises. Strength is determined by muscle cross-sectional area (related to the number of fibres in parallel) and leverage of the muscle that produces joint motion.

Further Reading and References

Basmajian, J.V., and C.J. de Luca. 1985. *Muscles alive: Their functions revealed by electromyography.* 5th ed. Baltimore: Williams & Wilkins.

Cowin, S.C., ed. 2001. *Bone mechanics handbook.* 2nd ed. Boca Raton, FL: CRC Press.

Jenkins, D.B. 2002. *Hollinshead's functional anatomy of the limbs and back.* 8th ed. Philadelphia: Saunders.

Levangie, P.K., and C.C. Norkin. 2001. *Joint structure and function: A comprehensive analysis.* 3rd ed. Philadelphia: Davis.

Nordin, M., and V.H. Frankel. 2001. *Basic biomechanics of the musculoskeletal system.* 3rd ed. Philadelphia: Lippincott, Williams & Wilkins.

Oatis, C.A. 2004. *Kinesiology: The mechanics of pathomechanics of human movement.* Philadelphia: Lippincott, Williams & Wilkins.

Van De Graaff, K.M. 2002. *Human anatomy.* 6th ed. Boston: McGraw-Hill.

CHAPTER 3

Basic Concepts of Anthropometry

*I*n this chapter we examine the field of anthropometry and discuss the measurement of, and variation in, human body size, shape, and composition. We also discuss the implications of these variations for movement capability.

The major learning concepts in this chapter relate to

- the definition of anthropometry,
- the basic equipment used for physical measurement,
- quantifying body size and shape by determination of the relationships between different body segments,
- the composition of the human body,
- describing physique through use of a shorthand method, and
- human variation in the musculoskeletal system and in physical dimensions.

Definition of Anthropometry

Anthropometry is the science that deals with the measurement of the size, proportions, and composition of the human body. In most cases it is the sizes that are directly measured, and these direct measurements can be combined to indicate the shape of the whole body or body segments (figure 3.1). Body composition typically involves using direct anthropometric results to predict the relative amount of a particular component in the whole body. The best-known example is the use of skinfold thicknesses to predict the percentage of fat in the body (see "Predicting Body Density and Body Composition" on p. 35).

One specialized branch of anthropometry of particular relevance to the study of human movement is kinanthropometry. **Kinanthropometry,** a term proposed by Canadian academic Bill Ross, has been defined by the International Society for the Advancement of Kinanthropometry as the "scientific specialisation dealing with the measurement of people in a variety of morphological perspectives, its application to movement and those factors which influence movement, including:

- components of body build
- body measurements
- proportions
- composition

- shape and maturation
- motor abilities and cardiorespiratory capacities
- physical activity including recreational activity as well as highly specialised sports performance.

Defined as such, kinanthropometry is a scientific specialisation closely allied to physical education, sports science, sports medicine, human biology, **auxology,** physical anthropology, **gerontology,** ergometry, and several medical disciplines" (www.isakonline.com).

Tools for Measurement

In anthropometry, **stadiometers** are used to measure height **(stature), anthropometers** to measure lengths of body segments, tapes to measure body segment circumferences, **bicondylar calipers** to measure bone diameters, **skinfold calipers** to measure thicknesses of skin plus **subcutaneous** fat, scales to measure masses, and so on.

Simple or complex mathematical manipulation can be used to derive indices to describe the shape of a body segment. The cross-sectional shape of the thorax **(thoracic index)** can be described by dividing the anteroposterior (from front-to-back)

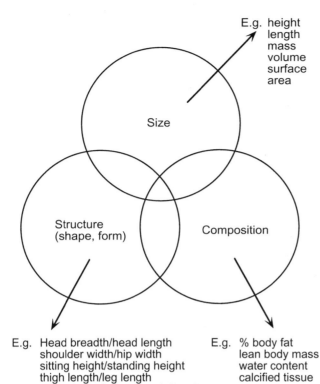

E.g. height
length
mass
volume
surface
area

Size

Structure
(shape, form)

Composition

E.g. Head breadth/head length
shoulder width/hip width
sitting height/standing height
thigh length/leg length
neck circumference/neck length

E.g. % body fat
lean body mass
water content
calcified tissue

Figure 3.1 Venn diagram indicating that the three components of anthropometry overlap with one another. Relative sizes indicate shapes of body segments, while simple measurements may be used to predict body composition.

diameter by the transverse (side-to-side) diameter. Statistical procedures are generally used to formulate prediction equations when the direct measurement of a parameter presents difficulties such as cost, time, or technical complexity.

In kinanthropometry, the basic measurement tools may be anthropometric equipment, but the researchers require important skills for interpretation of the results. Required knowledge includes the anatomy and biomechanics of the human musculoskeletal system. Because of the breadth of knowledge required by kinanthropometrists, this aspect is not emphasized here; however, examples are provided in chapter 5.

Body Size

The number of measurements of the dimensions of the human body that one could take is almost infinite. Anthropometry manuals recommend a core number of measurements with a longer list of optional measurements. In most cases, the manuals describe techniques to measure the size of the whole body and its various segments. Body size includes height and mass; segment sizes include lengths, circumferences, diameters, and so on. Individual segment masses cannot be directly measured, so these are predicted from previous research. As the techniques of anthropometry are critical to measurement **validity,** the measurer or experimenter must rigorously follow predetermined measurement protocols. Different manuals may describe different protocols for apparently the same body segment, so it is crucial that the measurer choose one particular protocol and adhere to it closely.

Determination of Body Shape

Body shape can best be described by the proportions between the sizes of different body segments. These proportions indicate the structure of the body. The relationship between different measurements may be termed *dimensionality.* The relationship between height and weight is a common example. In modern nutritional surveys, the **body mass index (BMI)** is often quoted. This is the mass (in kilograms) divided by the height (stature in metres) squared: $BMI = mass/ht^2$.

Many indices have been used to describe shape. Have you ever noticed that sometimes when people are standing up there are large differences among their heights but when they sit down the differences are very small? This observation is related to their body proportions. Some people have relatively long lower limbs and others relatively short limbs. To describe this relationship, the sitting height is divided by the standing height and multiplied by 100, so it is expressed as a percentage. If the lower limbs are investigated, the ratio between the lengths of the thigh (limb segment between hip and knee) and the leg (limb segment between knee and ankle; in biomechanics, known as the shank) can be calculated. Sprinters tend to have relatively short, muscular thighs and long "skinny" legs. The performance relationships of body proportions rely on mechanical knowledge for interpretation of the results and are consequently a topic of considerable interest for biomechanists.

The Tissues Composing the Body

One can contrast anthropometric models and anatomical models of the composition of the human body. Anatomical **dissection** has been used to describe the various components, separating the

IN FOCUS: PREDICTING BODY DENSITY AND BODY COMPOSITION

In the late 1980s, three articles were published by researchers on three continents, indicating the relationships between body measurements and density and percentage body fat. Researchers from the Department of Human Sciences, University of Technology, Loughborough, in England, and the Human Nutrition Unit, Istituto Nazionale della Nutrizione, Rome, took 21 body measurements from 138 male Italian shipyard workers. Examples of the mean values obtained were 1.73 m (5.7 ft) for height, 76.1 kg (167.8 lb) for mass, 10.4 mm (0.4 in.) for triceps skinfold thickness, 37.7 cm (14.8 in.) for calf girth, and 7.0 cm (2.8 in.) for bony width across the elbow. A research team from the School of Education at Flinders University of South Australia led by Dr. Bob Withers took 20 physical measurements from 135 female volunteers aged between 17 and 35 years. Examples of the mean values obtained were 1.66 m (5.4 ft) for height, 58 kg (128 lb) for mass, 16.5 mm (0.6 in.) for triceps skinfold thickness, 35 cm (13.8 in.) for calf girth, and 6.4 cm (2.5 in.) for bony width across the elbow. A group of seven researchers from five universities in the United States used underwater weighing to determine the body density of 310 subjects aged between 8 and 25 years. The subjects, from Illinois and Arizona, represented both genders and two major racial groups. As an example, the mean total body mass of the pubescent girls was 43.2 kg (95.2 lb).

The main aim of these projects was to compare the results of measuring percentage body fat by underwater weighing with the results obtained using existing prediction equations based on anthropometric measurements. Hardly any of the previous equations from the literature showed good correlations with the measured body densities. The results tended to confirm the general conclusion that equations developed to predict body density are population specific.

The authors of these studies therefore developed their own predictive equations using a statistical technique known as **stepwise multiple regression.** Variables included in the European equation for men were age and two skinfold thicknesses; the Australian equation for women included six skinfold thicknesses, two girth measurements, and one measurement of bone breadth. The American authors concluded that because chemical maturity has not been reached during puberty, adult prediction equations lead to an underestimate of body density and hence an overestimate of percent body fat. For the 59 American pubescent subjects, the sum of two skinfolds (triceps + calf or triceps + subscapular), plus an indication of physical maturation, was able to predict percent body fat reasonably well.

SOURCES

- Norgan, N.G., and A. Ferro-Luzzi. 1985. The estimation of body density in men: Are general equations general? *Annals of Human Biology* 12: 1-15.
- Withers, R.T., K.I. Norton, N.P. Craig, M.C. Hartland, and W. Venables. 1987. The relative body fat and anthropometric prediction of body density of South Australian females aged 17-35 years. *European Journal of Applied Physiology* 56: 181-90.
- Slaughter, M.H., T.G. Lohman, R.A. Boileau, C.A. Horswill, R.J. Stillman, M.D. van Loan, and D.A. Bemben. 1988. Skinfold equations for estimation of body fatness in children and youth. *Human Biology* 60: 709-23.

body into the six masses of skin, muscle, bone, nervous tissue and tissues from other organs, and fat (see table 3.1, "Proportions in the Human Body As Measured by Dissection of 25 Cadavers"). Anthropometric models seek to estimate body composition using noninvasive means. A two-component model, in common use at this time, seeks to separate lean tissue (such as bone and muscle) from fat tissue, given the realization that excess fat rather than excess lean tissue is a prime risk factor for premature death. Measurements of skinfold thickness have been introduced in an attempt to derive an indirect, noninvasive measure of body fatness (see "Predicting Body Density and Body Composition" above).

In the two-component anthropometric model, used in the calculation of body density, lean tissue and fat tissue each have an assumed **specific gravity.** Basically, lean tissue sinks in water because of its higher density, and fat floats on water. In the underwater weighing technique, which has been used as the criterion measure for the estimation of body composition, body weight in water is compared with body weight in air. Allowances

Table 3.1

Proportions in the Human Body As Measured by Dissection of 25 Cadavers

Males	Females
28.1% fat	40.5% fat
71.9% lean body mass	59.5% lean body mass
Skin = 8.5% lean body mass	Skin = 8.5% lean body mass
Bone = 20.6% lean body mass	Bone = 20.6% lean body mass
Muscle = 50% lean body mass	Muscle = 50% lean body mass
Undifferentiated tissues constitute the rest of the body	

Data from Clarys, Martin, and Drinkwater 1984.

are made for the volumes of air in the lungs and digestive tract. With use of this technique, calculations have shown that some elite power athletes have a negative percent body fat, which is biologically impossible and incompatible with life. This measurement error arises mainly because of the assumed specific gravity of nonfat tissue (about 1.1), which includes bone and muscle. The tissues composing the lean tissue component of the body, however, are variable in their densities and relative quantities. Also the specific gravity of the body varies between one region and another and may also vary within the one body segment. If underwater weighing can lead to incorrect estimates of body density, and therefore the percentage of body fat, it is not the valid criterion for development of predictive equations using skinfold measurements. Therefore, many researchers now recommend that skinfold measurements be reported directly and not be used in an equation that predicts percent body fat.

Radiographic techniques can be utilized in a three-component model of the human body. In **dual energy X-ray absorptiometry (DEXA),** a low-dose radiation source allows estimation of the masses of calcified tissue, fat, and nonfat tissue within the body. The modern imaging techniques of **computer-aided tomography (CAT)** and **magnetic resonance imaging (MRI)** have the potential to define proportions of many types of tissues within regions of the body.

Somatotyping As a Description of Body Build

In the past it was thought that body types were genetically determined and so did not change throughout life. Researchers now agree that body types do have a genetic component but that they are also a product of environment. **Somatotypes** were originally described as **genotypes,** but they are now considered **phenotypes** because of the environmental influences on body shape and the fact that somatotype can change throughout life. Anthropometric somatotypes use measurements to describe the shape and composition of the body at a particular point in time. Size must be reported separately.

The **Heath-Carter Anthropometric Somatotype** provides an example of most commonly used somatotyping methods. Its calculation uses 10 recorded anthropometric measurements (see "Measurements for Calculating the Heath-Carter Anthropometric Somatotype" on p. 37); but the protocol requires that each measurement be repeated at least once. Checking the **reliability** of measurements is a fundamental principle in any scientific experimentation.

In the Heath-Carter Anthropometric Somatotype, a person can be described by three numbers, each representing one component. The sum of three skinfolds from the arm and trunk is used to calculate the first component, which may be termed **endomorphy.** This component defines the relative fatness of the person, especially when it is related to the person's height. The second component, termed **mesomorphy,** represents musculoskeletal development. The skeletal development is estimated by the diameters of the bones at the elbow and knee. Muscular development is estimated by the circumferences of the arm and calf, after the estimated fat component of these segments has been accounted for. The bone diameters have more of a genetic determination than the segment circumferences, which are very much influenced by the type and quantity of physical activity. The third component is calculated from height and mass, using a cubic relationship. This component is termed **ectomorphy,** and a person

MEASUREMENTS FOR CALCULATING THE HEATH-CARTER ANTHROPOMETRIC SOMATOTYPE

1-4

Four skinfolds measured on the right-hand side of the body with skinfold calipers and expressed in millimetres

1. Triceps—on the back of the arm

2. Subscapular—just below the lowest part of the shoulder blade

3. Suprailiac—just above the bony projection of the pelvis at the approximate level of the umbilicus

4. Medial calf—inside of the widest part of the calf

5

5. Height measured with a stadiometer and expressed in millimetres (for calculation of mesomorphy) or centimetres (for calculation of ectomorphy)

6-7

Two bone breadths measured on both sides of the body with bicondylar calipers and expressed in centimetres

6. Humeral epicondylar—elbow width

7. Femoral epicondylar—knee width

8-9

Two limb segment circumferences measured on both sides of the body with an anthropometric tape and expressed in centimetres

8. Arm—elbow bent with biceps muscle maximally contracted

9. Calf—largest part of calf during standing normally on both feet

10

10. Weight—mass measured on scales and expressed in kilograms

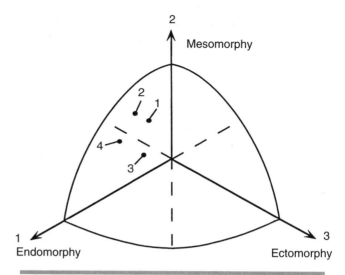

Figure 3.2 Somatochart indicating the somatotypes of four groups within the Canadian adult population. The four groups are (1) males younger than 30 years (3.5–5–2; endomorphy–mesomorphy–ectomorphy), (2) males older than 30 years (4–5.5–2), (3) females younger than 40 years (4.5–4–2.5), and (4) females older than 40 years (5–4.5–2).

Adapted from Bailey, Carter, and Mirwald 1982.

age of a normal man and woman, so males and females are not plotted on different somatocharts. An average man is more mesomorphic than the phantom whereas the average woman is more endomorphic than the phantom (see figure 3.2). A typical endomorph is the sedentary middle-aged man or a Sumo wrestler (figure 3.3a). A typical mesomorph is any elite power athlete, especially weightlifters, boxers, and wrestlers (figure 3.3b). Typical ectomorphs are ballet dancers and female gymnasts. High jumpers and marathon runners also tend to fit into this category (figure 3.3c). These examples indicate one aspect of kinanthropometry: defining relationships between body size, shape, and composition and human performance and then explaining the relationships.

Human Variation

Obviously there is a wide range of "normal" in almost all body measurements. Some individual differences may be explained by more general differences, such as differences between males and females, racial differences, age differences, and differences in types and amounts of activity. This short list implies two main sources of differences,

who rates highly on this component is light for his or her height.

Every person is rated on each of the three components and can be positioned on a special diagram called a **somatochart** (figure 3.2). At the centre of this triangular-shaped diagram is the unisex "phantom," which is the statistical aver-

Figure 3.3 Typical extreme somatotypes in different sports: *(a)* endomorph, *(b)* mesomorph, and *(c)* ectomorph.

genetic and environmental. Environmental factors include (a) the mechanical environment related to the size and frequency of muscular forces and (b) metabolic factors that may include deficiencies of certain nutrients in the diet. In other chapters we address differences related to age (chapter 4) and activity (chapter 5).

Human Variation in the Musculoskeletal System

Bone sizes obviously differ between individuals, but bone density differences can also be correlated with race. In North America, for example, it has been shown that African Americans have a higher bone density than Caucasians.

Peak bone mass in women is about 30% less than in men. This difference between the sexes is an example of sexual dimorphism, which in this case is due mainly to differences in physical size. Any differences in bone density between the sexes are small and are also site specific rather than general.

Muscle volume differences between males and females are also mainly related to differences in body size, but differences in muscle size also appear to be related to the presence of the different sex hormones and their interaction with physical activity. Generally the muscle volume of adult males is 40% more than that of adult females; however, this is not a large difference when muscle

volume is considered as a proportion of total body volume. Relative muscle masses in the lower limbs of males and females are similar, but there are larger differences in the upper limbs. Researchers who have analysed the tissues of the arm using CAT (which provides an image of the cross section of the arm) have found the average cross-sectional area of muscle plus bone to be 62 cm² (10 in.²) in a 30-year-old male compared with 35 cm² (5 in.²) in a female of the same age (see also "Relationships Among Thigh Circumference, Quadriceps Muscle Cross-Sectional Area, and Knee Extensor Strength," p. 39).

Human Variation in Physical Dimensions

There are many examples of **sexual dimorphism** in body dimensions, but often the overlap between the two sexes is so large that it would be difficult to predict the sex of an individual on the basis of body dimension measurements alone. Males tend to be taller and heavier than females, but these are not distinguishing features. At about 20 years of age, the ratios between males and females are 1.08:1 for height, 1.25:1 for weight, and 1.45:1 for lean body mass. These differences have important implications for both nutritional requirements and athletic performance.

The difference in shape between adult males and females is well known and is based mainly on dif-

IN FOCUS: RELATIONSHIPS AMONG THIGH CIRCUMFERENCE, QUADRICEPS MUSCLE CROSS-SECTIONAL AREA, AND KNEE EXTENSOR STRENGTH

Researchers from the Department of Anatomy and Human Biology, University of Western Australia, and Royal Perth Hospital examined thigh cross-sectional images produced by CAT in 15 males who had previously incurred knee injuries. The investigators compared uninjured and previously injured limbs (on average the knee injury had occurred seven years prior to testing). They also measured knee extension strength in a static situation, that is, during an isometric contraction (see figure 2.14).

Repeated measurements indicated that the measurements of muscle cross-sectional area were reliable. Cross-sectional areas of the knee extensor muscles on the previously injured side were 11% less than on the uninjured side. These sides were termed **"atrophic"** and "normal," respectively. The "atrophic" side was 21% weaker than the "normal" side. The relationship between knee extensor cross-sectional area and force varied between subjects and also changed for each person during a strengthening program. Analysis of muscle density indicated greater fat deposition in the "atrophic" muscles.

The authors discussed the clinical use of anthropometric measurements and advised that thigh circumference measured with a tape is not a valid indicator of muscle cross-sectional area. The circumference represents total amount of bone, muscle, and fat and thus does not allow distinction among these three components. Even an **anterior** thigh skinfold measurement may not provide a good indication of fat content because of the pattern of fat deposition within atrophied muscles.

SOURCE

- Singer, K.P., and B. Breidahl. 1987. The use of computed tomography in assessing muscle cross-sectional area, and the relationship between cross-sectional area and strength. *Australian Journal of Physiotherapy* 33(2): 75-84.

ferences in shoulder size. Measuring the width of the shoulders in centimetres and multiplying this by three, then subtracting the width of the pelvis, enables one to correctly identify almost 90% of the adult population as male or female. This relationship between shoulder and hip width, known as the **androgyny index,** is a good example of the use of proportions to describe shape and, in this case, emphasize differences.

Some sports (such as basketball) favour tall players, whereas jockeys are usually short because they must be very light. Figure 3.4 indicates the positions of a typical 200-cm (6.6-ft) basketball player and a 155-cm (5.1-ft) jockey relative to the young adult male population of Queensland in Australia. National Basketball Association centres are often more than six standard deviations above the average height. With regard to height comparisons, the average North American man is taller than 95% of Asian men, which probably explains the restricted interior space in the Japanese cars that were first exported during the 1960s. It is predicted that the height difference will be smaller in future generations. We explain the basis of this prediction in the next chapter on growth and maturation.

A number of proportions, including the sitting height/standing height ratio, also show racial differences. The group with the smallest sitting height/standing height ratio is the Australian

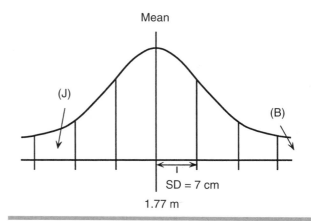

Figure 3.4 A curve indicating the normal distribution of self-reported heights among almost 100,000 young adult males in Queensland (aged 17-30 years). Heights of typical jockeys (J) and professional basketball players (B) are indicated. The normal distribution is drawn from data supplied by the Queensland Department of Transport (1985).

Aborigine; that is, people in this group have the longest lower limbs relative to their height. On average this figure is under 50% compared with over 53% for people from Southeast Asia. In a photographic atlas of Olympic athletes, Jim Tanner emphasized proportional differences between African American male athletes and Caucasians. The relatively longer lower limbs of the black athletes and other differences related to the lower limb are thought to partly explain differences seen in general performance in the 100-m (109-yd) sprint. While this example is appealing, one must realize that body dimensions are only one factor in performance and that many athletes do not conform to common perceptions of physical requirements. An example is the basketball player "Spud" Webb, who was able to win an NBA slam dunk competition even though he was only 170 cm (5 ft 7 in.) tall.

Statistical trends relate body size and shape to athletic performance, and certain generalizations follow from these trends. One statistician has argued that the taller athletes tend to perform better in almost all Olympic athletic events. Graphs of the relationships between height and performance certainly provide evidence in support of this concept.

Summary

Anthropometry involves the use of relatively simple equipment and techniques, following strict protocols, to take various measurements of the sizes of body segments. These measurements can be combined to indicate the shapes of different parts of the body, and simple measurements can be used to predict the more complex indicators of body composition. Kinanthropometry is an important aspect of functional anatomy because it applies the basic body measurements, and their mathematical combinations, to the movement situation; and knowledge of biomechanics and physiology aid the analysis and prediction of human movement.

Proportions between body measurements can be used to indicate the structure and the composition of the body. Indices are used to describe physique and are also used in public health surveys to determine risk factors for problems such as cardiovascular disease. Somatotyping is a shorthand way of defining shape and composition of the body. Endomorphy, based on skinfold thicknesses, indicates the relative amount of fat in the body; mesomorphy, based on bone diameters and muscle circumferences, indicates the relative development of the musculoskeletal system; ectomorphy, based on the basic measurements of height and mass, provides an idea of gross body shape. A person's somatotype is a phenotype, meaning that although it has a genetic base, it is highly dependent on lifestyle factors such as nutrition and activity levels. The concept of "normal" is important in that we must define "normal" so that we can describe "abnormal." The normal range may be different for males and females, different ethnic groups, and people of different ages.

Further Reading and References

Lohman, T.G., A.F. Roche, and R. Martorell, eds. 1988. *Anthropometric standardization reference manual.* Champaign, IL: Human Kinetics.

Norton, K.I., and T.S. Olds, eds. 1994. *Anthropometry and anthropometric profiling.* Sydney: Nolds Sports Scientific.

Norton, K.I., and T.S. Olds, eds. 1996. *Anthropometrica: A textbook of body measurement for sports and health courses.* Sydney: University of New South Wales.

Olds, T.S., and K.I. Norton. 1999. *Lifesize* (CD-ROM). Champaign, IL: Human Kinetics.

Ross, W.D., R.V. Carr, and J.E.L. Carter. 2003. *Anthropometry illustrated* (CD-ROM). Canada: Turnpike Electronic Publications.

CHAPTER 4

Musculoskeletal Changes Across the Life Span

*I*n this chapter we revisit basic concepts related to the musculoskeletal system and anthropometry, introduced in chapters 2 and 3, to focus on the topic of life span changes. The chapter addresses three major issues, describing the general process of physical growth, maturation, and aging; examining age-related changes in the skeletal, articular, and muscular systems; and examining changes in body dimensions throughout the life span.

Definitions of Auxology and Gerontology

Auxology and gerontology can be defined as the sciences of growth and aging, respectively. Of particular interest in auxology are the ages of onset of changes, the magnitudes of these changes, and the duration of the change period. All these factors are highly variable between individuals, but the order in which major changes occur is relatively constant. The knowledge available from auxology provides a basis for determining whether people are older or younger physically than their chronological age.

Gerontology is an increasingly important field of study as the percentage of aged individuals increases. One focus in gerontology, particularly important in the subdiscipline of functional anatomy, is the relative effects of aging per se, as opposed to the effects of inactivity, on the musculoskeletal system. It is necessary to determine the relative effects in order to estimate the probable impacts of physical activity programs.

The major learning concepts in this chapter relate to
- auxology and gerontology;
- modern medical imaging techniques;
- embryological development of the human musculoskeletal system;
- physical maturation during the postnatal years and physical changes related to aging;
- the various stages involved in the development of bone;
- the processes involved in the growth in length and width of bone;
- changes in bone composition throughout the life span;
- osteoporosis and its effects;
- the potential effects of repetitive muscular forces on developing bone;
- the development of joints and muscles;
- changes in body dimensions, size, and shape across the life span;
- intergenerational changes in body size and shape;
- physical differences between males and females at various stages of life; and
- methods of determining biological age.

Tools for Measurement

Growth involves measurable change in either the quantity of tissue or the size of body segments. Therefore, many large human growth and aging studies use the same tools that anthropometry uses (chapter 3). Most pregnant women have an **ultrasound (US) scan** at about 12 weeks of pregnancy, and the measurements taken by the obstetrician include the width of the skull and the length of the femur. These measurements are used to confirm the age of the **foetus.** During postnatal growth, physiological measurements may also be taken (see chapter 12).

Researchers are often interested in the skeletal development of children and the skeletal changes associated with aging. Studying these phenomena requires the use of radiological techniques. Skeletal age of children can be determined by hand/wrist radiographs, and the bone density of adults can be measured using computer-aided tomography (CAT) scans or dual energy X-ray absorptiometry (DEXA) scans.

Physical Growth, Maturation, and Aging

The major stages of the human life cycle are listed in table 4.1. About one-quarter of the human life span is spent growing. Humans have a long period of infant dependency followed by a unique extended childhood growth period between infancy and puberty. **Puberty** is the delayed period of rapid growth leading to sexual and physical maturity. Some claim that aging begins as soon as the peak in any given parameter is reached. This would imply a 40- to 50-year period of aging. Ideally this process should be as gradual as possible.

Embryological Development

Development of a human begins when the **ovum** is fertilized by a **spermatozoan,** forming a **zygote,** which is a single fertilized cell. The zygote quickly begins to divide; and during the first three weeks of development, three primary germ layers differentiate. These are the **ectoderm,** which eventually forms nervous tissue and the outer part of the skin; **endoderm,** which forms the internal linings of visceral organs of the body; and **mesoderm,** which forms the organs and tissues and structures of the musculoskeletal system and the inner layers of skin. Here we consider in detail only the mesoderm.

The foetal period is a time of rapid development of the mesoderm, characterized by a series of orderly and irreversible stages through which each organism passes (figure 4.1). From relatively undifferentiated cells and tissues, there is a progression of changes leading to highly specialized cells and tissues. By the end of the embryonic period, the **embryo** is about 35 mm (1.4 in.) long, measured from the top of the head to the buttocks; and all the fingers and toes can be seen. Most tissues differentiate during this period, and a number of landmark events occur.

The foetal period is a time of rapid growth of the tissues and structures that have formed during the preceding embryonic period. Growth of tissues can occur by an increase in the number of cells forming the tissue, called **hyperplasia,** or by an increase in the size of each cell, called **hypertrophy.** Hyperplastic growth is a feature of the foetal period.

The Postnatal Years

Birth, although it obviously indicates chronological age, is not a good indicator of biological age. Infancy, during which the baby learns to sit up,

Table 4.1

Stages in Human Development

Landmark events	Major stages	Periods within life cycle	Approximate ages
Conception	Prenatal	Pre-embryonic	Weeks 1-3 after conception
		Embryonic	Weeks 4-8 after conception
		Foetal	Months 3-9 after conception
Birth	Postnatal	Infancy	Years 1-2
		Childhood	Years 3-10
Peak height velocity		Puberty	Years 11-17
		Young adult	Years 18-40
		Middle-aged	Years 41-60
		Elderly adult	Years 61-
Death			

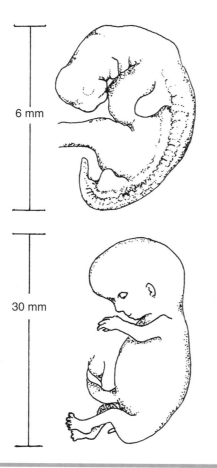

Figure 4.1 Illustrations of an embryo at 30 days and 55 days after conception, indicating sizes by the crown–rump length. At 30 days the upper limb buds are visible and the lower limb buds have begun to develop; by the end of the embryonic period, all essential external and internal structures are present.

crawl, stand, and walk, is a period of rapid motor development. During childhood, growth continues at a steady rate. Another period of rapid growth occurs during puberty, which is associated with the development of the secondary sexual characteristics.

The term **maturation** can be used to describe the physical changes that occur between birth and maturity. These internal processes appear to be genetically determined, in contrast to external environmental factors that affect maturation in the process of adaptation.

During early adulthood, many physical characteristics exhibit their peak. The aging process begins at this time and continues through the middle years to old age. One aim of physical activity is to delay some of the physical changes

associated with aging, as some of the so-called effects of age relate more to decreased physical activity than to an inbuilt aging mechanism (see chapter 12).

Age-Related Changes in the Skeletal and Articular Systems

For the musculoskeletal system, the major periods of growth are the foetal and pubertal stages. The early foetal growth period is a time of rapid multiplication of cells (hyperplasia), whereas growth during the pubertal period results mainly from development and enlargement of existing cells (hypertrophy). Development can also occur via the replacement of one type of tissue by another.

Stages in the Development of Bone

Adult human bones appear in a variety of shapes and sizes. Their basic shape is genetically determined, but their final shape is influenced greatly by the environment in which they develop. Environmental influences include mechanical factors, such as muscle forces acting on the developing bone, and metabolic factors including the supply of nutrients.

Typical long bones, such as the humerus in the arm and the femur in the thigh, develop via a process of **endochondral ossification** in which a cartilaginous model precedes the bone formation. The cartilage is eventually replaced by bone. The cartilage model initially is small, and its replacement by bone begins at an early stage of development. The primary ossification centres appear near the middle of the shaft of the future long bone at about eight weeks after fertilization, when the embryo is about 35 mm (1.4 in.) long (see figure 4.1).

The cartilage model of the future bone grows before it is replaced by bone. The replacement occurs in a series of stages. Initially, the cartilage model grows by two processes: an increase in the number of cartilage cells and then an increase in the size of each cell. Next, the gel-like matrix surrounding the cartilage cells is calcified. This hardening of the tissue surrounding the cells effectively cuts off their supply of nutrients as well as prevents the removal of metabolic waste products. These "imprisoned" cartilage cells eventually die so that the calcified cartilage has a honeycomb appearance. Invading blood vessels bring nutrients and bone-forming cells (osteoblasts) to the calcified cartilage to lay down new bone. This new bone has a disorganized appearance. If ossification were the final stage, a long bone would not develop its hollow structure. What happens instead is that some of the newly laid down bone is removed from internal sites by special bone-eroding cells. These osteoclasts work with osteoblasts during remodelling of the bone, causing the cortex to drift away from the central axis of the shaft during growth. Changes in shaft diameter and compact bone thickness can occur at any age by a remodelling process. Bone can respond to changes in mechanical stimuli by changing its structure.

One type of remodelling that occurs is the replacement of the originally laid down bone by compact bone that has an organized structure (figure 2.1). Most primary centres of ossification (in the developing shaft) appear before birth, whereas most secondary centres of ossification (in the developing ends of the bone) appear after birth. The times of appearance of these secondary centres of ossification vary widely. The centre in the femur near the knee is normally present at birth, but the centre in the **clavicle** (collarbone) closest to the **sternum** (breastbone) does not appear until about 18 years of age.

Growth in Length and Width of Bone

When both the shaft and ends of a long bone are developing, they are separated by a "growth plate" or **epiphyseal plate.** Within this region, all the stages of bone development can be seen under a microscope (see figure 4.2). Each zone in figure 4.2 represents a stage in the growth and then replacement of cartilage by bone; that is, the zones illustrated do not migrate downward, but rather retain their position and change their nature. This effectively increases the distance between the growth plates at each end of the bone. The process illustrated is continuous for almost 20 years. Growth in length of the long bones ceases when the process stops and epiphyseal plates "fuse" and disappear. Actually the growth stops because the cartilage cells no longer respond to hormonal influence.

Although growth in height ceases at a certain age, changes in thickness of the compact bone of the shaft and the density of spongy bone can occur at any time of life (see chapter 5). Growth in the thickness and diameter of long bones occurs by **appositional growth.** In this process, bone is normally added on the outside of the shaft and removed from the inside of the shaft.

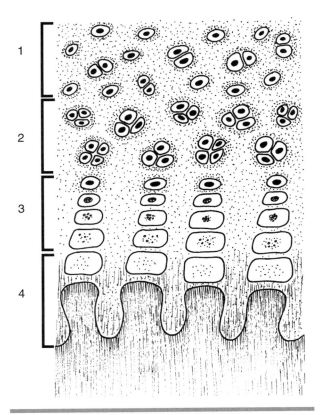

Figure 4.2 Zones in a section through the growth plate of a developing bone as seen under a light microscope. The zones, numbered from the end of the bone toward the shaft: (1) resting cartilage; (2) cartilage proliferation (hyperplasia); (3) cartilage maturation (hypertrophy); (4) cartilage calcification.

Skeletal Composition Changes Across the Life Span

The relative proportions of the inorganic salt crystals and the organic collagen change throughout the life span.

In a child, the flexibility of the bone is related to the large proportion of collagen. It is thought that the rapid growth of children's bone and the consequent amount of remodelling result in less than optimal mineralization. In a newborn baby, the skeleton accounts for about 13% of the total body weight, but about two-thirds of the skeleton is cartilaginous.

In a young adult the inorganic material is close to the optimal two-thirds mineralization, so the bone is strong and tough. Though the proportion of skeletal material in the mature adult is also about 13% of total body weight, just over 10% of the skeleton is cartilaginous. Of the bone, four-fifths is compact and one-fifth is spongy.

Eventually, aged bone becomes brittle. This change relates partly to the increased proportion of the inorganic salts; but of more significance is the reduced total mass of bone material in the aged musculoskeletal system. Bone mass decreases mainly because of a decrease in the number of trabeculae in spongy bone (see "Why Trunk Height Decreases With Aging" on p. 46), combined with resorption, and consequent thinning, of the trabeculae. In compact bone, the size of the **Haversian canals** (see figure 2.1) increases, and the cross-sectional area of the bone surrounding each Haversian canal decreases. This increased porosity of bone, in turn, causes decreased bone density (mass per unit volume of tissue). In an 80-year-old male the density of bone in the spine is often 55% of the value in a 20-year-old; in females who are elderly, it is often only 40% of its peak value. With aging, bone also becomes stiffer, partly because of the collagen cross-linking. A consequence of these mechanical changes is that aged bone can absorb less energy before it fractures, so it is more liable to fail when subjected to large forces.

Osteoporosis

For most tissues in the body, their mass at any time is determined by the balance between the simultaneous formation and destruction of the tissue. **Osteopenia** ("bone poverty") refers to reduced bone density due mainly to decreased bone synthesis, particularly in women during the postmenopausal years, when the dramatic decrease in the female hormone, estrogen, allows the rate of synthesis to be lower than the resorption rate. The initial effects are mainly to the spongy bone, but later the compact bone in the shafts of long bones also decreases in thickness.

According to the World Health Organization, **osteoporosis** ("bone porosity") is present when the bone density is more than 2.5 standard deviations below the average bone mass of a young adult woman. In general, the porosity of bone in the human femur doubles from 40 to 80 years of age, and people with osteoporosis show an even greater change. Thus osteoporotic bone is even less able to absorb energy before it fails than is normally aged bone. Moreover, even though a chemical analysis of osteoporotic bone material may indicate slightly

elevated levels of calcium per unit mass of bone, the overall mass of bone decreases so markedly that less calcium is stored in the skeleton overall.

During maturation, the density and calcium content of bones increase due to genetic predisposition aided by the interaction between dietary calcium and exercise. It is now thought that peak bone mass occurs at about 16 to 20 years of age in women, so some health authorities claim that osteoporosis is a paediatric disease and have promoted the idea of "bone banks" to help prevent osteoporosis. The idea is that if a person's lifestyle during maturation and early adulthood promoted maximum deposition of bone material, then, later in life, when withdrawals of bone material are inevitable, the structural integrity of the bones would be maintained longer. Financial advisers tell everybody to start saving as young as possible, and this concept is even more important for prevention of osteoporosis.

Osteoporosis is commonly associated with women, but the prevalence among men is now the same as it was in women about 25 years ago. This trend has been attributed to lifestyle changes, which include inadequate intake of calcium and less than optimal levels of physical activity. Both the personal lifestyle changes due to, and public health care costs of, osteoporosis are so great that they are considered further in chapter 22.

Bone Failure in Relation to Bone Development, Age, or Activity

The modern automotive industry attempts to protect passengers by enclosing them in a cage lined with material that is designed to cushion the forces involved in a collision. In similar fashion, the soft tissues of the human body, including fat and muscle, perform the role of a "crumple zone." But sometimes the energy involved in a mishap is enough to cause a fracture anyway. Bone failure may be related to bone development, age, or activity.

Certain types of fractures are associated with particular stages of development. The flexible bones of a child tend to splinter similarly to the way a branch does as it breaks off a growing tree; this type of fracture is called a "greenstick fracture." Late in life the fracture is more likely to be of the brittle type because of the increased stiffness of old bones. The bones of adults who are elderly are also more likely to fail because of the decreased bone mass and density in osteoporosis.

IN FOCUS: WHY TRUNK HEIGHT DECREASES WITH AGING

Researchers from the School of Physiotherapy at the Western Australia Institute of Technology (now Curtin University of Technology) and the Department of Anatomy at the University of Western Australia measured the dimensions of thin sagittal sections of the third lumbar vertebrae removed from 93 adult cadavers. The gender and age of death for each cadaver had been recorded. To calculate vertebral body bone density, the measured dry mass of each section was divided by the calculated volume of each speci-

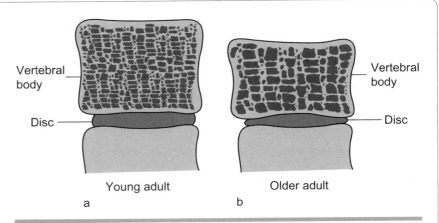

Figure 4.3 Age-related changes in dimensions of the vertebral body and intervertebral disks.

men. As another measure of bone density, the amount of light transmission through each section was analysed. Size and orientation of trabeculae were also analysed (see figure 4.3). To define the shape of each vertebral body section, the mid-vertebral height of the section was divided by the anterior vertebral height. Average height of each lumbar intervertebral disk was calculated as the mean of three measurements of disk height: anterior, middle, and posterior. To describe disk shape, the middle disk height was divided by the sum of the anterior and posterior disk heights.

All three methods of measuring bone density confirmed a loss of bone density (or increase in porosity) related to aging. The decline in the number of vertical trabeculae was not statistically significant, but there was a marked decrease in the number of horizontal trabeculae, especially in the females. The thicknesses of trabeculae were not measured. Shape measurements of the vertebral bodies indicated increasing concavity of the vertebral end plates with age for both sexes. Overall there was no decrease in disk height with aging, and in fact some individuals exhibited an increase in disk height. Evidence of disk degeneration was observed more often as age increased, but the group with degenerative disks was still a minority in the aged individuals. The authors concluded that the age-related structural changes of the lumbar vertebral column are mainly related to the hard tissues of the vertebral body as opposed to the soft tissues of the intervertebral disk (see figure 4.3). During aging, the vertebral end plates become concave, related to decrease in bone density of the vertebral body. Buckling of the vertical trabeculae occurs because of thinning or loss of the bracing horizontal trabeculae. The intervertebral disk becomes more convex but its average height is maintained.

The authors, suggesting that the results indicated the effects of normal aging, did not propose a definition of osteoporosis based on the measurements used. They used an engineering concept called Euler's theory to explain that the horizontal trabeculae are very important in that they act as "cross braces" to support the vertical trabeculae. The loss of horizontal trabeculae with aging leads to the collapse of vertical trabeculae; this in turn affects the shape of the end plates (at the top and bottom of the vertebral body), which become more concave. The authors cited this example of bone geometry as more important than mass per se. Further, they suggested that the increasing concavity of the vertebral bodies explains the loss of trunk height and increased curvature of the vertebral column observed in old age (figure 4.3). Radiological images of the lumbar spine may give the impression of disk space narrowing because the anterior and posterior disk heights may decrease.

SOURCES

- Twomey, L.T., and J.R. Taylor. 1987. Age changes in lumbar vertebrae and intervertebral discs. *Clinical Orthopaedics and Related Research* 224: 97-104.

- Twomey, L., J. Taylor, and B. Furniss. 1983. Age changes in the bone density and structure of the lumbar vertebral column. *Journal of Anatomy* 136: 15-25.

Specific fractures relate to certain ages. For example, fractures of the neck of the femur (commonly called hip fractures) are common in women in the elderly population. Forearm fracture incidence increases near the peak of the pubescent growth spurt. This is associated with increased porosity of compact bone as the remodelling space increases to provide calcium to the rapid-growth regions. A second increase in wrist fractures is seen in women who are elderly, related to the combination of osteoporosis and increased incidence of falls. In developing children, the cartilaginous growth plate may also fail, most often at the zone of maturation, or hypertrophy (see figure 4.2).

The development of bone has implications for potential injury to the musculoskeletal system. For example, injury to the cartilaginous growth plates can be produced by single excessive forces causing trauma or by repetitive compressive forces, or by tensile forces associated with muscle–tendon pull on an area of developing bone.

In sports medicine, specific injuries have been attributed to certain activities. One of the best examples is "Little League elbow," described in more detail in chapter 22. Other examples of the relationships between activity and types of growth plate problems are wrist injuries in gymnasts and tibial tuberosity avulsions in sports involving sprinting and jumping. Pain over the tibial tuberosity (just below the kneecap) in a prepubescent child is indicative of Osgood-Schlatter condition.

Effect of Various Factors on Range of Motion

Range of motion at joints is affected by joint structure and the mechanical properties of the tissues associated with the joints. A general perception is that joint range of motion decreases during life; and although this is the general trend, the rate of loss is not constant. Ranges of joint motion are very large in a newborn baby. As an example, the range of ankle **dorsiflexion** in a newborn is limited only by the contact of the top of the foot against the shin. Try this movement yourself to see how much flexibility you have lost since your birth.

Between the ages of 6 and 15 years, there is a general trend for joint range of motion to decrease in boys, whereas for girls the effects are variable and joint dependent. Girls are generally more flexible than boys during childhood and adolescence. Changes in joint ranges of motion between adolescence and young adulthood are variable and, as we will see in the next chapter, appear to be related to physical activity. Similarly, the general trend for joint flexibility to decrease during aging may not be completely explained by biological aging processes.

Aging people may experience **arthritis** ("inflammation of joints"), which markedly restricts range of motion. **Rheumatoid arthritis** involves inflammation of the synovial membrane, while **osteoarthritis** involves degeneration of the articular cartilage. In the total adult population, about three times as many women have osteoarthritis as do men. Thus there is some genetic determination of osteoarthritis, to which are added the effects of environmental factors. It would appear that the risk factors for rheumatoid arthritis are mainly related to a family history of the condition.

Age-Related Changes in the Muscular System

Mesoderm, which forms toward the end of the third week of development, will eventually form the muscle and connective tissues of the body. Some of the mesoderm is segmented such that it is in lumps on either side of the midline, and these lumps form the muscles of the trunk. Other mesoderm will form the muscles of the limbs, and this tissue migrates into the limbs when they first appear. Limb muscles migrate toward their final position during further development.

Certain precursor cells differentiate into **myoblasts** (muscle-forming cells) that fuse longitudinally to form the long muscles and muscle fibres we can observe.

The number of muscle fibres appears to be genetically determined so that large people often have more muscle fibres than persons who are smaller in stature. As muscles grow in length, the number of sarcomeres increases. Although direct evidence is sparse, many believe that the stimulus for growth of muscle is the bone growth that pulls on the attached ends of the muscle. Developmental growth occurs in both muscle length and cross-sectional area, and these factors may also be affected by activity. The effects of physical training on muscle are summarized in chapters 5, 9, and 11.

In an infant, muscle tissue accounts for about 25% of total body weight, but in a young adult the proportion of muscle is more than 40%. In terms of growth, muscle increases from about 850 g (2.3 lb) at birth to about 30 kg (66 lb) in a young adult male weighing 70 kg (154 lb).

Strength, which may be defined as the capacity to produce force against an external resistance, is related to the cross-sectional areas of the muscles producing the force. Muscle volume peaks at about 30 years of age and then gradually decreases. The atrophy of skeletal muscle associated with disuse is much faster than any aging-related atrophy. Muscle elasticity also decreases during aging so that muscles are stiffer and less extensible. This is one contributing factor to the loss of joint range of motion described earlier.

Changes in Body Dimensions Across the Life Span

Since size measurements can be reported directly, information on the average heights (figure 4.4) and weights of children is available. Height can be plotted against age, but plotting height gain against age more obviously illustrates the "growth spurt" (figure 4.4). Within this period is the **peak height velocity (PHV),** which is a major landmark in pubertal growth and a reference for many other changes. In boys, the relative amount of fat in the body may be decreasing during this period of rapid growth. The peak weight velocity is delayed relative to PHV by a few months. Peak strength velocity may be delayed for more than a year. Unfortunately, PHV and other peaks can be determined only retrospectively; they cannot be easily or accurately predicted. Age at PHV can be used to measure maturational stage, and this in turn can be used to determine the effects of early versus late maturation on adult body dimensions (see "The Pattern of Growth in Swedish Children" on p. 49).

Overall dimensions of the body such as stature and volume (related to body mass) do not increase at a constant rate during growth, nor do they change at a constant rate during aging. This idea is illustrated in figure 4.4 and is recognized in the common expression "growth spurts." If one has a specific interest in "growth spurts" and PHV, mentioned earlier, one can plot the height versus age as height gain per year against age, as illustrated in figure 4.4.

Combining Size Measurements to Provide Information About Shape

Galileo first described a square–cubic relationship that is seen in the human body. Assume that two objects have the same shape. If one object is twice as tall as the other, its cross-sectional area will be

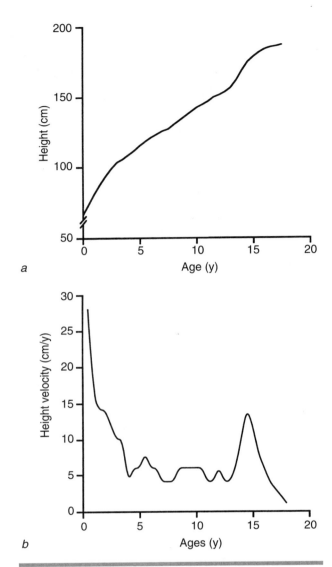

a

b

Figure 4.4 Two growth charts for the same individual derived from the same data. When *(a)* growth in height is expressed as height gain per year, *(b)* the peak height velocity (at 15 years) can be determined. These curves are derived from measurements taken by de Montbeillard of his son from the year of his birth in 1759 to 1777.

four times as large (squared relationship) and its volume will be eight times as large (cubic relationship). Remember that the area of a circle is πr^2 and the volume of a sphere is $4/3\pi r^3$. Let's see if these relationships are likely to describe a normal pattern of growth. Suppose that a baby is 50 cm (20 in.) and 3 kg (6.6 lb) when born, and assume that she grows to 150 cm (59 in.), that is, to three times longer. Assuming that volume is directly related to mass, we would expect her adult mass to be 3 mul-

IN FOCUS: THE PATTERN OF GROWTH IN SWEDISH CHILDREN

Commencing in 1955, 103 boys and 80 girls were followed from birth to adulthood. Patterns of maturation were analysed by researchers from the Department of Orthodontics, University of Lund, Malmo, and the Department of Pediatrics, University of Goteborg. The subjects were divided into three maturity groups according to their age at PHV. On average, the early-maturing girls were 10.7 years at PHV, and the boys were 12.6 years. The average maturers were 12.1 years (girls) and 14.0 years (boys) at PHV, and the late maturers were 13.6 years (girls) and 15.6 years (boys). Just over half of the subjects were classified as average maturers, while almost a quarter were early maturers and about a fifth were late maturers.

Between the ages of 5 and 14 years, the early-maturing girls were taller than the late maturers (maximum of 13.1-cm (5-in.) difference at 13 years); however, the three groups were very similar in height at 25 years of age. The early-maturing boys were taller than the late maturers between the ages of 12 and 15 years, but were actually shorter as adults. The early maturers were 11.8 cm (4.6 in.) taller than the late maturers at 14 years but were 6.5 cm (2.6 in.) shorter at 25 years of age. The late maturers were also taller than the average maturers at 25 years, by 4.2 cm (1.7 in.).

SOURCE

- Hagg, U., and J. Taranger. 1991. Height and height velocity in early, average and late maturers followed to the age of 25: A prospective longitudinal study of Swedish urban children from birth to adulthood. *Annals of Human Biology* 18: 47-56.

tiplied by 3 cubed kilograms. The calculated result of 81 kg (179 lb) is very heavy for a 150-cm-tall person. What does this tell us about growth? Basically it demonstrates that the relationship between height and weight during growth is not cubic and thus all dimensions are not changing at the same rate. In other words, the shape of the human body changes during growth. We can see this change in proportionality most clearly by considering the relative size of the head at different ages (figure 4.5). This illustrates that the brain and surrounding cranial bones are closer to adult size at an earlier age than other body parts are, demonstrating the concept of differential growth (that different parts of the body grow at different rates during different periods of development). Obviously human growth is not explained by a squared relationship either, because the baby girl would have grown to 150 cm (59 in.) and 27 kg (60 lb) in this situation!

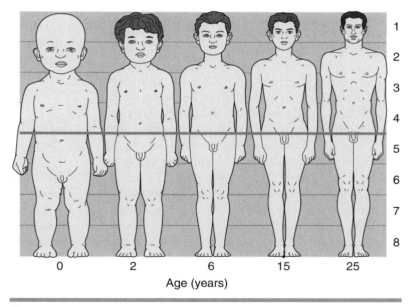

Figure 4.5 Relative sizes of the head at five different ages; the total body is scaled to the same height.
Adapted from Timiras 1972.

Secular Trend in Body Dimensions

You may be familiar with the idea that there are differences in size between people of different generations. It appears to be common for teenagers to be taller than their parents. This generational change, known as a **secular trend,** can be illustrated by differences in the heights of Australian schoolchildren from 1911 to 1976 as shown in figure 4.6. The figure illustrates a number of interesting changes that occurred in just over two generations. In 1911, 7- and 8-year-old boys and girls were similar in height, but in 1976 the boys were

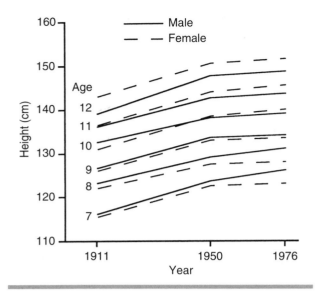

Figure 4.6 Secular trend in height of Queensland schoolchildren from 1911 to 1976 showing an increase in all age groups.
Adapted from Dugdale, O'Hara, and May 1983.

considerably taller. Note that 10-year-old boys in 1911 were taller than the girls, but this was reversed in 1976. Earlier maturation of girls is indicated by the height differences of the 12-year-old children. Twelve-year-old girls were still taller than boys of the same age; therefore there must be an age at which the girls overtake the boys. This was 11 years of age in 1911, but it was already 10 by 1950. The age of crossover between the male and female heights provides some indirect evidence for another good example of secular trend.

Studies from a number of countries show that the age of menarche, the age at which menstruation first occurs, has decreased by up to four years during the last century. Girls appear to be maturing earlier relative to the boys than they were in previous generations.

If the heights of 80-year-old women were compared to those of 20-year-old women, would this provide evidence of the secular trend in height of adult females? Unfortunately the answer is no, because there are confounding variables. One effect of aging is decreased stature due to two main factors. One is the collapse of the vertebrae as a result of decreased bone density (see "Why Trunk Height Decreases With Aging" on p. 46); the other is postural changes including exaggerated curvature of the thoracic region of the vertebral column and flexion at the lower limb joints. Some researchers have calculated that structural and postural changes account for a 4.3-cm (1.7-in.) decrease in stature,

with an extra 3.0-cm (1-in.) difference observed between 20- and 80-year-old adult females being attributable to the secular trend in height.

Growth Rates of Body Segments

An important concept in growth is the differential growth of tissues and body segments. Total body height, usually termed stature, is the total of lower limb length, trunk length, and head height. These three components do not grow at the same rate. Head height approximately doubles from birth to maturity, while trunk length approximately triples. The limbs grow faster; during maturation, the length of the upper limb increases fourfold while lower limb length increases fivefold.

Differential changes in the sizes of body segments mean changes in physique because the shape of the body is defined by proportions. The sitting height/standing height ratio changes from birth to adulthood; but the rate of change is not constant, and even the sign (direction) of the ratio changes. At 2 years of age the ratio is close to 60%. It then decreases to a minimum near puberty, indicating the relatively rapid growth of the lower limbs, after which it increases slightly because of some "catch-up" growth by the trunk. At the age of 9 to 10 years, the sitting height/standing height ratio is similar to the adult ratio of about 52%.

Growth Rates of Body Tissues

The various components of the human body are listed in table 3.1. Earlier in this chapter we noted that the proportion of the body made up of muscle tissue increases during growth and maturation. Consequently, the relative amounts of other tissues must decrease, indicating that the various tissues must grow at different rates. The differential growth in some body tissues is illustrated in figure 4.7. Notice again that nervous tissue, such as the brain, is closer to adult size at an earlier age than other tissues. The rapid increase in the reproductive tissues is associated with the development of secondary sexual characteristics and the "growth spurt" around PHV. As well as changes in muscle and bone, there are changes in major organs such as the brain and liver. In the newborn baby, 13% of the body weight is brain and 5% is liver. In the adult, these relative weights have decreased to 2% and 2.5%, respectively. Obviously there is a good correlation between the size of the skull and the size of the brain. The protruding abdomen observed during infancy and early childhood is indicative of the relatively large digestive organs, including the liver.

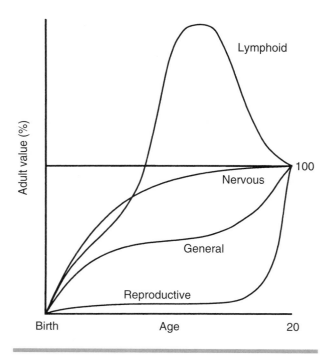

Figure 4.7 Differential growth in nervous, reproductive, and lymphoid tissue compared with the general growth of tissues.

Sexual Dimorphism in Growth

Figure 4.6 shows sexual dimorphism in growth. At any given age, the average of girls' heights is closer to their average adult height than is the case for boys. This phenomenon is particularly noticeable during the upper elementary and lower high school years, when girls are often taller than their male classmates because the female growth spurt usually occurs two years earlier than in the male. A rule of thumb is that girls reach half their adult height at about 2 years of age, while for boys this age is 2 1/2 years.

There are three main reasons for the tendency of adult males to be taller than adult females. First, because the pubescent growth spurt starts later in males (see earlier section on the pattern of growth in Swedish children), they are already taller when they enter the period of rapid growth. Second, the male growth spurt is more intense, meaning that the PHV is greater. Third, the growth spurt lasts longer in males. Of these factors, the predominant one is the delay in entering the growth spurt.

Sexual dimorphism is also seen in divergences in the average sitting/standing height ratio between boys and girls of the same age. During childhood the ratios are very similar, but from about 11 to 12 years onward the ratio is higher for girls. One explanation is that during the period of rapid growth the lower limbs of boys grow relatively faster than those of girls. During aging, the ratio between sitting and standing heights decreases again because of the shortening of the vertebral column, related to both structural and postural changes mentioned earlier.

Relative amounts of fat tend to be greater in females at all ages, but the major divergence between males and females occurs when girls enter puberty (figure 4.8). The size of the triceps skinfold

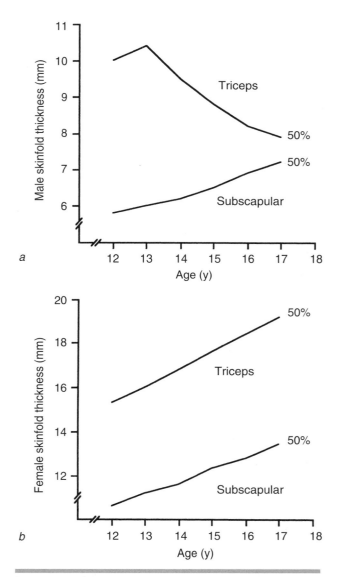

Figure 4.8 Changes in skinfold thicknesses during growth and maturation for 50th percentile Australian (a) males and (b) females. The triceps skinfold is measured at the back of the arm, and the subscapular skinfold is measured just below the shoulder blade.

Adapted from Court et al. 1976.

between the ages of 1 and 20 years exhibits different time courses for males and females. At 20, the skinfold thickness is greater than during infancy for females, but it is less than at 1 year of age for males. The minimum skinfold thickness is at 7 years of age for females; it then gradually increases, with almost a plateau for about two years during puberty. For males, there are two minima, at 7 and at 15 to 16 years of age. The second minimum occurs because of a decrease in skinfold thickness for about four years during puberty. Especially for females, the term "puppy fat" or "baby fat" is probably inappropriate, as there is no period during puberty when the skinfold thicknesses are decreasing.

Muscle and bone masses tend to be similar in males and females until about 13 years of age. At this age the hormonal influence on males is such that they have a greater increase in bone and, particularly, muscle mass.

Somatotype Changes During Growth, Maturation, and Aging

A longitudinal study of somatotypes of European males between the ages of 11 and 24 years indicated two stages of puberty. In the first stage, between about 11 and 15 years, endomorphy decreases because a decrease in skinfold thicknesses occurs at the same time as an increase in height. This increase in height is rapid, so the growth tends to be "linear," resulting in an increase in ectomorphy. The second stage of puberty is marked by an increase in mesomorphy, which results from relative increases in transverse diameters of the skeleton and increases in muscle volume. Ectomorphy decreases between 15 and 24 years of age because of the increased mass of the musculoskeletal system.

A study of almost 14,000 Canadian adults showed that males over 30 years of age are slightly more endo-mesomorphic than those under 30, and that there are similar differences between women over and under the ages of 40 years (see figure 3.2).

Methods of Determining Age

Of some interest is the concept of chronological age versus biological age. We all know children who are early maturers and teenagers who are late maturers. Masters athletes are pleased when told by their doctor that they have the body of a person 10 years younger. Obviously the time since birth is not the only method of representing age.

Examples of maturation include deciduous dentition ("baby teeth") followed by the permanent dentition, the gradual replacement of cartilage by bone, the onset of **menarche** (beginning of menstruation in girls), and the appearance of secondary sexual characteristics. These changes tend to occur in a similar order across individuals; the age of onset of the changes and the period of time required to complete the changes are more variable. Therefore stages in development can be used as a basis for determining biological age during maturation.

In young people, skeletal or bone age can be calculated by inspection of hand/wrist X-ray images. This site is the most suitable because the appearance of ossification centres and fusion of bony parts take place over many years in a fairly regular pattern. Bone age can be determined to the nearest month and then compared with chronological age to ascertain whether the person is ahead of or behind the normal developmental pattern. Clinically, bone age can be used, in conjunction with other measurements, to predict adult height.

Radiographic techniques are not the most suitable for use in large studies because of the ethical issues involved with exposure to X-radiation. It is preferable to use less potentially hazardous techniques.

Much of the musculoskeletal development that occurs around puberty is related to hormonal changes associated with sexual maturation. Therefore sexual maturity itself can be determined using stages proposed by Tanner. During the development of secondary sexual characteristics, all individuals pass through a common series of body changes that of course are different for males and females. The order of development is relatively constant, but the age of onset of the first stage of maturation and the duration of progress through all stages show wide variation. Genital and associated development is classified to determine which of the five Tanner stages best describes a particular person.

Sexual maturity rating has been used in male junior sports in North America because it relates to the musculoskeletal development induced by the increased levels of testosterone. In a collision sport such as American football, musculoskeletal development is important, and the school administrators in New York State during the early 1980s decided that no high school boy could play football until he had entered stage 3 on the sexual maturity rating. This recommendation was seen as an important sports medicine intervention, protecting boys who were perceived as most likely to incur injury during participation in the sport. Because sexual maturity rating involves estimation of genital development, it is psychologically invasive; and other less embarrassing methods are being developed that one hopes will be equally valid.

During aging, such parameters as bone density, range of joint motion, and muscle strength decrease; if these values are greater than expected for a person of a given age, the person could be described as biologically young for his or her chronological age. There are no formalized equations grouping parameters for people who are elderly as there are for children.

Summary

Most tissues and structures develop during the eight weeks after conception. The foetal period therefore involves growth of tissues and organs, mainly by increase in the number (hyperplasia) and size (hypertrophy) of cells. Maturation involves the changes that occur between birth and maturity. This period may vary depending on the specific physical parameter being considered. Most of the long bones of the limbs develop by a process of replacement of a cartilage model by bone, which takes about two decades. Around puberty, the cartilaginous growth plates may be susceptible to large muscle forces. During maturation, calcium is laid down in the bone; the peak amount may determine an individual's probability of becoming osteoporotic in later years. After birth, muscle mass increases over 30 times; and during maturation, the relative muscle mass almost doubles. Differential growth is an important concept: All tissues do not grow at the same rate at the same times, and

organs do not necessarily maintain their shape as they grow.

Aging involves decrements in a number of physical parameters, but some of the decrease in physical ability may relate to lifestyle changes rather than to natural biological aging processes. The quality of life, associated with general mobility and physical independence, will become more important as people live longer. Relative muscle mass and strength decrease in old age, and the negative effects on lifestyle can be severe. The prevention and treatment of osteoporosis and osteoarthritis are becoming increasingly important.

Each generation exhibits some physical differences from the previous generation. Children have been progressively getting taller for the last few generations. Unfortunately the incidence of osteoporosis has also been increasing rapidly so that now the incidence in men is the same as it was in women in the previous generation.

Further Reading and References

Eveleth, P.B., and J.M. Tanner. 1990. *Worldwide variation in human growth.* 2nd ed. Cambridge, New York: Cambridge University Press.

Shephard, R.J. 1997. *Aging, physical activity, and health.* Champaign, IL: Human Kinetics.

Shephard, R.J., and J. Parizkova, eds. 1991. *Human growth, physical fitness, and nutrition.* Basel: Karger.

Sinclair, D., and P. Dangerfield. 1998. *Human growth after birth.* 6th ed. Oxford, New York: Oxford University Press.

Tanner, J.M. 1989. *Foetus into man: Physical growth from conception to maturity.* 2nd ed. Ware, England: Castlemead.

CHAPTER 5

Musculoskeletal Adaptations to Training

The purpose of this chapter is to examine the changes in the skeletal, articular, and muscular systems and in overall body shape, size, and composition that occur as an adaptation in response to physical activity.

> The major learning concepts in this chapter relate to
> - the effects of activity and injury on bone;
> - the effects of activity on joints and their ranges of motion;
> - the effects of activity on muscle structure and function;
> - the effects of training and other lifestyle factors on body size, shape, and composition; and
> - the relationships between body sizes and types and different types of sports.

Effects of Physical Activity on Bone

Charles Darwin (1809-1882) recognized the relationship between physical activity and bone mass in his groundbreaking book *The Origin of Species,* first published in 1859. A modern version, abridged by the physical anthropologist Richard Leakey, includes the statement that "with animals the increased use or disuse of parts has a marked influence; thus in the domestic duck the bones of the wing weigh less and the bones of the leg more, in proportion to the whole skeleton, than do the same bones in the wild-duck; and this may be safely attributed to the domestic duck flying much less, and walking more, than its wild parents" (Darwin 1979, p. 50).

Human clinical cases and animal experimentation have shown similar adaptability of bone. The major weight-bearing bone in the human leg is the tibia, which is in contact with the femur at the knee. In cases in which the tibia is congenitally missing (not present at birth), the fibula has been surgically moved over to establish the necessary contact between it and the femur. The increases in size and strength of the fibula in this situation are quite spectacular.

According to one version of Wolff's law (p. 13) proposed by Sir Arthur Keith in a lecture to the Royal Society in London in 1921, "Every change

in the form and function of a bone or of their function alone, is followed by certain definite changes in their internal architecture, and equally definite secondary alterations in their external conformation, in accordance with mathematical laws" (Tobin 1955, p. 57). This translation of the law implied that bones sustain a maximum force with a minimum of bone tissue and that bone reorganizes to resist forces most economically.

Effects of Activity Level on Bone

Not all adaptation of bone is positive; certain circumstances may lead to maladaptation (see "How Does Intensive Exercise Affect the Bones of Growing Animals?" on p. 55). For example, clinical evidence from elite junior athletes (see chapter 22) suggests that intense physical activity can produce maladaptation in growing bones. Stress fractures may occur in young adults, especially when the onset of intense physical activity is rapid, as among recruits in the armed forces.

One can find a number of good examples of bone adaptation in response to external forces. It was realized soon after animals and humans went into space that they incurred a rapid reduction in bone mass, probably related to the decreased muscle activity in a zero gravity environment.

IN FOCUS: HOW DOES INTENSIVE EXERCISE AFFECT THE BONES OF GROWING ANIMALS?

Researchers from the Department of Anatomy at the University of Queensland studied the effects of a one-month intensive exercise program on the structural and functional properties of 17 pubescent male rats. An equal number of rats not specifically trained were used as controls for comparison with the experimental group.

For five days of each of the four weeks of the experiment, the experimental rats ran on a treadmill for 1 h per day and also swam for 1 h. It was estimated that the rats were working at about 80% of their maximal oxygen consumption. The researchers studied the structure and function of the major lower limb bones, measuring bone length and width and examining bone slices under a light microscope. Whole bones were also mounted in a specially manufactured torsion-testing machine to measure the strength of the bones when twisted at a physiological rate.

During the month, the experimental rats ate more than the controls; but at the end of the training period the experimental rats were lighter. The long bones of the hindlimb were shorter and lighter in the experimental animals. Also the epiphyseal plates were thinner. As a result of the intense exercise program the femur was not significantly affected, but the tibia exhibited a significant decrease in the amount of energy it could absorb before it fractured.

The authors postulated that the repetitive cyclical loading caused an accumulation of microcracks in the bone that resulted in maladaptation. The authors therefore questioned the proposal that exercise is always beneficial for young animals, indicating that the literature related to stress fractures in humans reflected the potential problem.

SOURCE

- Forwood, M.R., and A.W. Parker. 1987. Effects of exercise on bone growth: Mechanical and physical properties studied in the rat. *Clinical Biomechanics* 2: 185-190.

American and Russian scientists are still trying to develop the optimal exercise regimen to prevent loss of bone in space. Likewise when a limb is immobilized in a plaster cast, evidence of disuse osteopenia can be observed on X-ray images. Osteopenia, the mechanism by which bone mass decreases, is due to the imbalance between bone resorption and deposition. The result is that the porosity increases and therefore the bone is less dense, appearing less radio opaque on an X-ray image (bone absorbs more X-radiation than any other tissue except teeth). The bone loss is related to the absence of external forces on the bone, which are produced mainly by muscle activity. It would appear that bone has a genetically determined baseline mass that is less than that required for normal functioning. Therefore a certain level of physical activity is necessary for good bone health. Complete rest results in rapid loss of both bone and muscle mass.

Generally, lack of physical activity results in a loss of bone mass and increased activity results in a gain. Some of the best examples of the relationship between physical activity and bone mass involve comparisons between the humerus bone of the playing and nonplaying arms of professional tennis players. For example, radiological studies demonstrated not only that the bone was denser in the playing arm, but also that the diameter of the shaft was larger and the compact bone forming the shaft was thicker. Recent reanalyses of the original data have shown that the greatest observed increases occurred in individuals who began training at a young age, that is, prior to puberty.

These results have implications for young adults who are advised to optimise their bone density. Cross-sectional studies indicate that adults with active lifestyles have greater bone mass than sedentary adults, and better-controlled longitudinal studies indicate that exercise can cause an increase in bone mass. Even in postmenopausal women, exercise can add bone; exercise is certainly important in maintaining bone mass that would otherwise be lost through aging. Modest additions of bone can occur in aging individuals with adequate levels and types of exercise, but the biggest contribution of exercise is to reduce the bone loss that would occur with insufficient levels of physical activity.

Effects of Activity Type on Bone

Weight-bearing activities such as walking and running are associated with the addition of bone, but this effect may not be so great in other activities such as swimming and cycling. Elite swimmers have lower bone densities than other elite athletes; weightlifters have the highest bone density. As indicated in figure 5.1, the initial positive changes occur slowly whereas the initial response to decreased activity is rapid; therefore an exercise program must be regular and must be maintained over a long period of time to be beneficial. As noted in chapter 2, a remodelling cycle is about three months. About three or four remodelling cycles are required to produce a new steady state in response to increased physical activity. The type of exercise most beneficial for bone health is discussed further in chapters 11 through 13.

Bone Repair and Physical Activity

Our discussion thus far indicates that if exercise level gradually increases from an initial baseline level of activity, positive adaptation will result. If activity is too intense, however, there may be negative changes that could lead to injuries such as stress fractures. The following period of immobilization will result in loss of bone and decreased density so that the bones are weaker. During rehabilitation,

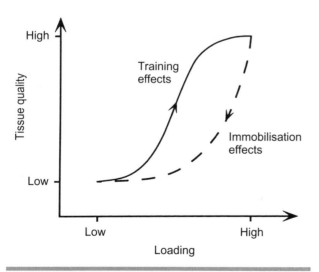

Figure 5.1 Possible courses of musculoskeletal tissue adaptation related to training (solid line) and immobilization (broken line). There is a delay in improvements related to training, but the negative effects of immobilization are almost immediate.

Adapted from Kannus et al. 1992.

after repair of the injured site, bone will regain its former properties. However, because the loss of strength is rapid in comparison with its return, it is important for coaches and sports medicine personnel to ensure that this scenario does not become a recurring cycle for an elite athlete. This concept applies to ligament sprains and muscle strains as well. Figure 5.1 indicates that there is an optimal level of activity for each person, and the challenge for prescribers of exercise is to determine this level on an individual basis.

Remember that bone is more like a tree than a piece of wood because it is a living, dynamic, metabolically active tissue with a good blood supply. It undergoes continual remodelling and can therefore recover from damage. The major stages during fracture healing are illustrated in figure 5.2. During a fracture, the bone and the surrounding tissues bleed when the blood vessels are ruptured. Subsequently a blood clot forms, and later, capillaries invade the clot, bringing cells and nutrients. A fibrous **callus** fills the space between the ends of broken bone; then a bony callus replaces the fibrous callus. Appropriate physical rehabilitation is important because the final strength of the bone will depend on remodelling of the bony callus, and this remodelling is affected by physical activity.

Effects of Physical Activity on Joint Structure and Ranges of Motion

Warm-up is usually recommended before any physical activity. Many believe that warm-up is beneficial only for cardiovascular reasons, including increasing the heart rate, but warming up also has positive effects on the musculoskeletal system.

Synovial Fluid, Articular Cartilage, and Ligaments

During cyclical exercise, such as jogging, the lower limb joints are continually moving within their ranges of motion. The short-term effect on articular cartilage is that it thickens and thus can better perform its function of force dissipation during dynamic activities. The cartilage thickens because it absorbs synovial fluid, and the flow of synovial fluid into and out of the cartilage improves the supply of nutrients and the removal of waste products of metabolism. Chronic exercise results in long-term thickening except when compressive

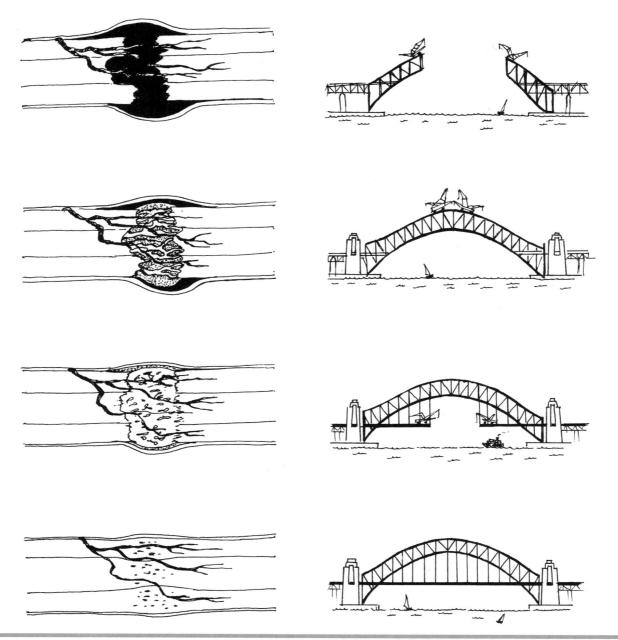

Figure 5.2   The major stages in bone healing compared with the construction of the Sydney Harbour Bridge, which was completed in 1932.

forces on the cartilage are excessive, as may occur during repeated downhill running or heavy workouts involving a great deal of jumping.

Normally the quantity of synovial fluid in a joint is very small. The knee has the largest joint cavity in the body, but it contains only about 0.2 to 0.5 ml of synovial fluid at rest. After a 1- or 2-km (0.6- or 1.2-mile) run, the volume may have increased by up to two or three times. One can argue that the synovial fluid can then better perform its functions

of lubrication and nutrition. Certainly short-term exercise makes synovial fluid less viscous and thus more like oil than grease. It may therefore be a better lubricant but does it sacrifice this property for the role of protection? Because the articular cartilage soaks up the "thin" synovial fluid, the joint is probably better protected by the thickened cartilage after a warm-up period.

Ligaments are passive stabilizing structures of joints, and the loads applied to them vary when a joint is moved through its full range of motion. The oscillating increase and decrease in tensile forces on ligaments during physical activity result in adaptations. The size of the ligament increases because of an increase in collagen synthesis, so the ligament becomes stronger. Moreover, an increase in cross-linking within the ligament makes it stiffer. There is evidence that endurance exercise has more positive effects on ligament strength than sprint training.

Degenerative Joint Disease and Exercise

Indications of degenerative joint disease, or osteoarthritis, are thinning of the articular cartilage and thickening of the thin layer of compact bone under the articular cartilage. It has been hypothesized that changes in this subchondral bone precede the cartilage thinning. Animal studies have shown that repetitive impulsive loading causes stiffening of the thin subchondral bone so that the cartilage may be more susceptible to damage. The human knee is among the joints commonly affected by osteoarthritis, which increases in incidence with age. To these possible genetic and aging factors may be added environmental factors such as obesity, which is associated with a greatly increased risk of osteoarthritis.

Although clinicians report that osteoarthritis is among the long-term effects of jogging, epidemiological studies of the relationship between exercise and degeneration of the articular cartilage have shown that regular runners do not experience an increased incidence of osteoarthritis. Recent studies indicate that the connection between exercise and later degenerative changes in the articular cartilage relates to previous synovial joint injury such as ligament sprains. Ligament damage may allow abnormal motion within the synovial joint, resulting in excessive localized loading.

Effects of Physical Activity on Muscle–Tendon Units

If you have experienced a major joint injury, you know how quickly muscle size decreases with disuse. A major goal of rehabilitation after such an injury is to increase the size and contraction force of the muscles involved. If you go to the gym for regular strength training, you expect to see increased muscle size; you would also hope

that the attached tendons would become stronger in order to withstand the greater muscle forces being produced.

Flexibility

Flexibility is a term commonly used to indicate ranges of joint motion. The term "joint range of motion" seems to indicate that some of the specific structures of synovial joints, illustrated in figure 2.6, restrict motion. In most cases, however, flexibility or lack of it depends on the muscle–tendon units that cross joints. In fact, most flexibility exercises stretch muscle–tendon units and not joint capsules or ligaments. Flexibility that relates to muscle–tendon length is positive. When some ligaments are stretched, however, the joint may be classified as *loose*, or *lax*. This characteristic is associated with increased risk of injury. Joint laxity may in fact require physical therapy or orthopaedic surgery.

Flexibility is often listed among the factors composing physical fitness, which gives the impression that flexibility is a general characteristic of an individual. In reality, flexibility is highly joint specific because it is related to normal activity patterns. The flexibility of ballet dancers provides an excellent example. A range of motion study in ballerinas showed that the range of external rotation of the hip was much greater than in the normal population but that the range of internal rotation was less than normal. This variability suggests that activity patterns in ballet, involving externally rotated postures of the hip, produce structural changes that adapt to the forces but that then restrict motion in the opposite direction.

Studies on teenage gymnasts and aged persons indicate that regular performance of flexibility exercises can prevent or even reverse normal decreases in ranges of joint motion. Certainly stretching exercises performed regularly for a period of weeks, months, or years can increase joint flexibility. When this increase occurs, how does the whole muscle respond to regular static stretching? Remember that whole muscles consist of both muscle and connective tissue. Does the number of sarcomeres in series forming each muscle fibre increase, or do the extensibility and elasticity of the connective tissue component increase? In other words, does a muscle grow longer or become more extensible as a result of flexibility exercises? The answer is probably that increased extensibility of the connective tissue is the predominant factor, although sarcomere number will adapt rapidly to any habitual length change.

A perception about weight trainers is that they are "muscle bound"—that their muscle development leads to a decrease in the range of joint motion. This general idea appears to be ill founded. With correct weight-training techniques, including flexibility exercises, range of motion can be maintained and even enhanced, as evidenced by the celebratory gymnastics of some Olympic weightlifters after winning a gold medal.

Understanding what is behind flexibility helps clarify why a general warm-up, even when combined with massage, seems to have little effect on joint flexibility. Clearly, specific joint flexibility exercises that stretch specific muscle–tendon units must be a separate part of the warm-up routine.

Strength

People who undertake a weight-training program often experience a rapid increase in measured strength, followed by a plateau effect and then further gradual increases. Electromyographic (EMG) studies have indicated that during the initial six to eight weeks, the muscular activation pattern becomes more efficient and thus the strength gains are related to improved control of the neuromuscular system. In the next phase, the size of the muscle fibres increases due to an increase in the number of myofibrils within each fibre. That is, an initial phase of neural adaptation (neurotrophic phase) is followed by a longer period of muscle hypertrophy (hypertrophic phase) in which the cross-sectional area of each muscle fibre increases. The major effect of weight training is hypertrophy of the muscle fibres as discussed further in chapter 11. It appears that increases in the connective tissue component of whole muscle parallel those of the muscle tissue during strength-training programs. Long-term effects of endurance exercise relate mainly to changes within the muscle fibre to make the muscle less fatigable, but there is also an increase in the relative amount of connective tissue within the whole muscle.

Tendon Adaptation

Exercise has a positive effect on tendon; but, as with other connective tissue structures, the rate of adaptation is much slower than that of

muscle. In tendon, collagen synthesis increases and the collagen fibres line up more regularly in a longitudinal direction. As with other tissues, training that is too strenuous may be injurious (figure 5.1). When muscle strains occur, they often happen at the junction between the muscle and its tendon.

Effects of Physical Activity on Body Size, Shape, and Composition

It was emphasized in chapters 3 and 4 that a person's somatotype is more of a phenotype than a genotype, implying that the somatotype can be modified by factors such as training. Although amount of body fat and its distribution are under a certain amount of genetic control, the combination of increased activity and reduced energy intake via the diet will decrease the fat content of the body. A person following such a regimen will become less endomorphic. All studies of athletes compared with the sedentary population have shown that athletes have less body fat and a larger lean body mass, partly because of an increased skeletal mass. Thus, exercise may not alter a person's ectomorphy rating because the decrease in endomorphy (fat) may be balanced by an increase in mesomorphy (muscle and bone). In other words, the skinfolds decrease, but the muscular circumferences of the arm and leg increase. If body mass does not decrease as a result of an exercise program, this does not necessarily mean that the program has been ineffective.

Role of Lifestyle Factors in Determining Physique

Athletes such as bodybuilders, weightlifters, and ballet dancers appear to have unusual body shapes that relate more to their adaptation to training than to differences determined before birth. When measurements of weightlifters and ballet dancers are compared with values for the normal population, interesting similarities and differences can be observed (figure 5.3). The measurements of bone sizes, such as circumferences of the wrist, knee, and ankle, are relatively normal. Remember that the bone in the shafts of the weightlifters is likely to be denser and heavier, but these parameters are not measured in normal anthropometric protocols. Width of the

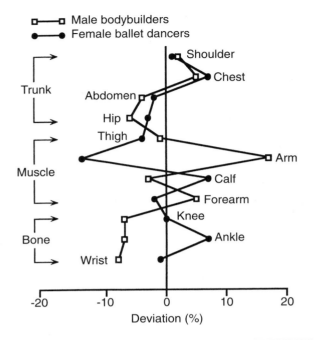

Figure 5.3 Differences in relative body segment circumferences between male bodybuilders and their reference population and between female ballet dancers and their reference population. Both reference populations are represented by the solid vertical line.

Reprinted from Katch et al. 1980.

expanded ends of long bones is less susceptible to training. Notice particularly that the muscular circumferences, such as of the biceps, are relatively large in weightlifters as a result of muscle hypertrophy. In contrast, they are relatively small in ballet dancers, related to the demands of their training, including type of physical activity (e.g., demands on the muscles of the arm are far less than on the muscles of the calf) and dietary intake. Training can affect physical and physiological factors at any time of life, and this aspect is a major focus of chapter 12.

Relationship of Body Sizes and Types to Sports

One good example of kinanthropometry is provided by the following quote: "Javelin throwers and gymnasts have practically identical somatotypes, although the javelin throwers, at 179.5 cm and 76.7 kg, are much bigger than the gymnasts,

at 167.4 cm and 67.1 kg" (Hebbelinck and Ross 1974, p. 545). After reading the next section of this book on **biomechanics** you should be able to explain these similarities and differences. One important aspect is the large difference in the size measurements, such as height and weight. Height is predominantly genetically determined and under normal conditions is not greatly affected by environmental factors. Therefore sport performance is affected by genetically determined factors as well as training and the physical adaptations to training. The physical dimensions that cannot be changed are often used in talent identification, a topic we explore more fully in part III of this book.

In a given sport, particular events may require mainly strength or endurance. In running, for example, the 100-m (109-yd) sprint and the marathon are at two ends of a strength–endurance continuum. In running, the athlete's own body weight must be transported, and runners tend to be lighter than the average person; they get progressively lighter as the distance of the event increases. Sprint and middle distance runners tend to be taller than normal, but long distance runners tend to be shorter than the average person.

Events such as the shot put require even greater strength than sprinting; height is also an advantage because it means that the shot can be released from a greater height (hence travel further for any particular velocity and angle of release). Shot-putters and discus throwers tend to be taller and heavier than other track and field athletes. Body weight is not such a limiting factor for these athletes because they do not have to transport their own body over a long distance or lift their centres of mass very far, so they tend to be very heavy. Also a large lean body mass would be advantageous for the development of power required in explosive movements.

Somatotypes also change as the requirements for strength and endurance change. In events that have a high strength component, such as the shot put and discus, the athletes are very mesomorphic. Athletes tend to move from this position on the somatochart downward and to the right as the endurance component increases (see "Are There Physical Differences Between Players in Different Sports?" on p. 61).

IN FOCUS: ARE THERE PHYSICAL DIFFERENCES BETWEEN PLAYERS IN DIFFERENT SPORTS?

A research group from the School of Education at the Flinders University of South Australia, led by Bob Withers, used the Heath-Carter Anthropometric Somatotype method (see chapter 3) to characterize 206 national standard sportsmen in 17 sports and 127 female representative squad members in 10 sports. Endomorphy was corrected for height. Age, height, and mass were reported separately.

The male athletes were described generally as balanced mesomorphs. Their mesomorphy ratings were above average, and the ratings for endomorphy and ectomorphy were both below average. When compared with an age- and sex-matched group of untrained Canadian males, the South Australian athletes were taller, lighter, less endomorphic, more mesomorphic, and more ectomorphic. The overall mean for relative body fat of the athletes was estimated to be 10%.

The most mesomorphic sportsmen were the powerlifters, and the least mesomorphic were the long distance runners. The field lacrosse players were more endomorphic than the Australian Rules football players and the gymnasts. The gymnasts were more mesomorphic than the track and field athletes when considered as a single group. The Rugby Union footballers and swimmers differed on all three components of the somatotype. The footballers were more endomorphic, more mesomorphic, and less ectomorphic. The tallest sportsmen were the basketball players and rowers, and the shortest groups were the powerlifters and the gymnasts. The heaviest group was the Rugby Union football players, and the lightest group was the gymnasts.

The authors compared their results with those of previous studies and provided a good kinanthropometric explanation of the differences between groups. The differences confirm the relationship between somatotype and relative requirements for strength and endurance (see figures 5.4 and 5.5).

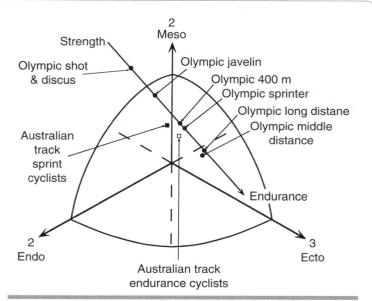

Figure 5.4 Somatochart illustrating the somatotypes of some Olympic track and field athletes and elite Australian cyclists. Notice the strength–endurance continuum from a predominance of mesomorphy to a predominance of ectomorphy. Note also that none of these elite athletes is endomorphic.

Data from Carter, Aubry, and Sleet 1982.

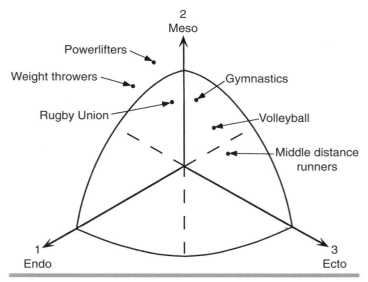

Figure 5.5 Somatochart illustrating the somatotypes of selected elite male South Australian sportspeople. Notice the same strength–endurance continuum as shown in figure 5.4.

Adapted from Withers, Craig, and Norton 1986.

(continued)

IN FOCUS: ARE THERE PHYSICAL DIFFERENCES BETWEEN PLAYERS IN DIFFERENT SPORTS? *(continued)*

Compared with a reference sample of 135 women, the female athletes were less endomorphic, slightly more mesomorphic, and more ectomorphic. Other studies have demonstrated a tendency for endomorphy to decrease as the level of physical activity increases. The basketball and netball players were taller than those playing soccer, lacrosse, cricket, hockey, and softball; and the softball players were lighter than the basketball players. There was a tendency for the netball, squash, and volleyball players to be less endomorphic and more ectomorphic than those playing badminton, basketball, cricket, hockey, lacrosse, soccer, and softball.

The authors emphasized that many factors other than body size and shape determine successful sport performance. This caveat is particularly important in the case of field games that require a range of skills, especially perception–action coupling, as well as various components of fitness.

SOURCES

- Withers, R.T., N.P. Craig, and K.I. Norton. 1986. Somatotypes of South Australian male athletes. *Human Biology* 58: 337-356.
- Withers, R.T., N.O. Whittingham, K.I. Norton, and M. Dutton. 1987. Somatotypes of South Australian female games players. *Human Biology* 59: 575-584.

Summary

Biological tissues in general respond to their environment. The tissues of the musculoskeletal system, in particular, respond to mechanical forces. The genetic baseline bone mass is much less than normal, so physical activity is required just to maintain normal levels. Appropriate exercise prescription can lead to increased bone mass, joint flexibility, and strength, all of which can be beneficial for general health and exercise performance.

The processes of adaptation of the various tissues of the musculoskeletal system occur at different rates and may involve various stages, such as improved neural control followed by muscle hypertrophy during a weight-training program. Training results in a change of somatotype as the relative amounts of the major tissues of the body alter in response to training. Generally, athletes have reduced amounts of fat and increased amounts of bone and muscle when compared with the general population. Excessive levels of activity can lead to tissue breakdown and a decrease in bone mass in women.

Further Reading and References

Bloomfield, J., T.R. Ackland, and B.C. Elliott. 1994. *Applied anatomy and biomechanics in sport.* Melbourne: Blackwell.

Darwin, C. [1859] 1979. *The illustrated origin of species.* London: Book Club Associates.

Delavier, F. 2001. *Strength training anatomy.* Champaign, IL: Human Kinetics.

Elliott, B., ed. 1998. *Training in sport: Applying sport science.* Chichester, New York: Wiley.

Hebbelinck, M., and W.D. Ross. 1974. Kinanthropometry and biomechanics. In *Biomechanics IV*, ed. R.C. Nelson and C.A. Morehouse (p. 545). Baltimore: University Park Press.

Tanner, J.M., R.H. Whitehouse, and S. Jarman. 1964. *The physique of the Olympic athlete. A study of 137 track and field athletes at the XVIIth Olympic Games, Rome, 1960: And a comparison with weight-lifters and wrestlers.* London: Allen and Unwin.

Tobin, W. 1955. The internal architecture of the femur and its clinical significance: The upper end. *Journal of Bone and Joint Surgery* 37A: 57-72.

PART II

MECHANICAL BASES OF HUMAN MOVEMENT

The Subdiscipline of Biomechanics

Biomechanics is the application of mechanics to the study of living systems. Having evolved from the days of Aristotle in 384 to 322 B.C., today biomechanics sits at the intersection of biology, medicine, and quantitative mechanics. Basic scientists use the principles of mechanics to understand the mechanisms underlying function in the healthy state. Clinicians, on the other hand, are more interested in formulating and solving problems related to injury and disease. The four chapters that follow show how the principles of biomechanics may be used to perform quantitative analyses of movement in the healthy state and to examine how musculoskeletal changes introduced by injury, training, growth and maturation, and aging affect motor performance. The focus, at all times, is at the systems level, although many biomechanists also work at the cellular and molecular levels, attempting to understand the mechanics of molecules and how it applies to the morphology and motility of cells.

Typical Questions Posed and Levels of Analysis

Why can some people jump higher than others? Why do some run faster and throw farther? These are questions typically addressed by sport biomechanists. Why do children and healthy older adults walk differently from healthy young adults? How does rupture of the anterior cruciate ligament (ACL) alter function of the remaining knee ligaments during activity? What are the best practices for reconstructing a torn ACL? How does leg muscle function change to compensate for the presence of spasticity resulting from cerebral palsy, stroke, or traumatic brain injury? How does a total hip replacement alter the mechanics of gait? These are all questions that concern the clinical biomechanist, orthopaedic surgeon, and physical therapist.

The approach taken to answer a question depends on the question being asked. For example, if one is interested in learning how high Michael Jordan can jump, one can obtain the answer by performing a relatively simple, noninvasive biomechanical experiment. A high-speed video system may be used to track the positions of surface markers mounted on various parts of the body, and these data can then be used to calculate the trajectory of the centre of mass during a jump. Alternatively, a device called a force plate may be used to record the time history of the vertical component of force exerted on the ground during the jump. A fundamental principle of **mechanics** called the impulse–momentum principle may then be applied to calculate the vertical velocity of the centre of mass at the instant of liftoff, and hence jump height.

On the other hand, if one is interested in understanding *why* Michael Jordan can jump higher than most people his size, then one needs a much deeper knowledge of the interaction between physiology and mechanics. Specifically, information about the architectural, physiological, and mechanical properties of Michael Jordan's leg muscles must be correlated with records of the forces developed by his leg muscles during the ground contact phase of a jump. One cannot obtain this information simply by performing a biomechanical experiment on Michael Jordan. Records of muscle forces, in particular, cannot be obtained by direct measurement, and so another approach involving computer modelling is often used. In this approach, the geometry and properties of the musculoskeletal system (i.e., muscles, tendons, ligaments, bones, and joints) are represented in a **model** of the body, and a problem is formulated to estimate those quantities that cannot be obtained directly from experiment. If the problem entails the calculation of muscle forces, the method most commonly used is inverse **dynamics.** Here, measurements of body motions and ground forces are used to determine the net muscle actions about the joints. A mathematical theory called **optimisation** is then applied to determine the forces developed by the individual leg muscles during the jump.

This section of the text concerns the mechanical bases of movement. Chapter 6 introduces the reader to the kinematics of movement, specifically the concepts of displacement, velocity, and acceleration of a body segment. Displacement and velocity, in particular, are needed to calculate mechanical energy in movement, which is the focus of chapter 7. The final two chapters of this part of the book present issues related to changes in biomechanical performance brought about by growth and development, aging, and injury. Chapter 8 discusses biomechanical changes that occur throughout the life span, and chapter 9 describes how the neuromuscular and musculoskeletal systems may be trained to combat both aging and injury.

Historical Perspectives

The history of science begins with the ancient Greeks. Aristotle was probably the first biomechanist. He wrote the first book related to movement, *De Motu Animalium (On the Movement of Animals)*, and actually viewed the bodies of animals as mechanical systems. Nearly 2,000 years later, Leonardo da Vinci would take up where Aristotle left off.

Da Vinci was born in 1452 and was a self-educated man. He became famous as an artist, but worked mostly as an engineer. His contributions to science and technology were prodigious by any measure, and he is credited with numerous inventions ranging from water skis to hang gliders. Da Vinci understood concepts such as the components of force vectors, friction, and how a falling object accelerates through the air. By combining his knowledge of human anatomy and mechanics, da Vinci was able to analyse relatively detailed systems of muscle forces acting on body parts.

The man regarded by many as the father of biomechanics is Giovanni Alfonso Borelli. Borelli was born in 1608 in Naples, Italy. He was the first to understand that the levers of the musculoskeletal system (i.e., moment arms) cause muscles to develop very large forces during physical activity. Using his intuition about the principle of **statics,** Borelli figured out the forces needed to hold a joint in equilibrium. He also determined the position of the centre of gravity of the whole body in a human and calculated inspired and expired air volumes.

There was little to show for the 250 years following Borelli until the appearance of the French physiologist, Etienne Marey. Marey was the first to actually film people moving. He constructed a camera to expose multiple images of a subject on the same photograph at set intervals. Marey and his colleague Eadweard Muybridge were the first to use high-speed film recordings to analyse movement. In 1887 Muybridge published his monumental, 11-volume *Animal Locomotion*, containing high-speed, sequential photographic analyses of human and animal gait. (Some of these historical motion photographs can be viewed at www.eastman.org/10_colmp/10_video/10_vid_muy.html.)

Professional Organizations

The major international organizations representing biomechanics are the International Society of Biomechanics (ISB), the American Society of Biomechanics (ASB), the European Society of Biomechanics (ESB), the Canadian Society of Biomechanics (CSB), and the Gait and Clinical Movement Analysis Society (GCMAS). These societies organize and hold major international scientific meetings regularly. Other professional organizations having major initiatives in biomechanics include the American Society of Mechanical Engineers, which has a Bioengineering Division, and the Society for Experimental Biology.

Further Reading and References

Carr, G. 1997. *Mechanics of sport.* Champaign, IL: Human Kinetics.

Enoka, R.M. 2002. *Neuromechanics of human movement.* 3rd ed. Champaign, IL: Human Kinetics.

McGinnis, P.M. 1999. *Biomechanics of sport and exercise.* Champaign, IL: Human Kinetics.

Panjabi, M.M., and A.A. White III. 2001. *Biomechanics in the neuromuscular system.* New York: Churchill Livingstone.

Zatsiorsky, V.M. 2002. *Kinetics of human motion.* Champaign, IL: Human Kinetics.

Some Relevant Web Sites

American Society of Biomechanics
[http://asb-biomech.org]

Biomechanics World Wide
[www.per.ualberta.ca/biomechanics/]

Canadian Society for Biomechanics
[www.health.uottawa.ca/biomech/csb/]

European Society of Biomechanics
[www.utc.fr/esb/]

Gait and Clinical Movement Analysis Society
[www.gcmas.org]

International Society of Biomechanics
[www.isbweb.org]

CHAPTER 6

Basic Concepts of Kinematics and Kinetics

This chapter introduces some fundamental concepts of kinematics and kinetics as they relate to human movement. We emphasize application of the basic laws of mechanics rather than the laws themselves. Vectors underlie much of what is known about mechanics, so we begin with a brief review of these quantities and illustrate how they may be used to describe forces and motion in human activity. We pay particular attention to displacement, velocity, and acceleration of a point on a rigid body; force and its relation to moment of force about a point in space; and application of free-body diagrams to analyses of force systems. The chapter concludes with a brief introduction to computer modelling and simulation.

The major learning concepts in this chapter relate to

- adding two vectors together to produce a resultant vector,
- calculating linear velocity (the time rate of change in displacement) and acceleration (the time rate of change in velocity) of a particle,
- calculating angular velocity (the time rate of change in angular displacement) and acceleration (the time rate of change in angular velocity) of a body segment,
- calculating velocity and acceleration of any point on a body segment,
- the degrees of freedom of a rigid body,
- the difference between external and internal forces and their relation to free-body diagrams,
- how static equilibrium analyses are used to determine unknown forces acting on body segments, and
- computer modelling of movement.

Vectors

Any physical quantity that is fully described by a number is called a **scalar**. Examples of scalars commonly encountered in biomechanical studies of movement are length and mass. A physical quantity that is described by its magnitude *and* direction is called a **vector**. Examples of vectors include the position, velocity, and acceleration of a point on a body (e.g., the **centre of mass**); a muscle force; the moment of force of a muscle about a joint; forces arising from contact of body parts with the environment (e.g., forces exerted by the ground on the feet in walking and running); and forces acting between the bones that meet at a joint (i.e., joint-contact forces). In this and subsequent chapters, any quantity that is underlined is a vector; otherwise, it is a scalar.

Vectors cannot simply be added to, subtracted from, and multiplied or divided by each other in the same way that scalars can. There are special rules for operating on vectors. To understand some of the most basic concepts of **kinematics** and kinetics, it is important to understand first how two vectors can be added together to produce a resultant vector. Consider moving a book from one position to another. As the book moves it experiences a **displacement** (change in position), which is represented by a vector $\underline{r}$ as shown in figure 6.1. The direction of $\underline{r}$ indicates the direction of the displacement of the book in moving from location 1 to location 2, and the actual **distance** moved is given by the magnitude of the displacement represented by $|\underline{r}|$. (The symbol $|\ |$ denotes the magnitude of the vector.) Let the displacement of the book in moving from location 2 to location 3 be represented by the vector $\underline{s}$ and the corresponding distance moved by $|\underline{s}|$. The resultant displacement of the book in moving from location 1 to location 3 (see figure 6.1) is then

$$\underline{q} = \underline{r} + \underline{s}. \qquad 6.1$$

In the special case in which the vectors $\underline{r}$ and $\underline{s}$ are perpendicular, the magnitude of the resultant displacement vector is given by $|\underline{q}| = \sqrt{|\underline{r}|^2 + |\underline{s}|^2}$; this follows from Pythagoras' theorem for a right-angle triangle. The direction of the resultant displacement is found from $\theta = \tan^{-1}\left(\dfrac{|\underline{s}|}{|\underline{r}|}\right)$, where θ is the angle formed between the resultant and the first displacement vector $\underline{r}$ and *tan* is an abbreviation for tangent (see figure 6.1). Vector addition is also often used to find the resultant of force systems (e.g.,

muscle forces) acting on a body part (see "How Much Force Do the Ankle Muscles Develop in Normal Walking?" below).

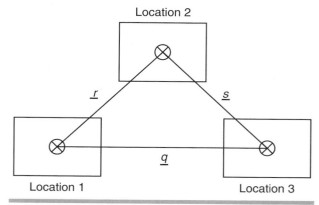

Figure 6.1   Diagram showing how vector addition may be used to calculate the displacement of a rigid body in space.

IN FOCUS: HOW MUCH FORCE DO THE ANKLE MUSCLES DEVELOP IN NORMAL WALKING?

A critical phase of the walking cycle is the so-called push-off phase. This is the time when the ankle plantar flexor muscles of the stance leg are heavily activated in order to push the centre of mass of the body forward and upward just prior to heel strike of the **contralateral** (swing) leg. The gastrocnemius (GAS) and soleus (SOL) muscles both exert forces to plantar flex the ankle during push-off. Both muscles insert on the foot via the Achilles tendon. The forces developed by these muscles have been estimated using sophisticated computer models of normal walking. In figure 6.2, the origin of the reference frame defined by the X- and Y-axes lies at the insertion of the Achilles tendon on the calcaneus. F_{GAS} and F_{SOL} are the forces developed in GAS and SOL, respectively. θ_{GAS} and θ_{SOL} are the angles that these muscle force vectors make with the vertical Y-axis. A recent analysis showed that GAS and SOL develop their peak forces at the same time during the push-off phase. Specifically, GAS generated a peak force of $F_{GAS} = 900\ N$, while the peak force in SOL was estimated to be $F_{SOL} = 2000\ N$. The line of action of each muscle was calculated from anatomical measurements

of its origin and insertion sites. At the instant under consideration, GAS's line of action was $\theta_{GAS} = 20\ deg$ relative to the vertical (Y-axis), while SOL's line of action was $\theta_{SOL} = 30\ deg$ relative to the vertical (see figure 6.2).

From this information, one may find the resultant peak force developed by the plantar flexor muscles. First, resolve each muscle force vector along the X- and Y-axes as follows:

$$F_{GASX} = F_{GAS}\sin\theta_{GAS}; \quad F_{GASY} = F_{GAS}\cos\theta_{GAS}$$

and

$$F_{SOLX} = F_{SOL}\sin\theta_{SOL}; \quad F_{SOLY} = F_{SOL}\cos\theta_{SOL}, \qquad 6.1.1$$

where F_{GASX}, F_{GASY} are the X and Y components of GAS's force vector and F_{SOLX}, F_{SOLY} are the X and Y components of SOL's force vector. Adding the X and Y components of each force gives

$$F_{RESX} = F_{GAS}\sin\theta_{GAS} + F_{SOL}\sin\theta_{SOL};$$

and

$$F_{RESY} = F_{GAS}\cos\theta_{GAS} + F_{SOL}\cos\theta_{SOL}, \qquad 6.1.2$$

(continued)

IN FOCUS: HOW MUCH FORCE DO THE ANKLE MUSCLES DEVELOP IN NORMAL WALKING? *(continued)*

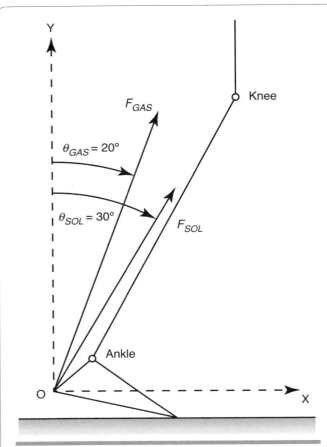

Figure 6.2 Schematic diagram of a human leg with vectors representing the forces acting in two of the ankle plantar flexor muscles: gastrocnemius (GAS) and soleus (SOL).

where F_{RESX} and F_{RESY} are the X and Y components of the **resultant force** developed by the ankle plantar flexors. After substitution of the magnitudes and directions of the force vectors already given, equation 6.1.2 becomes $F_{RESX} = 308 + 1000 = 1308 \; N$; $F_{RESY} = 846 + 1732 = 2578 \; N$. The magnitude of the resultant plantar flexor muscle force is then found using Pythagoras' theorem, $F_{RES} = \sqrt{F_{RESX}^2 + F_{RESY}^2} = 2891 \; N$, and the direction of the resultant muscle force is found from

$$\theta_{RES} = \tan^{-1}\left(\frac{F_{RESX}}{F_{RESY}}\right) = 27 \; deg \text{ (from the vertical; Y-axis).}$$

The peak forces developed by GAS and SOL during a maximum isometric contraction are known to be approximately 1700 N and 3000 N, respectively. Thus, the peak force developed by the ankle plantar flexors during normal walking is roughly 80% of the peak force developed by this muscle group during a maximum isometric contraction.

SOURCE

- Anderson, F.C., and M.G. Pandy, 2001, Static and dynamic optimization solutions for gait are practically equivalent. *Journal of Biomechanics* 34: 153-161.

Motion

Motion is the term used to describe the displacement, velocity, and acceleration of a body in space. Displacement is the change in the body's position as it moves from one location to another; **velocity** is the rate of change in the body's displacement over time; and **acceleration** is the rate of change in the body's velocity over time. The following sections provide simple examples to illustrate each of these three quantities as they relate to human movement in particular.

Motion of a Particle

Consider a man walking at his natural speed over level ground. Let the man's body be idealized as a point, say P, located at the centre of mass of the whole body (figure 6.3). Let O be any point fixed

on the ground and $\underline{r}(t)$ be the position vector from O to the centre of mass P at time t. The displacement of P in a finite time interval Δt is the change in position of P from time t to time $t + \Delta t$. That is, the displacement of P is given by $\underline{r}(t + \Delta t) - \underline{r}(t)$, where $\underline{r}(t + \Delta t)$ is the position of P at time $t + \Delta t$. Velocity is defined as the rate of change in position with respect to time. Thus, the average velocity of P in the interval Δt is given by

$$\underline{v} = \frac{\underline{r}(t + \Delta t) - \underline{r}(t)}{\Delta t}, \qquad 6.2$$

where $\underline{r}(t + \Delta t) - \underline{r}(t)$ represents the displacement of point P in the interval Δt. Similarly, if acceleration is the rate of change in velocity over time, the average acceleration of P relative to O in the time interval Δt is

$$a = \frac{v(t + \Delta t) - v(t)}{\Delta t}, \qquad 6.3$$

where $v(t)$ is the velocity of P at time t and $v(t + \Delta t)$ is the velocity of P at time $t + \Delta t$. The SI (Système Internationale) unit of position and displacement is metres (m), so velocity and acceleration are given in m/s and m/s^2, respectively. Equations 6.2 and

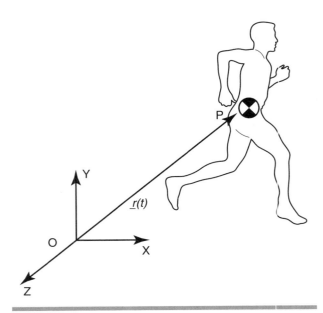

Figure 6.3 Centre of mass of the body idealized as a point P in space. O is the origin of the X,Y reference frame fixed on the ground.

6.3 may be used to estimate the velocity and acceleration of the centre of mass of the body in human locomotion (see "Velocities and Accelerations of Points on Bodies Can Be Estimated From Measurements of Position" below).

Angular Motion of a Rigid Body

A **rigid body** is an idealized model of an object that does not deform or change its shape; that is, the distance between every pair of points on the body remains constant. Consider again the case of a man walking on level ground, but now assume that one leg is represented by a rigid link as shown in figure 6.4. The centre of mass of the body is lumped at the tip of the pendulum, which coincides with the man's hip. Let $\theta(t)$ be the angular displacement of the leg at time t relative to a horizontal line fixed on the ground. The average **angular velocity** of the leg over a finite interval Δt is the rate of change of angular displacement of the leg,

$$\omega = \frac{\theta(t + \Delta t) - \theta(t)}{\Delta t}, \qquad 6.4$$

where $\theta(t + \Delta t)$ is the angular displacement at time $t + \Delta t$. The direction of the angular velocity vector, ω, is perpendicular to the plane of motion of the body; in figure 6.4, ω points out of the plane of the paper because the body is assumed to rotate

IN FOCUS: VELOCITIES AND ACCELERATIONS OF POINTS ON BODIES CAN BE ESTIMATED FROM MEASUREMENTS OF POSITION

Equations 6.2 and 6.3 can be used to estimate the velocity and acceleration of a point on a body part if the position of that point has been measured during the activity. Consider motion of the hip joint during walking. One can measure the position of the hip by placing a reflective marker on the surface of the skin at the approximate centre of the joint. Video cameras can then be used to track the changing position of the marker over time as the person walks. This type of measurement is routinely performed in movement biomechanics laboratories throughout the world. These laboratories are equipped not only with video cameras to measure the

changing positions and orientations of body parts but also with devices (called force plates) for measuring the forces exerted on the ground and instrumentation for monitoring and recording the pattern of muscle activity.

Table 6.1 gives the position of the hip joint in the fore–aft (X) direction at three successive instants during normal walking. The fore–aft velocity of the hip can be estimated using equation 6.2, $v = \frac{r(t + \Delta t) - r(t)}{\Delta t}$, where $r(t)$ is the position of the hip at time t, $r(t + \Delta t)$ is its position at time $t + \Delta t$, and Δt is the time step between any two successive instants during the gait cycle. For

(continued)

IN FOCUS: VELOCITIES AND ACCELERATIONS OF POINTS ON BODIES CAN BE ESTIMATED FROM MEASUREMENTS OF POSITION *(continued)*

the first two time points in table 6.1, $r(t) = 83.12$ cm, $r(t + \Delta t) = 85.69$ cm, and $\Delta t = 0.014$ s. Therefore, from equation 6.2, the velocity of the hip in the horizontal (fore–aft) direction is found to be $v = 183.6$ cm/s or $v = 1.84$ m/s. Similarly, using the data for the second two time points, the velocity of the hip is $v = 160$ cm/s or $v = 1.60$ m/s.

Table 6.1

Fore–Aft Position of the Hip in Normal Walking on the Level

T, time (seconds)	X position (centimetres)
0.272	83.12
0.286	85.69
0.399	87.77

Data from Winter 1990.

Most healthy young and middle-aged adults walk at a speed of 1.35 m/s when they are free to choose **speed, cadence,** and **step length.** This is the speed with which the centre of mass of the whole body moves during normal walking over level ground. Thus, the preceding calculations show that not all points on the body move with the same velocity during gait. Furthermore, the calculations show that the velocity of any given point on the body, such as the hip, changes over time. That is, all points on the body accelerate and decelerate during the gait cycle.

Indeed, equation 6.3 may be used to find exactly how much the hip accelerates (or decelerates) based on the data of table 6.1. From the results just obtained for the velocity of

the hip, one can write $v(t) = 1.84$ m/s, $v(t + \Delta t) = 1.60$ m/s, and $\Delta t = 0.013$ s. Using equation 6.3, the fore–aft acceleration of the hip is found to be $a = 18.5$ m/s².

It is important to realize that the values of velocity and acceleration calculated here are really average values; that is, one assumes by using equations 6.2 and 6.3 that the values of velocity and acceleration remain constant over the interval of time defined by Δt. It is also possible to obtain instantaneous values of the velocity and acceleration of a point on a body (i.e., the values of velocity and acceleration at a given point in time), but this requires knowledge of calculus and, more specifically, knowledge of the rules of differentiation.

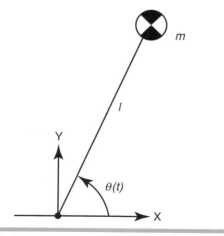

Figure 6.4 Diagram representing the human leg as an inverted single pendulum in normal walking on level ground. $\theta(t)$ is the angle that the leg makes relative to a horizontal line parallel with the ground at some time, t. l = length; m = mass of body.

counterclockwise relative to a horizontal line fixed on the ground. The average **angular acceleration** of the leg is then

$$\underline{\alpha} = \frac{\underline{\omega}(t + \Delta t) - \underline{\omega}(t)}{\Delta t}, \qquad 6.5$$

where $\underline{\omega}(t)$ is the angular velocity of the leg at time t and $\underline{\omega}(t + \Delta t)$ is the angular velocity at time $t + \Delta t$. The angular acceleration vector, $\underline{\alpha}$, is also directed perpendicular to the plane of motion of the leg (i.e., $\underline{\omega}$ and $\underline{\alpha}$ are assumed to have the same direction, out of the plane of the paper in figure 6.4). Note, however, that $\underline{\omega}$ and $\underline{\alpha}$ may not always have the same direction; for example, it may happen that the body is rotating in one direction (say counterclockwise), but slowing down (i.e., decelerating) as it rotates. In this case, $\underline{\omega}$ will be directed counterclockwise, but $\underline{\alpha}$ will be directed clockwise because the body is decelerating. The SI unit of angular displacement is radians (rad), and so angular velocity and angular acceleration are given in rad/s and rad/s², respectively. (Note that angular displacement, angular velocity, and angular acceleration have the same units in the Imperial and SI systems.)

Motion of a Point on a Rigid Body

To describe the motion of a rigid body in three-dimensional space, it is sufficient to consider the motion of a single point, such as the centre of mass, and the rotational motion of the body about that point. Consider a rigid body, say A, rotating about a fixed point C in the plane of the paper (figure 6.5). Because C is fixed relative to the ground, its absolute velocity is zero. In figure 6.5, $\underline{\omega}$ is the angular velocity vector of body A, and $\underline{v}$ is the velocity vector of any point P fixed on the body relative to C. Because C is fixed on the ground, $\underline{v}$ is also the velocity vector of P relative to O. The magnitude of the velocity of P is given by

$$v = \omega R, \qquad 6.6$$

where ω is the magnitude of the angular velocity of the body and R is the distance from C to P. The velocity vector of P relative to the ground lies in the plane of motion of the body and is directed perpendicular to the position vector from C to P as indicated in figure 6.5. Thus, $\underline{v}$ and $\underline{\omega}$ are also perpendicular to each other, because $\underline{\omega}$ is directed out of the page.

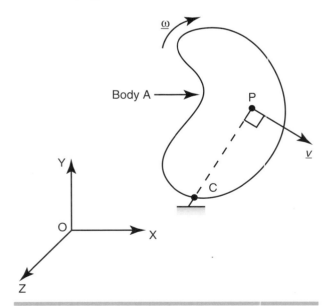

Figure 6.5   Diagram illustrating motion of a rigid body, body A, about a point C fixed in space. X, Y, Z is a ground-fixed reference frame with origin O. Because C is fixed relative to the ground, its absolute velocity is zero.

It happens more often that all points of a body are moving at each instant in time, rather than one point being fixed to the ground as shown in figure 6.5. Thus, point C usually has some velocity relative to the ground. In this case, the velocity of P relative to C is still given by equation 6.6; however, the velocity of C relative to some point fixed on the ground, say O, is no longer zero. One can find the velocity of P relative to O (the ground) by adding two vectors as follows:

$$^O\underline{v}^P = {}^O\underline{v}^C + {}^C\underline{v}^P, \qquad 6.7$$

where $^O\underline{v}^P$ is the velocity of P relative to O, $^O\underline{v}^C$ is the velocity of C relative to O, and $^C\underline{v}^P$ is the velocity of P relative to C. As described in the opening section of this chapter, one must follow special rules when operating on vector quantities, because a vector is defined by both its magnitude and its direction. The magnitude of the resultant velocity, $|^O\underline{v}^P|$, can be found using the cosine rule in trigonometry; thus:

$$\left|^O\underline{v}^P\right| = \sqrt{\left|^O\underline{v}^C\right|^2 + \left|^C\underline{v}^P\right|^2 - 2\left|^O\underline{v}^C\right|\left|^C\underline{v}^P\right|}\cos\alpha, \qquad 6.8$$

where $|^C\underline{v}^P| = \omega R$ from equation 6.6, and α is the angle formed between the vectors $^O\underline{v}^C$ and $^C\underline{v}^P$. Although the vector of P relative to O remains in the plane of motion of the body (i.e., perpendicular to the angular velocity vector of the body), its direction is not in general perpendicular to the position vector from C to P.

The acceleration of point P relative to any point O fixed on the ground follows from the equation defining the velocity of P relative to O. In the special case when one point C of the body is fixed relative to the ground, the magnitude of the acceleration of P relative to O is given by

$$\left|^O\underline{a}^P\right| = \sqrt{(\alpha R)^2 + (\omega^2 R)^2}, \qquad 6.9$$

where α is the angular of the body at any instant. The acceleration of P relative to O remains in the plane of motion of the body (i.e., perpendicular to the angular velocity vector), but it has two components indicated by the two terms on the right-hand side of equation 6.9. The first term, αR, is called the **tangential acceleration** and is directed perpendicular to the position vector from C to P; the second term, $\omega^2 R$, is called the **centripetal acceleration** and is directed radially inward from P to C. Note that the tangential and centripetal accelerations are always perpendicular to each other, and therefore we find the magnitude of the resultant acceleration simply by squaring each component, adding them together, and taking the square root of the result as indicated in equation 6.9.

Degrees of Freedom

The number of **degrees of freedom** of a rigid body is equal to the number of ways the body can move relative to some frame of reference such as the ground. Any rigid body has at most six degrees of freedom in three-dimensional space: three for the motion of any point on the body such as the centre of mass and three more for the rotations of the body about a line passing through that point. When the body is free to move in any manner whatsoever, it is said to have unconstrained motion.

A rigid body has fewer than six degrees of freedom whenever it is acted on by kinematic constraints. Kinematic constraints exist when forces act on the body so as to restrict the motion of the body at each instant in time. Consider, for example, a woman landing from a vertical jump as illustrated in figure 6.6. When the body is in the air, six generalized coordinates are needed to completely specify its position and orienta-

tion with respect to a reference frame fixed on the ground. Once the toes make contact with the ground, the number of degrees of freedom is reduced to four because the ground then applies a force at the toes that prevents the foot from slipping. Next, the heel comes into contact with the ground and the number of degrees of freedom is then reduced to three; only three coordinates are needed to fully specify the position and orientation of the shank, thigh, and trunk, because the foot is now firmly planted on the ground and is effectively fixed to it.

The concept of degrees of freedom is particularly important when one is characterizing joint motion. In the case of the human knee, for example, the femur and tibia are held together mainly by the cruciate and collateral ligaments. Because the ligaments are elastic, they stretch somewhat as the muscles apply forces to move the bones. Thus, the femur and tibia displace relative to each other so that the knee, in theory, has six degrees of freedom of joint movement. These six movements are often characterized in anatomical terms as three translations (i.e., linear movements) of the tibia relative to the femur (anterior–posterior translation, medial–lateral translation, and proximal–distal distraction) together with three rotations (i.e., angular movements) (flexion–extension, abduction–adduction, and internal–external rotation). Most of the other joints in the body have fewer than six degrees of freedom. The hip, for example, has only three degrees of freedom, all of which are rotations (i.e., flexion–extension, abduction–adduction, and internal–external rotation of the femur relative to the pelvis). There are no translational degrees of freedom at the hip because the femoral head sits tightly inside the acetabulum, preventing any relative displacement of the bones. Most joints of the fingers have just one degree of freedom, allowing only flexion and extension.

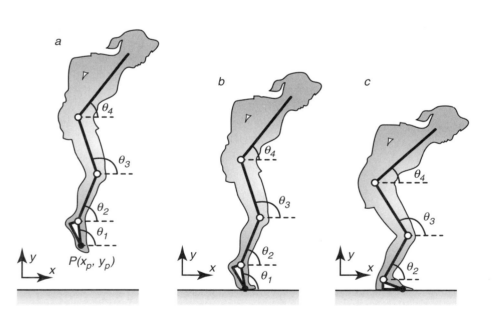

Figure 6.6   Diagram illustrating the number of degrees of freedom in a model for landing from a jump. All body segments are assumed to move only in the sagittal plane. X,Y is a reference frame fixed on the ground. Point P marks the tip of the toes segment in the model. (a) The jumper is airborne. (b) Only the toes are in contact with the ground. (c) The foot is flat on the ground.
Adapted from Zajac and Gordon 1989.

Force

Force is a measure of the amount of effort applied; it is a vector quantity, requiring both the magnitude and direction of the force to be given. The direction of a force is often referred to as the **line of action of the force.** The SI unit of force is the **newton** (N) (pound-force [lbf]).

Internal and External Forces

There are many kinds of forces in nature, but most of the forces encountered in biomechanical studies of movement fall into one of two categories: external forces and internal forces. A given body is subjected to an **external force** if the force is applied by another object. An **internal force** is any force applied by one part of a body on another part of the *same* body. Some types of forces are always external forces; for example, gravitational forces act on all parts of a body and are proportional to the body's mass. Other types of forces usually appear as internal forces—for example, the forces applied by the muscles and the forces transmitted by the bones that meet at a joint. One must use caution in distinguishing between external and internal forces because the distinction is intimately related to the definition of the system being analysed. Specifically, whether a force is treated as external or internal depends on how the boundary of the system is defined, which is determined by a concept known as the free-body diagram.

Free-Body Diagram

We can identify the external forces acting on a system by drawing a **free-body diagram.** Constructing such a diagram involves three steps. First, we must identify the system to be isolated. Second, we draw a sketch of the system isolated from its surroundings. Third, we draw and label as vectors all of the external forces acting on the system.

Which forces are external and which internal is not always immediately obvious. This fact is illustrated by the isokinetic knee extension exercise, commonly prescribed for maintaining thigh muscle strength subsequent to knee ligament injury or reconstruction. (Isokinetic exercise means that the joint is flexed and extended at constant speed.) The person sits comfortably in a Cybex or Biodex dynamometer with the torso and thigh strapped firmly to the seat (see figure 6.7a). The lower leg (shank plus foot) is strapped to the arm of the machine, which moves at a predetermined (constant) speed. A free-body diagram of the lower leg is shown in figure 6.7b. In this example, because the lower leg

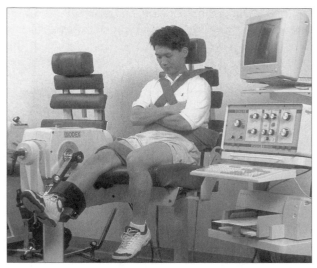

a

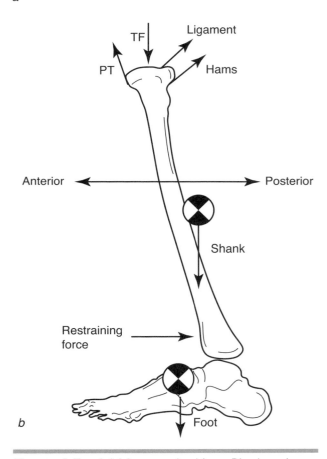

b

Figure 6.7 *(a)* Man seated in a Biodex dynamometer. The strap on the machine arm is attached just above the subject's ankle. *(b)* Free-body diagram showing the external forces acting on the lower leg (shank plus foot) during isokinetic knee extension exercise.

Free-body diagram courtesy of Dr. Kevin B. Shelburne. Copyright K.B. Shelburne.

is isolated from the thigh and the machine arm, any force that breaks the system boundary is treated as an external force and is therefore included in the free-body diagram. Thus, the forces applied by the quadriceps (via the patellar tendon, PT) and the hamstring (Hams) muscles are treated as external forces, while those applied by all the other muscles (e.g., soleus) are treated as internal forces and are therefore not shown in figure 6.7b because they do not break the imaginary system boundary. Note that the force acting between the femur and tibia (i.e., the tibiofemoral contact force, TF), the weight forces due to the shank and foot, the restraining force applied by the arm of the Biodex machine, and the forces induced in the knee ligaments (Ligament) also all break the system boundary and must therefore be included in the free-body diagram, as shown.

Moment of Force

The moment of force about a point is a measure of the **turning effect** of the force about that point. One may also think of the moment of a force about a point as a measure of the force needed to cause rotation of a body about that point. Moment is a vector, requiring specification of both magnitude and direction. Consider a force vector $\underline{F}$ and any point O as shown in figure 6.8. The magnitude of the moment vector about O is

$$M = F\,r, \qquad 6.10$$

where F is the magnitude of the force vector and r is the perpendicular distance from point O to the line of action of the force. It should be clear from equation 6.10 that the SI unit of moment is the **newton·metre** (N·m) (foot·pounds [ft·lb]).

Muscles develop force and cause rotation of body segments about the joints. The moment of a muscle force follows directly from equation 6.10, where r, the perpendicular distance between the line of action of the muscle and the axis of rotation of the joint, is called the **muscle moment arm.** Muscle moment arms are important because they describe how close to the axes of rotation of the joints the muscles pass. The closer a muscle is to the axis of rotation, the less effective it is in causing rotation of the body segments; that is, the smaller the moment arm, the smaller the moment of force (from equation 6.10) (see "How Much Force Can the Quadriceps Develop in a Maximal Contraction?" on p. 75).

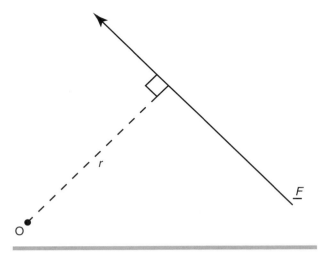

Figure 6.8   Force $\underline{F}$ acting about any point O in space. The force has a magnitude of F and a line of action as indicated in the diagram; r is the perpendicular distance from the line of action of $\underline{F}$ and point O.

Equilibrium

A body is in **equilibrium** if every point of the body has the same velocity. If the velocity and acceleration of every point on the body are zero, the body is said to be in **static equilibrium.** In mathematical terms, a body is in static equilibrium if the sum of all the external forces in the X, Y, and Z directions and the sum of all the moments of force about the X-, Y-, and Z-axes are simultaneously zero. That is,

$$\sum_{i=1}^{N} F_{xi} = 0\,; \quad \sum_{i=1}^{N} F_{yi} = 0\,; \quad \sum_{i=1}^{N} F_{zi} = 0 \quad 6.11.1$$

and

$$\sum_{i=1}^{N} M_{xi} = 0\,; \quad \sum_{i=1}^{N} M_{yi} = 0\,; \quad \sum_{i=1}^{N} M_{zi} = 0, \quad 6.11.2$$

where F_{xi}, F_{yi}, F_{zi} are the ith forces acting in the X, Y, and Z directions, respectively; and M_{xi}, M_{yi}, M_{zi} are the ith moments exerted about the X-, Y-, and Z-axes. All six conditions given in equations 6.11 must hold if static equilibrium is to apply. For example, a body may be at rest (zero velocity) but not in static equilibrium. Consider the simple case of a boy throwing a ball straight up into the air. When the ball reaches its maximum height its vertical velocity is zero, yet the ball continues to accelerate. Indeed, the vertical acceleration of the ball from the moment it leaves the boy's hand is 9.81 m/s² (32.2 ft/s²) if gravity is the only force acting while the ball remains in the air. In this case,

IN FOCUS: HOW MUCH FORCE CAN THE QUADRICEPS DEVELOP IN A MAXIMAL CONTRACTION?

The quadriceps is the strongest muscle in the body. This is demonstrated in the performance of an isometric contraction of the knee extensor muscles in a Cybex or Biodex dynamometer such as the one shown in figure 6.7. For peak force to be developed in the quadriceps, the hip must be flexed to approximately 60° and the ankle must be held roughly in the neutral (standing) position. The leg is strapped to the arm of the machine, which is then locked in place, so that the muscles crossing the knee contract isometrically (see figure 6.7a). Under these conditions, healthy young male subjects develop maximum moments in the range of 250 to 300 N·m (figure 6.9a). As one might expect, the maximum moment developed by the quadriceps varies with the knee flexion angle. In fact, peak knee extensor moment occurs when the knee is flexed to 60° in most people, and it decreases significantly as the knee is moved toward extension. This means that the knee extensor muscles are strongest—they exert the most leverage—when the knee is bent to 60° and that their leverage is severely compromised when the knee reaches full extension. Why?

To answer this question, consider the two factors that contribute to knee extensor moment: the moment arm of the knee extensor mechanism and the amount of force developed by the quadriceps (equation 6.10). The knee extensor

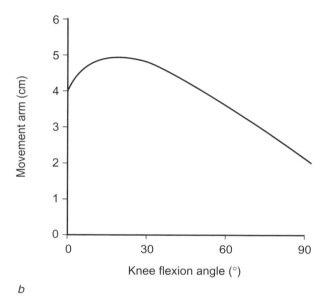

b

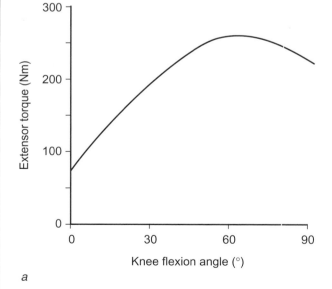

a

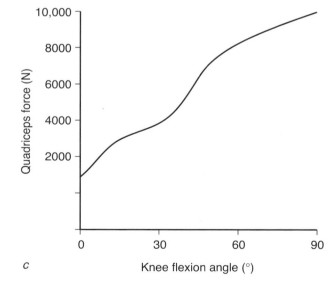

c

Figure 6.9 (a) Maximum isometric moment of force plotted against knee flexion angle for maximum isometric knee extension exercise, (b) the corresponding knee extensor moment arm, and (c) quadriceps force plotted against the flexion angle.
Reprinted from Shelburne and Pandy 1997.

(continued)

moment arm is actually the moment arm of the patellar tendon relative to the axis of rotation of the knee. The resultant force developed by the quadriceps is the vector sum of the forces in the separate heads of the vastus (i.e., medialis, lateralis, and intermedius) and the rectus femoris muscles. The curves shown in figure 6.9 were obtained from a computer model of the knee joint that was used to simulate maximum isometric contractions of the quadriceps muscles in healthy young males (Shelburne and Pandy 1997; Pandy and Shelburne 1998) (see figure 6.11). The quadriceps muscles in the model were given realistic physiological properties consisting of active and passive force–length curves. The elastic behaviour of the quadriceps tendon was also taken into account in the model.

The moment arm of the knee extensor mechanism is the perpendicular distance from the line of action of the patellar tendon to the axis of rotation of the knee (which is located roughly at the centre of the joint). This distance is determined by the shapes of the distal femur, proximal tibia, and patella and by the way the relative positions of these bones change during flexion and extension of the knee. The moment arm of the knee extensor mechanism has been measured in a number of studies using intact human cadaver specimens, and the result is given in figure 6.9b. The peak moment arm of the knee extensor mechanism is about 5 cm (2 in.) and occurs when the knee is flexed to about 20°. The minimum value is about 2 cm (0.8 in.) and occurs when the knee is flexed to 90°.

From equation 6.10, we can see that the moment of a muscle force is equal to the magnitude of the muscle force multiplied by the muscle's moment arm. Thus, dividing the muscle moment curve by the moment arm curve in figure 6.9 gives quadriceps force as a function of the knee flexion angle (figure 6.9c). Peak quadriceps force is about 9500 N for maximum isometric knee extension. Peak force occurs with the knee bent to 90°, and it decreases as the knee is moved toward extension. Quadriceps force decreases as the knee extends because the muscle then moves down the ascending limb of its force–length curve (i.e., muscle length decreases). Indeed, quadriceps force is minimum when the knee is fully extended because the muscle fibres are too short to develop much force in this position. Conversely, quadriceps force peaks when the knee is bent to 90° because this position corresponds with the length at which the maximum number of actomyosin cross-bridges are formed (as described in chapter 2).

The quadriceps exert the least leverage (i.e., smallest moment) when the knee is fully extended because these muscles cannot generate much force in this position. At full extension, the effectiveness of the quadriceps to cause rotation is determined by the force–length properties of these muscles and not by the geometry of the knee (i.e., the moment arm of the knee extensor mechanism). (The force that a muscle develops depends on the length at which the muscle is held. Near its resting length, the muscle develops maximum force when fully activated. At lengths much shorter and longer than its resting length, the muscle develops little force. The relation between the force that a muscle develops and the length at which it is held is called the force–length property of the muscle. See chapter 7.) At 60° of knee flexion, the effectiveness of the quadriceps is maximum because the force developed by the muscle and the moment of the extensor mechanism are both relatively large, and their product is then optimal. At 90° of knee flexion, quadriceps moment once again decreases, even though the force developed by the muscle continues to increase; compare force at 60° and 90° (figure 6.9c). In this case, quadriceps moment decreases because the moment arm of the extensor mechanism decreases. Thus, the effectiveness of the quadriceps at large knee flexion angles is governed by the geometry of the knee (i.e., the moment arm of the extensor mechanism).

SOURCES

- Pandy, M.G., and K.B. Shelburne. 1998. Theoretical analysis of ligament and extensor-mechanism function in the ACL-deficient knee. *Clinical Biomechanics* 13: 98-111.
- Shelburne, K.B., and M.G. Pandy. 1997. A musculoskeletal model of the knee for evaluating ligament forces during isometric contractions. *Journal of Biomechanics* 30: 163-176.

the ball is not in static equilibrium at the top of its flight even though it is momentarily at rest, because the net acceleration of the ball is not zero.

Consider once again the knee extension exercise illustrated in figure 6.7, but assume now that the exercise is performed with the muscles held isometric (i.e., at constant length). Assume also that movement of the lower leg is permitted only in the sagittal plane and that only the quadriceps muscles are being activated (i.e., the hamstring and gastrocnemius muscles are assumed to be fully deactivated). The free-body diagram for the isometric knee extension exercise is shown in figure 6.10 (compare with figure 6.7b). It is assumed that the knee is bent to 90°. The external restraining force, F_{ext}, acts perpendicular to the leg. The thigh is held horizontal as shown. F_Q is the magnitude of the resultant force developed by the quadriceps, which is applied via the patellar tendon to the leg. For the position shown, the quadriceps is assumed to be directed 15° anterior to the long axis of the leg. F_{Jx} and F_{Jy} are the X and Y components of the unknown articular contact force acting between the tibia and femur.

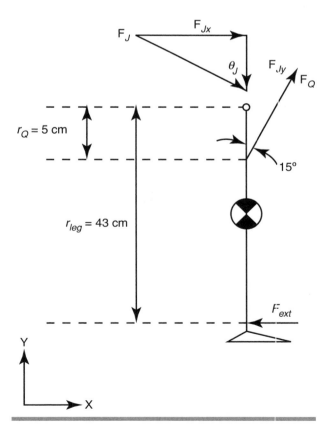

Figure 6.10 Free-body diagram showing the external forces acting on the lower leg during isometric knee extension exercise.

Equations 6.11 may be used to find the magnitude of the quadriceps force needed to hold the lower leg in static equilibrium during the exercise. Summing forces in the X and Y directions gives

$$\sum F_x = 0; \quad F_{Qx} + F_{Jx} - F_{ext} = 0 \qquad 6.12.1$$

and

$$\sum F_Y = 0; \quad F_{Qy} - F_{Jy} = 0, \qquad 6.12.2$$

where F_{ext} is the external restraining force applied by the machine arm, F_{Jx} and F_{Jy} are the X and Y components of the joint-reaction force acting between the femoral and tibial condyles, $F_{Qx} = F_Q \sin15°$ is the component of quadriceps force in the X direction, and $F_{Qy} = F_Q \cos15°$ is the component of quadriceps force in the Y direction (see figure 6.10).

Summing moments about the centre of the knee joint gives

$$\sum M_Z = 0; \quad F_{Qx}\, r_Q - F_{ext}\, r_{leg} = 0, \qquad 6.12.3$$

where r_Q is the distance from the insertion of the patellar tendon to the centre of the knee joint, and r_{leg} is the distance from the point of application of the external restraining force to the centre of the knee. Notice that the resultant joint-reaction force, F_J, does not appear in equation 6.12.3. The reason is that this force passes through the centre of the knee and therefore contributes nothing to the moment about the joint. If the magnitude of the external restraining force is known (F_{ext} is assumed to be 250 N here, which corresponds to the peak value typically exerted by young healthy men), the value of F_{Qx} can be found from equation 6.12.3 as follows: $F_{Qx} = F_{ext} \dfrac{r_{leg}}{r_Q} = 250 * \dfrac{43}{5} = 2150\ N$. Thus, the resultant force developed by the quadriceps is given by $F_Q = \dfrac{F_{Qx}}{\sin15°} = 8307\ N$.

The X and Y components of the joint-reaction force are then obtained from equations 6.12.1 and 6.12.2 as follows: $F_{Jx} = F_{ext} - F_{Qx} = -1900\ N$ and $F_{Jy} = F_{Qy} = F_Q \cos15° = 8024\ N$. Finally, the magnitude and direction of the resultant joint-reaction force are found from $F_J = \sqrt{F_{Jx}^2 + F_{Jy}^2} = 8246\ N$ and $\theta_J = \tan^{-1}\left(\dfrac{F_{Jx}}{F_{Jy}}\right) = 13°$, respectively. The magnitude of the quadriceps force is roughly 12 times greater than the weight of an average person (an average adult man weighs about 700 N). Indeed, muscles routinely develop very large forces in activity, with maximum isometric knee extension exercise being an extreme case. In normal walking, for example, the quadriceps develop peak forces as high as

1000 N, which is still larger than the average person's body weight.

Computer Modelling of Movement

Mathematical (computer) models are widely used today, mainly because they give information not easily obtainable from experiments. In biomechanical studies of movement, for example, muscle, ligament, and joint-contact forces cannot be measured noninvasively, so these quantities are often estimated using computer models of the body. Detailed models of the lower limb have been used to determine musculoskeletal forces in various activities, including walking, running, jumping, cycling, and throwing.

Figure 6.11 shows a model that was used to estimate muscle, ligament, and joint-contact forces in knee extension exercise. The lower limb was represented as a three-segment, four degree-

of-freedom kinematic chain. The hip and ankle were each modelled as a one degree-of-freedom hinge allowing only flexion and extension in the sagittal plane. The knee was modelled as a two degree-of-freedom joint allowing rolling and sliding of the tibia on the femur in the sagittal plane. The geometry of the distal femur, proximal tibia, and patella was based on digitised slices averaged across 23 cadaveric knees, and 11 separate bundles were used to model the geometry and mechanical properties of the cruciate ligaments, the collateral ligaments, and the posterior capsule. The model knee was then actuated by 11 musculotendinous units. Muscles included in the model and indicated in figure 6.11a are rectus femoris (RF), vastus medialis (VMED), vastus intermedius and vastus lateralis lumped together (VINT), biceps femoris long head (BFLH), biceps femoris short head (BFSH), semimembranosus (MEM), semitendinosus (TEN), and gastrocnemius (GAS). The pelvis and femur were fixed during the simulation. The ligament bundles included in the model knee were aAC (anteromedial bundle of the ACL), pAC (posterolateral bundle of the ACL), aPC (anterolateral bundle of the PCL), pPC (posteromedial bundle of the PCL), aMC (anterior bundle of the superficial MCL), iMC (intermediate bundle of the superficial MCL), pMC (posterior bundle of the superficial MCL), aCM (anterior bundle of the deep MCL), pCM (posterior bundle of the deep MCL), LCL (lateral collateral ligament), and pCap (posterior capsule). Note that the LCL inserts on the fibula, which is not shown in figure 6.11b.

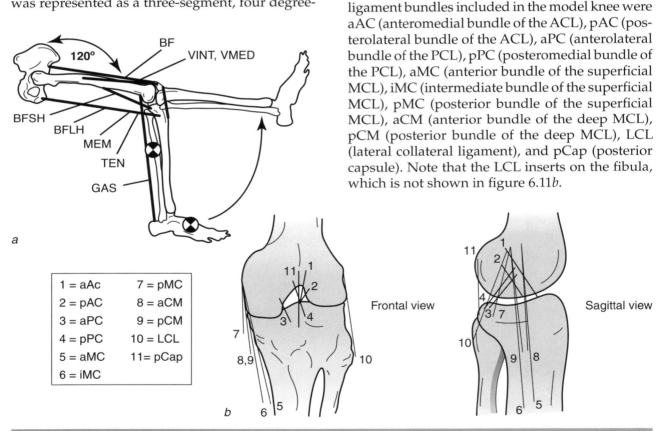

1 = aAc	7 = pMC
2 = pAC	8 = aCM
3 = aPC	9 = pCM
4 = pPC	10 = LCL
5 = aMC	11= pCap
6 = iMC	

Figure 6.11 Schematic of the model leg used to calculate the maximum isometric moment, moment arm, and quadriceps force developed during isometric knee extension exercise. (a) Muscles actuating the model leg; (b) shapes of the bones and locations of the knee ligaments assumed in the model.

Reprinted from Serpas, Yanagawa, and Pandy 2002.

The quadriceps muscles can exert up to 9500 N of force when fully activated and contracting isometrically (figure 6.12). Peak isometric force is developed when the knee is bent to 90°, and it decreases as the knee moves toward extension. Although the quadriceps force increases from full extension to 90° of flexion, the force transmitted to the ACL increases and then decreases as the knee is flexed from full extension (figure 6.12). The computer model predicts that the ACL is loaded only from full extension to 80° of flexion and that the resultant force in the ACL reaches 500 N at 20° of flexion, which is lower than the maximum breaking strength of the human ACL (2000 N). The calculations also show that load sharing within the knee ligament may not be uniform. For example, the force transmitted to the anteromedial bundle of the ACL (aAC) increases from full extension to 20° of flexion, where peak force occurs, and aAC force then decreases as knee flexion increases (figure 6.12, aAC). On the other hand, the force transmitted to the posterolateral bundle of the ACL (pAC) is quite different (compare aAC and pAC in figure 6.12). The forces transmitted to the MCL, LCL, and capsular structures are not shown, but their contributions are relatively small.

The forces exerted between the femur and patella and between the femur and tibia depend mainly on the geometry of the muscles that cross the knee. For maximum isometric extension, peak forces transmitted to the patellofemoral and tibiofemoral joints are around 11000 N and 6500 N, respectively (i.e., approximately 16 and 9 times body weight, respectively) (figure 6.13). As the knee moves faster during isokinetic extension, joint-contact forces decrease in direct proportion to the drop in quadriceps force. Quadriceps force decreases as the knee extends faster because the muscle then shortens more quickly; and, from the force–velocity property of muscle, an increase in shortening velocity leads to less force (see "Creatine Supplementation: To What Extent Does the Immediate Energy System Limit Exercise Performance?" in chapter 10, p. 127).

The computer model predictions of figures 6.12 and 6.13 have significant implications for the design of exercise regimens aimed at protecting injured or newly reconstructed knee ligaments. For maximum isolated contractions of the quadriceps, figure 6.12 shows that the ACL is loaded at all flexion angles less than 80°. Quadriceps-strengthening exercises should therefore be

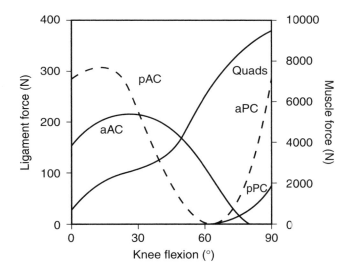

Figure 6.12 Resultant force developed by the quadriceps and the corresponding cruciate ligament forces calculated for maximum isometric knee extension.

Reprinted from Shelburne and Pandy 1997.

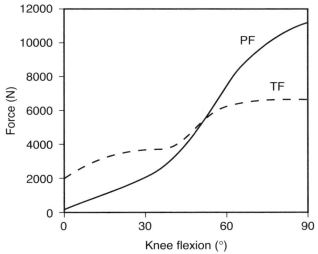

Figure 6.13 Resultant forces acting between the femur and tibia (TF) and between the femur and patella (PF) during maximum isometric knee extension exercise.

Reprinted from Pandy and Barr 2002.

limited to flexion angles greater than 80° if the injured or newly reconstructed ACL is to be protected from **stress.** In this region, however, very high contact forces can be applied to the patella (figure 6.13), so much so that limiting this exercise to large flexion angles may result in patellofemoral pain. This scenario, known as the "paradox of exercise," is the reason surgeons and physical therapists now prescribe so-called **closed-chain exercises,** such as squatting, for strengthening the quadriceps muscles subsequent to ACL injury or repair. (Closed-chain exercises are those in which an external force is applied to one or both feet during the exercise—e.g., the ground applies a force to both feet during squatting exercise. Open-chain exercises, on the other hand, are those in which no external force is applied to the feet—e.g., during isokinetic exercise, no force is applied directly to the foot.)

Summary

Any physical quantity that is described by its magnitude *and* direction is called a vector. Vector algebra is fundamental to analyses of human movement; for example, displacement, velocity, and acceleration vectors are used to study how body segments move under the action of force systems created by the actions of muscles and the environment. The number of degrees of freedom of a body segment is equal to the number of ways the body segment can move relative to some reference frame (typically the ground). Any body segment has at most six degrees of freedom: three for the translation of any point on the segment such as the centre of mass, and three for the rotation of the segment about three mutually perpendicular lines passing through that point. Body segments have fewer than six degrees of freedom when their motion is constrained by the actions of external forces (e.g., contact of the foot with the ground during walking). Free-body diagrams and the concept of equilibrium are used to analyse the resultant effect of force systems acting on body parts. Analyses based on the principle of static equilibrium are often used to estimate the magnitudes and directions of muscle, ligament, and joint-reaction forces present during activity.

Further Reading and References

Bedford, A., and W. Fowler. 1995. *Engineering mechanics: Dynamics.* Reading, MA: Addison-Wesley.

Bedford, A., and W. Fowler. 1995. *Engineering mechanics: Statics.* Reading, MA: Addison-Wesley.

Lieber, R.L. 1992. *Skeletal muscle structure and function: Implications for rehabilitation and sports medicine.* Baltimore: Williams & Wilkins.

Nordin, M., and V.H. Frankel. 1989. *Basic biomechanics of the musculoskeletal system.* 2nd ed. Philadelphia: Lea & Febiger.

Pandy, M.G. 2001. Computer modeling and simulation of human movement. *Annual Review of Biomedical Engineering* 3: 245-273.

Winter, D.A. 1990. *Biomechanics and motor control of human movement.* New York: Wiley.

CHAPTER 7

Basic Concepts of Energetics

One of the fundamental laws of mechanics, as elementary as Newton's Laws of Motion, is the **principle of work and energy.** Indeed, this principle, which states that the work done on an object is equal to the change in its kinetic energy, can be derived from Newton's famous Second Law ($F = m\,a$). The work done on an object is equal to the net force acting on the object multiplied by its displacement (change in position) as a result of the force. The mechanical energy of the object, its capacity to do work, may be visible as kinetic energy, gravitational potential energy, and (unless the object is rigid) elastic strain energy, which is stored as a result of the deformation created by the applied force.

This chapter describes the roles that kinetic energy, potential energy, and elastic energy play in human movement. While kinetic and potential energy are related to the mass of the body, elastic **strain energy** is associated with the properties of muscle and tendon tissue. Voluntary movement is made possible by the development of muscle force, which is fueled by chemical energy made available through the oxidation of foodstuffs. The last two sections of this chapter describe the energetics of muscle contraction and show how the metabolic cost of movement is governed by the interplay of kinetic energy, gravitational potential energy, and strain energy stored in the elastic tissues of muscle and tendon.

The major learning concepts in this chapter relate to

- calculating the kinetic energy of a body segment—the amount of mechanical energy a body segment possesses due to its motion;
- calculating the potential energy of a body segment—the amount of mechanical energy a body segment possesses by virtue of its height above the ground;
- total mechanical energy of a body segment—the sum of kinetic energy and gravitational potential energy;
- calculating the instantaneous power of a body segment—the rate at which work is done on the body segment;
- calculating elastic strain energy—the amount of mechanical energy stored in the elastic tissues of muscle and tendon during movement;
- describing qualitatively and quantitatively how metabolic energy is consumed during movement; and
- defining the efficiency of movement—the ratio of mechanical energy output to metabolic energy input.

Energy and Power

In this section we discuss three common forms of mechanical energy—kinetic energy, potential energy, and elastic strain energy—and show how to calculate each quantity. We also describe the concept of power, which is the rate of change of mechanical energy over time.

Kinetic Energy

The **kinetic energy** of a body is the amount of mechanical energy the body possesses due to its motion. Consider a rigid body translating and rotating in three-dimensional space (figure 7.1). Let the mass of the body be represented by m and the **moment of inertia** of the body about its centre of mass be given by Ic. (The moment of inertia of a body is a measure of the ability of the body to resist changes in its angular velocity. More simply, it may be thought of as a measure of a body's resistance to rotation. The moment of inertia depends on the point of the body about which it is calculated. It is minimal about the centre of mass of the body. The

Système International (SI) unit of moment of inertia is $kg \cdot m^2$ [$lb \cdot ft^2$].) At some time t during the body's motion, $\underline{v}$ represents the velocity of the centre of mass of the body and $\underline{\omega}$ is its angular velocity. The kinetic energy of the body is made up of two parts: the kinetic energy due to translation of the centre of mass, and the kinetic energy due to rotation about the centre of mass. The translational kinetic energy is given by

$$T_t = \frac{1}{2} m \underline{v}^2,\qquad 7.1$$

and the rotational kinetic energy is

$$T_r = \frac{1}{2} I_c \underline{\omega}^2.\qquad 7.2$$

Note that T_t and T_r are both scalars because any vector multiplied by itself is a scalar. That is, $\underline{v}^2 = \underline{v} \cdot \underline{v} = v^2$, and v^2 is a scalar. Thus, the total kinetic energy of the body is

$$T = \frac{1}{2} m \underline{v}^2 + \frac{1}{2} I_c \underline{\omega}^2,\qquad 7.3$$

and because $\underline{v}^2 = \underline{v} \cdot \underline{v} = v^2$ and $\underline{\omega}^2 = \underline{\omega} \cdot \underline{\omega} = \omega^2$, equation 7.3 can be rewritten more simply as

$$T = \frac{1}{2} m v^2 + \frac{1}{2} I_c \omega^2.\qquad 7.4$$

The SI unit of kinetic energy is the **joule** (which is equivalent to newton·metre [ft·lb]).

Potential Energy

Potential energy is the amount of mechanical energy a body possesses by virtue of its height above the ground. Consider the rigid body of figure 7.1 moving from position 1 to position 2 (see figure 7.2). The change in gravitational potential energy of the body is given by

$$U_g = m g(y_2 - y_1),\qquad 7.5$$

where y_1 is the vertical position (height) of the centre of mass of the body in position 1, y_2 is its height in position 2, and g is the gravitational acceleration constant near the surface of the earth (equal to 9.81 m/s² or 32.2 ft/s²). Potential energy is a scalar, and its SI unit is the joule (newton·metre [ft·lb]).

Total Mechanical Energy

The **total mechanical energy** of a rigid body is equal to the sum of its kinetic energy and gravitational potential energy,

$$E_T = \frac{1}{2} m v^2 + \frac{1}{2} I_c \omega^2 + m g (y_2 - y_1).\qquad 7.6$$

When the total mechanical energy of the system remains constant (i.e., the value of E_T does not change over time), energy is conserved, and the system is said to be **conservative.** In the real world, no system

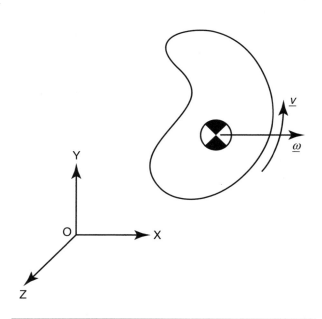

Figure 7.1 Rigid body rotating with an angular velocity $\underline{\omega}$ and translating with a linear velocity of its centre of mass of $\underline{v}$.

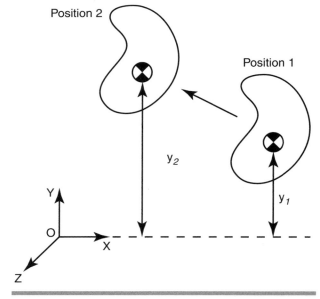

Figure 7.2 Displacement of a rigid body between two positions. The change in potential energy of the body is proportional to the displacement of the centre of mass from position 1 to position 2.

is completely conservative; some energy is always lost by virtue of the system's interaction with the environment. Take the pendulum commonly found in a grandfather clock. In order for the clock to keep perfect time, the pendulum bob must swing with exactly the same amplitude from cycle to cycle. The small amount of friction in the pendulum's bearings will act to decrease (by a very small amount) the amplitude of the pendulum's swing over time. To ensure that the pendulum swings with precisely the same amplitude from cycle to cycle, a small moment is exerted on the pendulum to compensate for the slowing effect of friction.

It is worthwhile looking at the fluctuations in kinetic and potential energy of the ideal pendulum (in the absence of friction) during each cycle of its swing, because this model illustrates many of the same biomechanical features evident in human walking. Figure 7.3 shows a single pendulum of length l and mass m. At any instant during its motion, the kinetic energy of the pendulum can be written as

$$T = \frac{1}{2}mv^2, \qquad 7.7$$

where v, the velocity of the tip of the pendulum, is given by $v = \omega l$, and ω is the angular velocity of the pendulum. Note that the pendulum has no rotational kinetic energy here, because its moment of inertia is assumed to be zero. The gravitational potential energy is given by

$$U = mg(l - l\cos\theta), \qquad 7.8$$

where the zero position is (arbitrarily) taken to be the lowest point reached by the pendulum during its motion. Thus, the total mechanical energy of the pendulum at any instant is

$$E_T = \frac{1}{2}mv^2 + mg(l - l\cos\theta). \qquad 7.9$$

As shown in figure 7.4, kinetic energy is maximal when the pendulum reaches its lowest point because its velocity is then maximal. Conversely, potential energy is minimal at the lowest point. Thus, kinetic energy is maximal when potential energy is minimal, and vice versa. The fluctuations of kinetic and gravitational potential energy are exactly equal and opposite in phase during each cycle because energy losses due to friction are neglected here. Thus, the total mechanical energy, E_T, of this ideal pendulum remains constant for all time (see figure 7.4).

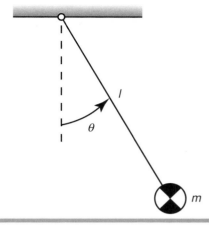

Figure 7.3 Diagram of an inverted single pendulum. All the mass of the pendulum is concentrated at its tip, and the moment of inertia of the pendulum is neglected. The hinge about which the pendulum pivots is assumed to be frictionless.

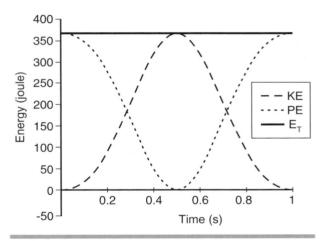

Figure 7.4 Kinetic energy (KE), gravitational potential energy (PE), and total mechanical energy (E_T) calculated for the inverted single pendulum model of figure 7.3.

The human body is not a conservative system; if it were, the total mechanical energy of the body would remain constant and no additional energy would be needed from the muscles to sustain movement. This is not the case, as muscles consume metabolic energy during contraction and do mechanical work to move the joints. Fluctuations in kinetic and potential energy of the whole body usually do not cancel each other during each cycle of movement, and so additional energy is needed to keep the body moving—energy that must be supplied by the muscles (see "Is Mechanical Energy Conserved in Walking?" on p. 84).

IN FOCUS: IS MECHANICAL ENERGY CONSERVED IN WALKING?

The changes in kinetic and potential energy of the centre of mass that occur during locomotion can be calculated from the measured force exerted on the ground. Force-measuring devices called force plates are used for this purpose. These devices can measure force very accurately to within one or two newtons, depending on whether the force is applied statically (constant in time) or dynamically (changing in time).

One can find the position and velocity of the centre of mass of the body by integrating the ground force recorded from the force plate. For example, an equation that relates the vertical component of the ground force to the vertical acceleration of the centre of mass can be written as follows:

$$F_{gy} = m\ddot{y}_{cm} + mg, \qquad 7.9.1$$

where F_{gy} is the measured vertical ground force, $\ddot{y}_{cm}$ is the vertical acceleration of the centre of mass, m is the mass of the body, and g is the value of gravitational acceleration at the surface of the earth. Solving equation 7.9.1 for the vertical acceleration of the mass centre and then integrating gives

$$\dot{y}_{cm} = \int_0^{t_f} \ddot{y}_{cm}\, dt \qquad 7.9.2$$

and

$$y_{cm} = \int_0^{t_f} \dot{y}_{cm}\, dt, \qquad 7.9.3$$

where $\ddot{y}_{cm} = (F_{gy} - mg)/m$ from equation 7.9.1; $\dot{y}_{cm}$ and y_{cm} are the vertical velocity and position of the centre of mass, respectively; and t_f defines the interval of time over which the integration is carried out. Equations 7.9.2 and 7.9.3 give the change in vertical velocity and vertical position of the centre of mass, respectively. A similar analysis can be done to find the horizontal velocity and position of the centre of mass. Substituting these quantities into equation 7.6 then gives the change in total mechanical energy of the centre of mass over one cycle, where ω in equation 7.6 is taken to be zero because the centre of mass is a point (i.e., only rigid bodies can rotate and have angular velocities; points on bodies can only translate).

If the preceding analysis is carried out for walking at a normal speed, the fluctuations in kinetic and gravitational potential energy are found to be nearly equal in magnitude and opposite in

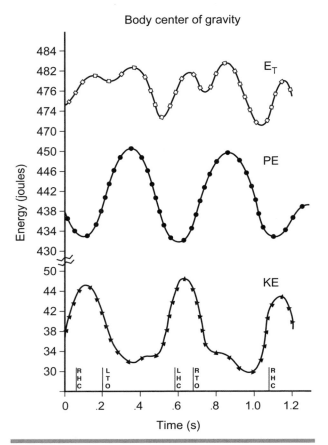

Body center of gravity

Figure 7.5 Kinetic energy (KE), gravitational potential energy (PE), and total mechanical energy (E_T) calculated for normal walking on level ground. RHC = right heel contact; RTO = right toe-off; LHC = left heel contact; LTO = left toe-off.

Reprinted from Winter 1979.

phase (see figure 7.5 and compare with results shown in figure 7.4; in figure 7.5, all energies were calculated from force plate data recorded from humans walking at their self-selected speeds (i.e., at speeds of approximately 1.35 m/s)). In walking, the centre of mass reaches its highest point at midstance, and gravitational potential energy is therefore maximal at this point. Almost all of the kinetic energy in walking is due to the forward velocity of the body; the changes in vertical velocity are much smaller, so the fluctuations in vertical kinetic energy are negligible. The forward velocity of the body is maximal when the body is at its lowest point, when both legs are in contact with the ground; and it is least when the body is at its highest point, at midstance. Thus, kinetic energy is maximal during double support and minimal in midstance, precisely opposite to the pattern of changing gravitational potential energy.

When the curves representing kinetic and potential energies in figure 7.5 are added numerically (which is possible because energy is a scalar quantity), the fluctuations in total mechanical energy of the centre of mass are seen to be relatively small. This result indicates that the mechanical energy of the body is nearly conserved when people are free to choose the speed at which they walk. That is, at the speed at which metabolic energy consumption is mini-

mized, there is almost a complete exchange of kinetic and gravitational potential energy, similar to what the ideal single pendulum model of walking predicts.

SOURCES

- Cavagna, G.A., H. Thys, and A. Zamboni. 1976. The sources of external work in level walking and running. *Journal of Physiology* 262: 639-657.
- Farley C.T., and D.P. Ferris. 1998. Biomechanics of walking and running: Center of mass movements to muscle action. *Exercise and Sport Science Reviews* 26: 253-285.

Power

Power is the rate of doing work. The work done by any external force $\underline{F}$ acting on a mass during a very small (infinitesimal) displacement $d\underline{r}$ is $\underline{F} \cdot d\underline{r}$. To find the power P, which is a scalar quantity, we divide the work done by the interval of time, dt, during which the displacement occurs. Thus,

$$P = \frac{\underline{F} \cdot d\underline{r}}{dt} = \underline{F} \cdot \underline{v}, \qquad 7.10$$

where $\underline{v}$ is the velocity of the point on the rigid body at which the force is applied. The analogous relation for rotational motion is

$$P = \frac{\underline{M} \cdot d\theta}{dt} = \underline{M} \cdot \underline{\omega}, \qquad 7.11$$

where $\underline{M}$ is the net moment applied to the body and $\underline{\omega}$ is the angular velocity of the body. In SI units, power is expressed in **watts** (which is equivalent to newton·metres/s [ft·lb/s]).

Power can be positive or negative, depending on whether it is being transferred to or taken from the mass. A muscle produces energy when it contracts concentrically (i.e., shortens) against an external load; in this case the value of muscle power is taken to be positive. Conversely, muscles that contract eccentrically (i.e., lengthen) absorb energy, and the value of muscle power is taken to be negative here (see also figure 2.14).

If muscle force and the rate of muscle shortening (or lengthening) are known, equation 7.10 can be used to estimate the power produced (or absorbed) by the muscle during contraction (see "Modelling the Mechanics of Muscle Contraction" on p. 86). Indeed, equations 7.10 and 7.11 are the basis on which detailed analyses have been performed to determine how muscles contribute power to the

body segments in various tasks, including jumping, pedaling, and walking (see "Which Muscles Are Most Important to Vertical Jumping Performance?" on p. 87).

Elastic Strain Energy

Potential energy may also be stored in an elastic rather than a gravitational form. In explosive movements like running and jumping, considerable amounts of strain energy may be stored in the elastic tissues of muscle and tendon. If some of this stored elastic energy can be returned to the skeleton, less metabolic energy will be needed to keep the body moving. Storage and utilization of elastic strain energy may therefore reduce the amount of metabolic energy consumed by the muscles.

As explained in "Is Mechanical Energy Conserved in Walking?" on page 84, calculations based on force-plate measurements of human walking show that changes in forward kinetic energy are nearly perfectly out of phase with changes in gravitational potential energy so that the total mechanical energy of the centre of mass is kept nearly constant during a step. The reason is that the centre of mass is highest in midstance, when the forward kinetic energy is least, and lowest in double support, when the forward kinetic energy is greatest. The opposite is true for running, where changes in forward kinetic and gravitational potential energy are substantially in phase, leading to large changes in the total mechanical energy of the mass centre during each step. Thus, in running, the centre of mass is lowest in midstance when the forward kinetic energy is least, and it is highest during the flight phase when the forward kinetic energy is greatest.

IN FOCUS: MODELLING THE MECHANICS OF MUSCLE CONTRACTION

In the late 1930s, muscle physiologist A.V. Hill postulated a conceptual model of muscle contraction. This theoretical model integrates three of the most important force-producing properties of muscle: the force–length and force–velocity properties plus muscle's active state. In the late 1950s, Sir Andrew Huxley proposed the first complete mechanistic model to explain how the actin and myosin filaments interact with each other in the development of muscle force. This theory has come to be known as the sliding filament theory of muscle contraction. Although Hill's model cannot explain the mechanisms by which force and energy are produced during a contraction, it yields much insight into the mechanophysiological relationships between length, velocity, activation, and force.

Figure 7.6 is a model of a musculotendon actuator that is often used in theoretical studies of movement. The musculotendon actuator is represented as a three-element muscle in series with tendon. The mechanical behaviour of muscle is described by a Hill-type contractile element (CE), which models the muscle's force-length and force–velocity properties; a series elastic element (SEE), which models muscle's active stiffness; and a parallel elastic element (PEE), which models muscle's passive stiffness. The active stiffness is thought to arise from the cross-bridges formed by actomyosin binding, so this property is present only when the muscle is activated. The passive stiffness resides in the material properties of the collagen molecules that are the building blocks of each muscle fibre. Because tendon (and, for that matter, ligament)

is composed of collagen as well, its force–length property is qualitatively similar to that of passive muscle.

Passive muscle develops a force that is proportional to its stretch. This follows from the fact that the behaviour of the muscle is fully described by its force–length curve in the passive state. However, when muscle is activated, the force it develops depends on the instantaneous values of its length, velocity, and activation level. In the model shown in figure 7.6, all muscle fibres are assumed to insert at an angle, α, on tendon. Thus, pennation angle, α, is assumed to be constant. F^{MT} is the force developed by the whole musculotendon actuator, and this is the force transmitted to tendon. Four muscle-specific parameters are needed to specify the behaviour of the model: the peak isometric force developed by the muscle and the corresponding fibre length and pennation angle, plus the resting length of the tendon.

A first-order differential equation can be derived to describe the dynamics of the musculotendon actuator shown in figure 7.6 (Zajac 1989):

$$\frac{d F^{MT}}{dt} = f(F^{MT}, l^{MT}, v^{MT}, a(t)); \quad 0 \le a(t) \le 1. \quad 7.11.1$$

Equation 7.11.1 indicates that the time rate of change of musculotendon force, $\frac{d F^{MT}}{dt}$, depends on musculotendon length, l^{MT}; musculotendon velocity, v^{MT}; muscle activation level, α; and musculotendon force, F^{MT}. If we know the values of musculotendon length, musculotendon velocity, activation level, and musculotendon force at one time instant, we can integrate equation 7.11.1 to find the value of musculotendon force at the next time instant. Thus, if the trajectories of musculotendon length, velocity, and muscle activation are known for all times and the value of musculotendon force is given at the initial state (i.e., at time $t = 0$), we can find the time history of musculotendon force by integrating equation 7.11.1 for the duration of the motor task. The problem of integrating a differential equation given a forcing function and a set of initial conditions is referred to as the forward-dynamics problem in biomechanics.

SOURCES

- Pandy, M.G. 2001. Computer modeling and simulation of human movement. *Annual Review of Biomedical Engineering* 3: 245-273.

- Zajac, F.E. 1989. Muscle and tendon: Properties, models, scaling, and application to biomechanics and motor control. *CRC Critical Reviews in Biomedical Engineering* 19: 359-411.

- Zajac, F.E., and M.E. Gordon. 1989. Determining muscle's force and action in multi-articular movement. *Exercise and Sport Science Reviews* 17: 187-230.

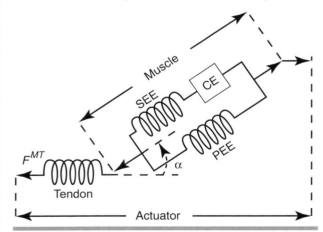

Figure 7.6 Schematic representation of a model of musculotendon actuation. Three components are used to model the mechanical behaviour of muscle: a contractile element (CE), a series elastic element (SEE), and a parallel elastic element (PEE). Tendon is represented as a nonlinear spring.

Reprinted from Zajac 1989.

IN FOCUS: WHICH MUSCLES ARE MOST IMPORTANT
TO VERTICAL JUMPING PERFORMANCE?

Because muscle force cannot be measured noninvasively in people, computer models are often used to estimate muscle force in tasks such as walking, running, jumping, and cycling. The model of figure 7.7 was used to simulate maximum-height jumping in humans. The skeleton was represented as a four-segment, four degree-of-freedom, planar linkage, joined to the ground at the toes and articulated at the ankle, knee, and hip by frictionless hinge joints. The head, arms, and torso were lumped together and represented as one rigid body. Thus, the effect of arm swing was neglected in the model simulations. The model skeleton was actuated by eight lower extremity musculotendinous units, each unit modelled as a three-element muscle in series with tendon (see "Modelling the Mechanics of Muscle Contraction" on p. 86). Muscles included in the model were gluteus maximus (GMAX), rectus femoris (RF), hamstrings (HAMS),

vasti (VAS), gastrocnemius (GAS), soleus (SOL), other uniarticular plantar flexors (OPF), and tibialis anterior (TA).

A mathematical theory called optimisation was used to simulate the biomechanics of maximum-height squat jumping. In this task, the body begins from a static squatting position with the hip, knee, and ankle all flexed to 90°. Comparison of the model results with kinematic, force plate, and muscle electromyographic data obtained from experiments showed that the model reproduced the major features of the ground contact phase of the jump. The simulation results were then analysed to determine how muscles accelerate and contribute power to the body segments during jumping.

Equations similar to equations 7.10 and 7.11 were used to calculate the power developed by the muscles and the amount of energy subsequently transferred to the skeleton during the ground contact phase of a maximum-height squat jump. Specifically, power was calculated by multiplying musculotendon force by musculotendon contraction (shortening or lengthening) velocity (see equation 7.10). In figure 7.8, UPF represents the combined power of the uniarticular

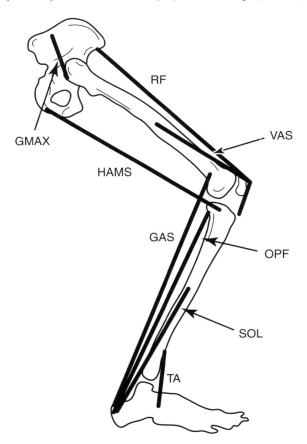

Figure 7.7 Diagram showing a musculoskeletal model of the body that was used to simulate a maximum-height squat jump.

Reprinted from Pandy, Zajac, Sim, and Levine 1990.

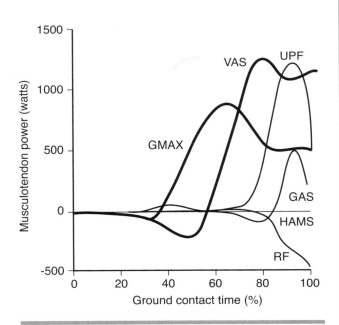

Figure 7.8 Mechanical power generated by the musculotendon actuators in the model of figure 7.7 during the ground contact phase of a maximum-height squat jump.

Reprinted from Pandy and Zajac 1991.

(continued)

ankle plantar flexors in the model (i.e., OPF and SOL). The area under each curve is equal to the total energy developed or absorbed by each actuator.

Vasti and gluteus maximus muscles were found to be the major energy producers, the prime movers, of the lower limb in vertical jumping. These muscles contributed most significantly to the total energy made available for propulsion (note the area under each curve in figure 7.8). However, in the final 20% of ground contact time, just before liftoff, the ankle plantar flexors (soleus and gastrocnemius) also contributed significantly to the total energy delivered to the skeleton.

Figure 7.9 is a plot of the total instantaneous power delivered to each body segment during the jump. The total area under each curve is equal to the total mechanical energy (kinetic plus gravitational potential energy) of each body segment at the instant the body leaves the ground. A large proportion of the total energy developed by the muscles was delivered to the trunk segment. In fact, the combined energy of the thigh, shank, and foot amounted to only 30% of the total energy available at liftoff. The remainder (approximately 70%) of the input muscle energy was transferred to the trunk. This is not a surprising result given that the trunk segment represents approximately 70% of the total body mass.

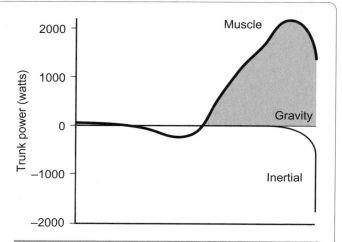

Figure 7.10 Contributions of all the muscles (Muscle), gravitational forces (Gravity), and inertial forces (Inertial) to the instantaneous power of the trunk for the ground contact phase of the jump.

Reprinted from Pandy and Zajac 1991.

The model simulation results also showed that muscles dominate the instantaneous power of the trunk segment for most of the ground contact phase of the jump (figure 7.10). (In figure 7.10, the total power delivered to the trunk segment (head, arms, and trunk lumped together) is indicated by the shaded region.) Only near liftoff do centrifugal forces (i.e., forces arising from motion of the joints) become so important that they dominate the power delivered to the trunk. Centrifugal forces are large only near liftoff because the velocities of the joints increase greatly just before the body leaves the ground. According to the model calculations, gravitational forces contribute little to trunk energy in maximum-height jumping.

Figure 7.11 shows the relative contributions of individual leg muscles to trunk power during the jump. The shaded region is the total power delivered to the trunk by all the muscles in the model. The dashed line is the contribution of all the biarticular muscles in the model (hamstrings and rectus femoris). The area under each curve represents the total energy delivered by that muscle (or group of muscles) to the trunk segment at the instant the body leaves the ground.

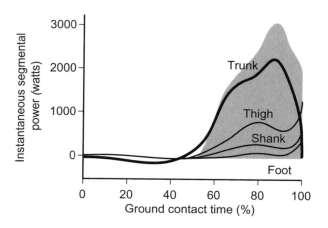

Figure 7.9 Instantaneous power delivered to each body segment in the model during the ground contact phase of the simulated squat jump.

Reprinted from Pandy and Zajac 1991.

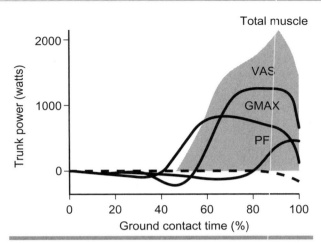

Figure 7.11 Contribution of each muscle to the total power delivered to the trunk segment during the ground contact phase of the simulated squat jump. VAS is the contribution of vasti; GMAX is the contribution of gluteus maximus; and PF is the contribution of all the ankle plantar flexors (SOL, OPF, and GAS) in the model (see figure 7.7).

Reprinted from Pandy and Zajac 1991.

Of all the muscles, vasti and gluteus maximus contributed most significantly to the power delivered to the trunk (figure 7.11). The energy delivered by these muscles amounted to nearly 90% of the total energy delivered to the trunk during the jump. Thus, vasti and gluteus maximus appear to be the most important muscles for maximum-height jumping. The ankle plantar flexors, soleus and gastrocnemius, are also important, but only in the last 20% of ground contact. Near liftoff, the ankle plantar flexors (PF) deliver as much power to the trunk as either vasti or gluteus maximus. Finally, the model simulation results also suggest that the biarticular muscles are relatively unimportant to overall jumping performance.

SOURCES

- Pandy, M.G., and F.E. Zajac. 1991. Optimal muscular coordination strategies for jumping. *Journal of Biomechanics* 24: 1-10.
- Zajac, F.E. 2002. Understanding muscle coordination of the human leg with dynamical simulations. *Journal of Biomechanics* 35: 1011-1018.

Two very different mechanisms explain this difference in mechanical energy between walking and running. The almost complete transfer of potential and kinetic energies in walking means that not all the energy required to lift and accelerate the centre of mass has to come from the muscles—storage and utilization of gravitational potential energy allow the muscles to do less work, as is the case in the pendulum. In running, however, an exchange of gravitational and forward kinetic energy is not possible because the centre of mass goes through an increase in height and speed at the same time during a step. The large fluctuations in total mechanical energy during each cycle suggest that the pendulum model of walking does not apply to running. Nonetheless, metabolic energy is still saved, not by the exchange of gravitational potential and kinetic energy, but instead by storage and utilization of elastic strain energy.

Although we cannot calculate the amount of elastic energy stored in muscle and tendon precisely from force-plate measurements, we can estimate it if we know the efficiency with which muscles convert chemical energy into mechanical work. The kangaroo perhaps best exemplifies an animal that is able to exploit the benefits of elastic

energy storage. A 40-kg (88-lb) kangaroo has an Achilles tendon measuring approximately 1.5 cm (0.6 in.) in diameter and 35 cm (14 in.) in length. It is likely, then, that substantial quantities of elastic energy are stored in the tendons of these animals following impact with the ground. Indeed, at a speed of 30 km/h (18.6 miles/h), it has been estimated that the contractile machinery supplies only 1/3 of the energy required to lift and accelerate the centre of mass while the legs are on the ground. The remaining 2/3 of the energy required for each hop is thought to come from strain energy stored in the animal's tendons. Although this fraction is likely to be much smaller in a person who is running, some estimates put it as high as 1/2; that is, one-half of the mechanical energy needed to lift and accelerate the centre of mass during ground contact, plus the energy needed to move the limbs, is supplied by strain energy stored in the elastic tissues of muscle and tendon.

More quantitative estimates of elastic energy storage require knowledge of the mechanical properties of muscle and tendon and of the forces they exert. Although all biological tissues exhibit nonlinear force–extension curves, there is always a region in which extension varies linearly with force. Consider the elastic behaviour of tendon as represented by a simple linear spring. Let the stiffness of the tendon (spring) be given by k^T and the amount of stretch be Δl^T. The elastic strain energy stored in the tendon spring is then

$$U_s = \frac{1}{2} k^T (\Delta l^T)^2. \qquad 7.12$$

This equation has been used to estimate the amount of strain energy stored in the elastic tissues of muscle and tendon when humans jump to their maximum height.

Metabolic Energy Consumption

When a muscle is stimulated, heat is liberated, and the amount of heat produced can be measured by the temperature change within the muscle. If there is also a change in length during a contraction, then mechanical work is done by the muscle. According to the **First Law of Thermodynamics,** the total rate of energy production during muscle contraction, $\dot{E}$, is equal to the rate at which heat is liberated, $\dot{H}$, plus the rate at which work is done to move the external load, $\dot{W}$. Thus,

$$\dot{E} = \dot{H} + \dot{W}. \qquad 7.13$$

The rate at which mechanical work is done is equal to the power developed by the muscle. Power, in turn, is given by the force exerted by the muscle multiplied by its shortening (or lengthening) velocity (see equation 7.10). The rate of heat production is more difficult to estimate. This quantity depends on a number of factors related to the mechanics and physiology of muscle contraction, including muscle length, mass, contraction velocity, activation level, muscle fibre recruitment rate, and muscle fibre type. At least qualitatively, though, the rate at which heat is produced during a contraction can be estimated from four quantities: activation heat, maintenance heat, shortening heat, and resting heat.

When muscle is excited, a relatively large amount of heat is produced early in the contraction. This **activation heat** reflects the energetics of calcium release and re-accumulation, and it can account for as much as 25% to 30% of the total energy consumed during a contraction.

The continued heat production observed once steady state has been reached is called the **maintenance heat.** This portion of the heat rate is thought to represent the chemomechanical events associated with muscle contraction and relaxation (see chapter 2). The maintenance heat accounts for the largest fraction of the total heat liberated during contraction, and its magnitude is a strong function of muscle's force–length property (see "How Much Force Can the Quadriceps Develop in a Maximal Contraction" on p. 75 in chapter 6).

Shortening heat was first discovered by A.V. Hill in the mid- to late 1930s. Hill defined this quantity as the difference between the heat liberated by a muscle when it shortens and the heat liberated by the same muscle when it contracts isometrically. The shortening muscle liberates more energy because it does external work to move a load and also because it liberates more heat. From his early experiments on frog muscle, Hill concluded that the amount of shortening heat was directly proportional to the distance the muscle shortened. Shortening heat represents only a small (often insignificant) fraction of the total energy consumed during contraction.

All muscles produce a little heat as a consequence of being alive. **Resting heat** is the heat released as chemical energy, derived from oxidation of foodstuffs, consumed in the resting state. The rate of heat production in the resting state (i.e., sitting quietly or lying down) is about 1.04 J/(kg·s). Interestingly, the cost associated with quiet standing is some 30% higher, approximately 1.5 J/(kg·s).

As soon as a person begins to walk, a great increase in energy expenditure occurs, reflecting the metabolic cost to the muscles of moving the body against gravity and of accelerating and decelerating the limbs. A large number of experimental studies have shown that metabolic energy expenditure increases with walking speed in the manner indicated (see figure 7.13). The relationship is expressed fairly well by a parabolic equation of the form

$$\dot{E} = 32.0 + 0.005\, v^2, \qquad 7.14$$

where $\dot{E}$ is the rate of metabolic energy consumption expressed in J/(kg·s) and v is walking speed. The constants in equation 7.14 were determined on the basis of data obtained from nearly 100 men and women who walked on level ground at a frequency and step length of their choosing.

Dividing equation 7.14 by walking speed, v, gives the rate at which metabolic energy is consumed per unit distance traveled, or

$$\dot{E}_d = \frac{32.0}{v} + 0.005\, v, \qquad 7.15$$

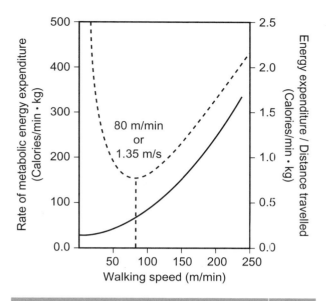

Figure 7.12 Metabolic energy expenditure plotted as a function of walking speed. The rate of metabolic energy consumption increases parabolically with walking speed (solid line). When the rate of metabolic energy consumption is normalized by the distance traveled, an optimal walking speed is predicted at roughly 80 m/min or 1.35 m/s (dashed line).
Reprinted from Ralston 1976.

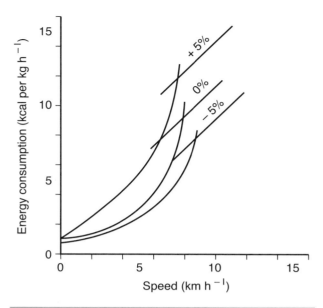

Figure 7.13 Rate of metabolic energy consumption plotted against speed for walking and running on level ground (0%), on a 5% incline (5%), and on a 5% decline (–5%).
Reprinted from Margaria 1976.

where $\dot{E}_d$ is expressed in J/(kg·m). Importantly, while $\dot{E}$ increases proportionally with walking speed, $\dot{E}_d$ has a well-defined minimum at about 1.3 m/s (figure 7.12). As shown in figure 7.12, the minimum in metabolic energy expenditure predicted by equation 7.15 coincides with that obtained from oxygen consumption measurements made on people.

Energy consumption rate increases as a curvilinear function of speed for walking, but it becomes a straight-line function of speed for running (figure 7.13). At a given speed, the rate of energy consumption increases as the slope of the ground increases, and it decreases as the slope decreases, reaching a minimum at a gradient of –10% (figure 7.13). For gradients steeper than –10%, energy consumption increases again because of the postural changes necessary for continued locomotion and the significant braking action needed from the muscles.

Efficiency of Movement

Muscles create and absorb mechanical energy by shortening and lengthening during flexion and extension of the joints. When mechanical energy is absorbed, muscles are said to do negative work; this is the work done by a muscle when it is developing an active force at the same time that it is being lengthened by some externally applied force. If the muscle is shortening as it develops a force, it is said to do positive work. The efficiency with which a muscle operates under these conditions can be defined as

$$efficiency = \frac{mechanical\ work\ done}{metabolic\ energy\ consumed},\quad 7.16$$

where the mechanical work done *on* the muscle is regarded as negative work, while that done *by* the muscle is positive work.

The efficiencies of walking and running on negatively and positively sloped surfaces have been calculated and are plotted in figure 7.14. The lines emanating from the zero point on the graph represent constant efficiency of transport. Walking up steeper and steeper inclines requires more and more energy for the same distance traveled. Notice that the curve for walking up inclines approaches the limit of 25% efficiency. This represents the maximum efficiency with which a muscle can convert chemical energy into the mechanical work needed to move the centre of mass. For walking

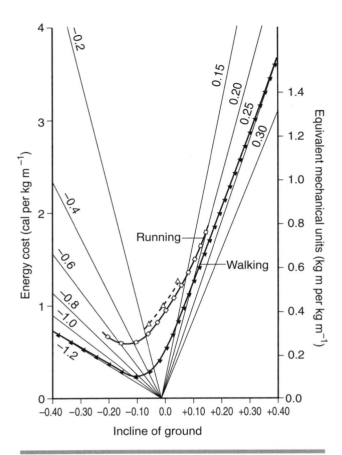

Figure 7.14 Metabolic energy consumed per unit distance moved plotted against inclination of the ground. The data were obtained from humans walking and running on a treadmill whose orientation to the ground was varied.
Reprinted from Margaria 1976.

down inclines, however, the limit of efficiency is given to be –120%. This negative value of efficiency is explained by the fact that the leg muscles do mostly negative work (i.e., absorb mechanical energy) during walking down a steep incline.

Notice that the curve for running lies above the curve for walking at the same speed. In walking, metabolic energy is saved through the conversion of gravitational potential energy into forward kinetic energy, as explained by the single pendulum model. Although some amount of elastic strain energy is stored and reutilized in each cycle of running, the results of figure 7.14 suggest that this mechanism is less efficient than the pendulum-like mechanism present in walking. That is, in relation to the cost of transport, storage and utilization of elastic energy in running are not as efficient as the conversion of gravitational potential energy into forward kinetic energy in walking.

Summary

As we will see in chapter 10, muscles metabolise high-energy phosphate molecules to produce chemical energy and to do mechanical work. Roughly 75% of the total metabolic energy consumed during activity is dissipated as heat; the remaining 25% is used to perform useful mechanical work (i.e., to support the body and to swing the limbs back and forth during locomotion). Mechanical energy is visible in two forms: kinetic energy and potential energy. Kinetic energy is the energy the body possesses due to its motion; it is a function of the rate of change of position (i.e., velocity) of the body at each instant. Potential energy may be stored in its gravitational form, as strain energy in the elastic tissues of muscle and tendon, or both. Storage and utilization of elastic strain energy can increase the efficiency with which muscles convert chemical energy to mechanical work.

Further Reading and References

Alexander, R.McN., and G. Goldspink. 1977. *Mechanics and energetics of animal locomotion*. London: Chapman and Hall.

Bedford, A., and W. Fowler. 1995. *Engineering mechanics: Dynamics*. Reading, MA: Addison-Wesley.

Margaria, R. 1976. *Biomechanics and energetics of muscular exercise*. Oxford: Clarendon Press.

McMahon, T.A. 1984. *Muscles, reflexes, and locomotion*. Princeton, NJ: Princeton University Press.

Rose, J., and J.G. Gamble. 1994. *Human walking*. 2nd ed. Baltimore: Williams & Wilkins.

CHAPTER 8

Biomechanics Across the Life Span

The major learning concepts in this chapter relate to
- the kinematics and kinetics of normal walking,
- major differences in the gait patterns of children and healthy young adults, and
- major differences in the gait patterns of elderly persons and healthy young adults.

As life expectancy continues to increase, so does the population of elderly adults. Currently, one of every 10 persons is 60 years or older, and about one in every 100 is 80 years or older. The Population Division of the United Nations predicts that by 2050 the proportion of persons older than 60 years will be one in five and by 2150 one in three. Likewise the number of persons 80 years or older is predicted to rise to around 3% of the population by 2050 and to just under 10% of the population by 2150 (for more details visit http://www.un.org/esa/socdev/agewpop.htm).

Maintaining independent function is an important goal for older adults, as people in this group are prone to fall-related injuries. Independence in activities such as shopping, using public transportation, and visiting friends requires adequate muscle strength, adequate flexibility (i.e., joint ranges of motion), and adequate neuromotor performance (i.e., reaction times and proprioception). The incidence of falling is known to increase significantly when musculoskeletal performance is degraded by aging.

This chapter describes how the natural aging process affects biomechanical performance of motor skills. Because the ability to walk is critical to maintaining an independent lifestyle, the discussion centres on the effects of aging on the biomechanics of gait. Less is known about the effects of maturation on movement in the very young. However, some researchers have investigated how growth and development affect gait mechanics in young children, and the results of these studies are also summarized here.

Biomechanics of Normal Walking

To appreciate the changes that occur in walking during the normal life span, it is necessary first to understand the biomechanics of gait in healthy young adults. The gait cycle is usually defined as the time between successive heel strikes of the same leg. Because normal walking is very nearly symmetric, one-half of the gait cycle is defined from heel strike of one leg to heel strike of the other (figure 8.1). The normal gait cycle is usually divided into two distinct phases: stance and swing. The stance phase, which is the time from heel strike to toe-off of one leg, can be divided into three subphases: initial double support, single-limb stance, and second double support. The swing phase is then the portion of the cycle from toe-off to heel strike of the same leg, and it is also usually divided into three subphases: initial swing, midswing, and terminal swing. There are two periods of single and double support in each gait cycle: each double-support phase occupies 12% of the cycle time, and each single support occupies 38%. Each leg is in contact with the ground for 62% of the gait cycle (stance) and spends 38% of the time in the air (swing).

Much of what is currently known about the biomechanics of normal walking is based on the

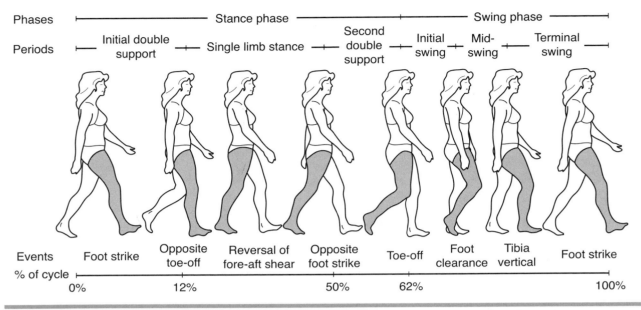

Phases	Stance phase				Swing phase		
Periods	Initial double support	Single limb stance		Second double support	Initial swing	Mid-swing	Terminal swing

| Events | Foot strike | Opposite toe-off | Reversal of fore-aft shear | Opposite foot strike | Toe-off | Foot clearance | Tibia vertical | Foot strike |
| % of cycle | 0% | 12% | | 50% | 62% | | | 100% |

Figure 8.1 Schematic diagram illustrating the major events of the normal gait cycle. Adapted from Vogler and Bove 1985.

results of gait analysis experiments performed over the last 50 years. The quantities recorded in these experiments are the relative movements of the joints, the pattern of force exerted on the ground, and the sequence and timing of leg muscle activity. The relative movements of the joints are usually obtained using high-speed video cameras that track the three-dimensional positions of reflective markers mounted on the body parts; ground forces are recorded using force plates that measure the three components of force (anterior–posterior, vertical, and medial–lateral) exerted by the leg on the ground; and muscle activations are usually monitored noninvasively through attachment of electromyography (EMG) electrodes to the surface of the skin.

Kinematics of Normal Gait

Kinematic measurements have shown that the centre of mass describes a smooth sinusoidal path when projected on the sagittal plane (figure 8.2). The total amount of vertical displacement (figure 8.2b) is 5 cm (2 in.) when healthy young adults walk at their preferred speeds. The peaks of the vertical oscillations occur at midstance of the supporting limb, and the troughs occur at roughly the middle of the double-support phase. The curve is very smooth and fluctuates regularly between its peaks and troughs. Interestingly, at midstance, when the vertical displacement of the centre of mass is greatest, the top of the head is actually

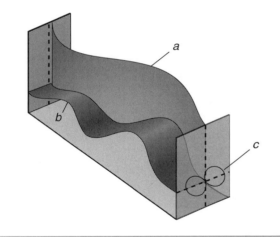

Figure 8.2 Schematic diagram showing the displacements of the centre of mass of the body in (a) the transverse plane, (b) the sagittal plane, and (c) the frontal plane. The data represent normal gait.
Reprinted from Rose and Gamble 1994.

lower than the standing height of the person. Thus, people are slightly shorter when they walk than when they stand fully erect. This explains why one is able to walk freely through a tunnel that is exactly the same height as one's standing height without the fear of bumping one's head.

The centre of mass is also displaced laterally in the transverse plane (i.e., in a direction perpendicular to the sagittal plane) (figure 8.2). The amplitude of the path traced in the transverse plane (figure 8.2*a*) is also about 5 cm (2 in.), but its frequency is only one-half that of the vertical oscillations. When viewed from the back, the centre of mass is seen to move up and down and to swing from side to side during each cycle. Indeed, when the curves describing the vertical and lateral displacements are projected onto the frontal plane, the resulting curve resembles the number "8" lying on its side (figure 8.2*c*).

The upper body moves forward throughout the gait cycle, but its speed does not remain constant. It moves fastest during the double-support phase and slowest when the supporting leg is near midstance. The trunk twists about its long axis as the shoulder and pelvis rotate in opposite directions. The arms also swing in opposition to the movements of the legs; thus, the right arm and right side of the shoulder move forward in unison with the left leg and left side of the pelvis. The total displacements of the shoulder and pelvis are relatively small, measuring 7° and 12°, respectively, during the course of a full cycle. Similar to the displacement of the centre of mass, the trunk rises and falls twice and moves from side to side once during each cycle. In addition to twisting about its long axis in the transverse plane, the pelvis also tips slightly backward and forward (in the sagittal plane) and from side to side (in the frontal plane).

The hip flexes and extends once during the gait cycle (figure 8.3). The limit of hip flexion is reached near midswing, after which the hip remains flexed to this position until initial contact. Peak hip extension is reached before the end of the stance phase (initial swing).

The knee shows two flexion and extension peaks during each gait cycle. Flexion–extension movements of the knee remain relatively small during the stance phase compared to swing. The knee is nearly fully extended at initial contact, and it flexes slightly (to no more than 20°) during loading response. It extends again during midstance and terminal stance, so much so that it is nearly fully extended just prior to opposite heel strike. Knee flexion increases rapidly during pre-swing and continues well into initial swing, reaching a peak of nearly 60°. The knee then extends rapidly during midswing and terminal swing, and becomes nearly fully extended once again in preparation for initial contact.

The ankle is very close to the neutral position (i.e., its position during normal standing) at the time of initial contact, after which it plantar flexes rapidly to bring the foot flat on the ground (figure 8.3).

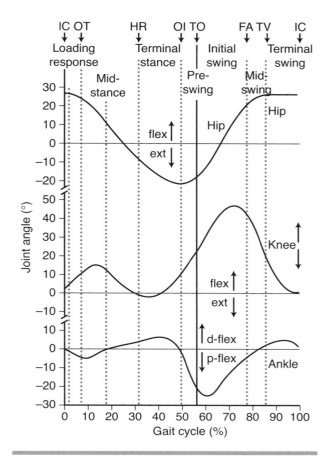

Figure 8.3 Mean sagittal plane joint angles measured in healthy young adults walking at their preferred speeds. Data are plotted for the right hip (flexion positive), right knee (flexion positive), and right ankle (dorsiflexion positive). The gait cycle events are shown at the top of the figure. IC = initial contact; OT = toe-off of opposite leg; HR = heel rise; OI = initial contact of opposite leg; TO = toe-off; FA = feet adjacent; TV = tibia vertical.

Reprinted from Whittle 1996.

During midstance, the tibia rotates forward relative to the foot, moving the ankle into a dorsiflexed position. Just before opposite heel strike, the ankle plantar flexes rapidly, which serves to lift the centre of mass into the air and smooth the transition from single to double support. During swing, the ankle dorsiflexes in order to provide toe clearance for the freely swinging leg. Once the toes have cleared the ground, the ankle maintains a roughly neutral position in preparation for initial contact.

Although the major movements of the lower limb joints occur in the sagittal plane, joint displacements in the transverse and frontal planes

About 50 years ago, a group of scientists at the University of California at Berkeley proposed six kinematic mechanisms to explain how movements of the lower limb joints contribute to the pathway of the centre of mass when humans walk at their preferred speeds. These six mechanisms, termed the major determinants of gait, are hip flexion, stance knee flexion, ankle plantar flexion, pelvic rotation, pelvic list, and lateral pelvic displacement. These six determinants form the basis of much of our understanding of the kinematics of normal walking.

The simplest form of walking is **compass gait,** in which all joints except the hips are locked in the anatomical position. Flexing and extending the hips means that the centre of mass moves on the arcs of circles, whose radii correspond to the lengths of the legs. To produce reasonable step lengths, the centre of mass must rise and fall much more in the compass gait than it does in normal walking. Apart from the fact that the vertical displacement of the centre of mass is too large, the main objection to compass gait is the discontinuity (i.e., the sharp corner) that it produces in the path of the centre of mass during the transition from single- to double-leg stance (see dashed line in figure 8.4).

The second determinant permits bending of the knees in the sagittal plane; at normal speeds of walking, the amplitude of knee flexion during stance is roughly 20°. Stance knee flexion flattens the arcs defined by compass gait, as does the third determinant, ankle plantar flexion. However, ankle plantar flexion also eases the transition from double- to single-leg stance by producing a smoother trajectory of the centre of mass in the vicinity of the two intersecting compass-gait arcs. In normal walking, the ankle plantar flexes by more than 30° during terminal stance.

Two of the remaining three determinants—pelvic rotation and pelvic list—further flatten and

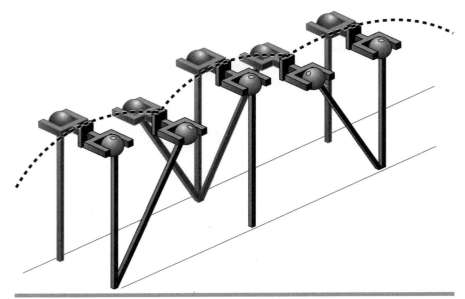

Figure 8.4 Diagram depicting the compass gait using only hip flexion and extension. The path of the centre of mass is a series of arcs.

Reprinted from Saunders, Inman, and Eberhart 1953.

smooth the arcs traced by the centre of mass. Pelvic rotation permits movement of the pelvis in the transverse plane, which allows for an increase in step length; the amplitude of transverse pelvic rotation is about 8° on either side of the central axis of the body. The fifth determinant, pelvic list, is the downward movement of the pelvis in the frontal plane, brought about by abduction of the hip. Transverse pelvic rotation raises the points of intersection between two compass-gait arcs, while pelvic list further flattens the trajectory by lowering the centre of mass as the pelvis tilts from the weight-bearing leg to the leg that is about to enter swing. The sixth determinant, lateral pelvic displacement, concerns movement of the centre of mass in the transverse plane. As support is alternated from one leg to the other, the centre of mass is displaced from side to side. Shifting the centre of mass from side to side is made possible by transverse rotation of the tibia about the subtalar joint.

SOURCE

- Saunders, J.B., V.T. Inman, and H.D. Eberhart. 1953. The major determinants in normal and pathological gait. *Journal of Bone and Joint Surgery* 35-A: 543-558.

are also important for producing the sinusoidal displacement of the centre of mass shown in figure 8.2. Indeed, three of the six major determinants of gait are associated with movements of the bones in the transverse and frontal planes (see "The Major Determinants of Normal Gait" on p. 96). The pelvis rotates as much as 20° in the transverse plane and lists up to 10° in the frontal plane. The tibia also rotates about its long axis, courtesy of the subtalar joint at the ankle. Pelvic rotation and pelvic list reduce the vertical displacement of the centre of mass and therefore lower the metabolic cost of transport in each step. Ankle inversion–eversion also keeps the metabolic cost of walking low by allowing the centre of mass to pass closer to the hip of the supporting leg in midstance.

Muscle Actions During Normal Gait

Kinematic and ground force measurements may be used to calculate the net muscle moments exerted about the joints in gait. This procedure is known as the *inverse dynamics* method. The pattern of muscle moments for normal walking is illustrated in figure 8.5. Shortly after initial contact, the ankle dorsiflexors exert a small moment to decelerate the foot and prevent it from slapping the ground (figure 8.5, Ankle; 0-5% of the gait cycle). Once the foot is placed flat on the ground, the quadriceps contract eccentrically and exert an extension moment at the knee, which peaks just after the opposite leg leaves the ground (figure 8.5, Knee; 5-20% of the gait cycle). Thus, the main function of the quadriceps in early single-leg stance is to limit stance knee flexion (i.e., accelerate the knee into extension).

Knee extension in early stance (loading response) is also brought about by the action of the ankle plantar flexors. The soleus muscle exerts a plantar flexor moment at the ankle that restrains forward rotation of the tibia and simultaneously extends the knee. This plantar flexor moment increases through midstance and peaks during terminal stance, just before the opposite leg makes contact with the ground (figure 8.5, Ankle; 10-50% of the gait cycle). The peak moment created by the ankle plantar flexors in terminal stance is often referred to as push-off. Subsequent to opposite heel strike, the plantar flexors (with assistance from the hip flexors, iliopsoas) continue to lift and accelerate the leg forward (figure 8.5, Hip; 50-60% of the gait cycle).

Perhaps the most distinguishing feature of the swing phase is the lack of muscle activity in the lower limb (compare the joint moments in figure 8.5 at the ankle, knee, and hip during stance and

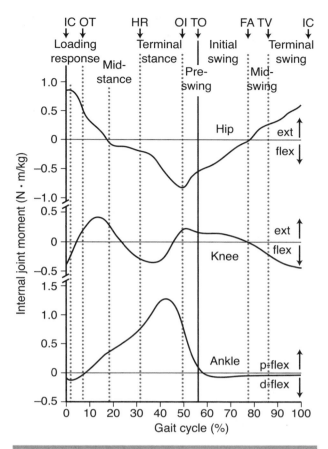

Figure 8.5 Mean sagittal plane net muscle moments applied at the lower limb joints during one cycle of normal walking. Data are shown for the right hip (extensor moment positive), right knee (extensor moment positive), and right ankle (dorsiflexor moment positive). The gait cycle events are shown at the top of the figure. IC = initial contact; OT = toe-off of opposite leg; HR = heel rise; OI = initial contact of opposite leg; TO = toe-off; FA = feet adjacent; TV = tibia vertical.

Reprinted from Whittle 1996.

swing). Very little muscle activity is needed in swing because the leg behaves very much like a double pendulum, swinging freely through the air under the action of gravity and inertia alone. Only small moments must be exerted at the hip (extensor) and knee (flexor) to decelerate the leg in preparation for heel strike.

Breakthroughs in computer processing power are also allowing researchers to use sophisticated computer models of the body to learn more about function of individual muscles during gait. As described in chapters 6 and 7, because models can be used to estimate muscle forces during activity,

There is great range in the complexity of computer models used to study human gait. The simplest of all models is the inverted single pendulum, which has been used to explain the changes in kinetic and potential energy that occur when humans walk at their preferred speeds. At the other end of the spectrum, models of exceeding complexity have been built to provide more knowledge about how muscles coordinate movement of the body segments during the gait cycle.

One of the most complex models ever used to simulate the biomechanics of walking is shown in figure 8.6. The model is able to simulate the three-dimensional movements of the body parts during walking because it includes all six major determinants of normal gait: hip flexion, stance knee flexion, ankle plantar flexion, pelvic rotation, pelvic list, and lateral pelvic displacement (see "The Major Determinants of Normal Gait" on p. 96). The skeleton is represented as a 10-segment, 23 degree-of-freedom mechanical linkage. The first six degrees of freedom define the position and orientation of the pelvis relative to the ground. (The lines with arrows represent joint axes of rotation in figure 8.6. The numbers appearing beside the arrows describe the sequence of body movements or degrees of freedom defined in the model. Thus, #1-6 define the six degrees of freedom (three translations and three rotations) of the pelvis in the model.) The head, arms, and torso are lumped together and represented as a single rigid body, which articulates with the pelvis by means of a three degree-of-freedom ball-and-socket joint located roughly at the level of the third lumbar vertebra (#7-9). Each hip is modelled as a ball-and-socket joint having three degrees of freedom: hip flexion (#10 and #17), abduction–adduction (which permits pelvic list) (#11 and #18), and internal–external rotation (which permits pelvic rotation) (#12 and #19). Each knee is modelled as a single degree-of-freedom hinge joint, allowing only flexion and extension (#13 and #20). Each ankle subtalar complex is represented as a universal joint with two degrees of freedom: ankle plantar flexion (#14 and #21) and subtalar inversion–eversion (#15 and #22). Each foot is represented by two segments in the model: a

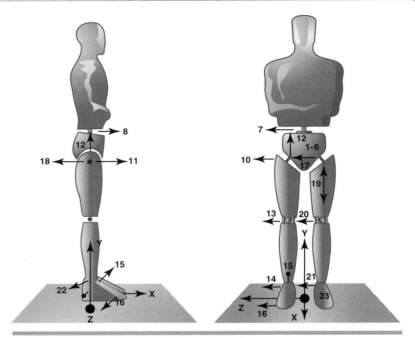

Figure 8.6 Schematic diagram of the three-dimensional model used to simulate one full cycle of normal walking.
Reprinted from Anderson and Pandy 2001.

hindfoot and a toes segment, hinged together by a single degree-of-freedom metatarsal joint (#16 and #23). Five springs with damping are placed under the sole of each foot to simulate the interaction of the foot with the ground.

Relative movements of the body segments in the model are controlled by 54 muscles: 24 muscles per leg plus 6 abdomen and back muscles. Each muscle is given realistic force–length and force–velocity properties, as well as series and parallel elastic elements with active and passive stiffness properties (see "Modelling the Mechanics of Muscle Contraction" on p. 86 in chapter 7). Tendon is modelled as an elastic material. Ligament action is included in the model in order to prevent the joints from hyperextending during a simulation. The delay between the incoming neural excitation and muscle activation (force) is also taken into account.

To calculate muscle forces during walking, the cost of transport during walking (i.e., metabolic energy per unit distance moved) was minimized. Metabolic energy was estimated by summing the heat liberated during muscle contraction and the mechanical work done by the muscles to move the joints (see chapter 7). One full cycle of walking (i.e., single- and double-leg support) was simulated by solving a dynamic optimisation problem using high-performance parallel computing. The joint angles, ground forces, and muscle activation patterns predicted by the dynamic optimisation solution were similar to measurements obtained from gait analysis experiments performed on five healthy young adult male subjects.

A general principle that has emerged from analyses of **computer simulations** of movement is that muscles can accelerate joints they do not span. In other words, a muscle can contribute to the acceleration of a body segment without actually touching it. The physical explanation is as follows. When a muscle touches a body segment, it can apply a force to that segment, which is then transmitted up and down the multilink chain via the contact forces acting at the joints. Thus, a muscle like vasti, which originates on the femur and inserts on the tibia, can accelerate not only the femur and tibia, but also the trunk and the foot, because vasti induces contact forces at all of the joints (i.e., ankle, knee, and hip) simultaneously.

This *principle of muscle-induced accelerations* is nicely illustrated by the behaviour of the soleus muscle during the midstance phase of the gait cycle. As mentioned previously, soleus is activated in midstance to restrain forward rotation of the tibia about the ankle. As this occurs, however, soleus also accelerates the knee into extension with great vigour. Thus, soleus, which crosses only the ankle joint, may accelerate the knee into extension as much as, if not more than, it accelerates the ankle into plantar flexion.

The principle of muscle-induced accelerations was recently used to show how muscles provide support to the centre of mass (i.e., contribute to the pattern of vertical force exerted on the ground) during normal walking. From foot-flat to just after opposite toe-off, the model gluteus maximus (hip extensor), vasti (knee extensor), and gluteus medius/minimus (hip abductors) contribute most significantly to the vertical ground force (figure 8.7a). These muscles are responsible for the first peak seen in the vertical ground force during early single-leg stance. Gravity and gluteus medius/minimus contribute nearly all of the support needed during midstance (figure 8.7b) while the ankle plantar flexors, soleus and gastrocnemius, are responsible for the second peak just prior to opposite heel strike (figure 8.7c).

Note that figure 8.7 also explains how specific leg muscles contribute to the vertical acceleration of the centre of mass during the stance phase of walking. The reason is that the

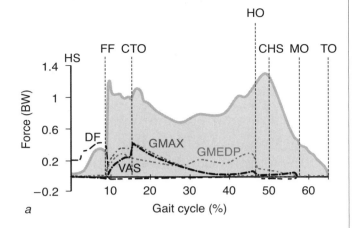

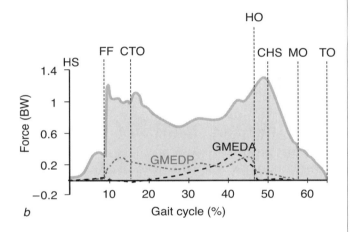

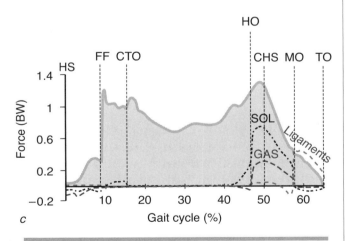

Figure 8.7 Major contributions made by the leg muscles during normal walking. The results were obtained from the model illustrated in figure 8.6.

Reprinted from Anderson and Pandy 2003.

(continued)

**IN FOCUS: USING COMPUTER MODELS TO STUDY
MUSCLE FUNCTION IN GAIT** *(continued)*

vertical ground force is directly proportional to the vertical acceleration of the centre of mass. Specifically, the vertical ground force is equal to mass of the body multiplied by the vertical acceleration of the centre of mass, plus the weight of the body, which is a constant. The reason the model could provide such insight is that the model simulations were able to predict how much force each muscle develops at each instant during the gait cycle. Such information cannot be obtained at present from gait analysis experiments alone.

SOURCES

- Anderson, F.C., and M.G. Pandy. 2001. Dynamic optimization of human walking. *Journal of Biomechanical Engineering* 123: 381-390.
- Anderson, F.C., and M.G. Pandy. 2003. Individual muscle contributions to support in normal walking. *Gait and Posture* 17: 159-169.
- Pandy, M.G. 2001. Computer modeling and simulation of human movement. *Annual Review of Biomedical Engineering* 3: 245-273.

analysis of the model simulations provides understanding of muscle function at a much deeper level than is presently possible with gait analysis experiments alone (see "Using Computer Models to Study Muscle Function in Gait" on p. 98).

Gait Development in Children

Gait matures in a relatively short period of time. Most children begin walking within three months of their first birthday. Heel strike and reciprocal swinging of the extremities are usually established by the age of 1.5 years. The kinematic, kinetic, and muscle EMG patterns observed in adults are fully developed in children by the age of 7 years.

Understanding the mechanics of walking in children is hampered by the fact that gait experiments are often difficult to perform in the very young. Measurements of ground forces are difficult to obtain in children under 2 years of age because these children take such small steps and usually contact the same force plate with successive strides. Muscle EMG data are also difficult to obtain for ethical reasons. Fine-wire electrodes, which are needed to access deep-lying muscles of the leg and abdomen, have yet to be used on very young children.

Nonetheless, a clear picture of at least the five major determinants of mature gait has emerged (contrast these with the six major determinants of normal gait described in "The Major Determinants of Normal Gait" on p. 96).

- The first determinant, duration of single-leg stance, increases steadily from about 32% in 1-year-olds to about 38% in 7-year-olds, with the most rapid increase occurring between the ages of 1 and

2.5 years. The duration of the single-support phase during normal walking in adults is typically 39%.

- The second determinant, walking speed, increases steadily from 0.64 m/s at age 1 year to 1.14 m/s at age 7 years. The preferred speed of walking in 7-year-olds is still significantly lower than the average walking speed of 1.34 m/s recorded in adults.

- The third determinant, cadence (i.e., the rate at which the legs swing back and forth), decreases steadily with age. Small children walk with a fairly high cadence (short gait cycle time), with a mean at age 1 year of approximately 176 steps/min. At age 7 years, cadence decreases to around 145 steps/min, which is still well over the average of 113 steps/min measured in adults.

- The fourth determinant, step length, increases almost linearly with age between 1 and 4 years and thereafter continues to increase, but at a slower rate. Step length in 1-year-olds is typically around 0.2 m (0.7 ft), whereas 7-year-olds take steps that are as much as 2.5 times bigger (approximately 0.5 m or 1.6 ft). The mean step length during normal walking in adults is around 0.7 m (2.3 ft).

- The fifth determinant, **step width,** decreases with age. In figure 8.8, step width is expressed as pelvic width divided by ankle spread, with ankle spread being the distance between the two ankles when both legs are in contact with the ground. Thus, as age increases, step width decreases, so the ratio of pelvic span to ankle spread increases as shown. The increase in pelvic span/ankle spread is most pronounced between the ages of 1 and 3.5 years. The support base is roughly 70% of pelvic width in 1-year-olds, falling to 45% in 3.5-year-

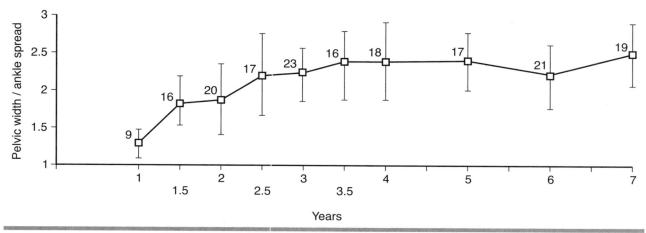

Figure 8.8 Change in the ratio of pelvic width to ankle spread with age in young children. As pelvic width/ankle spread increases, the support base (i.e., step width) decreases because ankle spread decreases. The numbers next to the points on the curve are the numbers for each age group. The vertical bars are plus and minus one standard deviation.
Reprinted from Sutherland, Olshen, Cooper, and Woo 1980.

olds. After the age of 3.5 years, the ratio of pelvic span to ankle spread remains more or less constant until the age of 7 years. The average value in adults walking at their preferred speeds is around 30%.

Thus, very young children (i.e., 1-year-olds) walk with high cadence and take short steps to move their centres of mass at relatively slow speeds. The fact that children take shorter steps than adults during independent walking is only partly explained by the fact that they have much shorter legs. Indeed, step length increases linearly with leg length between the ages of 1 and 7 years. The immature child, however, takes shorter steps than his or her leg length would allow. In connection with this, young children also spend less time supporting their centres of mass with one leg (i.e., the swing phase occupies a smaller proportion of the gait cycle in children than in adults; see figure 8.9).

There are three possible reasons for the lesser step length and duration of single-leg stance in young children compared to adults. First, balancing skills are not fully developed, which poses a threat to stability. Second, intermuscular control (i.e., muscle coordination) has yet to be learned. Third, the ankle plantar flexors are still too weak to provide the necessary thrust to lift and accelerate the body forward in late stance. With maturation and growth, step length and walking speed increase, while cadence decreases; these three determinants reach normal adult values only around the age of 15 years.

Other important kinematic and kinetic features that distinguish walking in the very young from normal walking in adults are as follows:

- Very young children (1-year-olds) have no heel strike (i.e., initial contact with the ground is made with the entire foot).

- The degree of hip extension is reduced, and the hip does not remain flexed for as long as it does during the swing phase of normal walking (see figure 8.9).

- Young children do not extend their knees very much during single-leg stance (figure 8.9).

- Children keep their legs externally rotated at all times during the gait cycle.

- Reciprocal arm swing is absent from the gait patterns of the very young.

One-year-olds show some increase in knee flexion at initial contact but very little knee extension during midstance compared with adults. This behaviour is thought to be a result of inadequate ankle plantar flexor muscle force. During normal walking in adults, knee extension in stance is brought about by the action of the ankle plantar flexors combined with some amount of quadriceps force (as described in the previous section). In particular, the ankle plantar flexors act to restrain forward rotation of the tibia during loading response and midstance. Because the plantar flexors are activated less during stance

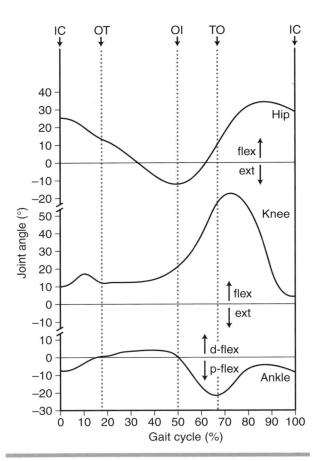

Figure 8.9 Mean sagittal plane angles measured for the lower limb joints in young children. Data are plotted for the right hip (flexion positive), right knee (flexion positive), and right ankle (dorsiflexion positive). The gait cycle events are shown at the top of the figure. IC = initial contact; OT = toe-off of opposite leg; OI = initial contact of opposite leg; TO = toe-off.
Reprinted from Whittle 1996.

in 1-year-olds than in adults, excessive forward rotation of the tibia results, leading to an increase in knee flexion. To compensate, the quadriceps are activated for a longer period of time during stance phase in 1-year-olds.

Gait Changes in Older Adults

Gait kinematics are different in healthy older adults compared to healthy young adults when subjects walk at their preferred speeds. Muscle strength and balance seem to play a major role in these changes, and the metabolic costs of walking are affected.

Changes in Gait Kinematics

Cross-sectional studies have shown that preferred and maximal walking speeds remain constant until the seventh decade (60-70 years); thereafter, preferred walking speed declines at a rate of 12% to 16% per decade, while maximal walking speed decreases at a rate of 20%.

Healthy older persons walk more slowly than healthy young adults because they take shorter steps, not because they move their legs at a slower rate (i.e., by decreasing cadence). Many studies have shown that healthy elderly adults have a reduced step length and therefore spend less time with one foot in contact with the ground (i.e., single-support time is shorter). This also means that older persons spend more time with both legs on the ground (i.e., double-support time is longer) compared to their younger counterparts. For example, stance time has been found to increase from 59% in 20-year-old men to 63% in 70-year-old men. The concomitant increase in double-support time is 18% to 26% from the healthy young to the healthy old. Cadence, however, remains nearly unchanged with advancing age.

Changes also occur in gait kinetics as age increases beyond the seventh decade. The most consistent finding reported in the gait literature is a decrease in ankle plantar flexor function, specifically joint torque and power developed during the push-off phase. Other changes related to muscle function have also been reported—for example, an increase in hip extensor and hip flexor torque and power, as well as a decrease in knee extensor torque and power. These changes, though, appear to be more variable and may be a function of the design of the experimental studies reported (e.g., the pool of participants recruited for the individual studies and the conditions under which the studies were conducted, including whether or not walking speed was controlled). Healthy older adults also exert smaller forces on the ground when they walk. Specifically, peak anterior–posterior forces are smaller, as is the peak vertical force exerted on the ground during single-limb stance. Smaller ground reaction forces are consistent with the knowledge that older adults prefer to walk more slowly.

Causes of Gait Changes in Aging

Why do older people walk more slowly than younger people? The reasons are not fully clear, but muscle weakness and impaired balance seem to be the prime suspects underlying the decline in walking speed with age. Qualitatively, balance is the ability to control the position of the centre of mass during movement, so a reduction in step length (and therefore walking speed) may be an appropriate strategy for increasing stability by reducing the duration of single-limb support. Weaker muscles develop less force, so less power is available for support and propulsion. Weakness of the hip extensor muscles may reduce joint power early in the stance phase, leading to a decrease in walking speed. Walking speed may also be compromised by weak ankle plantar flexors, as these muscles are responsible for providing a large thrust, both upward and forward, in terminal stance (see "Using Computer Models to Study Muscle Function in Gait" on p. 98). Thus, weak ankle plantar flexors (specifically soleus and gastrocnemius) may reduce the power delivered to the skeleton during push-off, causing older adults to walk more slowly by taking smaller steps.

In one study, three-dimensional biomechanical analyses were performed on 26 older adults (average age of 79) and 32 young adults (average age of 26). All participants walked at their preferred cadence and step length (i.e., their natural walking speed). The older subjects chose step lengths 10% shorter than those of the young subjects mainly because they had significantly reduced ankle plantar flexion in terminal stance (i.e., just prior to opposite heel strike). Furthermore, the older subjects exerted less plantar flexor torque and developed less power at the ankle during terminal stance than the younger subjects. The older subjects also developed more flexor power at the hip in terminal stance. These results suggest an adaptation in older adults that manifests itself as a redistribution of motor patterns in the lower limb during gait.

The ankle plantar flexors are believed to have three roles in normal walking, but precisely what these muscles do during the gait cycle remains unknown. Most researchers would agree that the major function of the ankle plantar flexors in single-limb stance is to contract eccentrically (i.e., lengthen) in order to control forward rotation of the tibia during midstance. Because the plantar flexors are almost fully activated and contract concentrically (i.e., shorten) in terminal stance, just prior to opposite heel strike, some scientists believe that their role in this portion of the gait cycle is to lift and accelerate the centre of mass forward (see "How Much Force Do the Ankle Muscles Develop in Normal Walking" on p. 67 in chapter 6). However, some researchers have argued that the ankle plantar flexors work with the hip flexors in terminal stance, immediately after opposite heel strike, to accelerate the leg forward and upward in preparation for swing. So, although the function of these muscles appears to be well defined in early stance, their role in late stance is open to

question. If healthy elderly adults use their hip flexor muscles more in order to compensate for a reduction in the power developed by their ankle plantar flexors, then this suggests that the main role of the ankle plantar flexors in late stance is to lift and accelerate the leg forward in preparation for swing, rather than to propel the centre of mass forward just prior to opposite heel strike.

Metabolic Cost of Walking in Older Adults

Recall from chapter 7 that when people are permitted to choose their cadence and step length, they select a walking speed that minimizes metabolic energy consumed per unit distance traveled. This most economical speed is approximately 1.3 m/s in healthy young adults (see figure 7.12). As the speed of walking deviates from this value, the cost of transport (i.e., metabolic energy consumed per unit distance traveled) increases. If healthy elderly persons walk more slowly than their younger counterparts, does this imply that they also consume less metabolic energy to move their centres of mass a unit distance (one meter)? The answer is no; in assuming lower preferred walking speeds, healthy older adults actually show higher aerobic demands per unit distance walked than healthy young adults.

In one study, oxygen consumption was measured as 30 healthy young adults ranging in age from 18 to 28 years, and 30 healthy older adults ranging in age from 66 to 86 years, walked at various speeds (including their preferred speed) on a treadmill. Starting with the treadmill at low speed, walking speed was slowly increased until each person subjectively identified the speed that was most comfortable. Both the young and older adults showed the familiar U-shaped speed–energy curve when metabolic energy consumption was normalized by distance walked (figure 8.10). The minimum cost of transport occurred at the same speed (1.34 m/s) in both groups. This minimum, however, was significantly higher (by 7%) in the older adults than in the younger ones. These two results show that a single speed–energy curve like that of figure 7.12 does not describe the metabolic cost of walking in healthy young and older adults, and that the effect of aging is represented by a vertical shift upward in the speed–energy curve.

What explains the age-related difference in the relationship between metabolic cost and walking speed shown in figure 8.10? The answer appears to be the higher cost of generating muscle force in older people. It is well known that both muscle

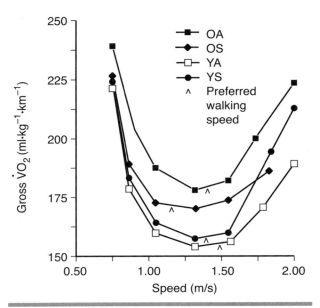

Figure 8.10 Gross oxygen consumption normalized to body mass plotted against walking speed for young and older adults. The preferred walking speed for each group is indicated. OA = old active; OS = old sedentary; YA = young active; YS = young sedentary.

Reprinted from Martin, Rothstein, and Larish 1992.

mass and muscle strength decrease beginning in the fourth decade. There is also evidence from studies in mice that the force generated per unit cross-sectional area of muscle is roughly 20% lower in old muscle than in young muscle. It would appear, therefore, that healthy elderly adults need to recruit a greater number of motor units per muscle in order to generate the muscle force needed for a given motor task. It may also be that people who are elderly need to recruit a higher number of fast-twitch fibres to move their centre of mass at the required rate; and fast-twitch fibres are known to be less economical in terms of energy consumption than slow and intermediate muscle fibre types. Thus, to walk at their preferred speed, older adults must use more oxygen, which implies that metabolic energy consumption is not the criterion optimised during gait.

Exercise and Aging

There is a strong correlation between walking speed and level of mobility. Elderly persons who walk very slowly and who have poor balance are prone to falling. Indeed, very slow walkers are usually homebound. Normal walking requires adequate levels of muscle strength as well as adequate

neuromuscular control and joint proprioception. As noted earlier, walking speed begins to decline in the seventh decade; and this is usually a sign of deterioration in strength, control, proprioceptive abilities, or a combination of these. Interventions such as resistance strength training may reverse, or at least retard, the decline in walking speed with age and may therefore decrease the risk of falling.

Various combinations of strength and **neuromuscular training** have been tried, and their effects on walking speed are well documented. In one study, 34 subjects who were 75 years and older underwent an exercise protocol consisting of flexibility, balance, and resistance strength training three times a week for 12 weeks. Flexibility consisted of stretching exercises performed while sitting, followed by calf and hamstrings stretches and finally spinal extension exercises. Resistance training included isometric hip abduction and knee extension exercises performed to fatigue, with resistance set at roughly 80% of maximum voluntary muscle contraction. Balance exercises included shifting weight from one foot to the other in the medial–lateral and anterior–posterior directions, as well as performing simple tai chi movements for 5 to 10 min each day.

Walking speed and step length both correlated with knee extensor (quadriceps) muscle strength in healthy older adults. The results show that below a knee extensor moment of roughly 50 N·m, walking speed and step length both decrease with progressive quadriceps weakness (figure 8.11). Above this limit, the correlation is poorer (i.e., the slopes of the lines in figure 8.11 are lower beyond 50 N·m), implying that the limit of walking speed is then more closely related to the strengths of the other leg muscles.

Significant improvements in both muscle strength and walking speed were obtained from the training program. Isometric knee extension strength increased by more than 30%, and walking speed increased by 8%. The improvement in walking speed, in particular, is clinically significant in view of the fact that gait velocity declines by roughly 15% per decade after the age of 60 years. These results are not universal, however, as others have found no improvement in walking speed when subjects aged 60 to 71 years undertook moderate resistance and balance exercises. The difference in these findings may be explained by differences in subject characteristics. Subjects who did not improve their walking speed were much stronger to begin with than those who did. Thus, resistance and balance training improved muscle strength in these subjects by less than 10%, which

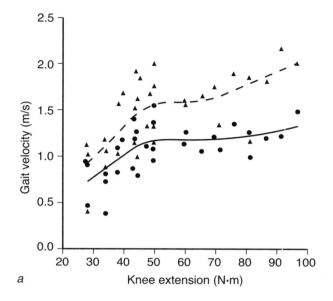

a

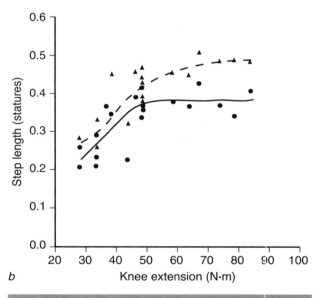

b

may explain the imperceptible change in walking speed. This in turn implies that walking speed improvements are more likely to occur in older adults who have significant leg muscle weakness (i.e., the frail elderly).

Summary

Normal gait is characterized by six kinematic mechanisms known as the major determinants of gait: hip flexion, stance knee flexion, ankle plantar flexion, pelvic rotation, pelvic list, and lateral pelvic displacement. These mechanisms explain how movements of the lower limb joints contribute to the sinusoidal displacement of the centre of mass when humans walk at their preferred speeds. Very young children (below the age of 7) and healthy older adults walk differently than healthy young adults. The development of mature walking is characterized by changes in the duration of single-leg stance, walking speed, cadence, step length, and step width. The durations of single-leg stance, step length, and walking speed are all lower in very young children than in healthy young adults; cadence and step width are both higher. The trends are similar in older adults. Step length, walking speed, and the duration of single-leg stance all decrease with advancing age, while step width increases to widen the base of support. Regular intervals of strength and neuromuscular training may help to maintain walking speed in older adults, but the effects on step length are currently unknown.

Figure 8.11 *(a)* Relationship between walking speed and maximum isometric knee extensor moment (quadriceps strength) in older adults. The solid line and circles are data for subjects walking at their preferred speeds. The dashed line and triangles are data for subjects walking at much faster than their preferred speed. *(b)* Relationship between step length (normalized by height in statures) and maximum isometric knee extensor moment (quadriceps strength) in older adults. The circles are data for subjects walking at their preferred speeds. The triangles are data for subjects walking at much faster than their preferred speed. The solid and dashed lines are the lines of best fit through the respective data sets.

Reprinted from Judge, Underwood, and Gennosa 1993.

Further Reading and References

Malina, R.M., and C. Bouchard. 1991. *Growth, maturation, and physical activity.* Champaign, IL: Human Kinetics.

Rogers, M.A., and W.J. Evans. 1993. Changes in skeletal muscle with aging: Effects of exercise training. In *Exercise and Sport Science Reviews*, ed. J.O. Holloszy (pp. 65-102). Baltimore: Williams & Wilkins.

Rose, J., and J.G. Gamble. 1994. *Human walking.* 2nd ed. Baltimore: Williams & Wilkins.

Schultz, A.B. 1992. Mobility impairment in the elderly: Challenges for biomechanics research. *Journal of Biomechanics* 25: 519-528.

Sutherland, D.H., R. Olshen, E.N. Biden, and M.P. Wyatt. 1988. *The development of mature walking.* London: MacKeith Press.

Whittle, M.W. 1996. *Gait analysis: An introduction.* Oxford: Butterworth-Heinemann.

CHAPTER 9

Biomechanical Adaptations to Training

The major learning concepts in this chapter relate to

- the effects of training on muscle properties, particularly muscle strength and muscle contraction speed;
- how training can change neuromuscular control of movement, specifically muscle coordination;
- how neuromuscular control and therefore muscle function can change in response to injury; and
- the effects of muscle strength and muscle contraction speed on motor performance.

*T*he use of training methods to improve muscular function is widespread. Specialized training programs are designed not only to improve performance but also to prevent injuries. Training can change both the contractile properties of muscle (i.e., strength and contraction speed) and the ability of the nervous system to control muscular function (i.e., coordination). Resistance training increases a person's ability to produce maximum muscle force, while plyometric training, in the form of dynamic depth jumps, can enhance the ability to rapidly develop force. Plyometric training is also used as an adjunct in jump-training programs to improve neuromuscular and proprioceptive control. This chapter describes how the neuromuscular and musculoskeletal systems respond to various training methods instituted for the purpose of improving performance and decreasing the incidence of injuries in sport. Experimental and theoretical results are also presented to show how biomechanical performance may be altered through manipulation of changes in strength, speed, and control.

Muscular Adaptations to Training

Training can change both muscle strength and muscle contraction speed. Strength may be increased through an increase in the net neural drive to the muscle, an increase in muscle size, or both. Muscle contraction speed may be altered through change in the shape of a muscle's force–velocity curve, change in the value of its intrinsic maximum shortening velocity, or both.

Two of the most common methods of increasing muscle strength are isometric and isokinetic training. Both types of exercises are performed on a device called a dynamometer. A dynamometer is an electromechanical machine containing a speed-controlling mechanism that accelerates to a preset speed when an external force is applied by a body part to the machine's arm. Once the preset speed has been reached, a loading mechanism creates an equal and opposite force that balances the applied force (see figure 6.7a).

Performance of an isometric contraction provides a measure of the muscle's maximum strength. In isometric knee extension, for example, the participant is asked to maximally activate the quadriceps while the machine arm is prevented from rotating. The maximum force produced by the muscle depends on the length at which the muscle is held. Thus, one can obtain the muscle's maximum isometric force–length curve by measuring the muscle's maximum isometric force at various flexion angles of the knee (see figure 6.9).

Early in a training program (two to eight weeks), strength gains are more attributable to neural adaptations than to increases in muscle size. The reason is that the ability to activate all of the available motor units is enhanced in the early phases of training, when subjects are still learning how

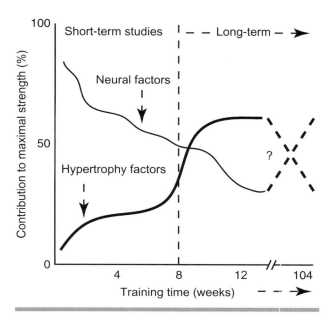

Figure 9.1 Contributions of neural factors and hypertrophy factors to maximal muscle strength plotted against training time.
Reprinted from Kraemer, Fleck, and Evans 1996.

to exert force effectively. As shown in figure 9.1, a dramatic increase in adaptation of neural factors occurs over the first 6 to 10 weeks of a resistance training program. However, as the duration of training extends beyond 10 weeks, muscle hypertrophy begins to contribute more than neural adaptations to the strength gains. Eventually, muscle hypertrophy reaches a plateau, and additional exercise produces no further change in maximum muscle strength.

Interestingly, strength training can also change the rate at which muscles shorten and lengthen over time (i.e., contraction speed). Isokinetic knee extension training performed at relatively high speed has been shown to change the shape of the force–velocity curve for quadriceps, with relatively higher quadriceps force obtained at high velocities after training for four weeks. Training can also increase a muscle's maximum shortening velocity. The maximum shortening velocity of the thumb muscle, adductor pollicis, was found to increase by 21% after a 10-day training program consisting mainly of isokinetic exercise. Increasing a muscle's maximum shortening velocity (v_{max}) increases the power developed by the muscle, because increasing v_{max} increases the area under the muscle's force–velocity curve. Changing the shape of the muscle's force–velocity curve by increasing the force that the muscle develops at high veloci-

ties also increases the power developed by the muscle, because this too increases the area under the muscle's force–velocity curve (see "At What Speed Must a Muscle Shorten to Develop Maximum Power?" on p. 108).

Neuromuscular Adaptations to Training

Besides increasing the neural drive to muscle, training can also change the way muscle action is coordinated during a motor task. Current interest in developing training programs to improve neuromuscular control is driven by the need to reduce muscle, ligament, and joint injuries in sport.

Training to Prevent Anterior Cruciate Ligament Injury

One particularly important example relates to the disproportionate number of anterior cruciate ligament (ACL) injuries sustained by female athletes in sports characterized by running, jumping, and cutting manoeuvres. In fact, the incidence of serious knee injury is reported to be approximately six times higher in female athletes than in their male counterparts. Many factors, both intrinsic and extrinsic, have been identified as predictors of ACL injuries in the female athlete. Intrinsic risk factors include lower extremity malalignment, decreased intercondylar notch width at the knee, increased knee joint laxity, and hormonal influences. Extrinsic risk factors include imbalance in quadriceps and hamstring muscle strength and inadequate neuromuscular control. While sports medicine research has yet to isolate the cause of ACL injury in females, many studies have shown that the incidence of ACL injuries is reduced through participation in training programs that focus on teaching neuromuscular control.

Neuromuscular training programs designed to prevent ACL injury are based on the dictum that knee stability and function are improved when postural equilibrium, intermuscular control, and leg muscle strength are all enhanced. One example is the jump-training program developed at the Sports Medicine and Orthopaedic Center in Cincinnati, Ohio. This program emphasizes improving both muscle strength and neuromuscular control by incorporating stretching, plyometric exercises, and weightlifting. The overall goals are to decrease landing forces by teaching neuromuscular control of the leg muscles during landing, and to increase strength of the leg musculature through weight

IN FOCUS: AT WHAT SPEED MUST A MUSCLE SHORTEN TO DEVELOP MAXIMUM POWER?

Maximum-height jumping, throwing, and pedaling are all activities in which performance is determined by the ability to generate maximum muscle power. The speed at which a muscle develops maximum power is not immediately obvious because this value is determined by the shape of the muscle's force–velocity curve, which is nonlinear. Indeed, it is not an easy feat to deduce this property by experiment.

The procedure involves a quick-release experiment performed using the muscle–lever system shown in figure 9.2. The muscle is taken out of the body soon after death and immersed in a salt bath to ensure that the contractile machinery is still able to function. Tendon is usually separated from the musculotendinous preparation because the **compliance** of the tendon will affect the force developed by the muscle. One end of the muscle belly is clamped to the workbench (ground), and the other end is attached to the lever as shown. A weight is also hung from one end of the lever as indicated. An electromagnetic device is used to prevent the lever from rotating as the muscle is initially stimulated to its maximal force. At some point in time, the electromagnet is released, and the lever is then free to rotate. The direction the lever rotates in depends on the relative magnitudes of the forces applied by the hanging weight and the pull of the muscle. If the hanging weight is greater than the maximum isometric force developed by the muscle (for the length at which the muscle is contracting), the lever will rotate counterclockwise as shown in figure 9.2 and the muscle will lengthen (i.e., contract eccentrically). Conversely, if the isometric force developed by the muscle is greater than the force exerted by the hanging weight, the lever will rotate clockwise and the muscle will shorten (i.e., contract concentrically). In either case, the muscle undergoes an **isotonic** (constant force) **contraction,** as the hanging weight is the only external force applied.

Unfortunately, the results of the quick-release experiment tend not to be dependable when muscle is forcibly stretched. As actomyosin bonds are broken, the muscle incurs some damage during stretching. If the hanging weight is made

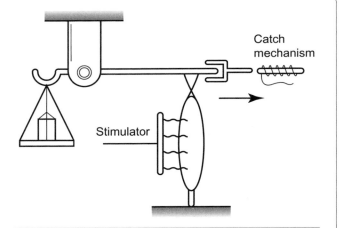

Figure 9.2 Schematic diagram of instrumentation used to estimate a muscle's force–velocity curve in a quick-release experiment. When the electromagnetic catch mechanism on the right is released, the muscle shortens or lengthens depending on whether the weight in the pan on the left is lighter or heavier than the isometric force developed by the muscle.

Reprinted from McMahon 1984.

large enough, the muscle will eventually yield as it is stretched; this phenomenon is indicated by the somewhat flat line obtained in the lengthening region of the force–velocity curve shown in figure 9.3 (solid line). Note that the muscle develops more force when it is stretched (i.e., in lengthening). Note also that the slope of the curve is discontinuous in the transition between shortening and lengthening, the slope in lengthening being roughly four times greater (solid line).

The experiment works well, however, as long as the muscle is permitted to shorten against the externally applied load. For each weight, the shortening velocity of the muscle can be measured from its length trajectory in time. Specifically, the length change of the muscle is recorded as it shortens against the force of the hanging weight. The slope of the length–time trajectory is calculated at a point when the muscle length change is no longer instantaneous. (As soon as the electromagnet is released, the muscle undergoes an instantaneous change

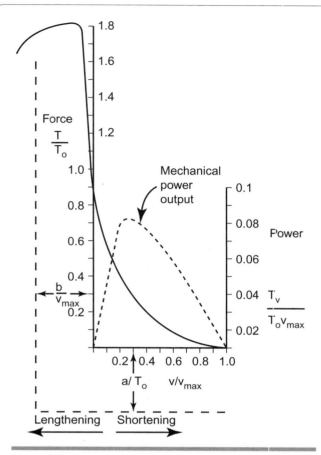

Figure 9.3 Force–velocity curve for muscle as fitted by the hyperbola given in equation 9.1.1 (solid line). The dashed line is the corresponding power developed by the muscle in the shortening region. Peak power occurs at about 1/3 of the muscle's maximum shortening velocity (see text).

Reprinted from McMahon 1984.

in length, which is due to the stiffness of the cross-bridges—the so-called series elastic property of muscle; see figure 7.6.) The slope of the length–time trajectory is taken as the shortening velocity of the muscle for the load in question. Thus, different muscle shortening velocities can be found, depending on the amount of weight placed in the pan.

The solid line in figure 9.3 was derived by fitting a curve to the data points obtained from such a series of quick-release experiments (first reported by A.V. Hill in 1938). The relation describing this

curve is a hyperbola and is known as Hill's force–velocity equation. The equation takes the form

$$(F^M + a)(v^M + b) = (F_o^M + a)b, \qquad 9.1.1$$

where F^M is the force developed by the muscle, F_o^M is the muscle's maximum isometric force at a given length, v^M is the shortening velocity of the muscle, and a and b are constants that vary with the temperature of the muscle. Interestingly, the shape of the force–velocity curve is the same for almost all muscles. The only exception appears to be insect flight muscle, for which the force–velocity data do not fit the curve predicted by equation 9.1.1.

One can find the maximum shortening velocity of the muscle, v_{max}^M, by extrapolating the experimental data back to the point where force is zero (i.e., the intercept of the velocity axis in figure 9.3). This defines the (maximum) velocity with which the muscle will shorten when no external load is applied. It is often regarded as a constant property of the muscle, although some evidence exists to suggest that the value of v_{max}^M depends on muscle length and muscle activation level. Nonetheless, maximum shortening velocity is a good indicator of the distribution of fibre type within a given muscle: Muscles with low maximum shortening velocities (e.g., $v_{max}^M = 3$ muscle lengths per second for human muscle) can be expected to be composed of predominantly slow-twitch, oxidative (type SO) fibres; those with intermediate values of v_{max}^M (i.e., $v_{max}^M = 10$ muscle lengths per second) will have mixed, fast-twitch, oxidative glycolytic (type FOG) fibres; and those with high values of v_{max}^M (i.e., $v_{max}^M = 20$ muscle lengths per second) will have predominantly fast-twitch, glycolytic (type FG) fibres.

Because power is the product of force and velocity, one can find muscle power by solving equation 9.1.1 for muscle force, F^M, and multiplying the result by the muscle's shortening velocity. Thus,

$$power = F^M v^M = \frac{v^M(b F_o^M - a v^M)}{(v^M + b)}. \qquad 9.1.2$$

The dashed line in figure 9.3 is obtained when equation 9.1.2 is plotted against muscle shortening velocity. Maximum muscle power can then be found using the rules of differentiation from

(continued)

IN FOCUS: AT WHAT SPEED MUST A MUSCLE SHORTEN TO DEVELOP MAXIMUM POWER? *(continued)*

calculus. Specifically, equation 9.1.2 is differentiated with respect to shortening velocity, and the result is then set equal to zero. Solving the resulting equation for shortening velocity gives the velocity at which the muscle develops maximum power. It can be shown that, in general, maximum muscle power occurs at roughly $0.3\, v_{max}^{M}$. That is, muscles must shorten at approximately 1/3 of their maximum shortening velocity if they are to develop maximum power in each stroke. This result is used to design the gearing on bikes so that maximum power may be delivered to

the crank at all times, especially during riding up and down hills.

SOURCES

- Hill, A.V. 1938. The heat of shortening and the dynamic constants of muscle. *Proceedings of the Royal Society (London), Series B* 126: 136-195.
- McMahon, T.A. 1984. *Muscles, reflexes, and locomotion.* Princeton, NJ: Princeton University Press.

---

training. The program has three distinct phases: stretching exercises, followed by jump training, and finally resistance strength training.

The stretching exercises focus on muscles of the leg, particularly soleus, quadriceps, and hamstrings, but attention is also paid to the muscles of the upper limb including latissimus dorsi, pectoralis major, and biceps. Weight-training exercises include abdominal curls, leg presses, calf raises, bench presses, and forearm curls. The jump-training protocol emphasizes proper jumping and landing techniques with special attention to body posture at ground contact, knee stability, and soft landings (i.e., toe-to-heel landings with the knees bent).

Evaluating the Effectiveness of Injury Prevention Training

Researchers evaluated the efficacy of this training program in relation to jumping performance and landing technique, as well as its effectiveness in decreasing the incidence of ACL injuries in female athletes. High school female volleyball players were trained for six weeks using the protocol just outlined. The training sessions, lasting approximately 2 h per day, took place three days per week on alternating days. The control group for the study included untrained male subjects matched for height, weight, and age. Peak landing forces resulting from a volleyball block jump decreased by 22%; knee abduction and adduction muscle moments decreased by approximately 50%; and hamstring-to-quadriceps muscle moment

ratio increased by as much as 26%. Also, the peak hamstring-to-quadriceps moment ratio was significantly higher in trained females compared to untrained female athletes but was similar to that measured in the male controls.

The results of prospective studies support the notion that neuromuscular training can reduce the incidence of knee injuries in sport. When the Cincinnati training protocol was implemented on 1,263 high school volleyball, soccer, and basketball players, untrained females demonstrated a knee injury rate 3.6 times higher than that for trained females and 4.8 times higher than that for untrained male controls. The incidence of knee injury in the untrained group was 0.43 per 1,000 player exposures compared to 0.12 in the trained group. The injury rate for the male controls was 0.09 per 1,000 exposures. In all, only 14 serious knee injuries were sustained among the 1,263 athletes who participated in the training program. The decreased injury rate in the trained athletes was attributed to an increase in dynamic stability of the knee after training. Knee joint stability was increased after training through correction of the imbalance between quadriceps and hamstring muscle strength and through increase in the ratio of hamstring-to-quadriceps strength in the female athletes.

This finding is important because it shows that the ratio of hamstring-to-quadriceps muscle strength is an important factor in preventing ACL injury. The ACL functions to prevent excessive anterior (forward) translation of the tibia relative to the femur during activity. When the quadriceps muscles are activated, they develop a force that is transmitted to the tibia by means of the extensor mechanism composed of the quadriceps tendon, the patella, and the patellar tendon. The force in the patellar tendon acts to pull the tibia forward because it points anteriorly relative to the long axis of the tibia. Thus, a force developed by the quadriceps muscles serves to translate the tibia anteriorly, which strains the ACL. Hamstring muscle action reduces ACL strain as these muscles pass posterior to the knee and insert on the back of the tibia. Thus, the hamstrings can protect the ACL by creating a backward pull on the tibia, thereby decreasing the force transmitted to the ligament (see "How Can Hamstring Muscle Contractions Protect the ACL From Injury?" on p. 112).

As an example, consider figure 9.4, which is a free-body diagram showing all the forces that typically act on the leg during activity. PT represents the resultant force applied by the patellar tendon; Ligament is the resultant force of the cruciate and collateral ligaments; Hams is the resultant force applied by all the hamstring muscles; Ground is the resultant ground reaction force; and TF is the resultant compressive force acting between the femur and tibia. Each force may be resolved into a shear component that acts in either the anterior or posterior direction, and an axial component that acts along the long axis of the leg. Anterior shear forces tend to move (translate) the tibia anteriorly (forward) relative to the femur, and therefore will strain the ACL. Posterior shear forces, on the other hand, tend to move the tibia backward relative to the femur, and so protect the ACL from strain; however, posterior shear forces may simultaneously strain the posterior cruciate ligament (PCL). Neuromuscular training can increase knee joint stability not only by increasing the ratio of hamstring-to-quadriceps muscle strength, but also by fine-tuning the control exerted over the hamstrings.

Biomechanical Adaptations to Injury

Adaptations are also common following injury and surgical treatment. One interesting example in the orthopaedics literature is a phenomenon known as "quadriceps-avoidance gait." This particular adaptation occurs in anterior cruciate ligament-deficient (ACLD) patients, that is, patients who have incurred a complete rupture of their ACL. These patients usually have difficulty with movements involving either lateral thrusts (e.g., cutting from side to side) or twisting (i.e., rotation of the femur relative to the tibia in the transverse plane).

Gait analysis has been done on ACLD patients during various activities of daily living, including level walking; jogging; walking up and down stairs; and even running, cutting, and pivoting. Kinematic and force-plate data were used to estimate the net muscle moments exerted at the knee in all three planes of movement: flexion–extension, abduction–adduction, and internal–external rotation. Interestingly, the greatest functional changes between ACLD patients and normal subjects occurred in normal walking on level ground. The ACLD patients exhibited a significantly lower than normal net extensor moment during the midstance

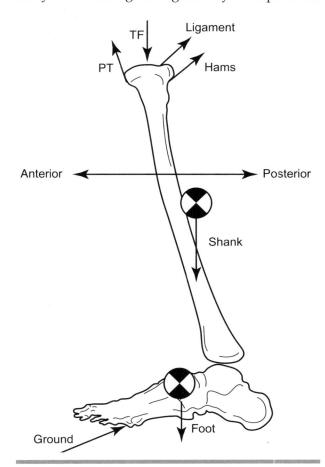

Figure 9.4 Forces that typically act on the lower leg during activity. See also "How Can Hamstring Muscle Contractions Protect the ACL From Injury?" on page 112.

IN FOCUS: HOW CAN HAMSTRING MUSCLE CONTRACTIONS PROTECT THE ACL FROM INJURY?

Rupture of the ACL is one of the most common and debilitating injuries sustained by athletes. Approximately one in every 1,000 people experiences an injury to their ACL. Deleterious effects resulting from ACL injury include degenerative changes inside the knee consistent with osteoarthritis, as well as injury to the remaining passive structures, particularly the collateral ligaments, joint capsule, and menisci, all of which provide secondary restraint to anterior translation of the tibia. The risk of recurring injury is increased in the ACL-deficient (ACLD) knee because of the significant increase in anterior translation that occurs when the ACL is absent. For example, injury to the medial collateral ligament (MCL) is common because an increase in anterior tibial translation means that the MCL will be stretched by a proportionately larger amount. An increase in anterior tibial translation may also cause tearing of the menisci, which serve to protect the underlying cartilage from increased stress.

One effective method of protecting the ACL from injury is the use of hamstring co-contraction. Because the hamstrings insert on the back of the tibia, these muscles are able to apply a posterior pull to the leg, which decreases the amount of anterior translation at the knee. The model shown in figure 6.11 (p. 78) has been used to explain the effect of hamstring muscle action on ACL force in the intact knee and on anterior stability of the ACLD knee. The analyses were performed on isometric and isokinetic knee extension because physical therapists commonly prescribe this exercise as a means of preserving thigh muscle strength subsequent to ACL injury and repair. Figure 9.5a shows the results obtained when, at each flexion angle of the knee, the quadriceps muscles in the model were fully activated and the activation levels of the hamstring muscles were varied from 0% to 100%.

For maximum isolated contractions of the quadriceps in the intact knee, the peak force transmitted to the ACL is around 500 N. Furthermore, the ACL is loaded in the range 0° to 75° of knee flexion if the hamstring muscles remain inactive throughout this task. Activation of the hamstrings decreases both the peak force borne by the ACL and the flexion range over which the ligament is loaded (see figure 9.5a). As the level of hamstring co-contraction is increased, ACL force

decreases in proportion; however, the effect is not linear, as indicated in figure 9.5b, which shows ACL force plotted against hamstring co-contraction force over the full range of knee flexion angles. In fact, the relationship between hamstring co-contraction and ACL force depends on a third variable: the flexion angle of the knee. Both the model and experimental results have shown that hamstring co-contraction is more effective when the knee is bent than when it is

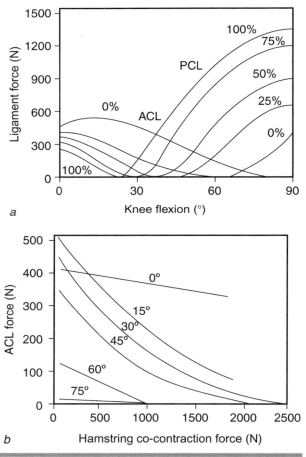

a

b

Figure 9.5 Effect of hamstring co-contraction on anterior cruciate ligament (ACL) and posterior cruciate ligament (PCL) forces during isometric knee extension exercise. (a) Ligament force plotted against knee angle for various levels of hamstring activation; (b) ACL force plotted against hamstring co-contraction force at various knee angles.

Reprinted from Pandy and Shelburne 1997.

straight. With the knee fully extended, hamstring co-contraction has hardly any effect on lowering the force transmitted to the ACL (figure 9.5, 0°). However, with the knee bent to angles greater than 15°, activating the hamstrings together with the quadriceps decreases ACL load significantly (compare curves for 15°, 30°, and 45° in figure 9.5 with that for 0°).

Hamstring co-contraction reduces ACL force by changing the balance of shear forces applied to the leg. The ACL is stretched and bears force whenever the tibia is translated anteriorly (forward) relative to the femur. Anterior translation of the tibia is ultimately a result of a net anterior shear force applied to the leg. Thus, the pattern of ACL loading during knee extension exercise is explained by the balance of shear forces acting on the leg. Figure 9.6, *a* and *b*, shows the anterior–posterior shear forces applied to the leg during isometric knee extension exercise performed with and without hamstring co-contraction, respectively. The results were obtained from the two-dimensional model of the knee shown in figure 6.11. All shear forces acting on the leg were resolved perpendicular to the long axis of the leg. In the figure, Total (thick black line) represents the net shear force applied to the leg, which was found by summing the contributions from the patellar tendon force, (PT, dashed line), the tibiofemoral compressive force (TF, dotted line), and the restraining force (Restraint, dot-dash line). Hams represents the shear force applied to the leg by the resultant force developed by hamstrings. Total is equal and opposite to the net shear force applied by all the ligaments and the capsule in the model. The tibiofemoral compressive force (TF) applies an anterior shear force to the leg at all knee flexion angles because the tibial plateau slopes roughly 8° from front to back in the knee. The patellar tendon applies an anterior or a posterior shear force to the leg depending on its orientation relative to the long axis of the leg. The orientation of the patellar tendon relative to the long axis depends, in turn, on the flexion angle of the knee.

At all but very small knee flexion angles, ACL force decreases with increasing hamstring co-contraction because the net anterior shear force applied to the leg decreases (compare Total in figure 9.6, *a* and *b*). The net anterior shear force decreases because the hamstring muscles apply a posterior force to the leg. When the knee is near full extension, however, the hamstrings all meet the tibia at relatively small angles, so they

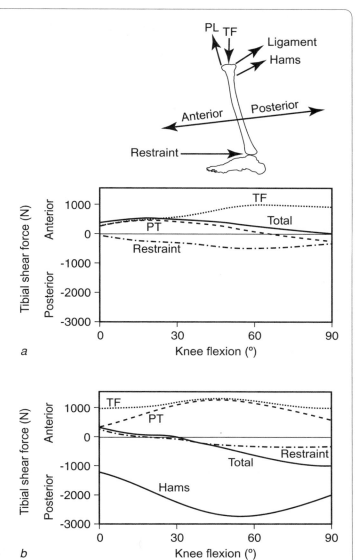

Figure 9.6 Anterior–posterior shear forces applied to the leg during isometric knee extension exercise. (*a*) Results for isolated maximum contractions of the quadriceps muscles (i.e., with the hamstring and gastrocnemius muscles turned off in the model); (*b*) results for simultaneous contractions of the quadriceps and hamstring muscles.

Adapted from Pandy and Shelburne 1997.

can no longer apply a large posterior shear in this position. Thus, the hamstrings cannot pull the tibia backward far enough to unload the ACL when the flexion angle of the knee is small.

The same mechanism acts to limit excessive anterior translation of the tibia and increase

(continued)

stability of the ACLD knee. A large number of experimental studies have shown that anterior stability of the knee is compromised when the ACL is absent (this phenomenon is often referred to as "giving way"). Figure 9.7 shows the effect of ACL deficiency on anterior translation of the tibia in maximum isometric knee extension exercise. The data points in the figure are the mean of measurements obtained from normal subjects (top) and ACLD patients (bottom) during maximum contractions of the quadriceps muscles. The shaded areas define plus and minus one standard deviation from the mean; 0 mm represents the

"neutral position" of the knee, which is the point where the tibia and femur contact each other when no external forces are applied to the leg and all the muscles are relaxed. The results show that peak anterior translation of the tibia is significantly greater in the ACLD knee (mean of 15 mm or 0.6 in.) than in the intact knee (mean of 10 mm or 0.4 in.). The reason is that the ACL restrains forward movement of the tibia relative to the femur in the intact knee; thus when it is absent, anterior translation of the tibia is much greater.

The model in figure 6.11 has also been used to study the effect of hamstring co-contraction on anterior tibial translation (ATT) during isokinetic extension exercise in the ACLD knee. The model simulations show that ATT is inversely related to hamstring co-contraction at all knee extension speeds (figure 9.8). Peak ATT decreases by more than 1 cm (0.4 in.) at all speeds when the hamstrings are fully activated. Again, peak ATT decreases as hamstring activation increases because the posterior shear force provided by these muscles overwhelms the anterior shear force supplied by the patellar tendon and the shear force due to the compressive force developed as a result of contact between the tibia and femur (see figure 9.4).

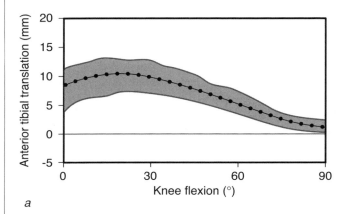

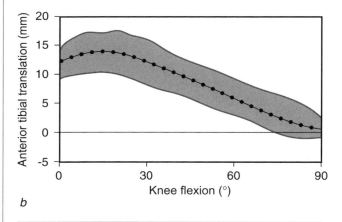

Figure 9.7 Anterior translation of the tibia relative to the femur in (a) the anterior cruciate ligament (ACL)-intact knee and (b) the ACL-deficient (ACLD) knee during knee extension exercise.

Adapted from Pandy and Shelburne 1998.

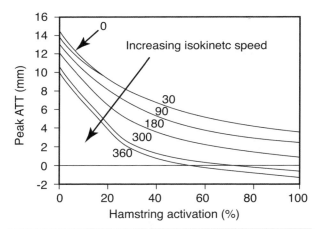

Figure 9.8 Peak anterior tibial translation (ATT) versus hamstring co-contraction level for isokinetic knee extension at 0°, 30°, 90°, 180°, and 300°/s in the anterior cruciate ligament-deficient knee.

Adapted from Yanagawa, Shelburne, Serpas, and Pandy, 2002.

Although the mechanism by which the hamstrings limit ATT in the ACLD knee is now well understood, ultimately the usefulness of this mechanism depends on whether training programs can be developed to teach ACLD patients to activate their hamstrings at appropriate intervals during an activity. Otherwise, excessive ATT over time may cause damage to the menisci and surrounding passive structures, resulting in further degeneration of the joint.

There is now some evidence to suggest that anterior knee stability may be improved in ACLD patients through alteration of their muscle recruitment patterns during activity. One study addressed the effects of strength training versus neuromuscular training on muscle reaction times of the quadriceps, hamstring, and gastrocnemius muscles during isokinetic exercise. Neuromuscular training, which focused on improving coordination of the lower extremity muscle groups that stabilize the knee during activity, significantly decreased the reaction times of the knee flexor muscles. Thus, the time to peak knee flexor torque decreased, allowing the flexor muscles, particularly hamstrings, to exert more control over anterior translation of the tibia at the knee.

SOURCES

- Pandy, M.G., and K.B. Shelburne. 1997. Dependence of cruciate-ligament loading on muscle forces and external load. *Journal of Biomechanics* 30: 1015-1024.
- Pandy, M.G., and K.B. Shelburne. 1998. Theoretical analysis of ligament and extensor-mechanism function in the ACL-deficient knee. *Clinical Biomechanics* 13: 98-111.
- Wojtys, E.M., L.J. Huston, P.D. Taylor, and S.D. Bastian. 1996. Neuromuscular adaptations in isokinetic, isotonic, and agility training programs. *American Journal of Sports Medicine* 24: 187-192.
- Yanagawa, T., K.B. Shelburne, F. Serpas, and M.G. Pandy. 2002. Effect of hamstrings muscle action on stability of the ACL-deficient knee in isokinetic exercise. *Clinical Biomechanics* 17: 705-712.

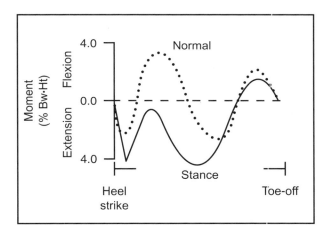

Figure 9.9 Net moment exerted about the knee by normal subjects and by anterior cruciate ligament-deficient patients during walking on level ground. Knee extensor moment is normalized by body mass and height.

Adapted from Andriacchi, Natarajan, and Hurwitz 1997.

portion of the gait cycle (figure 9.9). One interpretation of these findings is that the ACLD patients walk by reducing the demand on their quadriceps during stance (i.e., the quadriceps muscles are activated less in the ACLD patients, giving rise to a quadriceps-avoidance gait pattern).

It is important to realize here that a lower net extensor moment at the knee does not necessarily mean that the moment exerted by quadriceps is lower. Because the net moment at the knee is a combination of quadriceps and hamstring muscle action, a lower extensor moment could result from an increase in the flexor moment applied by hamstrings. However, muscle electromyography results tend to support the interpretation that quadriceps moment is lower because the activation level of quadriceps is lower in the ACLD patients.

Gait analysis results also showed that the ACLD patients reduced their net extensor moment by 25% during jogging compared with a 100% reduction in walking. However, the net extensor moment in stair climbing was the same in ACLD patients and normal subjects. Taken together, these results suggest that the observed quadriceps avoidance in ACLD patients is not a result of quadriceps weakness but instead due to a change in neuromuscular control.

Dependence of Motor Performance on Changes in Muscle Properties

Strength and quickness are critical determinants of performance in explosive movements like jumping, throwing, and sprinting. Together, these two factors determine the amount of power developed instantaneously during a task (i.e., muscle power is given by the product of muscle force and muscle contraction speed; see "At What Speed Must a Muscle Shorten to Develop Maximum Power?" on p. 108). As discussed earlier in this chapter, resistance strength training can alter both muscle force and fibre contraction speed. This section describes the effects that changes in the contractile properties of muscle have on motor performance—specifically, how biomechanical performance depends on muscle strength and muscle fibre contraction speed.

Vertical jumping is one of the most heavily studied motor tasks. Many scientists have studied jumping with the aim of learning more about the biomechanics and control of this task and, specifically, about how performance may be influenced by training. In a recent jump-training program that incorporated stretching, plyometric exercises, and weightlifting, female high school volleyball players demonstrated a mean increase of 3.8 cm (1.5 in.) in jump height after training. This represented an almost 10% increase in jump height as a result of the six-week period of training. Even larger increases in vertical jumping performance have been documented in the literature. One extreme example is the 1984 U.S. Olympic gold medal volleyball team, which showed a 10-cm (4-in.) increase in jump height after two years of jump training.

Using Computer Modelling to Study Vertical Jumping Performance

While it is gratifying to learn that jumping performance can be increased significantly by strength and neuromuscular training, it is difficult, if not impossible, to explain why. Noninvasive measurements of biomechanical performance cannot pinpoint the factor or factors responsible for the increase in jump height. The reason is that several properties of the neuromuscular and musculoskeletal systems change simultaneously during the training regimen. Alternatively, computer models may be used to study the relationships between training effects and performance; that is, a model of the neuromusculoskeletal system may be used to predict how changes to specific parameters affect the performance of a motor task.

Models of the body similar to the one shown in figure 7.7 (p. 87) have been used to study how changes in muscle strength, muscle contraction speed, and motor unit recruitment affect jump

height. The values of each of these parameters in the model were increased by amounts consistent with results obtained from strength-training programs. For example, the peak isometric strength and maximum shortening velocity of each muscle in the model were increased by 20%, while the activation level of each muscle was increased by 10%. The three training effects were first applied to all muscles simultaneously, and then to the ankle plantar flexors, knee extensors, and hip extensors separately. In this way, the modelling results were used to determine whether it is better to train all the leg muscles simultaneously or to isolate specific muscle groups such as the quadriceps (knee extensors) and gluteus maximus (hip extensors).

Insights Into the Effects of Training Provided by Computer Models

Increasing the peak isometric force of all the leg muscles in the model by 20% produced the largest increase in jump height. In this case, jump height increased by 7 cm (2.8 in.), or 5% of the value obtained prior to the simulated training effect. Increasing the maximum shortening velocity of each muscle by 20% or the activation level of each muscle by 10% increased jump height by only 4 cm (1.6 in.), or 3% of the value obtained for the nominal (untrained) model. When all three training effects were introduced simultaneously, jumping performance increased by nearly 17 cm (6.7 in.), or 12% of the value calculated for the untrained model. Thus, training programs that increase strength, fibre contraction speed, and motor recruitment (i.e., activation level) of all the leg muscles simultaneously are most beneficial to overall jumping performance.

The modelling results also suggest that training the knee extensors is better than training either the ankle plantar flexors or the hip extensors. When peak isometric force and maximum shortening velocity of the quadriceps were increased by 20%, and quadriceps activation was simultaneously increased by 10%, jump height increased by almost 10 cm (4 in.), or 7% of the value calculated for the untrained model. The same changes made to either the ankle plantar flexors or the hip extensors produced increases in jump height of only 3 cm (1.2 in.), or 2% of the value obtained for the untrained model.

The latter result is a little puzzling in light of the fact that the model calculations had shown previously that the quadriceps and gluteus maximus are the major energy producers, the prime movers, of the body in vertical jumping (see "Which Muscles Are Most Important to Vertical Jumping Performance?" on p. 87 in chapter 7). It should be noted here that each time a change was introduced to the model, a new optimal pattern of muscle activations was found by re-solving the optimisation problem for a maximum-height jump. Thus, one interpretation of the result is that what matters most in terms of performance is the ratio of quadriceps to gluteus maximus muscle strength, not the absolute strength of these muscles. In other words, jumping performance is most sensitive to a change in the ratio of knee extensor to hip extensor muscle strength. Increasing quadriceps strength by 20% increases the ratio of knee extensor to hip extensor muscle strength in the model, whereas increasing gluteus maximus muscle strength decreases this ratio. The same line of reasoning may explain why increasing ankle plantar flexor muscle strength by 20% leads to an increase in jump height of just 3 cm (1.2 in.) in the model.

Training for strength is also better than training for quickness or speed. Figure 9.10 shows the effects of increasing muscle strength and muscle contraction speed on vertical jump height as predicted by the four-segment, eight-muscle, sagittal plane model of the body described in chapter 7 (see figure 7.7). An increase in muscle strength was simulated in the model by simultaneously increasing body weight, since muscle strength increases in proportion to muscle mass. Thus, figure 9.10 represents changes in body strength-to-weight ratio rather than changes in muscle strength alone. Also, jumping performance is normalized in these results by dividing by the value of jump height calculated for the untrained model. Similarly, body strength-to-weight ratio and muscle contraction speed are each normalized by dividing through by the value of body strength-to-weight ratio and muscle contraction speed in the untrained model, respectively.

The simulation results show that the slope of the line predicted for changes in body strength-to-weight ratio is twice as large as that obtained when changes in muscle fibre contraction speed are made. Thus, muscle strength has a greater effect on vertical jump height than muscle fibre contraction speed, even when the accompanying increase in body mass is taken into account.

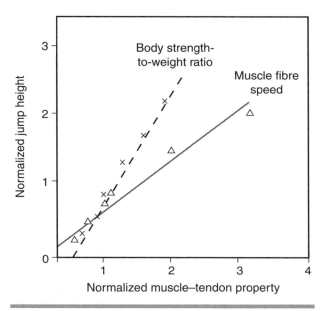

Figure 9.10 Effect of changes in muscle strength and muscle fibre contraction speed on vertical jump height. The results were obtained from the model shown in figure 7.7. Jump height is normalized by dividing by the performance predicted for the untrained (nominal) model. Adapted from Pandy 1990.

Another important lesson from the modelling studies is that musculoskeletal changes must be accompanied by appropriate changes in neuromuscular control; otherwise, the expected improvement in motor performance will not occur. In vertical jumping, if the pattern of muscle activations remains unchanged subsequent to strength training, jump height actually decreases relative to that in the untrained state. In the simulated jump shown in figure 9.11, the strength of the knee extensor muscles has been increased by 20%, but the control exerted over the joints is the same as that calculated for the untrained model before the training effect was introduced. When the pattern of muscle activations was not optimised to match the changes introduced to the neuromusculoskeletal model, the body left the ground prematurely (i.e., the centre of mass was at a lower height at liftoff than is optimal for a maximal jump).

One consequence of not optimising the pattern of muscle activations (i.e., the controls) is that a larger fraction of the total work produced by the muscles goes into rotating the body

Knee extensors 20% stronger, not optimised
Vertical velocity at liftoff: 2.45 m/s
Jump height: 0.31 m

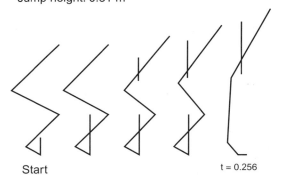

a   Start   t = 0.256

Knee extensors 20% stronger, not optimised
Vertical velocity at liftoff: 2.73 m/s
Jump height: 0.43 m

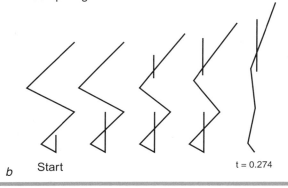

b   Start   t = 0.274

Figure 9.11 Stick figures showing the effect of changing quadriceps muscle strength in a model of jumping without re-optimisation of the control. At liftoff, the knee was hyperextended and the hip was not extended enough.
Reprinted from Bobbert and van Soest 1994.

segments rather than accelerating the centre of mass upward. In a maximum-height jump, approximately 90% of the total work done by the leg muscles is used to propel the centre of mass upward. This number is closer to 80% when muscle coordination (i.e., the sequence and timing of muscle activations) is not optimal. So, even though jumping performance depends heavily on muscle strength, and to a lesser extent on muscle fibre contraction speed, optimal performance is also intimately related to neuromuscular control.

For this reason, and as discussed at the beginning of this chapter, jump-training programs now focus on manoeuvres that blend muscle strength with neural control.

Summary

Training can change both muscle strength and fibre contraction speed. Strength may be increased through an increase in motor unit recruitment (net neural drive to the muscle), an increase in muscle fibre size, or both. Contraction speed may be altered through change in the shape of a muscle's force–velocity curve, change in the value of its intrinsic maximum shortening velocity, or both. Early in a training program (two to eight weeks), increases in muscle strength are brought about by neural adaptations rather than by increases in muscle size. Training can also change the way muscle action is coordinated during activity (i.e., neuromuscular control). Neuromuscular training programs are usually designed to improve stability (balance and coordination) and proprioception (joint position sense), in addition to muscle strength. Training is vital not only for improving biomechanical performance but also for preventing injuries during sport.

Further Reading and References

Andriacchi, T.P., R.N. Natarajan, and D.E. Hurwitz. 1997. Musculoskeletal dynamics, locomotion, and clinical applications. In *Basic orthopaedic biomechanics*, ed. V.C. Mow and W.C. Hayes (pp. 37-68). Philadelphia: Lippincott-Raven.

Kraemer, W.J., S.J. Fleck, and W.J. Evans. 1996. Strength and power training: Physiological mechanisms of adaptation. In *Exercise and Sport Science Reviews*, ed. J.O. Holloszy (pp. 363-397). Baltimore: Williams & Wilkins.

McMahon, T.A. 1984. *Muscles, reflexes, and locomotion*. Princeton, NJ: Princeton University Press.

Pandy, M.G. 1990. An analytical framework for quantifying muscular action during human movement. In *Multiple muscle systems: Biomechanics and movement organization*, ed. J. Winters and S.L. Woo (pp. 653-662). New York: Springer-Verlag.

PART III

PHYSIOLOGICAL BASES OF HUMAN MOVEMENT

The Subdiscipline of Exercise Physiology

*E*xercise physiology is a subdiscipline of human movement studies that focuses on the physiological responses to exercise. Physiology can be defined as the study of the body's functions; hence, **exercise physiology** is the study of the responses of body function to exercise. Exercise physiology focuses on both the acute (immediate) responses to exercise and the chronic (long-term) adaptations to physical activity.

Exercise physiology draws on knowledge and techniques from a variety of speciality areas including other subdisciplines in human movement studies, physiology, biochemistry, nutrition, endocrinology (the study of hormones and their functions), histology (the study of microscopic structure of tissues and cells), and cell and molecular biology.

Applications of Exercise Physiology

As "Some Applications of Exercise Physiology" (p. 120) reveals, exercise physiology knowledge and skills are used in varied situations including sport, physical fitness, health promotion and preventive medicine, exercise rehabilitation, the workplace, school health and physical education, and the research laboratory.

Exercise principles are used in designing effective assessment and training programs to enhance fitness and performance of athletes. Talent identification involves physiological and anthropometric profiling to identify young people with potential to excel in specific sports. Exercise physiology forms the basis of physical activity prescription for healthy adults, as well as for special populations such as children; persons who are elderly; cardiac patients; and individuals with conditions such as diabetes, pregnancy, or arthritis.

Regular exercise also has a role in preventing and rehabilitating injury, including sporting or occupational injuries. Exercise physiology principles are used to assess the physical demands of some occupations, such as firefighting, and to develop programs to help workers achieve the needed fitness. Coaches and physical education teachers working with children use exercise physiology to design training programs specifically for the young athlete. Finally, research in exercise physiology focuses on understanding the mechanisms underlying the physiological responses and adaptations to exercise.

In this part of the book, we discuss and apply exercise physiology principles across a wide spectrum. Chapter 10 discusses basic principles of exercise metabolism and how the body provides the energy needed for muscles to produce

SOME APPLICATIONS OF EXERCISE PHYSIOLOGY

Sport
- Fitness profiling of athletes
- Talent identification
- Designing sport-specific training programs
- Evaluating training program effectiveness

Health Promotion and Preventive Medicine
- Physical activity for general health
- Exercise prescription in specific conditions, such as obesity, diabetes, and pregnancy
- Corporate and community fitness programming and health promotion

Exercise Rehabilitation
- Exercise programming in patient groups, such as those with heart and blood vessel diseases (e.g., ischemic heart disease, hypertension), respiratory diseases (e.g., asthma, chronic obstructive pulmonary disease), metabolic conditions (e.g., diabetes, obesity), musculoskeletal conditions (e.g., arthritis, low back pain), and neurological diseases (e.g., multiple sclerosis, Parkinson's disease)
- Injury rehabilitation (e.g., work or sport related)
- Fitness programming to prevent reinjury

Work Site
- Assessment of physical fitness demands of specific occupations, such as firefighting, military, and police
- Exercise programming to meet occupational physical fitness standards
- Prevention of occupational injuries, such as back injury

School Health and Physical Education and Sport
- Specialist teaching in health and physical education
- Coaching in community-based sport for children
- Adapting sport for children

Research Laboratory
- Physiological, biochemical, and hormonal responses to exercise
- Mechanisms of training adaptations
- Mechanisms responsible for health benefits of physical activity

force and move the body. In chapter 11, we extend these principles to describe how the body adapts to exercise training (e.g., run faster or farther, lift more weight) and present recommendations for exercise prescription. Chapter 12 discusses how exercise capacity and responses to exercise training change over the life span. Finally, in chapter 13, we extend and apply these principles to the role of exercise in enhancing health and preventing or treating prevalent diseases.

Typical Questions Posed and Levels of Analysis

Exercise physiology can be studied at many different levels, from the whole-body to the molecular level. At the whole-body level, the exercise physiologist might ask whether a lifetime of exercise can increase life span. At the systemic level, exercise physiology aims at understanding the effects of regular exercise on the cardiovascular system—for example, how exercise helps prevent heart disease. Understanding the effects of exercise on the pattern of fat (adipose tissue) distribution and implications for prevention of obesity represents work at the tissue level. At the cellular and subcellular level, the exercise physiologist may study the mechanisms responsible for muscular hypertrophy following strength training. Finally, work at the molecular level might focus on the extent to which elite sport performance is determined by genetics.

Historical Perspectives

Just as the applications and levels of analysis of exercise physiology are multifaceted, so too are its historical origins. The ancient Greeks and Romans recognized many exercise physiology principles—for example, that muscles hypertrophy (grow larger) in response to strength training and that regular physical activity is important to good health.

Spanish physician Christobel Mendez was the first to publish a printed book on the topic, in 1553, titled *Book of Bodily Exercise*. The European interest in exercise and health continued well into the 19th century and spread to the United States, where early concern focused on ill health among city dwellers. With the rise of university sport, particularly rowing in mid-19th-century England, "trainers" (equivalent to today's coaches) began to apply physiological principles to the training of athletes.

Modern exercise physiology has its origins in the fields of physiology, biochemistry, and nutrition. Work during the early part of the 20th century centred on understanding energy metabolism and efficiency of movement during prolonged steady-rate exercise. A.V. Hill in England was among the first to study the metabolic responses to maximal exercise in athletes. Hill's work directly influenced the Harvard Fatigue Laboratory in the United States, which generated much of the foundation of modern exercise physiology, eventually expanding the application of exercise physiology to sport, medicine and health, physical education, and the workplace.

In the 1960s, exercise physiology research adopted a biochemical and subcellular focus with reintroduction by Bergstrom of the needle biopsy technique, which allows the exercise physiologist to explore the metabolic responses of muscle cells to exercise. For example, the importance of muscle glycogen (glucose) stores and dietary carbohydrate to exercise performance could be identified only by use of the biopsy procedure.

Since the 1970s, exercise physiology research has been firmly placed within human movement or exercise sciences. At the same time, with recent advances in cell and molecular biology, exercise physiology has incorporated many of the tools of modern biomedical research. Exercise physiology now encompasses the study of the physiological and metabolic adaptations to exercise, from basic research at the molecular and cellular level through to applied research aimed at improving sport performance in athletes or the health of entire communities.

Professional Organizations and Training

The three major international organizations relevant to exercise physiology are the International Federation of Sports Medicine (FIMS), the International Council of Sport Science and Physical Education (ICSSPE) and the American College of Sports Medicine (ACSM). Details of these organizations are provided toward the end of chapter 1.

In addition to these international organizations, there are many regional and national associations representing the professional interests of sport and exercise physiology practitioners. Web sites of some associations are listed at the end of this introduction.

Exercise physiologists are often members of other professional associations with specialized interest in various aspects of basic physiology, physical activity, or health. These might include associations devoted to research and applications in physiology (e.g., American Physiological Society), rehabilitation (e.g., American Association of Cardiovascular Pulmonary Rehabilitation), resistance training (e.g., National Strength and Conditioning Association), or health promotion (e.g., American Public Health Association). It is not unusual to belong to several different professional associations at the same time, depending on one's training, interests, and occupation.

Further Reading and References

Tipton, C.M. 1998. Contemporary exercise physiology: Fifty years after the closure of the Harvard Fatigue Laboratory. *Exercise and Sport Science Reviews* 26: 315-340.

U.S. Department of Health and Human Services. 1996. *Physical activity and health: A report of the Surgeon General* (pp. 11-57). Atlanta: U.S. Department of Health and Human Services, Centers for Disease Control and Prevention, National Center for Chronic Disease Prevention and Health Promotion.

Some Relevant Web Sites

Active Australia
[www.activeaustralia.org]

American College of Sports Medicine (ACSM)
[www.acsm.org]

American Heart Association
[www.americanheart.org]

American Society of Exercise Physiologists (ASEP)
[www.asep.org]

Australian Association for Exercise and Sports Science (AAESS)
[www.aaess.com.au]

British Association of Sport and Exercise Sciences (BASES)
[www.bases.org.uk/newsite/home.asp]

British Heart Foundation
[www.bhf.org.uk]

Canadian Society for Exercise Physiology (CSEP)
[www.csep.ca]

Centers for Disease Control and Prevention (U.S.) Physical Activity and Health Initiative
[www.cdc.gov/nccdphp/sgr/npai.htm]

European College of Sport Science (ECSS)
[www.dshs-koeln.de/ecss/HTML/]

Health Canada
[www.hc-sc.gc.ca]

Healthy People 2010 (U.S.)
[www.healthypeople.gov]

Heart and Stroke Foundation of Canada
[www.heartandstroke.ca]

Hong Kong Association of Sports Medicine and Sports Science
[www.fmshk.com.hk/hkasmss/]

International Council of Sport Science and Physical Eduction (ICSSPE)
[www.icsspe.org]

International Federation of Sports Medicine (FIMS)
[www.fims.org/fims/frames.asp]

Sport and Recreation New Zealand (SPARC)
[www.hillarysport.org.nz]

Summer Active (Canada)
[www.summeractive.canoe.ca]

World Health Organization
[www.who.int/hpr/physactiv]

World Heart Federation
[www.worldheart.org]

CHAPTER 10

Basic Concepts of Exercise Metabolism

*O*ne important question asked by exercise physiologists is "What factors limit performance?" That is, what prevents a runner from running faster, what causes the athlete to become fatigued, what limits how much weight a powerlifter can lift?

Exercise capacity is determined by how much energy the muscle cell can produce and how quickly this energy can be made available to the contractile elements within skeletal muscle. For example, maximum sprinting speed can be maintained for only 100 to 200 m (109 to 219 yd) (or 10-20 s at world record pace). After the first 10 to 20 s, running pace slows because the muscle cells cannot maintain the required rate of energy supply. Fatigue occurs when the rate of energy demand exceeds the rate of production in skeletal muscle.

If we know what limits exercise capacity in a particular activity, we can use this knowledge to improve performance. For example, it is well known that depletion of stored glucose (**glycogen**) from skeletal muscle causes fatigue in long-duration events such as marathon running. It is now common practice for endurance athletes to consume a high-carbohydrate diet to enhance muscle glycogen stores and delay the onset of fatigue during training and competition.

In this chapter we examine a range of fundamental concepts related to exercise metabolism, discussing how energy for exercise is produced in muscle, how energy metabolism during exercise is measured, how different physical activities require different amounts of energy, and how energy metabolism is affected by dietary manipulation.

The major learning concepts in this chapter relate to

- muscle as a unique tissue,
- how energy for exercise is produced,
- how the body provides a continual supply of oxygen to the working muscles for sustained exercise,
- how exercise capacity is measured,
- different types of human skeletal muscle cells,
- the amount of energy required for different activities, and
- the role of diet in energy metabolism and exercise performance.

Production of Energy for Exercise

As the body goes from rest to maximal exercise, the metabolic rate in human skeletal muscle can increase by up to 50 times. The muscle must be able to match the rates of energy supply with demand, that is, to produce energy at the needed rate. It is metabolically inefficient for cells to produce more energy than needed, and muscle cells do not store vast quantities of energy. During sustained exercise, chemical energy must be continually supplied to the working muscles.

Production of Adenosine Triphosphate

Muscular work requires transfer of chemical energy to mechanical energy. This chemical energy is supplied in the form of **adenosine triphosphate (ATP).** Adenosine triphosphate is called a "high-energy phosphate" molecule because energy is released upon cleavage of its terminal phosphate bond (figure 10.1*a*). The ATP splits into adenosine diphosphate (ADP) and inorganic phosphate (P_i) in a reversible reaction. The energy released when this bond is cleaved is required for cross-bridge interaction between the thin and thick filaments of skeletal muscle resulting in production of force in muscle (see figure 2.11). During exercise and recovery, ADP can be re-phosphorylated to ATP provided sufficient chemical energy is supplied via metabolic pathways. Muscle cells contain all the necessary equipment to continually resynthesize ATP from ADP and P_i.

Three energy systems resynthesize ATP: (1) the **immediate energy system,** sometimes called the stored energy, high-energy phosphagen, or ATP-PCr system; (2) the anaerobic system, sometimes called the **anaerobic glycolytic system** or **lactic acid** system; and (3) the aerobic or oxidative system (table 10.1). The three systems combine to make ATP available for all types of exercise from very short bursts such as a 10-m (11-yd) sprint to sustained activity such as a triathlon. As described in this section, all three systems operate simultaneously, and their relative contributions to ATP resynthesis depend on the intensity and duration of exercise.

The Immediate Energy System

The immediate energy system provides stored chemical energy in the form of **phosphocreatine (PCr),** another high-energy phosphate molecule (figure 10.1b). Stored PCr provides an immediate source of energy for exercise. At the start of exercise, PCr is rapidly degraded to creatine plus P_i, donating its phosphate group to ADP to make ATP (figure 10.1c). PCr is the major fuel source at the onset of any exercise and for brief exercise lasting less than 30 s such as sprinting or jumping (table 10.1).

Phosphocreatine is rapidly depleted during exercise—by 65% after 30 s and by nearly 100% at 1 min of maximal exercise (figure 10.2). The PCr is not replenished until exercise stops; complete replenishment requires 6 to 7 min of rest at the end of exercise. As will be discussed in chapter 11, effective sprint training must consider the time course of PCr replenishment during recovery between exercise bouts. "Creatine Supplementation: To What Extent Does the Immediate Energy System Limit Exercise Performance?" on page 127 addresses whether increasing muscle PCr stores enhances performance in intense exercise.

The Anaerobic Glycolytic System

The anaerobic glycolytic system provides the major source of ATP for maximal exercise lasting between 20 s and 3 min (table 10.1). For example, during 30-s sprinting, anaerobic glycolysis provides 60% to 65% of the needed ATP. As the name implies, this system produces energy anaerobically (without oxygen) using glucose as the fuel (substrate). The glucose comes from muscle glycogen stores and to a limited extent from blood glucose; glycogen is the

Table 10.1

The Three Energy Systems

	Immediate (stored energy, high-energy phosphagen, or ATP-PCr)	Anaerobic glycolytic (anaerobic or lactic acid)	Oxidative (aerobic)
Substrate(s)	ATP (adenosine triphosphate), PCr (phosphocreatine)	Glycogen or glucose	Glycogen or glucose, fat, protein
Relative rate of ATP production	Very fast	Fast	Slower
Duration at maximal pace	0-30 s	20-180 s	>3 min
Limiting factors	PCr depletion	Lactic acid accumulation	Glycogen depletion
Examples of activities	Power- and weightlifting	Longer sprints	Endurance events
	Short sprints	Middle distance team sports	Team sports
	Jumping, throwing	Ball games (soccer, basketball, rugby)	Ball games (soccer, field hockey, Australian rules football)

a Energy for muscular work

b PC splitting to resynthesise ATP

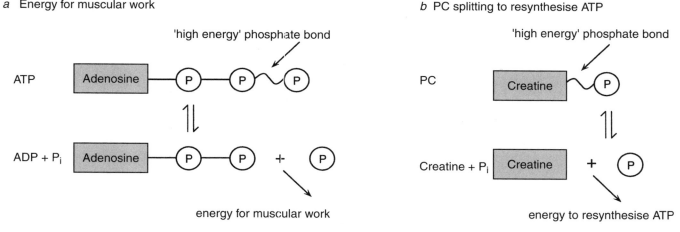

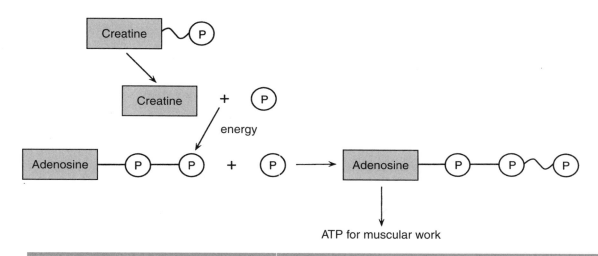

c At onset of exercise

Figure 10.1 Schematic representation of "high-energy" phosphagens. *(a)* Adenosine triphosphate (ATP) is split into adenosine diphosphate (ADP) + inorganic phosphate (P_i). *(b)* Phosphocreatine (PCr) is split into creatine + P_i. *(c)* At the start of exercise and during brief intense exercise, PCr provides most of the energy for ATP production.

storage form of glucose, consisting of a polymer of many glucose molecules. Lactic acid is a by-product of anaerobic glycolysis, and this system is sometimes called the lactic acid system.

One molecule of glucose, a six-carbon sugar, is degraded via a series of 13 steps to two molecules of pyruvic acid (figure 10.3). Pyruvic acid can then either be converted to lactic acid or enter the Krebs cycle (also known as the tricarboxylic acid cycle and the citric acid cycle). Two ATP molecules are produced anaerobically for each molecule of glucose converted to two pyruvic acid molecules.

The Oxidative System

Oxidative production of ATP occurs in the mitochondria, membrane-bound subcellular organelles found in most cells. Pyruvic acid, fatty acids, and amino acids are further degraded via the Krebs cycle, producing carbon dioxide, electrons, and hydrogen ions (H^+) (figure 10.3). The carbon dioxide diffuses out of muscle cells into the blood and is transported to the lungs where it is exhaled. The electrons and hydrogen ions enter the electron transport chain, a series of enzymes that eventually combine the electrons, hydrogen ions, and oxygen

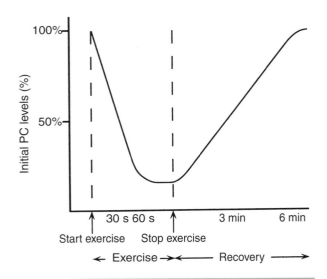

Figure 10.2 Phosphocreatine (PCr) depletion and replenishment. Up to 6 to 7 min are required to fully replenish muscle PCr stores after depletion by intense exercise.

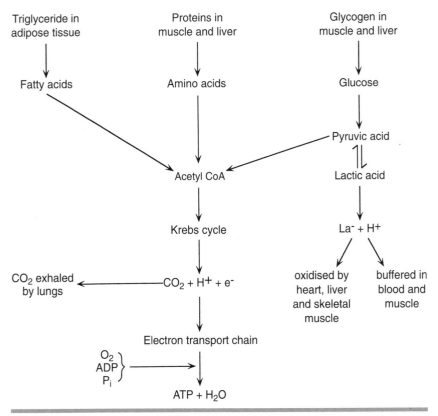

Figure 10.3 Simplified scheme of energy metabolism. Fats (fatty acids), proteins (amino acids), and carbohydrate (glycogen or glucose) can all be metabolised to produce adenosine triphosphate (ATP). Carbohydrates can be metabolised aerobically or anaerobically, but the other two require oxygen. The end products are carbon dioxide, lactic acid, ATP, and water.

to produce water. This transfer of electrons and hydrogen ions provides chemical energy to resynthesize ATP from ADP and P_i. Continued production of ATP via the electron transport chain requires a constant supply of oxygen to the muscle cell.

The Three Energy Systems As a Continuum

The three energy systems operate as a continuum; each system is always functioning, even at rest. What varies is the relative contribution each system makes to total ATP production at any given time (figure 10.4).

Even in the extremes of activity, such as marathon running or brief sprinting, all three systems are used. For example, in a marathon, the immediate system provides ATP at the start of the race; the anaerobic system provides much of the needed ATP for the first few minutes until the oxidative system reaches steady state. The anaerobic system also provides a significant amount of ATP during the race, for example, for uphill running and for sprinting at the end. Over the course of the entire marathon, the oxidative system provides most of the ATP needed.

The Fueling of ATP Production by Fats, Proteins, and Carbohydrates

As just discussed, ATP can be synthesized via metabolism of fats, proteins, and carbohydrates. It is important to recognize that these nutrients are not transformed into ATP. Rather, the body breaks down these nutrients to release energy from their chemical bonds, which is then used to synthesize ATP (figure 10.3).

Carbohydrate (i.e., glucose) can be used to produce ATP either anaerobically or aerobically. In contrast, fats in the form of fatty acids and proteins in the form of amino acids can be used to produce ATP only via the aerobic pathway. At any given time, the body metabolises a mixture of these nutrients to

IN FOCUS: CREATINE SUPPLEMENTATION: TO WHAT EXTENT DOES THE IMMEDIATE ENERGY SYSTEM LIMIT EXERCISE PERFORMANCE?

The immediate energy system provides ATP for brief, intense exercise such as sprinting. Muscle stores of PCr can be depleted by all-out exercise lasting 30 s; this is thought to limit performance in such events. Increasing the amount of muscle PCr may enhance performance in power events. Creatine occurs naturally in food, mainly meat and fish, and is also synthesized by the liver. If we could increase muscle creatine stores, and if doing so enhanced performance in power events, we would know the extent to which performance is limited by PCr levels.

Muscle takes up ingested creatine from the blood and uses it to synthesize PCr. Muscle PCr levels can increase by only 10% to 20%, however, and any excess is excreted. The effectiveness of supplementation depends on the subject, amount ingested, and type of exercise. Creatine supplementation does not affect endurance exercise or maximum strength, because neither is limited by PCr depletion. Creatine supplementation works best for repeated high-intensity exercise such as resistance training and multiple sprints.

In one recent study, team handball players were randomly assigned to a creatine-supplemented (20 g creatine per day for five days) or placebo group (Izquierdo et al. 2002). A variety of muscular fitness measures were made before and after the five days. Athletes in the supplemented group showed increases in body mass (0.6 kg (1.3 lb)), muscular endurance, average power output in an anaerobic test, and performance in repeated sprints or jumping. There were no changes in aerobic capacity or **1-repetition maximum (1RM)** strength. The authors concluded that the main effects of creatine supplementation were enhanced muscular endurance and average power output in repeated rather than single efforts. The increase in body mass most likely resulted from water retention by the muscles.

In another recent study, active but untrained males performed short sprints (6 s) on a cycle ergometer repeatedly over 80 min (Preen et al. 2001). One group ingested 20 g/day creatine and another group ingested a placebo for five days. When the cycling test was repeated after five days, the supplemented group showed higher total work and power output over the 80 min compared with the placebo group. This was true for a range of recovery times, from 24 to 84 s, between repeat sprints.

These studies suggest that short-term creatine supplementation enhances performance of repeated bouts of intense exercise. Creatine supplementation may not help performance in a single event, but it allows the athlete to train harder by enhancing PCr resynthesis between intervals. Although creatine is not currently banned in sport, there is some concern about long-term health consequences of supplementation, which are not completely known.

SOURCES
- Izquierdo, M., J. Ibanez, J.J. Gonzalez-Badillo, and E.M. Gorostiaga. 2002. Effects of creatine supplementation on muscle power, endurance, and sprint performance. *Medicine and Science in Sports and Exercise* 34: 332-343.
- Preen, D., B. Dawson, C. Goodman, S. Lawrence, J. Beilby, and S. Ching. 2001. Effect of creatine loading on long-term sprint exercise performance and metabolism. *Medicine and Science in Sports and Exercise* 33: 814-821.
- Terjung, R.L., P. Clarkson, E.R. Eichner, P.L. Greenhaff, P.J. Hespel, R.G. Israel, W.J. Kraemer, R.A. Meyer, L. Spriet, M.A. Tarnopolsky, A.J.M. Wagenmakers, and M.H. Williams. 2000. ACSM roundtable: The physiological and health effects of oral creatine supplementation. *Medicine and Science in Sports and Exercise* 23: 706-717.

produce ATP. However, the relative contribution of each nutrient to ATP production varies with exercise intensity and, thus, the metabolic rate.

At rest and during low-intensity exercise, fatty acids and glucose are used in approximately equal amounts as substrates for ATP production. As exercise intensity increases, ATP production progressively relies more on glucose and less on fatty acids. During maximal exercise, the muscles metabolise primarily glucose, derived from muscle glycogen. Amino acids usually contribute little to ATP resynthesis (<5%) during moderate exercise. Metabolism of amino acids may provide up to 20% of energy production after several hours of prolonged exercise in which glucose supply to the muscle is severely limited.

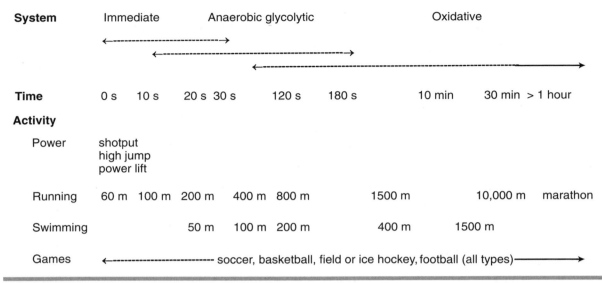

System	Immediate	Anaerobic glycolytic				Oxidative		

Time	0 s 10 s	20 s 30 s	120 s	180 s	10 min	30 min > 1 hour
Activity						
Power	shotput high jump power lift					
Running	60 m 100 m	200 m	400 m 800 m		1500 m	10,000 m marathon
Swimming		50 m	100 m 200 m		400 m	1500 m
Games	◄-------------------------- soccer, basketball, field or ice hockey, football (all types)————————►					

Figure 10.4 The energy system continuum. The relative contribution of each system to ATP resynthesis depends on exercise duration and intensity.

Lactic Acid—Friend or Foe?

Lactic acid is produced as a by-product of anaerobic glycolysis. Excess lactic acid in the form of the lactate ion (explained next) is transported across the muscle cell membrane into the blood and circulated throughout the body. During maximal exercise, lactic acid concentration may increase 15-fold, from resting levels of 1 to 2 mmol/L up to 30 mmol/L in muscle and 15 mmol/L in blood.

Excess lactic acid is associated with muscular fatigue. Lactic acid rapidly dissociates into a lactate anion and free hydrogen ion (H^+) (figure 10.3). An increase in H^+ concentration increases the acidity (lowers the pH) of muscle and blood. Tissues and blood contain substances that partially, but not fully, buffer the increased acidity. The anaerobic glycolytic system is sensitive to changes in acidity, and the decrease in pH slows the anaerobic pathway. Thus, excess lactic acid accumulation resulting from anaerobic glycolysis inhibits further ATP production. This is perceived as fatigue, or the inability to maintain exercise pace. This inhibitory effect is a protective response, since excess acidity can lead to cell death.

During and after exercise, excess lactic acid diffuses from the working muscles and is circulated via the blood to tissues such as the heart, liver, and other muscles. Lactic acid can be converted back to pyruvic acid and degraded via oxidative metabolism to produce ATP in these tissues (figure 10.3). Thus, excess lactic acid produced via anaerobic glycolysis can become a fuel for further ATP production in skeletal muscle.

After the end of exercise, excess lactate is also reconverted in the liver back to glucose, which can then be used to resynthesize glycogen depleted during exercise. It takes approximately 20 to 40 min to fully remove lactic acid produced during maximal exercise. The rate of lactic acid removal is faster during active compared with passive recovery. The best form of active recovery is light activity, such as slow jogging at approximately 30% to 60% maximum pace. During active recovery, the working muscles use the excess lactic acid as a fuel for ATP production. It is important that the pace of recovery be low enough so that more lactic acid is not produced. As discussed in the next chapter, interval training programs should consider the rates of lactic acid accumulation and removal during exercise.

Oxygen Supply During Sustained Exercise

The aerobic energy system provides most of the ATP for sustained exercise lasting longer than 3 min, and about 20% to 30% of ATP for all-out exercise lasting 30 to 60 s. If oxygen supply is insufficient to sustain ATP production, the muscles must increasingly rely on the anaerobic system with consequent buildup of lactic acid and inhibition of further ATP production.

Oxygen consumption ($\dot{V}O_2$) is an important measure of energy expenditure during exercise. A standard curve of oxygen consumption during exercise has several components (figure 10.5). During

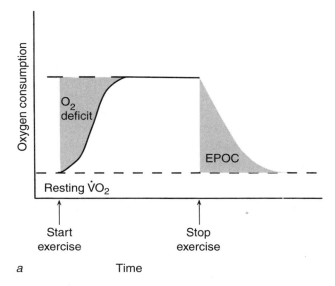

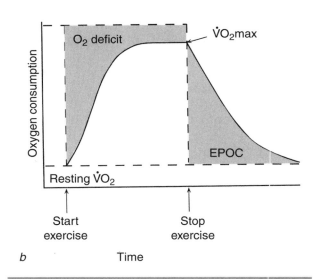

Figure 10.5 Oxygen consumption during exercise. *(a)* During submaximal steady-rate exercise, $\dot{V}O_2$ reaches a plateau. *(b)* During maximal exercise, oxygen consumption continues to increase until $\dot{V}O_2$max is achieved. During supramaximal exercise, adenosine triphosphate above that produced by oxidative metabolism is generated via anaerobic glycolysis.

the initial few minutes of exercise, oxygen uptake is not sufficient to provide all the energy needed, and the body is said to go into **oxygen deficit.** During this time, ATP is supplied primarily by the two anaerobic systems—stored phosphagen (PCr) and anaerobic glycolysis. The oxygen deficit occurs because it takes the cardiorespiratory system a few minutes to adjust to the increased energy demand at the onset of exercise. This is felt as discomfort during

the first few minutes of exercise, especially in unfit people. After the first few minutes comes a "second wind," as the oxidative system adjusts to provide the needed energy at the given rate of exercise.

During submaximal exercise of constant intensity, $\dot{V}O_2$ reaches a plateau ("steady state") at which oxygen uptake is sufficient to provide the ATP needed (figure 10.5a). Theoretically, exercise could proceed at this rate indefinitely. During exercise of continually increasing intensity, $\dot{V}O_2$ continues to increase with work rate until a maximal value and maximum exercise capacity are reached (figure 10.5b). Oxygen consumption will plateau at a value termed $\dot{V}O_2$**max** or maximum oxygen consumption, described in detail further on. (The dot over the V in $\dot{V}O_2$max indicates a volume rate, that is, volume per unit time, e.g., L/min.) During supramaximal exercise (above $\dot{V}O_2$max), ATP above that produced by oxidative metabolism is generated via anaerobic glycolysis, producing high lactate levels.

At the end of exercise, oxygen uptake does not return to resting levels immediately but takes some time to return to pre-exercise levels. This slow return of oxygen uptake after the end of exercise is called the **"excess postexercise oxygen consumption"** or EPOC (figure 10.5, *a* and *b*). In the past, EPOC was called the **"oxygen debt,"** a term implying that the oxygen consumed after exercise is used to "repay" the deficit occurring at the onset of exercise. We know now, however, that EPOC is more complex than a simple repayment of an anaerobic deficit. The excess oxygen goes toward removing lactate and resynthesizing muscle stores of glycogen, PCr, and ATP. Extra oxygen is needed for ATP production by the heart, since it is working hard to pump blood to muscles and skin, and also because elevated body temperature raises metabolic rate throughout the body. The size of EPOC varies with exercise intensity and duration. Intense prolonged exercise is associated with a large EPOC and a longer time (up to several hours) for oxygen consumption to return to baseline levels.

$\dot{V}O_2$max As an Indicator of Endurance Exercise Capacity

$\dot{V}O_2$max represents the maximum amount of oxygen an individual can consume per minute during exercise. The more oxygen consumed, the higher the capacity for ATP production via aerobic metabolism. Thus, $\dot{V}O_2$max gives an indication of endurance exercise capacity or the ability to continue exercising for a long time; it is also called **aerobic power.**

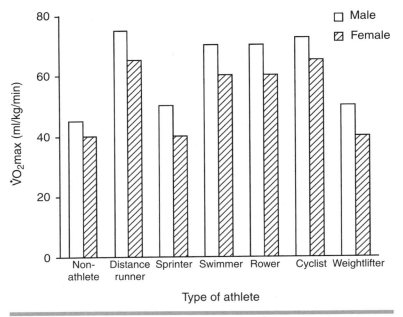

Figure 10.6 $\dot{V}O_2$ max values for young adult athletes. Average $\dot{V}O_2$ max levels for nonathletes and high-performance male and female athletes in various sports.

Due to a combination of heredity and training, elite endurance athletes generally have high $\dot{V}O_2$ max values (figure 10.6). Approximately 40% to 50% of $\dot{V}O_2$ max is genetically determined. In addition, $\dot{V}O_2$ max may increase by up to 40% with aerobic exercise training. Elite endurance athletes are, to a certain extent, genetically predisposed toward excelling in endurance events; see "Heredity and Exercise Capacity: How Much of Physical Performance Is Genetically Determined? (Or Are Great Athletes Born That Way?)" on page 148 in chapter 11. However, although a high $\dot{V}O_2$ max indicates broad potential for endurance exercise, $\dot{V}O_2$ max is not by itself the best predictor of exercise performance. That is, the individual with the highest $\dot{V}O_2$ max value will not necessarily be the top performer in endurance events. Other factors, such as training, motivation, skill, mechanical efficiency, and the ability to maintain exercise for long periods at a high percentage of maximum capacity, contribute to performance.

The effects of training on $\dot{V}O_2$ max are discussed in the next chapter, and changes with age and sex differences in $\dot{V}O_2$ max are discussed in chapter 12.

Measurement of Exercise Capacity

Exercise capacity can be precisely measured in the laboratory using various work monitors or ergom-

eters ("work meters") that determine the amount of work performed and energy expended during exercise. Exercise capacity can also be estimated or measured directly in sport-specific field tests. Standardized testing procedures have been developed to measure different types of exercise capacity, for example, **endurance exercise capacity,** muscular strength and power, and anaerobic work capacity.

Aerobic or Endurance Exercise Capacity

$\dot{V}O_2$ max is a measure of aerobic power, whereas endurance exercise capacity is a performance measure, such as the maximum time an individual can exercise at a given speed or the total amount of work that can be accomplished in a given time. $\dot{V}O_2$ max is assessed in the laboratory, where precise equipment can sensitively monitor oxygen consumption during exercise. $\dot{V}O_2$ max can be measured during any mode of exercise, but most frequently the mode is stationary running on a motorized treadmill (figure 10.7a) or cycling on a cycle ergometer. Sport-specific ergometers have also been developed to simulate activities such as rowing, kayaking, cross-country skiing, and swimming.

The mode of exercise selected to test $\dot{V}O_2$ max should be specific to the athlete's training. For example, a cycling test provides a more accurate measure of $\dot{V}O_2$ max for trained cyclists than a treadmill test. Measuring $\dot{V}O_2$ max on the treadmill is most common for the average person who is generally comfortable with walking and jogging, and for athletes in sports involving running.

In a $\dot{V}O_2$ max test, the participant starts to exercise at a comfortable pace, after which exercise intensity increases progressively until he or she can no longer continue to exercise at the given pace (this is called "volitional fatigue"). Exercise intensity is increased through increases in the speed or incline of the treadmill (or both), increases in pedal resistance on the bicycle ergometer, or increases in exercise pace for other types of ergometers.

Throughout the $\dot{V}O_2$ max test, the participant breathes into a two-way, lightweight mouthpiece or mask connected via a long tube to gas analysers. The amounts of oxygen, carbon dioxide, and total volume of air breathed are measured throughout the test and then used to calculate oxygen

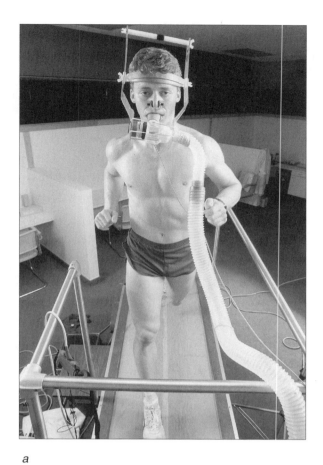

a   *b*

Figure 10.7   Exercise tests: *(a)* treadmill test of $\dot{V}O_2$max and *(b)* cycle ergometer test for anaerobic power and capacity.
Cycle ergometer photo courtesy of Michael Hinchey, The University of Queensland.

consumption. $\dot{V}O_2$max is defined as the highest value of oxygen consumed during exercise and is usually achieved during the final minute of exercise, just before volitional fatigue.

$\dot{V}O_2$max is expressed either as absolute volume in litres of oxygen consumed per minute (L/min) or, when adjusted for body mass, as millilitres of oxygen consumed per kilogram body mass per minute (ml $\cdot$ kg^{-1} $\cdot$ min^{-1}). $\dot{V}O_2$max in absolute terms (L/min) is closely related to body size, with larger individuals having higher absolute $\dot{V}O_2$max values than smaller individuals. Adjusting for body mass takes into account differences in body size and more accurately reflects endurance exercise capacity.

Anaerobic Exercise Capacity

Anaerobic power and capacity refer to exercise capacities in activities requiring energy production by the immediate and anaerobic glycolytic sys-tems, or brief intense exercise. **Anaerobic power** is the maximum or peak power, expressed in watts (W), that can be achieved in an all-out exercise test; it occurs within the first 2 to 3 s. **Anaerobic capacity** represents the total amount of work, expressed as kilojoules (kJ), that can be accomplished in a specified time, usually 30 to 60 s.

Anaerobic power and capacity are important in many sports and activities requiring rapid and powerful movement, such as sprinting or jumping. Several procedures—some general and others sport specific—are used for measuring anaerobic power and capacity. Among the most commonly used are 10- and 30-s cycle ergometer tests in which the participant pedals as fast as possible on a special cycle ergometer equipped with a work monitor to measure power and work (figure 10.7*b*). Other general tests include vertical jumping, sprinting, or stair climbing as measures of explosive power, and specific tests for team game sports.

Why Measure Exercise Capacity?

Precise measurement of aerobic and anaerobic power or anaerobic capacity allows evaluation of an athlete's current state of fitness and training program effectiveness. Because certain types of exercise capacity are very much related to genetics, testing can also be used in talent identification for some sports (see chapter 22).

$\dot{V}O_2$max provides an accurate and reproducible means of standardizing exercise intensity for research or exercise prescription. Exercise intensity can be set at a given percentage of maximum, say at a work rate eliciting 70% of $\dot{V}O_2$max. The absolute work rate may vary between people, but each individual will be working at a rate relative to his or her own ability. For example, athletes often train at a high percentage of maximum aerobic power (80-100% $\dot{V}O_2$max), whereas a lower exercise intensity (50-85% $\dot{V}O_2$max) is recommended for the average healthy nonathlete (see chapter 11).

$\dot{V}O_2$max is a good way to determine initial fitness level from which to develop an individual exercise prescription and to monitor progress. It is frequently measured in the clinical setting to assess exercise tolerance. For example, cardiologists measure $\dot{V}O_2$max and the heart's response to exercise in heart patients. These measures provide the cardiologist with a good indication of the extent of heart disease as well as the patient's progress during treatment.

Anaerobic power and capacity are important to many sports, especially those requiring explosive movements, such as sprinting, and game-type sports. Accurate assessment of these factors is important in talent identification and in developing and evaluating effective training programs.

The Cardiorespiratory System and Oxygen Supply During Exercise

The cardiorespiratory system consists of the lungs, the breathing tubes that carry air to the lungs (the trachea, bronchii, and smaller branches), and the heart and blood vessels (figure 10.8). Oxygen, which composes approximately 20.9% of inspired air, diffuses from the smallest air sacs in the lungs (alveoli) to blood contained in capillaries surrounding the alveoli. Gas exchange between the alveoli and capillaries is so rapid that, even during maximal exercise, arterial blood is almost fully saturated with oxygen. Blood is carried via the pulmonary veins to the heart, then via the arterial

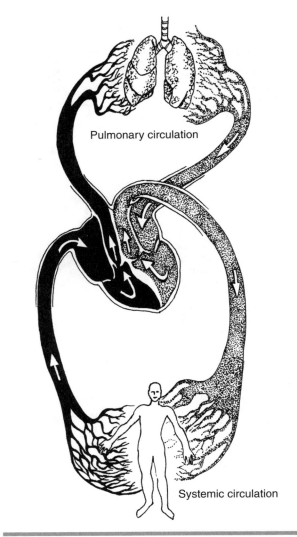

Figure 10.8   A schematic diagram of the circulatory system.

system through progressively smaller arteries to capillaries where gas exchange occurs within tissues. Oxygen delivery via the circulatory system to the working muscles is a major limiting factor for endurance exercise performance.

Cardiovascular Responses to Exercise

As most athletes know, even before the onset of exercise, heart rate and respiration increase (figure 10.9). This "anticipatory" response is due to neural input into centres controlling the heart and respiration. In general, heart rate, blood flow to the heart and skeletal muscles, and respiratory rate increase proportionately with increasing exercise intensity.

Heart rate increases linearly with increasing work rate up to a maximum value, which is age dependent (figure 10.9*a*). **Maximum heart rate** can be estimated by the equation 220 minus age, although there is large individual variation at any given age. Heart rate is controlled by input from the central nervous system and responds to changes in blood acidity (pH), oxygen and carbon dioxide content, and temperature as well as body movement.

Stroke volume (SV) is the volume of blood pumped by the heart with each contraction. During exercise, SV increases as the heart contracts more forcefully, ejecting more blood with each contraction. In untrained persons, SV increases early in exercise and reaches peak levels at 40% to 60% of $\dot{V}O_2$max. In endurance-trained athletes, however, SV continues to increase with exercise intensity (figure 10.9*b*).

Cardiac output ($\dot{Q}$) is the volume of blood pumped throughout the body per minute. $\dot{Q}$ is a function of both heart rate (HR) and SV ($\dot{Q}$ in L/min = HR in beats/min $\times$ SV in ml/beat). As with heart rate, cardiac output increases linearly with increasing work rate and reaches a plateau at maximum exercise capacity. During maximum exercise,

cardiac output may reach values approximately four to eight times resting levels (figure 10.9*c*). Cardiac output represents the ability of blood to circulate and to deliver oxygen to the working muscles, and is a major limiting factor for endurance exercise capacity. Cardiac output and oxygen consumption are linearly related; that is, a high cardiac output is needed for a high $\dot{V}O_2$max.

Oxygen extraction by skeletal muscles also increases during exercise (figure 10.9*d*). At rest, tissues extract only about 25% of the oxygen contained in blood, but during maximal exercise the skeletal muscles extract 75% to 85% of the oxygen in blood. Thus, during exercise, increased blood flow to, and oxygen extraction by, the working muscles ensure adequate oxygen delivery.

Minute ventilation, volume of air brought into the lungs per minute, is a function of both respiratory rate and the depth of each inspiration, both of which increase during exercise. Ventilation increases linearly with work rate up to about 55% to 75% of $\dot{V}O_2$max, after which it rises disproportionately with increasing work rate (figure 10.9*e*).

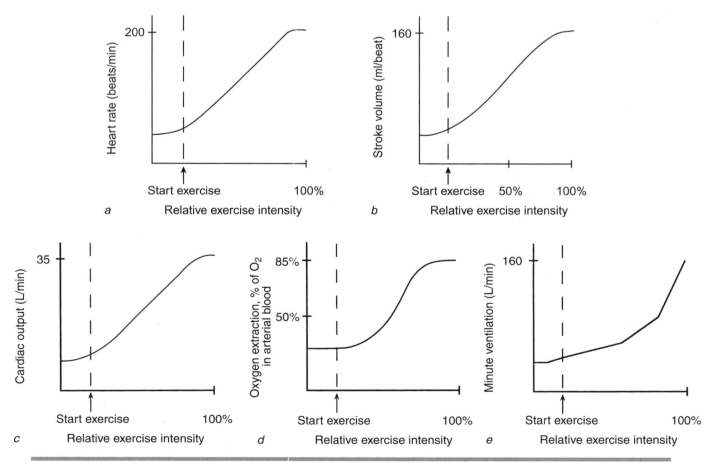

Figure 10.9   Cardiorespiratory system responses to exercise. *(a)* Heart rate, *(b)* stroke volume, *(c)* cardiac output, *(d)* oxygen extraction by skeletal muscle, and *(e)* minute ventilation.

During maximal exercise, ventilation may increase to 20 to 25 times above resting values. The large increase in ventilation ensures that blood flowing through the lungs is almost fully saturated with oxygen, even at maximum exercise. Thus, ventilation is not considered a limiting factor to exercise capacity, although fatigue of respiratory muscles may contribute to general fatigue during high-intensity exercise.

The combination of all the cardiorespiratory changes just discussed ensures that during exercise, oxygen supply closely matches oxygen use and energy demand in the skeletal muscles. At exercise intensities above $\dot{V}O_2$max, oxygen supply is insufficient to fully meet the demand for oxygen and ATP within skeletal muscle, and the excess energy is supplied via the nonoxidative pathways.

Distribution of Blood Flow During Exercise

During exercise the increased cardiac output is not uniformly distributed throughout the body. Rather, blood is redirected through the circulation so that, with increasing exercise, proportionally more blood flow goes to working muscles (figure 10.10). At rest (figure 10.10a), only about 20% of blood flow goes to skeletal muscles, with most going to the brain and internal organs (viscera). At the onset of exercise, blood flow to the viscera is reduced by narrowing of small arteries (constriction). At the same time, small arteries in the working muscles and skin open (dilate), increasing blood flow to these areas. Circulation to the heart increases in proportion to heart rate and exercise intensity. During submaximal exercise (figure 10.10b), about 50% to 60% of blood flow may be directed to working muscles and about 10% to the skin. During maximal exercise (figure 10.10c), nearly 80% of cardiac output may go to the working muscles. This redirection of blood flow is important to providing sufficient oxygen, glucose, and fatty acids to the muscles and heart and to removing carbon dioxide, lactic acid, and heat from the working muscles.

Human Skeletal Muscle Cells

Human skeletal muscle cells (called muscle fibres) are not homogeneous. The different fibre types provide specialization in muscle function, permitting muscle fibres to adapt to a wide range of demands. It is thought that muscle cells cannot adapt completely to the full range of demands imposed by intense endurance and strength training. The specialization of fibre type allows certain fibres to adapt optimally to one type of demand, say by increasing in size and glycolytic capacity in response to strength training, while another fibre type can adapt optimally to another set of demands, say by increasing oxidative capacity in response to endurance training.

Muscle Fibre Types

In the human and many other mammals are three types of muscle fibres, each adapted for specialized function. Muscle fibre types are classified according to their physiological, biochemical, and histological properties (see table 10.2).

Muscle fibres can be classified according to

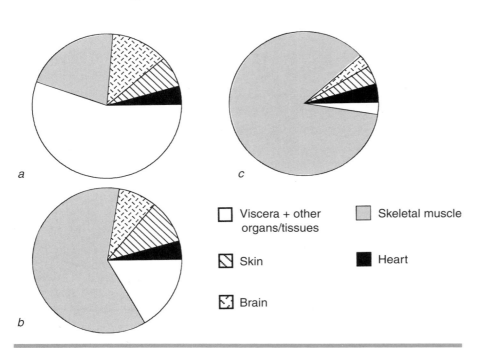

Figure 10.10 Blood flow redistribution during exercise. (a) Rest, (b) submaximal exercise, and (c) maximal exercise. Relative proportions are expressed as a percentage of cardiac output.

□ Viscera + other organs/tissues

▨ Skin

▨ Brain

▨ Skeletal muscle

■ Heart

Table 10.2

Human Skeletal Muscle Fibre Types

Characteristic	Slow-twitch (ST)	Fast-twitch (FT)a	FTb
Fibre size	Small	Large	Large
Contraction speed	Slow	Fast	Fast
Force	Low	High	High
Glycolytic capacity	Low	High	High
Oxidative capacity	High	Moderately high	Low
Capillary supply	High	Moderately high	Low
Fatigue resistance	High	Moderate	Low

their relative speed of contraction as slow-twitch (ST) or fast-twitch (FT). Slow-twitch fibres are also known as type I and FT as type II. The ST fibres contract within 100 ms, whereas FT fibres contract within 50 ms. Muscle fibres can also be differentiated by the predominant energy system used to produce ATP. Glycolytic fibres have a high capacity for anaerobic (glycolytic) metabolism, whereas oxidative fibres have a greater capacity for aerobic metabolism. Fast-twitch fibres are primarily glycolytic, and ST fibres are primarily oxidative. Combining information about contraction speed and metabolism allows us to classify human skeletal muscle fibres into three types:

- Type I or slow oxidative (ST) fibres
- Type IIA or fast oxidative glycolytic (FTa) fibres
- Type IIB or fast glycolytic (FTb) fibres

Each muscle fibre is innervated by only one motor neuron (nerve to the muscle cell), and it is the motor neuron that determines fibre type. Each motor neuron may innervate up to several hundred muscle fibres; one motor neuron and all the muscle fibres it innervates are called the motor unit (figure 2.12). Because only one motor neuron innervates a motor unit, all muscle fibres within a motor unit are of the same fibre type.

In the average human, skeletal muscles are composed of approximately 50% ST and 50% FT fibres; about 25% of muscle fibres are type FTa and 25% type FTb. Most human skeletal muscles consist of a mixture of muscle fibre types, although particular muscles may contain a higher percentage of one fibre type. Athletes in particular sports may exhibit different fibre type distribution compared with the average individual.

Slow-twitch fibres are well suited to activities requiring low force generated over a long time. Muscles used primarily for posture and endurance activities (e.g., the soleus muscle, the deeper muscle in the calf) contain a high proportion of ST fibres. Muscles used for forceful contractions (e.g., the gastrocnemius, the superficial muscle in the calf) generally contain a higher percentage of FT fibres. The various fibre types are mixed within the muscle so that a microscopic view of a muscle in cross section reveals a "mosaic" of the different fibre types as shown in figure 10.11. The sample depicted was histochemically stained to delineate the three fibre types: type I (ST or slow oxidative), type IIA (FTa or fast oxidative glycolytic), and type IIB (FTb or fast glycolytic) (see table 10.2). Capillaries (C) appear as small round profiles between skeletal muscle fibres.

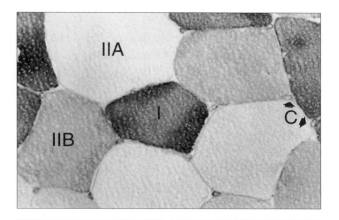

Figure 10.11 Muscle biopsy—photomicrograph (photograph through a light microscope) of a cross section of human skeletal muscle from the vastus lateralis (thigh) muscle.

Photomicrograph courtesy of Dr. P.R.J. Reaburn.

As table 10.2 shows, ST fibres are smaller, are slower to contract, and are not capable of generating as much force as FT fibres. Slow-twitch fibres are fatigue resistant; that is, they can continue to contract repeatedly without fatigue. These fibres contain many mitochondria and are surrounded by several capillaries, ensuring a generous supply of oxygen. Thus, ST fibres have a high capacity for oxidative metabolism and are used primarily for endurance activities. It is also believed that, during intense exercise, ST fibres take up and metabolise lactic acid produced by FT fibres.

FTb fibres are the largest, fastest, and most forceful of the three fibre types. FTb fibres have a low oxidative but high anaerobic glycolytic capacity and are capable of producing large amounts of lactic acid; these fibres fatigue easily, that is, they are not fatigue resistant.

FTa fibres exhibit properties of both ST and FTb fibres. They resemble FTb fibres in that they are large, fast, capable of forceful contraction, and high in glycolytic capacity. FTa fibres are also similar to ST fibres in that they have more mitochondria, a moderate capillary supply, and higher oxidative capacity than type FTb fibres. FTa fibres are used in activities such as competitive rowing, swimming, sprinting, and moderate-intensity weightlifting.

Muscle Fibre Type and Exercise Capacity

The different types of muscle fibres allow muscles to respond to a wide variety of exercise demands—from finely controlled movements requiring little force to fast and powerful movements requiring maximum force.

Activation of Fibre Types During Exercise

When a skeletal muscle contracts, not all muscle fibres are activated or "recruited" to produce force. Muscle fibres are activated in proportion to the amount of force required. Muscle fibre recruitment follows a pattern, called the "size principle," in which smaller ST fibres are activated first, followed by FTa and then type FTb fibres; FTb fibres are activated during forceful contractions requiring above 70% of maximum muscular force (figure 10.12). Slow-twitch fibres are activated during all contractions, and although FT fibres may provide most of the force during near-maximal contractions, ST fibres are still activated and contribute to force production. This principle can be applied to the design of exercise training programs. For example, low-intensity endurance (e.g., walking or slow jogging) exercise recruits mainly ST

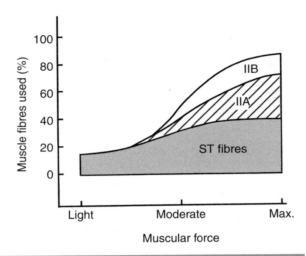

Figure 10.12 Skeletal muscle fibre recruitment during exercise. Smaller slow-twitch (ST) fibres are activated first, followed by larger fast-twitch fibres (FTa and then FTb fibres); FTb fibres are activated during forceful contractions requiring above 70% of maximum muscular force.
Adapted from Wilmore and Costill 1994.

fibres; higher-intensity endurance exercise must be performed in order to recruit and thus induce adaptations in FTa and FTb fibres. Similarly, high-intensity weight training, near maximum muscular effort, will recruit FTb fibres (as well as ST and FTa fibres). The next chapter covers the application of muscle fibre recruitment patterns to training for sport.

Skeletal Muscle "Fibre Typing"

In humans, muscle "fibre typing" is performed on a small (20 mg, about the size of a grain of rice) sample of muscle obtained by muscle biopsy technique. A small (1 cm or 1/2 in.) incision is made through the skin and underlying tissue down to the muscle layer. A thin muscle biopsy needle is inserted into this incision, and the muscle sample is removed along with the needle. The incision is then closed with a suture. There are no pain nerve endings in skeletal muscle, and the skin incision is performed under local anaesthesia. The entire procedure, performed under sterile conditions, has become quite common in exercise physiology research. The athlete can usually resume training within a day of having a muscle biopsy. However, because obtaining a muscle biopsy is an invasive procedure, it is not generally used for routine testing of athletes and remains primarily a research tool for understanding the metabolic responses to exercise.

Once obtained, the muscle sample is frozen to preserve its structure and then cut in cross section using specialized equipment; the cylindrical muscle fibres appear as circular profiles. Sections are stained with special dyes to visualize muscle fibre properties, such as glycogen content or muscle proteins.

Importance of Muscle Fibre Type to Sport Performance

As shown in figure 10.13, the relative proportions of fibre types differ between different types of athletes. For example, high-performance distance runners tend to have about 80% ST and 20% FT fibres, whereas elite sprinters and power athletes have 60% to 75% FT and 25% to 40% ST fibres. These differences imply that muscle fibre type is important to elite performance in specific sports. However, as with $\dot{V}O_2$max, top performance is related to many factors. Muscle fibre type gives only a broad indication of potential in those sports at the extremes of the energy system continuum, such as distance running, sprinting, or powerlifting. A mixture of fibre types is advantageous in other sports requiring use of all energy systems, such as soccer, basketball, or tennis, as well as skill-based activities such as golf.

The question whether skeletal muscle fibre type and number can be changed with exercise training is discussed in chapter 11.

Energy Cost of Activity

The energy cost of exercise depends on the activity, its intensity or pace, the exerciser's mechanical efficiency or technique, and—for nonsupported activities—the individual's body mass. Environmental factors such as temperature, wind, or terrain may also influence the energy cost of activity.

As described in chapter 7, mechanical efficiency is the energy cost used to accomplish a specific amount of work; the human body is at most 25% efficient. Individuals vary in their mechanical efficiency and the energy cost of their exercising at a certain pace; that is, a mechanically efficient person uses less energy to exercise at a given pace than someone less mechanically efficient. In some activities (e.g., swimming, rowing), mechanical efficiency or technique is an important determinant of performance. The term "economy of movement" denotes the oxygen cost of a particular exercise, such as the volume of oxygen consumed during running at a given pace. Economy of movement and mechanical efficiency are related, since oxygen consumption is an indirect measure of energy use.

For activities in which body mass is supported such as cycling or swimming, energy cost is relatively independent of body mass. However, for activities in which body mass is not supported, such as walking or running, the energy cost increases with body mass. The energy cost is highest in activities that use the entire body or large muscle groups, such as running or swimming.

Knowledge about the energy cost of activity can be used in several ways. For example, many people exercise to control body mass (weight); knowing how much energy is expended in certain activities is useful in prescribing the appropriate amount of exercise for fat loss. The energy cost of training must also be considered in the planning of diets for athletes. Athletes, especially endurance athletes, often have difficulty consuming sufficient energy and carbohydrate to meet the high metabolic demands of training. Measurement of the energy cost of exercise is important to understanding the physiological responses and regulation of metabolism during exercise.

Importance of Diet to Energy Metabolism and Exercise Performance

Most athletes and coaches are aware of the importance of diet to exercise performance and the special dietary needs of athletes. For example, athletes need to consume large amounts of energy and carbohydrate to fuel prolonged exercise, and

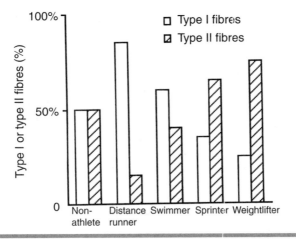

Figure 10.13 Skeletal muscle fibre type distribution in nonathletes and various types of athletes.

Data from McArdle, Katch, and Katch 1996.

sport teams often employ dietitians to provide nutritional advice for athletes.

Why Athletes Need a High-Carbohydrate Diet

The availability of certain fuels, especially carbohydrate, is important to exercise capacity. As shown in figure 10.14, exercise capacity is directly related to both pre-exercise muscle glycogen stores and the amount of dietary carbohydrate consumed. A low-carbohydrate diet is associated with low muscle glycogen stores and poor exercise capacity, whereas a high-carbohydrate diet increases muscle glycogen stores and exercise capacity.

Muscle glycogen stores can be depleted by as little as 40 min of intense prolonged exercise, such as distance running. Fatigue occurs at the point of muscle glycogen depletion. Once muscle glycogen stores are depleted, glucose is no longer available to the muscle and the body must rely almost entirely on fatty acids, and to a limited extent amino acids, for ATP production. Because of the different chemical composition of fats and glucose, about 7% less energy is produced per given amount of oxygen with fat as a substrate compared to glucose (table 10.3). Once glycogen is depleted and fat becomes the predominant substrate, exercise pace slows because less ATP is produced. Thus, glycogen

Table 10.3

Energy Released per Litre of Oxygen Consumed

Substrate	Energy per litre of oxygen, kcal (kJ)
Carbohydrate	5.05 (21.2)
Fat	4.70 (19.7)
Protein	4.82 (20.2)

depletion is associated with an inability to maintain the rate of exercise and with the perception of fatigue; in marathon running, this point is called "hitting the wall."

Muscle glycogen stores can also be depleted by prolonged periods of interval exercise in which anaerobic glycolysis is the predominant ATP-producing system. For example, 30 s of all-out sprinting may deplete 25% of muscle glycogen, and ten 1-min sprints may deplete 50% of muscle glycogen. Sports in which glycogen depletion may occur include soccer, football, basketball, and any activity requiring repeated high-intensity sprinting over a prolonged period. Swimmers, rowers, and weightlifters training over extended periods of time may also experience fatigue due to glycogen depletion. Thus, dietary carbohydrate is important to many athletes, not just those participating in continuous, prolonged events.

Once glycogen is depleted from muscle, it takes 24 to 48 h to fully restore glycogen levels. Athletes who train intensely each day may thus be chronically glycogen depleted and have difficulty maintaining training and competitive performance.

Fortunately, one adaptation to intense training is that muscle stores more glycogen. Moreover, muscle glycogen stores can be further increased by consumption of a high-carbohydrate diet. Muscle glycogen replenishment depends on the type of diet, type of carbohydrate, and how soon a meal is consumed after exercise. Glycogen repletion is generally faster if high-carbohydrate food is consumed as soon as possible after exercise (see "Replacing Muscle Glycogen After Exercise: Does the Type of Food Matter?" on p. 139).

Athletes who train for several hours each day should regularly consume a high-carbohydrate diet. For most athletes, a high-carbohydrate diet consists of 60% to 80% of daily energy intake as carbohydrate, equivalent to 6 to 8 g carbohydrate per kilogram of body weight per day or 420 g (15 oz) to 560 g (20 oz) carbohydrate per day for a 70-kg (154-lb) person. During times of very intense

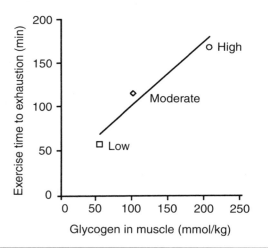

Figure 10.14 Dietary carbohydrate and exercise performance. Increasing dietary carbohydrate enhances muscle glycogen stores. Higher muscle glycogen content increases exercise time to exhaustion by delaying the point of glycogen depletion and, thus, the onset of fatigue.

Reprinted from Wilmore and Costill 2004.

IN FOCUS: REPLACING MUSCLE GLYCOGEN AFTER EXERCISE: DOES THE TYPE OF FOOD MATTER?

Intense endurance exercise or repeated high-intensity exercise can deplete muscle glycogen stores, leading to fatigue. To help replace muscle glycogen after exercise, many athletes consume a high-carbohydrate diet. Carbohydrates are contained in many foods, and athletes need specific information about whether certain types of carbohydrate foods are better than others for replenishing muscle glycogen.

Glycemic index (GI) is a measure of the impact of a food on blood glucose level; GI is determined through measurement of the blood glucose response in a fasted individual after ingestion of a specific amount of carbohydrate food. Foods with a high GI, such as bread or mashed potatoes, elicit a higher blood glucose and insulin response after ingestion. In contrast, ingestion of foods with a low GI, such as rolled oats or legumes, elicits a lower blood glucose and insulin response. High-GI foods cause a larger increase in blood insulin and glucose levels, which is thought to enhance glucose uptake and glycogen synthesis by muscle.

In one study (Burke, Collier, and Hargreaves 1993), well-trained male cyclists performed exercise to deplete muscle glycogen on two occasions. For 24 h after each exercise bout, the cyclists consumed either high-GI or low-GI foods. Blood glucose and insulin levels, as well as muscle glycogen storage, were higher with consumption of the high-GI foods. These results suggest that muscle glycogen stores are replenished faster when high-GI foods are eaten in the 24 h after intense endurance exercise. Thus, athletes needing to replace muscle glycogen should consume moderate- to high-GI foods during the day after intense exercise.

Adding protein to carbohydrate may enhance glycogen replacement after depletion (Zawadzki, Yaspelkis, and Ivy 1992). Roy and Tarnopolsky (1998) measured the rate of glyco-gen synthesis in 10 healthy men who performed resistance training to exhaustion. Immediately and 1 h after exercise, one group drank a high-carbohydrate drink and another group drank a combined carbohydrate–protein–fat drink; the two drinks contained the same total amount of energy. The rate of muscle glycogen resynthesis after exercise was similar in the two groups. It was concluded that the rate of muscle glycogen synthesis is sensitive to both the amount of carbohydrate and the total amount of energy contained in foods consumed immediately after exercise that depletes muscle glycogen.

Taken together, these studies indicate that the rate of muscle glycogen replenishment after exercise depends on the types of foods ingested in the first few hours of recovery. Glycogen replenishment is enhanced by high-energy, high-GI carbohydrate foods with or without high protein. Both the total amount of carbohydrate and total amount of energy ingested are important factors. These studies have greatly helped athletes and dietitians determine the best foods to eat during recovery after muscle glycogen-depleting exercise.

SOURCES
- Burke, L.M., G.R. Collier, and M. Hargreaves. 1993. Muscle glycogen storage after prolonged exercise: Effect of the glycemic index of carbohydrate feedings. *Journal of Applied Physiology* 75: 1019-1023.
- Roy, B.D., and M.A. Tarnopolsky. 1998. Influence of differing macronutrient intakes on muscle glycogen resynthesis after resistance exercise. *Journal of Applied Physiology* 84: 890-896.
- Zawadzki, K.M., B.B. Yaspelkis, and J.L. Ivy. 1992. Carbohydrate-protein complex increases the rate of muscle glycogen storage after exercise. *Journal of Applied Physiology* 72: 1854-1859.

training for more than 2 h per day, or during the taper before major competition, this may increase to 9 to 10 g carbohydrate per kilogram per day.

One recommendation is that, when an athlete plans further exercise within the next 6 to 12 h, he or she should consume 1 g of carbohydrate per kilogram body weight within the first 30 min after the first bout of exercise, and then again every 2 h up to the next exercise session. This practice is especially important in multi-event competitions in which athletes compete in several events within a few days. Easily digested foods such as sports drinks or soft drinks, fruits, breads, wheat cereals, and glucose confectionery are effective.

Do Athletes Need Extra Protein?

Many athletes, especially sprinters and weightlifters, consume a high-protein diet and supplement their diets with protein or amino acid powders or drinks, believing that the extra protein will increase muscle growth or hypertrophy. It is true that athletes need more protein in the diet than the average nonathlete. However, a well-balanced diet is more than adequate to meet the protein needs of virtually all athletes.

For the average healthy nonathlete, the recommended daily intake of protein is 12% to 15% of daily energy consumption, or 0.8 g protein per kilogram per day. For a 70-kg (154-lb) person, this is about 56 g (2 oz) of protein per day. (The protein content of certain foods is listed in table 10.4.) On an absolute protein basis, both power/strength and endurance athletes need about 50% to 100% more protein per day than the nonathlete. The general recommendation is that strength/power/speed athletes should consume 1.5 to 2.0 g protein per kilogram body weight per day, and endurance athletes should consume 1.5 to 1.6 g protein per kilogram per day.

However, because athletes expend large amounts of energy during training, they also consume at least 50% more energy per day than the nonathlete. Athletes can easily meet their additional protein needs through a well-balanced diet that contains 12% to 15% of daily energy as protein, simply by increasing the total amount of food consumed. These protein needs of athletes are depicted in table 10.5. To provide the recommended 1.5 to 2.0 g protein per kilogram per day, the 70-kg (154-lb) athlete would require approximately 105 to 140 g (3.7 to 4.9 oz) protein per day. A daily diet of 3,750 kcal (15,750 kJ) containing 15% of total energy as protein would provide sufficient carbohydrate and protein to meet the energy and protein needs of any athlete.

Excess dietary protein is not incorporated into muscle but is excreted by the kidney or used to synthesize fat. Neither is desirable, since excess protein excretion places an extra burden on the kidneys and excess fat impairs performance in athletes. High-protein diets and protein supplements are also costly. In addition, a high-protein diet usually does not have sufficient carbohydrate to fuel extended training sessions. By emphasizing protein at the expense of carbohydrate in the diet, the athlete may be unable to train at a high intensity for long. Moreover, muscular hypertrophy may be compromised because inadequate energy is available to fuel protein synthesis in the muscle cell.

Importance of Replacing Water Lost During Exercise

Although water is rarely thought of as a nutrient, it is an essential part of the diet. Because the human body is relatively inefficient, most (70-80%) of the energy produced during physical activity is not used to perform work but appears as heat. The body can withstand only a relatively narrow range of core temperature. To prevent core temperature

Table 10.4
Protein Contained in Some Foods

Food	Protein (grams)	Serving size
Milk	9	250 ml (8 oz)
Cheese	5-7	30 g (1 oz)
Steak	26-34	100 g (3.5 oz)
Fish	18-26	100 g
Chicken	30-34	100 g
Wholemeal bread (whole-grain)	3.5	1 slice
Cereal	2-6	30 g
Nuts	4	25 g
Egg	6	1 egg
Fruit	1	1 piece
Vegetables	1-2	30 g

Compiled from Burke and Deakin 1994; Inge and Brukner 1986; McArdle, Katch, and Katch 1996; Stanton 1988.

Table 10.5

Meeting Protein Needs Through the Diet

	Nonathlete	Athlete
Total energy consumed, kcal (kJ) per day	2,500 (10,500)	3,750 (15,750)
Body mass, kg	70	70
Percent energy intake as protein	12% to 15%	12% to 15%
Energy intake as kcal (kJ) per day	300 (1,260) to 375 (1,575)	450 (1,890) to 562 (2,362)
Grams of protein per day*	300 kcal/4.2 = 71 to 375/4.2 = 89	450/4.2 = 107 to 562/4.2 = 134
Grams of protein per kilogram body mass per day	71/70 = 1.0 to 89/70 = 1.3	107/70 = 1.5 to 134/70 = 1.9

*1 g protein yields 4.2 kcal when metabolised by the body.

from rising dangerously during exercise, heat produced during exercise is transferred from muscles to blood, then through the circulation to the blood vessels in the skin.

The primary mechanism for heat loss during exercise is evaporation of sweat. Depending on the individual, exercise intensity and duration, and environmental conditions, sweat rates may vary between 0.5 and 3.0 L/h (between 1 pint and 6.3 pints/h). Even with evaporative cooling, an athlete's core temperature may increase from the normal 37° C (98.6° F) up to 39.5° C (103.1° F) during intense prolonged exercise.

During prolonged exercise, redistribution of water within the body and loss of body water via sweating may reduce blood volume. Losing body water equivalent to 4% to 5% of body mass may adversely affect thermoregulation and exercise capacity. The cardiovascular system adjusts to this loss of blood volume by increasing heart rate to offset the concomitant decline in stroke volume and cardiac output. Thus, a given exercise will cause greater stress on the cardiovascular system when performed in a hot versus a more moderate environment.

If prolonged exercise in the heat continues without replacement of body water, blood volume may drop significantly and the body may be unable to lose excess heat. Body temperature may increase dangerously, above 42° C (107.6° F). Heat illness or heatstroke, in which the athlete's cardiovascular and thermoregulatory systems are severely impaired, may occur. Heatstroke is life threatening if not treated promptly.

Athletes should consume water at regular intervals during prolonged exercise, especially when exercising in the heat. The general recommendation is to drink approximately 500 to 1,000 ml (1 to 2 pints) plain water 1 h before exercise, an additional 250 to 500 ml (1/2 to 1 pint) 20 min before exercise, and then 250 ml every 15 min during exercise. Plain water is sufficient for exercise up to 60-min duration. Addition of glucose and electrolytes (e.g., sports drinks) is recommended for intense exercise lasting longer than 60 min, when muscles may become depleted of glycogen. A modest amount of glucose (about 6%) in solution will help improve performance without compromising water replacement in very long events. A low concentration of electrolytes promotes faster absorption of water.

Rehydration after exercise is especially important, and it may take several hours to completely replace water lost during exercise. Thirst is a poor indicator of the need for fluids or amount of fluid lost via sweating, and the athlete must make a conscious effort to fully replace fluids before the next bout of exercise. A simple guide to the amount of fluid lost during exercise is the difference between pre-exercise and postexercise body weight. Whenever possible, the athlete should drink enough fluid to return body weight to pre-exercise values before the next bout of exercise.

Summary

Exercise physiologists study limitations to exercise capacity—questions such as what causes fatigue during intense or prolonged exercise and what are the limits to human performance. Exercise capacity may be limited by the rate of energy production within muscle cells; when ATP demand exceeds the rate of production, fatigue occurs and exercise pace must slow. Skeletal muscles produce ATP via three different systems—the immediate (ATP-PCr), anaerobic glycolytic (anaerobic or lactic acid), and oxidative (aerobic) systems. These three systems ensure that ATP is available for all types of exercise, from short intense bursts to prolonged exercise. For prolonged exercise, oxygen must be continually supplied to the working muscles. When oxidative metabolism cannot produce sufficient ATP, the muscles rely more on anaerobic glycolysis, with subsequent buildup of lactic acid, causing muscle fatigue. Although $\dot{V}O_2$max (maximum oxygen consumption) is a good general indicator of endurance exercise capacity, there are better predictors, in particular the ability to exercise for a long time at a high percentage of $\dot{V}O_2$max without buildup of lactic acid. Human skeletal muscle cells are of three types: ST fibres are small with high oxidative capacity; FTb fibres are large with high glycolytic capacity; FTa fibres are large and have moderately high oxidative and high glycolytic capacity. In general, fibre type cannot be changed through exercise

training. Diet is important to exercise capacity, and athletes from most sports need to consume a balanced diet that is high in carbohydrate to ensure adequate energy supply for exercise. Replacement of water during and after exercise is important to exercise performance and health.

Further Reading and References

Burke, L., and V. Deakin. 2000. *Clinical sports nutrition.* 2nd ed. Sydney: McGraw-Hill.

Gore, C.J., ed. 2000. *Physiological tests for elite athletes.* Champaign, IL: Human Kinetics.

Heyward, V.H. 2002. *Advanced fitness assessment and exercise prescription.* 4th ed. Champaign, IL: Human Kinetics.

Houston, M.E. 2001. *Biochemistry primer for exercise science.* 2nd ed. Champaign, IL: Human Kinetics.

Wilmore, J.H., and D.L. Costill. 2004. *Physiology of sport and exercise.* 3rd ed. Champaign, IL: Human Kinetics.

CHAPTER 11

Physiological Adaptations to Training

The major learning concepts in this chapter relate to
- what limits exercise performance,
- responses of different energy systems to different types of training,
- adaptation of the muscular system after strength training,
- exercise training effects on muscle fibre number and fibre type,
- basic principles of training,
- training for cardiovascular endurance,
- exercise for health-related fitness,
- methods of strength training, and
- causes of muscle soreness.

The human body has a tremendous ability to adapt over time in response to repeated bouts of exercise, resulting in an increased physical work capacity. These long-term adaptations are called chronic or training adaptations.

The purpose of exercise training is to induce metabolic and structural adaptations in muscle that delay the onset of fatigue. Compared with the untrained individual, the trained athlete can perform more work, or exercise at a faster pace or for a longer time, before the onset of fatigue.

Effective training to enhance exercise capacity must take into account the different energy systems used to produce adenosine triphosphate (ATP) in skeletal muscle, discussed in the previous chapter. This is true whether the goal of training is to improve performance in an athlete, to enhance health in the average individual, or to treat a patient with a particular illness. As we shall see in this chapter, the outcomes of any training program depend very much on the type of training undertaken.

Muscular fitness is equally important to performance in virtually every sport; thus effective training programs must also attempt to induce structural change in the muscles that will result in increased strength, power, and endurance.

Training-Induced Metabolic Adaptations

One of the challenges for exercise physiologists is to identify the relative contributions of the various energy systems to a particular activity and to use this information to develop training programs that maximize adaptations and thus performance. The following are examples of how different types of training result in different types of change:

- Muscle glycogen stores increase with endurance training. As discussed in the previous chapter, muscles fatigue when depleted of their glycogen stores by endurance events or activities requiring repeated high-intensity exercise over an extended time (e.g., team game sports). Increasing pre-exercise glycogen stores can delay glycogen depletion and thus the onset of fatigue. In other words, the endurance-trained or team game athlete will be able to exercise longer before glycogen depletion causes fatigue.

- Muscle phosphocreatine (PCr) stores may be increased by power and short sprint training. As discussed in the previous chapter, PCr provides the major means of ATP production during short, maximal exercise. Since PCr is rapidly depleted, increasing PCr stores via training enables more

ATP to be synthesized before depletion during short explosive activities.

• An endurance-trained athlete begins to sweat earlier and sweats more during exercise than an untrained person. This allows the athlete to avoid a precipitous rise in body temperature, which could compromise performance, during long-duration events.

Factors Limiting Exercise Performance

Factors that limit exercise capacity and performance are closely related to the predominant energy system(s) used during a particular activity. For example, factors causing fatigue or limiting sprinting performance are different from those limiting performance in endurance events (see "Summary of Causes of Fatigue During Exercise" below).

For power and speed activities lasting only a few seconds, such as sprinting or jumping, performance is limited by the ability to recruit fast-

SUMMARY OF CAUSES OF FATIGUE DURING EXERCISE

Brief, high intensity (<1 min)
• PCr depletion
• Moderate to high lactate levels
• Disturbance of chemical gradients across cell membrane

Longer, high intensity (1-7 min)
• PCr depletion
• High lactate levels
• Disturbance of chemical gradients across cell membrane

Prolonged, moderate to high intensity (10-40 min)
• Moderate lactate accumulation
• Partial glycogen depletion
• Dehydration
• Disturbance of chemical gradients across cell membrane

Very prolonged (>40 min)
• Glycogen depletion
• Dehydration
• Increased body temperature
• Low blood glucose levels
• Disturbance of blood amino acid levels

twitch (FT) muscle fibres to generate maximal force and power, as well as by the ability to maintain balance and coordination while generating such high muscle force. In brief high-intensity movement in which maximal power is exerted for up to 20 to 30 s, such as 100- to 200-m (109- to 219-yd) running or 50-m (55-yd) swimming, performance is related to the limited amount of ATP and PCr stored in the muscles. As discussed in the previous chapter, PCr provides a rapid means of ATP synthesis but is soon depleted during maximal effort. Given that PCr is resynthesized only after exercise, ATP must be supplied by the two other systems (anaerobic glycolysis, oxidative metabolism). Elite sprinters are characterized by an ability to use PCr to resynthesize ATP at a faster rate than non-elite sprinters.

During maximal exercise lasting between 30 s and 2 to 3 min, such as 400- to 800-m (437- to 875-yd) running or 100- to 200-m (109- to 219-yd) swimming, anaerobic glycolysis is the major source of ATP synthesis. As discussed in chapter 10, anaerobic glycolysis provides ATP at a relatively fast rate but is inhibited and thus limited by lactic acid accumulation. Performance in very intense exercise is thus limited by the ability to buffer (neutralize) and remove excess lactic acid from the muscle. Fatigue also occurs as a result of disturbance of the chemical and electrical gradients across the muscle cell membrane, due to changes in the intra- and extracellular distribution of electrolytes such as potassium.

Middle distance events lasting between 3 and 10 min, such as 1,500- to 3,000-m (1,640- to 3,280-yd) running, 400- to 800-m (437- to 875-yd) swimming, and 4,000-m (4,374-yd) cycling, are limited by a combination of lactic acid accumulation, moderate glycogen depletion, and disturbance of electrolyte distribution between the muscle cells and extracellular environment.

Performance in longer events lasting between 10 and 40 min, such as 10-km (6.2-mi) running or 1,500-m (1,640-yd) swimming, is limited by a combination of factors including moderate lactic acid accumulation and partial glycogen depletion, dehydration, and disturbance of the chemical and electrical gradient across the muscle cell membrane.

In very long events lasting more than 40 min, such as road cycling or distance running, performance is limited by a combination of factors: glycogen depletion, dehydration, increased body temperature, low blood glucose levels, and changes in the ratios of amino acids in the blood.

Dehydration and increased body temperature increase demands on the cardiovascular system to direct blood flow to the skin and working muscles. Low blood glucose levels and altered blood amino acid ratios contribute to central nervous system fatigue, especially the sensation of fatigue.

Activities that rely on all the energy systems over an extended time, such as team games (e.g., basketball or soccer), are limited by a combination of factors similar to those described for long-duration events. For example, glycogen depletion may occur after repeated high-intensity sprinting required in many stop–start sports such as tennis or basketball. The extended duration of a match (60-90 min) may also increase body temperature, lower blood glucose levels, and cause dehydration.

Because training adaptations of a particular energy system lead to improvement in performance in events using that system, it is imperative to understand which energy systems are predominant in which sports and the interplay among these systems. To design an effective training program, then, one must tailor its metabolic demands to appropriately challenge the energy systems used by the sport in question. The following sections deal with responses to the demands of various types of training.

Immediate and Anaerobic System Changes After Strength and Sprint Training

Table 11.1 summarizes the metabolic changes in the anaerobic pathways occurring after sprint and strength training. Sprint and strength training increase PCr, ATP, and glycogen stores within

muscle fibres, especially FT fibres. These increases facilitate a higher power output in short-duration exercise through an increased capacity for ATP resynthesis via the breakdown of PCr and through anaerobic glycolysis.

The activities of the anaerobic glycolytic enzymes also increase with training, enhancing the amount of ATP generated by anaerobic glycolysis. In addition, sprint training and high-intensity strength training increase muscle glycogen storage, enhancing glycogen availability during repeated sprints and power (e.g., jumping) or strength events (e.g., weightlifting).

The muscle's capacity to generate and to tolerate high levels of lactic acid during maximal exercise also increases with sprint training. The increase in lactic acid production observed after sprint training is due to the enhanced rate of breakdown of glucose to lactic acid via anaerobic glycolysis. Exercise can continue despite high lactic acid levels because of an increased capacity to buffer (neutralize) the hydrogen ions (H^+) that dissociate from lactic acid and to remove lactic acid from muscle cells.

Muscle fibre size increases, especially in FT fibres, with strength, power, and sprint training; this is called hypertrophy. Because larger muscles contain more cross-bridges capable of generating force, strength and power output increase. High-intensity strength, power, and sprint training also enhance synchronization of motor unit recruitment, increasing the number of active muscle fibres contributing to force generation. Finally, the strength- or power-trained muscle fibre exhibits less disturbance of its internal (intracellular) environment during exercise; that is, cellular levels of

Table 11.1

Metabolic, Physiological, and Structural Adaptations to Sprint and Strength Training

Adaptation	Consequence
Increased muscle adenosine triphosphate (ATP) and phosphocreatine	More ATP at start of exercise
Increased muscle glycogen	More glycogen available, delaying onset of fatigue
Increased anaerobic enzymes	More ATP synthesized via glycolysis
Increased lactic acid buffering	Increased capacity to tolerate high lactic acid levels
Increased muscle fibre size	Increased force generation, muscular strength and power
Increased motor unit synchronization	Greater strength and power
Less disturbance of muscle cell chemical gradient, especially potassium and calcium	Increased force generation and strength

ions such as potassium and calcium are better retained during exercise, allowing for continued tension generation by muscle fibres.

Changes in Aerobic Metabolism After Endurance Training

Metabolic adaptations resulting from endurance training are summarized in table 11.2. Generally, six weeks of endurance training will elicit a 20% to 40% increase in $\dot{V}O_2$max due to changes in both the cardiovascular system and skeletal muscle cells.

The activity of mitochondrial enzymes controlling the Krebs cycle and electron transport chain greatly increases, often by more than 100%. Oxygen uptake and oxidative production of ATP increase in slow-twitch (ST) fibres, as well as in FTa fibres provided that training is at a pace that recruits these fibres. Increased capacity for oxidative metabolism means that less lactic acid is produced via anaerobic glycolysis during endurance exercise.

The capacity of skeletal muscle fibres to utilize fatty acids to produce ATP increases after endurance training. This means that, at any given submaximal exercise intensity, trained muscle uses more fatty acids and less glycogen to produce ATP. Increased capacity to use fatty acids spares glycogen stores, delaying glycogen depletion and the onset of fatigue during prolonged exercise. In addition, muscle stores of glycogen increase after endurance training, providing more glycogen for prolonged higher-intensity aerobic exercise. The combination of extra glycogen and slower rate of its use allows the muscles to work longer before depletion of glycogen and the onset of fatigue.

The capacity of ST muscle fibres and other tissues to remove lactic acid from the blood increases following endurance training. This occurs because muscles have a higher oxidative capacity and can use lactic acid to produce ATP. (Remember, lactic acid can be used as a fuel or substrate by the Krebs cycle and electron transport chain.) Enhanced removal of lactic acid keeps blood and muscle lactic acid levels low during exercise, preventing early fatigue.

Endurance training also increases blood capillary numbers within skeletal muscle, especially around ST and FTa muscle fibres. This increased blood supply enhances the delivery of oxygen to, and removal of carbon dioxide and lactic acid from, working skeletal muscle fibres.

Endurance-trained skeletal muscles contain more myoglobin than untrained muscles. Myoglobin is an iron-containing protein that transports oxygen through the skeletal muscle cell. Increased myoglobin content enhances oxygen delivery from the cell periphery to the site of oxidative metabolism in the mitochondria.

Table 11.2

Metabolic Adaptations to Endurance Training

Adaptation	Consequence
Increased $\dot{V}O_2$max	Greater endurance performance
Increased muscle glycogen	More work before onset of fatigue
Increased mitochondrial enzymes	Increased oxidative capacity
Increased use of fats as substrate	Less reliance on glycogen, less glycogen depletion
Enhanced lactic acid removal and oxidation	More work before onset of fatigue
Increased lactate threshold	More work before onset of fatigue
Increased capillary number	More blood, oxygen, and substrates delivered to muscle and more lactate and carbon dioxide removed from muscle
Increased oxygen extraction by muscle	More oxygen available for adenosine triphosphate production
Increased muscle myoglobin content	More oxygen delivered to mitochondria

It is important to note that these metabolic adaptations are specific to the type of training and recruitment pattern of muscle fibre types. Only endurance training will increase the oxidative capacity of skeletal muscle. Similarly, only high-intensity speed or power training will increase intramuscular stores of PCr and ATP.

Metabolic changes occur only in those muscle fibres recruited during activity. For example, only ST fibres show changes in oxidative capacity after low-intensity endurance training; training must be of higher intensity to recruit and train FTa and FTb fibres. Similarly, resistance (weight) training must include work above 70% of maximum strength to recruit and cause adaptations in FTb fibres. Sprinting recruits predominantly ST and FTa fibres, and these fibres exhibit the most profound changes after sprint training. Lower-intensity endurance training does not change muscle fibre size, but higher-intensity endurance training induces hypertrophy of ST and FT fibres.

Endurance Training-Induced Changes in the Cardiorespiratory System

The cardiorespiratory system also adapts to endurance training. These adaptations enhance oxygen delivery to skeletal muscle and the muscle's ability to use oxygen to produce ATP during exercise. A summary of these changes is provided is table 11.3.

Oxygen Consumption

Resting oxygen consumption is related to body size and, in the absence of large changes in body size or muscle mass, remains essentially unchanged after endurance training. Submaximal oxygen consumption at the same absolute exercise pace also remains unchanged after training unless there is a change in mechanical efficiency; efficiency may change in sports such as swimming in which technique improves with training. In contrast, $\dot{V}O_2$max may increase by 20% to 40% after endurance training.

The extent to which $\dot{V}O_2$max improves during endurance training depends on initial fitness level and previous training, genetics, age, and the type of training program. In general, previously unfit people show large relative gains in $\dot{V}O_2$max with training because they are farther from their genetically determined upper limit of $\dot{V}O_2$max. Heredity is also an important factor affecting improvement in $\dot{V}O_2$max; some people respond with larger improvements in $\dot{V}O_2$max than others, even when performing the same training program; see "Heredity and Exercise Capacity: How Much of Physical Performance Is Genetically Determined? (Or Are Great Athletes Born That Way?)" on page 148. The type of training is also important—higher intensity as well as frequent and longer-duration training induces larger and faster improvements in $\dot{V}O_2$max. An improved $\dot{V}O_2$max after training is due to a combination of metabolic changes, such as increased oxidative capacity and blood supply within muscle, and cardiovascular system changes that enhance delivery of blood and oxygen to working muscles.

Table 11.3

Cardiorespiratory System Responses to Endurance Training

Adaptation	Consequence
Increased $\dot{V}O_2$max	Increased endurance performance
Decreased resting and submaximal heart rate	Less work done by the heart
Increased resting and exercise stroke volume	Increased cardiac output during maximal exercise
Increased maximal cardiac output	Increased blood and oxygen delivery to muscles
Increased blood volume, red blood cell number, and haemoglobin content	Increased oxygen delivery to muscles
Increased oxygen extraction from blood	Increased oxygen delivery to mitochondria
Decreased blood viscosity	Easier movement of blood throughout body, increased cardiac output during maximal exercise

IN FOCUS: HEREDITY AND EXERCISE CAPACITY: HOW MUCH OF PHYSICAL PERFORMANCE IS GENETICALLY DETERMINED? (OR ARE GREAT ATHLETES BORN THAT WAY?)

Genetics is important for some sports. For example, basketball and volleyball players tend to be taller, and gymnasts smaller, than average. How important is genetics to other aspects of exercise capacity such as $\dot{V}O_2$max, muscular strength and power, and anaerobic capacity?

Claude Bouchard and associates have for many years studied twins and families to find out how genetics might determine exercise capacity and training adaptations. Identical (monozygotic, MZ) twins share identical genetic makeup; nonidentical (dizygotic, DZ) twins are no more similar than siblings in terms of their genetic makeup, but, having been born and raised together, share similar environmental influences. If exercise capacity is genetically determined, then identical twins should perform more similarly in tests of exercise capacity than nonidentical twins.

Endurance and anaerobic exercise capacities were measured in siblings and in both nonidentical and identical twins. On all physiological variables measured, MZ twins were more closely related than DZ twins or siblings; the genetic effect was estimated to range from 40% to 70% depending on the variable measured. For example, genetics was found to determine about 40% of $\dot{V}O_2$max and 70% of total work accomplished in 90 min of exercise. This means that about half of endurance exercise performance can be accounted for by genetic influence. At least half of anaerobic exercise capacity also seems to be related to heredity.

Not only are anaerobic and aerobic exercise capacity influenced by heredity, but the capacity to improve with training also is genetically determined. In a training study on twins and siblings, about 60% of the improvement in $\dot{V}O_2$max after endurance training could be attributed to genetic influence.

The HERITAGE family study is a large study involving hundreds of families in the United States and Canada focusing on the genetic influence on exercise responses. It shows strong family influence, ranging from 30% to 70%, on other physiological variables such as ventilatory threshold and submaximal exercise capacity.

There is also a strong family influence on how much physiological variables and endurance exercise capacity increase after training. In contrast, some variables do not seem to show much familial influence, such as skeletal muscle fibre type.

These studies indicate that although exercise performance is to some extent genetically determined, this is relevant only for the elite athlete and should not be viewed as discouraging wide participation in sport and physical activity. Only a few sports are at the extremes of physiological demands, such as distance running, sprinting, or weightlifting, in which performance is determined mainly by physiological factors. Most sports require a combination of physiological capacity, skill and technique, optimal training, and motivation. Moreover, regardless of genetic predisposition, all individuals can improve exercise capacity and health with regular exercise. These data also indicate that effective exercise prescription requires individualization because of differences in training responsiveness between individuals.

SOURCES

- Bouchard, C., M.R. Boulay, J.-A. Simoneau, G. Lortie, and L. Perusse. 1988. Heredity and trainability of aerobic and anaerobic performances: An update. *Sports Medicine* 5: 69-73.

- Bouchard, C., E.W. Daw, R. Rice, L. Perusse, J. Gagnon, M.A. Province, A.S. Leon, D.C. Rao, J.S. Skinner, and J.H. Wilmore. 1998. Familial resemblance for $\dot{V}O_2$max in the sedentary state: The HERITAGE family study. *Medicine and Science in Sports and Exercise* 30: 252-258.

- Bouchard, C., and T. Rankinnen. 2001. Individual differences in response to regular physical activity. *Medicine and Science in Sports and Exercise* 33(Suppl): S446-S451.

- Perusse, L., J. Gagnon, M.A. Province, D.C. Rao, J.H. Wilmore, A.S. Leon, C. Bouchard, and J.S. Skinner. 2001. Familial aggregation of submaximal aerobic performance in the HERITAGE family study. *Medicine and Science in Sports and Exercise* 33: 597-604.

Heart Rate

It is not unusual for an endurance athlete to have a resting heart rate of 30 to 40 beats/min; the normal resting heart rate in a nonathlete is about 60 to 70 beats/min. This decrease in resting heart rate is called training bradycardia (slowing of heart rate). During submaximal exercise at the same absolute work rate, heart rate is lower in the trained than in the untrained individual. Because of changes in stroke volume (discussed next), the heart is able to pump the same amount of blood with fewer contractions. Thus, the same exercise is less demanding on the endurance-trained heart. The decreases in resting and submaximal heart rate result from changes in the heart's neural control. In contrast to resting and submaximal heart rate, maximal heart rate is more related to age and remains essentially unchanged, or possibly reduced, by training.

Stroke Volume

Stroke volume, the volume of blood pumped with each heartbeat, is higher at rest and at all exercise intensities after training. Endurance training over a period of years increases the size and strength of the heart's ventricles, increasing the amount of blood that can be pumped with each contraction. At rest and during submaximal exercise, the higher stroke volume coincides with a lower heart rate, resulting in the same cardiac output. In addition, in the endurance-trained person, stroke volume continues to increase with increasing exercise intensity, whereas it plateaus at a relatively low work rate in untrained individuals (see figure 10.9b). The high stroke volume in well-trained endurance athletes is an important contributor to high cardiac output, $\dot{V}O_2$max, and endurance exercise capacity.

Cardiac Output

Cardiac output, or the volume of blood pumped each minute, is a product of stroke volume and heart rate. Resting cardiac output is related to body size and does not change with endurance training. During submaximal exercise at the same work rate, cardiac output is unchanged or somewhat lower in endurance-trained than in untrained persons. In contrast, maximal cardiac output may double after endurance training, due to the large increase in stroke volume.

Oxygen Extraction

The ability of skeletal muscle to extract oxygen from blood depends on blood flow and the muscle's oxidative capacity. Endurance training greatly increases oxidative capacity, mitochondrial density, and myoglobin content, especially in ST and FTa muscle fibres, and the ability of muscles to extract oxygen during submaximal and maximal exercise.

Blood Composition

Other cardiovascular system adaptations include changes in the composition of blood that enhance blood and oxygen delivery to the working muscles and the removal of by-products such as lactate and heat from the working muscles. Endurance training increases the total volume of blood in the body, number of erythrocytes (oxygen-carrying red blood cells), and content of haemoglobin (oxygen-carrying protein within erythrocytes). These changes provide for increased oxygen delivery to the working muscles. In addition, blood becomes less viscous due to a proportionally greater increase in the volume of solution (plasma) compared with the increase in erythrocyte number. Lower blood viscosity means less resistance to blood flow, which enhances blood delivery to the working muscles during exercise and reduces the work of the heart. These adaptations also increase the reservoir of fluid available for thermoregulation during exercise in the heat.

Together, these cardiovascular system adaptations greatly enhance oxygen delivery to the working muscles. The increase in oxygen delivery, coupled with the higher oxidative capacity of skeletal muscle, means that the body relies more on oxidative metabolism and less on lactic acid-producing anaerobic metabolism during exercise. In addition, these changes improve the capacity to use fat for ATP production, which helps delay glycogen depletion and the onset of fatigue during prolonged exercise. Thus, the capacity for endurance exercise greatly increases in response to training that uses the oxidative energy system.

Endurance Training-Induced Respiratory Changes

Due to adaptations in the respiratory muscles, the maximum minute ventilation, or maximum volume of air inspired each minute, increases with endurance training. Although ventilation does not usually limit endurance capacity, increased ventilatory capacity means that the respiratory muscles are less likely to fatigue during endurance exercise in the trained person. Moreover, the increased volume of air inspired and expired enhances the lung's ability to rid the body of carbon dioxide and to buffer lactic acid produced during exercise.

Endurance Training-Induced Changes in the Lactate Threshold

The amount of lactic acid produced during exercise is a function of exercise intensity and state of training. As can be seen from a curve of blood lactic acid level plotted against exercise work rate (figure 11.1), blood lactic acid level does not increase linearly with increasing work rate. Rather, the level remains low until a certain exercise intensity and then increases exponentially.

The point at which blood lactic acid level begins to increase has been given many names, including the **lactate** (or lactic acid) **threshold, anaerobic threshold,** and **onset of blood lactic acid accumulation (OBLA).** The lactate threshold generally occurs at 50% to 65% of $\dot{V}O_2$max in untrained people and 70% to 85% of $\dot{V}O_2$max in endurance-trained people. Thus, compared with untrained individuals, endurance-trained athletes can exercise at a higher intensity before blood lactic acid begins to accumulate.

The lactate threshold indicates the exercise intensity that can be maintained primarily via oxidative metabolism without significant lactic

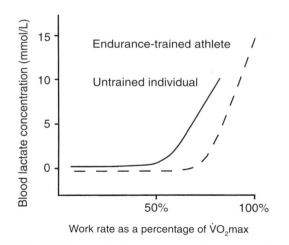

Figure 11.1 Lactate threshold. General trend of blood lactate concentration plotted against relative work rate, expressed as a percentage of $\dot{V}O_2$max. Endurance training shifts the lactate threshold toward a higher relative percentage of aerobic power, so the athlete can exercise at a higher intensity before blood lactic acid accumulates.

acid accumulation. This threshold represents the exercise intensity that can be maintained at the upper limit of aerobic metabolic capacity without a major contribution of the anaerobic system. Below this threshold, ATP production occurs without significant buildup of lactic acid; above the threshold, ATP production relies on an increasing contribution from anaerobic metabolism. Recall from chapter 10 that lactic acid produced during high-intensity exercise can cause fatigue by increasing muscle acidity and slowing the rate of ATP production via anaerobic glycolysis. Thus, the lactate threshold represents the intensity of exercise below which an individual can, theoretically, maintain exercise indefinitely without fatigue.

Although $\dot{V}O_2$max is a good general indicator of endurance exercise capacity, the exercise intensity or pace that coincides with the lactate threshold is a better predictor of performance among elite endurance athletes than $\dot{V}O_2$max. The endurance athlete who can maintain a faster pace of exercise with lower levels of lactic acid will be the better performer. Although the test is complex, regular measurement of the lactate threshold can provide an important means of evaluating an athlete's response to a training program.

Muscular System Changes After Strength Training

Elite powerlifters, sprinters, and bodybuilders are excellent examples of the tremendous capacity of skeletal muscle to adapt to strength training. Muscular fitness is important to performance in virtually every sport, and most athletes incorporate some type of muscular fitness exercise into the training regime.

Muscular fitness refers to the various components of muscle function, including muscular strength, power, and endurance. Muscular strength is defined as the maximum force that can be produced by a muscle or muscle group in a single movement. **Muscular power** represents the rapid application of force, as in a quick, explosive movement such as sprinting, throwing, or jumping; muscular power is a function of both speed and strength, or the product of force multiplied by speed. **Muscular endurance** is another important aspect of muscular fitness, and is defined as the ability of muscle to generate force repeatedly or continuously over time. Muscular endurance is important to performance in many sports, such as swimming, rowing, kayaking, and cycling, and to injury prevention. Muscular strength, power, and endurance are all interrelated, since each is a function of the maximum amount of force that can be applied by a particular muscle group or in a specific movement. However, increasing one component does not necessarily lead to increases in another. That is, simply increasing muscular strength will not necessarily enhance muscular power; each component must be specifically trained according to the desired outcome.

Muscular Strength

As noted in previous chapters, muscular strength is closely related to the size or cross-sectional area of muscle, as well as to muscle fibre type distribution, neural factors, and hormones such as the male hormone testosterone that stimulate muscle growth after periods of strength training. Muscular strength may increase by 20% to 100% over several months of **resistance training.**

Strength increases because of a complex interaction of many factors, including neural, structural, and metabolic adaptations in skeletal muscle. The relative contribution of each of these adaptations to strength gains varies by individual. For example, in older persons, the capacity for muscle hypertrophy may be limited and strength gains may depend more on neural factors. (See also chapter 12.)

As with anaerobic and endurance training, muscles adapt to strength training in a way that is specific to the intensity and volume of training. According to the size principle of motor unit recruitment, the largest and strongest FTb fibres are recruited for forceful contractions above 70% of maximum strength. Thus, strength-training programs must include high-intensity lifts to induce adaptations in FTb fibres and to maximize strength gains.

Muscle Hypertrophy

Muscle hypertrophy begins after approximately six to eight weeks of strength training, and is the major contributor to continued strength gains after this time. As noted in chapter 5, muscle hypertrophy occurs via increases in the average diameter of muscle fibres and the amount of connective tissue between muscle fibres. The increase in muscle fibre size occurs because of increased number of myofibrils and contractile filaments so that there are more cross-bridges generating force. Both protein synthesis and degradation occur at faster rates in trained than in untrained

fibres. However, in trained muscle fibres the rate of synthesis exceeds that for degradation, resulting in a net increase in skeletal muscle protein. The increased connective tissue between muscle fibres is important to preventing injury in the stronger, more forceful muscles.

Muscles also hypertrophy in a pattern specific to the intensity and volume of strength training and the muscle fibre type used. Low-intensity strength training, which recruits mainly ST and FTa fibres, induces hypertrophy only in these fibres. FTb fibres are the largest and strongest, with the highest capacity for hypertrophy (twice that of ST fibres). Maximum hypertrophy occurs when FTb fibres are recruited through training at high intensity. Maximum hypertrophy also seems to require high-volume strength training.

Metabolic Adaptations

Metabolic changes induced by intense strength training include increases in ATP, PCr, and glycogen content in FT fibres (table 11.1). The activity of several key enzymes involved in PCr breakdown and production of ATP in FT fibres also increases. Together these metabolic changes provide more ATP and PCr at a faster rate so that the muscle can generate more force in brief, maximal muscle actions.

Neural Adaptations

Neural adaptations account for much of the early strength gains in the first six to eight weeks of training, before muscle hypertrophy occurs. Strength training enhances synchronous recruitment and reduces inhibition of motor units. Motor units do not generally fire at the same time but instead fire in sequence or asynchronously. "Synchronous" recruitment of motor units results in a summation of force because more muscle fibres are activated at the same time. Strength training appears to relieve some of the natural inhibition of motor units. Under normal conditions, each motor neuron receives input from several other neurons, some of which are inhibitory. Reduced inhibition by strength training permits more motor units to become active and thus generate force. In addition, strength training appears to reduce activation of antagonist muscles (which oppose the action of a specific muscle group) and to increase activation of synergist muscles (which assist the muscles causing movement). The first four weeks of a strength-training program also improve skill and coordination in lifting heavy weights, both of which contribute to early improvements in strength.

Muscular Power and Endurance

Muscular power relates to the ability to generate a large amount of force rapidly; that is, power requires both strength and speed. The amount of muscular force that can be generated depends on several factors, including the speed of contraction. Essentially, force and speed of contraction are inversely related; for a given muscle, a faster movement reduces the amount of force generated. Thus, maximal power reflects a trade-off between speed and force. Maximum muscular power does not occur at maximum force but is achieved in the midrange of the force–velocity curve, at about 30% to 50% of maximum force. Changes in muscular power are closely related to the movement pattern and velocity of movement during training; these should match, as closely as possible, the demands of the sport. High-speed power training improves muscular power when measured at high speed, but does not necessarily improve muscular power or strength at slower speeds of movement. Similarly, training at slower speed may not improve muscular power in fast movements. It appears that these effects result to a great extent from neural changes in response to strength and power training, as just discussed.

Muscular endurance, or the ability to maintain force or repeated contractions over time, is related to strength. Increasing strength also improves muscular endurance because, after strength training, a lower proportion of maximum force is required to maintain a given submaximal force. In other words, the same submaximal force can be maintained over time with recruitment of fewer motor units, and thus less fatigue, after training.

Training and Muscle Fibre Number or Type

It is generally accepted that human skeletal muscle fibre number and fibre type are genetically determined and cannot be changed appreciably through normal activity such as exercise. However, recent research suggests that there is some "plasticity" (ability to change) in the structural and metabolic characteristics of muscle fibre types.

Evidence for increased fibre number (hyperplasia) is stronger in experimental animal models than in humans. Hyperplasia has been induced by intense, high-volume strength training in experi-

mental animals. Research on humans also provides indirect evidence for hyperplasia after intense, high-volume strength training. For example, some experienced weightlifters and bodybuilders have very large muscles that contain only average-size muscle fibres. It has been proposed that high-volume, high-intensity strength training over many years may stimulate skeletal muscle fibres to divide, which increases the number of fibres, each remaining about average in size. However, even if hyperplasia does occur, its contribution to gains in strength are minimal compared with muscle fibre hypertrophy and changes in neural factors, the two predominant mechanisms responsible for increases in strength after resistance training.

Fibre type is determined by the motor neuron. Fibre type transition, the changing of one fibre type to another, does not occur under most training circumstances. Recent research suggests, however, that some types of training may induce limited changes in the physiological and metabolic characteristics of muscle fibres. Several months of heavy resistance or power training (e.g., weightlifting, sprinting) may cause FTb muscle fibres to more closely resemble FTa fibres. This occurs because of changes in the metabolic characteristics, such as an increase in oxidative capacity and capillary number. The FTa fibres retain the characteristics of FTb fibres, such as size, speed, and strength, but are more fatigue resistant; the changes can be considered beneficial adaptations to enhance the muscle's ability to generate and maintain force. These changes are reversible if training stops. There is no evidence that ST fibres can become FT fibres, even with very heavy training.

Basic Principles of Training

Certain basic principles are common to all types of training, from conditioning at the highest (elite) level of sport competition to programming for the average adult who exercises for the health benefits.

Specificity

Skeletal muscle responds specifically to the physiological and metabolic demands of exercise. Thus, training must reflect the specific energy demands of the activity. For example, only endurance-type training will increase the oxidative capacity of skeletal muscle, and only strength training will induce hypertrophy in FT muscle fibres. The concept of specificity also extends to movement patterns and neural adaptations; and training should, as far as possible, simulate the speed, force, and timing of the activity being trained for.

Training Variables

One can manipulate several variables when designing an exercise training program: type of exercise (mode), duration or volume, intensity or pace, and frequency. The choices regarding each variable depend on the desired outcomes. Someone interested in health-related fitness has more choice in terms of these variables than an athlete training for a particular sport. For example, one can achieve health-related fitness using different modes of exercise (e.g., aerobic dance, swimming, walking), whereas an athlete would spend most training time using the specific mode of his or her sport. Moreover, one can achieve health-related fitness with much lower intensity, frequency, and volume of training than needed to excel at a high level in sport.

Overload

The body adapts to the physiological and metabolic demands of training over time. Training loads must progressively increase to induce continued improvement in exercise capacity. Athletes frequently experience a plateau in performance if training loads are not continually adjusted to accommodate recent training adaptations and increased work capacity. Overload is normally achieved by altering the combination of frequency, intensity, time (duration), and type of activity.

Individualization

Training adaptations are best achieved with individualized programs specific to the needs of the athlete. Although general principles of training apply under most situations, it is important to recognize that the rate or extent of adaptations may vary between individuals. As discussed later in this chapter, individualization is important for preventing overtraining syndrome.

Reversibility

Training adaptations last only as long as the physiological and metabolic demands of exercise continue, that is, as long as training is maintained at a certain intensity and volume. Detraining, a reversal of training adaptations, begins within days of cessation of training. Once a certain level of fitness is achieved, a minimum amount of regular exercise is needed to maintain training adaptations. In general, maintaining training adaptations requires less exercise than is needed to induce the initial changes. Provided that training intensity is maintained, training volume may be decreased by up to 50% with little loss of fitness over the short term (e.g., two to three weeks).

Periodization

Athletes frequently train in weekly, monthly, or seasonal cycles, varying the intensity and volume of training to achieve specific goals. Short cycles (one to two weeks) are called "microcycles," and longer periods (two weeks to two months) are called "macrocycles" (figure 11.2). In the high-performance athlete, performance plateaus at the end of each microcycle, requiring increased training volume, intensity, or both in the next microcycle. Periodization ensures that the athlete reaches peak form at crucial times, such as major competition. It also allows for variety in training to prevent boredom, injury, and overtraining and permits continued adjustments of the training program in response to the athlete's progress or other factors such as injury or illness.

For team game athletes, periodization allows the athlete and team to focus on various aspects of the

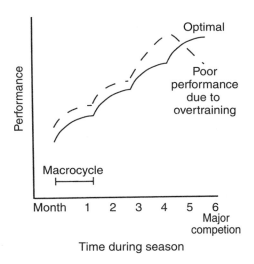

Figure 11.2 Training microcyles. Theoretical model of training macrocyles, each approximately six weeks long. Ideally, performance should peak during major competition. Excessive training early in the season may cause overtraining and a decline in performance late in the season and during major competition.

sport at different times in the season. For example, in sports such as soccer or basketball, early season training may emphasize conditioning for aerobic fitness and muscular hypertrophy and strength. As the season progresses, training will focus less on endurance capacity and strength and more on speed, power, skill, strategy, and teamwork.

Overtraining

Overtraining refers to excessive training leading to prolonged fatigue, frequent illness, and poor performance. The athlete is often unable to maintain training loads and to perform at the standard expected. Overtraining often results from increasing training volume or intensity too rapidly and not allowing adequate recovery between training sessions. Prolonged rest and a reduction in training loads over several weeks or months are necessary to restore performance in an overtrained athlete.

Optimal training must balance all of the principles discussed previously. Indeed, training of the top athlete is both an art and science, based on scientific principles underlying training adaptations along with the coach's intuition and observations regarding how an individual athlete adjusts to overload induced by training. Optimal performance depends on maximizing the positive training adaptations while minimizing the potential negative effects of excessive training loads.

Continuous and Interval Training

Training may involve continuous or interval exercise. As their names imply, continuous training is performed without any rest breaks, while **interval training** involves alternating intervals of exercise and rest. Table 11.4 summarizes the advantages and disadvantages of each type of training.

Continuous and interval training can be employed to develop different aspects of fitness. For example, short interval training involves repeated high-intensity intervals at maximal pace and induces adaptations in the immediate energy system; in contrast, longer intervals at lower intensity primarily train the aerobic energy system.

Continuous Training

Continuous training may be further defined by exercise intensity. Lower-intensity continuous training is usually in the range of 70% to 80% $\dot{V}O_2$max for athletes, and 50% to 60% $\dot{V}O_2$max for those seeking health-related fitness. Lower-intensity continuous training can be used in a variety of situations, including development of health-related fitness for the average adult or during early season aerobic training in many sports. Higher-intensity

Table 11.4

Types of Training

Type	Advantages	Disadvantages
Continuous, constant pace	Is time efficient Trains cardiovascular and muscular endurance Routine is easy to follow Can be sport specific (e.g., to distance running) Chance of injury less because of lower intensity	Athletes may need higher intensity Can be monotonous May not be specific to some activities (e.g., team sports)
Continuous, varied pace	Adds variety of pace Allows training at higher intensity	May not be sport specific Increased risk of injury
Interval	Adds variety of pace and duration Can be very sport specific	Requires more time Increases risk of injury
Circuit	Trains cardiovascular and muscular endurance and strength Can be very sport specific	Improvements in endurance and strength not maximal Requires access to equipment

continuous training, above 85% $\dot{V}O_2$max for the athlete and 60% $\dot{V}O_2$max for the nonathlete, is generally recommended only for those who are already physically fit. The athlete often exercises at or near race pace in this type of training.

Continuous training need not be at a constant pace but can vary within a given exercise session. Shorter bursts of higher-intensity exercise train both the anaerobic glycolytic and aerobic energy systems; the longer periods of slower exercise induce adaptations primarily in the aerobic system, enhancing removal of lactic acid produced during higher-intensity exercise. Varying the pace of training adds variety to continuous training, which may become repetitive and monotonous. Varying the intensity of training within a continuous session can add specificity for team game athletes in sports such as soccer or basketball, in which the athlete may run continuously but at various speeds throughout a match or game.

Interval Training

Interval training involves alternating periods (intervals) of exercise and rest. As shown in table 11.5, the length of both the exercise and rest intervals is manipulated to induce specific training adaptations.

Interval training offers several advantages over continuous training. For competitive athletes, the duration and intensity of exercise intervals can be varied to train specific energy systems, and training may be performed at or even above race pace. Training at or above race pace seems to be important to inducing neural adaptations such as increased synchrony of motor unit recruitment. Because of frequent rest intervals, interval training permits the athlete to perform more exercise than in continuous

training. Repeated, high-intensity exercise intervals also improve lactate buffering and tolerance. Moderate-intensity interval training is an effective way to gradually introduce exercise to the untrained adult interested in health-related fitness. There is less risk of overuse injury, muscle soreness, and an abnormal response in moderate interval training compared with higher-intensity continuous training. In addition, exercise is often perceived as less intense and therefore more comfortable. One disadvantage of interval training is that, because of the rest periods, interval sessions are typically longer than continuous training sessions.

There are three general categories of interval training, each corresponding to the predominant energy system used:

• *Short interval training,* or anaerobic interval training, consists of work intervals lasting 5 to 30 s. Short interval training relies predominantly on the immediate energy system and is used to develop muscular strength, muscular power, and speed; exercise is performed at or above race pace. Since PCr levels may be depleted by more than 50% after 30-s maximal exercise, rest intervals should be long enough (3 to 6 min) to restore PCr. Moderately high levels of lactic acid may accumulate in skeletal muscle and blood after repeated 20- to 30-s high-intensity exercise intervals.

• *Intermediate interval training,* or anaerobic–aerobic interval training, consists of work intervals lasting 30 s to 2 min and is performed at high intensity (>90% of race pace). Intermediate interval training relies on PCr breakdown and anaerobic glycolysis for energy production, and to some extent on aerobic metabolism. Muscle and blood lactic acid levels are very high at the end of each work interval, and

Table 11.5

Interval Training Variables

Variable	Short interval	Intermediate interval	Long interval
Work interval	5-30 s	30-120 s	2-5 min
Intensity/pace	>95% race pace	90-95% race pace	60-90% $\dot{V}O_2$max
Rest interval	Three to five times work interval	Two to three times work interval	One to two times work interval
Major energy system	Immediate	Anaerobic glycolysis	Oxidative
Major effects on:	Speed and power	Speed and power, muscular endurance, lactic acid buffering and tolerance	Cardiorespiratory endurance, muscular endurance, lactic acid removal

Compiled from Rushall and Pyke 1990.

iods (several minutes) are needed to
els and to remove lactic acid.

l training, or aerobic interval train-
......sts of work intervals lasting 2 to 5 min;
this type of training relies primarily on the aerobic
system for ATP production. Long intervals induce
changes in the oxidative capacity of muscle, especially ST and FTa fibres. If the training is performed
at a higher intensity, at 70% to 90% of $\dot{V}O_2$max, FTb
fibres are recruited and the anaerobic glycolytic
system is also taxed, producing moderately high
lactic acid levels. Higher-intensity long interval
training enhances lactic acid removal and oxidation.

Training for Cardiovascular Endurance

The optimal type of cardiovascular endurance
training program depends on the individual's
objectives and desired outcomes. Obviously, an
older person seeking health benefits and a high-
performance distance runner require different
programs. Thus, exercise programs should be
individualized, although certain basic principles
apply to all types of endurance training.

There appears to be a minimum threshold in
terms of exercise intensity, frequency, and duration
to improve aerobic exercise capacity. To improve
$\dot{V}O_2$max, the average healthy young adult needs
to exercise for at least a 15-min duration, at a
minimum of 60% $\dot{V}O_2$max, at least three times per
week. Improvements may occur at a lower exercise
intensity (even as low as 30% to 40% $\dot{V}O_2$max) in
older, less fit individuals or in persons with disease
(see chapter 13).

In general, improvements in $\dot{V}O_2$max and
endurance exercise capacity are related to the total
amount of exercise performed. Exercise intensity,
frequency, and duration may be manipulated
in many ways to increase aerobic power and
endurance exercise capacity. However, according to the principles of specificity and motor unit
recruitment, training intensity and duration for
the endurance athlete should approximate those
required in competition. There is some debate as
to the optimal amount, intensity, and frequency of
exercise needed to improve physical fitness and
health. We discuss this at length further on. Some
health benefits, such as loss of body mass, reduced
blood pressure, and lower risk of heart disease,
occur without significant changes in physical fit-

ness. Improving physical fitness (i.e., increasing
$\dot{V}O_2$max or endurance exercise capacity) requires
slightly more exercise, in the range of 50% to 85%
$\dot{V}O_2$max for 20 to 60 min, three to five times per
week. The elite athlete requires a far higher level
of exercise to improve physical fitness and performance—daily training for several hours at an
intensity greater than 85% of $\dot{V}O_2$max.

Methods of Strength Training

Some type of muscular strength training should be
included in all fitness programs, from programs for
strength and power athletes to those for individuals who exercise for the health benefits. Of course,
athletes in sports requiring strength and power,
such as weightlifting, bodybuilding, and sprinting,
rely heavily on resistance training. However, many
other athletes also benefit from strength training,
especially those in sports requiring a high level of
muscular endurance such as rowing, swimming,
and cycling. Strength training also helps prevent
and rehabilitate sport-related injuries.

Muscular strength exercises can enhance
health-related fitness. Moderate-intensity resistance training confers health benefits such as
favourable changes in glucose tolerance, body
composition, and blood lipids (fats), which are
related to heart disease (see chapter 13). Moderate resistance training may also help prevent and
treat some types of lower back pain and other
conditions such as arthritis and osteoporosis (see
chapter 13); maintain lean body mass, muscular
strength, and mobility with aging (see chapter 12);
and improve muscular "tone" and body shape, an
important aesthetic consideration for many adults
who exercise.

Types of Muscle Contractions

As we have seen in earlier chapters, the two general types of muscle contractions are static and
dynamic (figure 11.3). In a static, or isometric,
contraction, the muscle generates force but does
not change length because resistance to that force
is greater than the force generated by the muscle.
Isometric contractions act primarily to stabilize
joints during movement. Isometric strength
training is primarily used when joint mobility is
limited, for example, after an injury or in a joint
affected by arthritis.

In a dynamic contraction, the muscle's length
changes while it generates force against a movable
resistance. Dynamic contractions, as we have noted

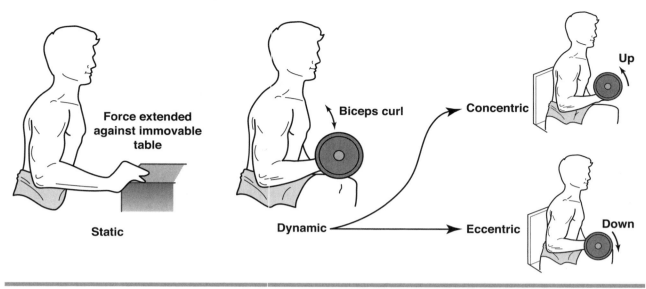

Figure 11.3 Static and dynamic muscle contractions in the biceps brachii.

previously, may be either concentric (the muscle generates force while shortening) or eccentric (the muscle generates force while lengthening). Most force-producing movements occur via concentric contraction—for example, the propulsive action in running, jumping, or throwing. Eccentric contractions are used to stabilize or decelerate the body especially when counteracting gravity. Two common examples are landing after a jump, in which the quadriceps muscles on the front of the thigh lengthen to absorb the impact and prevent falling, and slowing the arm movement at the end of throwing, in which the rotator cuff muscles at the back of the shoulder lengthen to decelerate the arm and prevent loss of balance and shoulder injury.

Types of Strength Training

Strength or resistance training programs may include static or dynamic contractions. Given that gains in strength are specific to the type of program, resistance-type activities should match the needs of the sport and desired outcomes. As already indicated, isometric training has limited applications, primarily because the gains in strength are limited to the particular joint angle at which training occurs.

Dynamic strength training may be further classified by the type of resistance against which force is applied. **Iso-inertial** (sometimes called **isotonic**) training is performed against a constant resistance, for example, a barbell or fixed-weight (pin weight) machine. **Isokinetic** training is performed with a constant speed of movement against variable resistance. Specialized equipment, such as the Cybex or KinCom isokinetic dynamometers, which control the velocity of movement, is required for isokinetic training. The machine controls the speed of movement while offering resistance equal to the force being applied; maximal force is measured at each joint angle throughout the range of motion. The speed may be preset to simulate movement of a particular activity.

The various types of resistance training (isometric, isotonic, and isokinetic) have advantages and disadvantages. Isometric training is low cost and low risk, but gives minimal gains in strength. Iso-inertial training can be very sport specific and yields excellent improvements in strength, but may increase the risk of injury and muscle soreness because of the eccentric component. Isokinetic training gives excellent strength gains with minimal risk of injury but is less sport specific, since movement is generally limited to simple flexion and extension only at certain joints, and natural movements rarely occur at constant speed. In addition, many movements in sport occur at speeds faster than an isokinetic dynamometer can provide. Isokinetic training is often used in the clinical setting to assess muscular strength throughout the range of motion, to identify weakness within a certain muscle group or imbalance between muscle groups, and to rehabilitate an injury. Isokinetic training is also used widely in the research laboratory.

Training to Improve Muscular Strength and Endurance and to Induce Hypertrophy

As in other forms of training, adaptations to strength training are specific to training program variables such as the type of training, method of measurement, training intensity, and duration. Improvement in strength also depends on the individual's age, fitness level, and initial strength. In general, training using one type of muscle action improves strength only when measured in a similar manner. That is, iso-inertial strength training improves strength only when measured iso-inertially, and isokinetic strength training improves strength when measured isokinetically at a speed of movement similar to that used in training. Thus, strength-training programs must be specific to the desired outcomes. Training program variables that can be manipulated include the following:

- The number of repetitions or times a weight is lifted
- Sets, or groups, of repetitions
- Training volume or total amount of work performed, which is the product of the number of repetitions and the number of sets
- Intensity or resistance

Intensity can be expressed two ways. Resistance can be prescribed as a percentage of maximal strength, or the weight that can be lifted only once, referred to as 1-repetition maximum (1RM). For example, if an athlete's maximum strength in the bench press is 50 kg (110 lb), a set might consist of five repetitions at 70% of 1RM; resistance would be 70% of 50 kg, or 35 kg (77 lb), to be lifted five times. This method requires initial measurement of maximum strength (1RM) for each muscle group, which may be inconvenient or contraindicated for some people. Resistance training intensity may also be prescribed in terms of the weight or resistance that can be lifted for a specified number of repetitions. For example, a set may consist of 10RM, which is the maximum weight the athlete can lift only 10 times. Although this method does not require the person to perform maximum lifts, it does require trial and error to estimate resistance for each muscle group.

One must consider the pattern of skeletal muscle fibre type recruitment and specific energy systems in prescribing strength-training programs specific to the desired outcomes. Table 11.6 gives examples of resistance training programs designed to improve muscular strength, power, and endurance; improve fitness; and induce hypertrophy. Gains in maximum strength require recruitment of the larger FTb fibres, which occurs above 70% of maximum force. Muscle hypertrophy requires moderate- to high-intensity, high-volume training for several weeks or more. Optimal development of muscular power requires hypertrophy and development of strength first, followed by work emphasizing speed and the explosive application of force that simulates the actual movement pattern. Power athletes generally follow a periodized program of resistance training, beginning with six to eight weeks of training emphasizing muscle hypertrophy, followed by a further six to eight weeks focusing on muscular strength, and then power training using movement patterns specific to the sport.

The Role of Eccentric Muscle Actions in Strength Training

Gains in strength are related to the mode and muscle actions used in training. Concentric-only training generally improves strength only when measured concentrically, with little effect on

Table 11.6

Prescription for Strength Training

To develop	Repetitions	Sets	Intensity	Rest between sets
Maximal strength	2-6	3-6	High	>3 min
Muscular hypertrophy	8-12	3-6	Moderate	>3 min
Muscular endurance	15-25	1-4	Moderate	<1 min
Muscular power	2-6	3-5	Moderate resistance, fast, explosive	>3 min
Health-related fitness	8-20	1-2	Moderate	<1 min

Compiled from Wilmore and Costill 1999.

eccentric strength. Most common forms of resistance training, such as free weights or pin-weight machines, include both concentric and eccentric muscle actions. Recent research suggests that maximum gains in strength require both types of muscle actions. Although eccentric muscle actions are more likely to induce injury to muscle fibres and muscle soreness (discussed next), it is believed that some degree of cellular damage leads to gains in muscular strength and muscle hypertrophy. Most serious strength and power athletes incorporate both types of muscle actions in their resistance training.

Causes of Muscle Soreness

All athletes, and most people who exercise, have at some point experienced muscle soreness following heavy, unaccustomed training sessions. There are two forms of postexercise muscle soreness—an immediate form appearing during or soon after exercise, and a delayed form that appears 12 to 48 h after exercise. Both forms cause muscle weakness; that is, a muscle is not capable of producing as much force when soreness is present.

The acute or immediate type of soreness feels like a painful or burning sensation in the affected muscles. This sensation and associated weakness are caused by buildup of lactic acid (discussed in the previous chapter) that increases the muscle's acidity. This form of soreness is temporary, and the sensation of soreness disappears after a few minutes or hours.

The second type of muscle soreness is called **delayed-onset muscle soreness (DOMS)** because it begins the day after exercise and may last for up to several days. Delayed-onset muscle soreness is more likely to occur and to be more severe after exercise involving eccentric muscle actions. For example, prolonged downhill walking or running causes DOMS in the quadriceps muscles (on the front of the thigh) because these muscles act eccentrically to counteract the force of gravity. It is generally believed that muscle soreness occurs because fewer muscle fibres are activated during eccentric than in concentric muscle actions. The high force generated by each activated muscle fibre is thought to cause damage to that muscle cell's interior structure as well as to the surrounding connective tissue.

Over the years, several mechanisms have been advanced to explain DOMS. Contrary to popular belief, DOMS is not related to lactic acid accumulation. For example, lactic acid levels are not elevated in the types of exercise most likely to cause DOMS, such as downhill walking. Muscle spasms during and after exercise have also been proposed as a mechanism for DOMS, although there is little experimental evidence to support this concept. The most likely cause of DOMS is damage to both individual muscle fibres and connective tissue surrounding the fibres resulting from intense, prolonged, or eccentric exercise. Damage to muscle fibres and connective tissue initiates an inflammatory response, causing influx of inflammatory cells from the circulation, tissue swelling, and the sensation of pain.

The skeletal muscle damage associated with DOMS is temporary, lasting usually less than one week. Muscle fibres are rapidly repaired, and there is no evidence for permanent tissue damage associated with DOMS. As discussed earlier, it is believed that some level of tissue breakdown and repair may be important in initiating training-induced adaptations in skeletal muscle, in particular gains in muscle size and strength. One interesting feature of DOMS is that a single session of eccentric exercise provides protection for up to several months against a subsequent occurrence of DOMS within the same muscle (see "Can Delayed-Onset Muscle Soreness Be Prevented?" on p. 160).

Exercise for Health-Related Fitness

As the term implies, health-related fitness involves **physical activity** performed primarily for health benefits rather than to improve performance or physical work capacity. Health-related fitness encompasses an ability to perform the physical tasks of daily living as well as reducing the risk of diseases associated with a sedentary lifestyle, for example heart disease or obesity as discussed in chapter 13.

ACSM and USSG Exercise Guideline Summaries

People interested in their health often ask, "What is the optimal amount of exercise that I should do to remain healthy?" Implied is another question—"What is the minimum amount of exercise I need to do to remain healthy?" These seem to be simple questions, but the answers are somewhat complex; the optimal and minimal amounts of

IN FOCUS: CAN DELAYED-ONSET MUSCLE SORENESS BE PREVENTED?

Anyone who has exercised intensely knows that muscles can become very sore after unaccustomed exercise. Understanding the mechanisms responsible for DOMS will help us develop ways to either avoid DOMS or minimize the adverse effects.

One interesting feature of DOMS is that prior training using eccentric muscle actions protects the same muscle against damage after subsequent exercise. This prompts the question of how long this protective effect lasts. A recent study suggests that the protective effect may last for up to 6 months, but not as long as 12 months (Nosaka et al. 2001). Young adult males performed maximal eccentric exercise of the elbow flexor muscles of the upper arm, resulting in DOMS. Subjects repeated the same exercise 6, 9, or 12 months later, and indicators of muscle cell damage and DOMS were assessed after the second bout. All indicators of muscle cell damage were significantly lower, suggesting a protective effect, when eccentric exercise was repeated 6 months later; some indices were lower when exercise was repeated 9 months later. However, there was no protective effect 12 months later. This study suggests that a single maximal bout of eccentric exercise protects against muscle damage for 6 to 9 months but that this protective effect is lost by 12 months.

It appears that eccentric exercise induces inflammation within damaged skeletal muscle, causing inflammatory cells to migrate from the blood into the injured area; this is a normal response to injury or inflammation in any part of the body. Once activated, these inflammatory cells release substances that cause swelling and the perception of pain in the affected area.

In a study on mice, inflammatory cells appeared in damaged muscle cells 6 hours to 3 days after eccentric muscle actions (Pizza et al. 2002). There was less damage after eccentric actions in muscles that had been passively stretched two weeks before than in muscles that performed eccentric actions without the prior stretching. Despite the reduced response after prior stretching, some inflammatory cells appeared in muscles after eccentric actions, suggesting that these cells may help protect muscle cells against subsequent damage. Thus, some level of inflammation may be needed to protect against excessive muscle damage by subsequent eccentric exercise.

Taken together, these studies suggest that some degree of damage and inflammation may be needed to protect against damage by subsequent eccentric exercise. This protection may last up to 9 months, but only in the muscles used in the initial exercise.

SOURCES

- Nosaka, K., K. Sakamoto, M. Newton, and P. Sacco. 2001. How long does the protective effect on eccentric exercise-induced muscle damage last? *Medicine and Science in Sports and Exercise* 33: 1490-1495.
- Pizza, F.X., T.J. Koh, S.J. McGregor, and S.V. Brooks. 2002. Muscle inflammatory cells after passive stretches, isometric contractions, and lengthening contractions. *Journal of Applied Physiology* 92: 1873-1878.

exercise needed to achieve health benefits vary according to the individual's goals, health status, fitness level, and age.

In recent years, recommendations by two influential organizations have provided guidance on the amount and types of physical activity needed to achieve health benefits; many other national and international organizations have followed with their own recommendations. Since 1978, the American College of Sports Medicine (ACSM) has regularly issued position statements on the recommended amount of exercise for healthy adults, based on extensive review of many years of research on the health benefits of exercise. The most recent (2000) ACSM recommendations, outlined in "Summary of American College of Sports Medicine Recommendations for Exercise for Health" (p. 161), focus on the appropriate type and amount of exercise suitable for maintaining health in the normal adult population.

SUMMARY OF AMERICAN COLLEGE OF SPORTS MEDICINE RECOMMENDATIONS FOR EXERCISE FOR HEALTH

- Any activity using large muscle groups that can be maintained for a prolonged period of time, such as walking, jogging, cycling, swimming
- Intensity from 40/50% to 85% of $\dot{V}O_2$max or 55/65% to 90% of age-predicted maximal heart rate
- Duration from 20 to 60 min total each day; minimum of 10-min bouts
- Frequency of three to five times per week
- Inclusion of moderate-intensity strength training of the major muscle groups two to three times per week

Compiled from American College of Sports Medicine 2000.

In 1996, the landmark **U.S. Surgeon General's** (USSG) **report,** *Physical Activity and Health,* clearly linked regular physical activity with a variety of health benefits (described in more detail in chapter 13). The USSG recommendations are outlined in "Summary of U.S. Surgeon General's Recommendations for Physical Activity". Before discussing these two sets of recommendations in more detail, it is important to note that, despite some differences, the recommendations are more similar than dissimilar and that both sets provide a starting point for people to develop their own physical activity programs to meet their specific needs.

p 179

SUMMARY OF U.S. SURGEON GENERAL'S RECOMMENDATIONS FOR PHYSICAL ACTIVITY

- Everyone over age 2 should accumulate at least 30 min of endurance-type physical activity on most, preferably all days, of the week.
- Physical activity should be at least of moderate intensity; additional health benefits can be achieved by more vigorous activities, or by being active for more than 30 min per day.

- People with heart, lung, metabolic, or other chronic diseases should first be evaluated by their physician and should be given an appropriate physical activity program.
- Previously inactive men over age 40 and women over age 50, and those at risk of cardiovascular disease, should consult their physician before performing vigorous activity.
- Strength-developing activities, such as resistance training, should be performed at least twice per week, using the major muscle groups of the legs, arms, trunk, and shoulders.

Adapted from U.S. Department of Health and Human Services 1996.

Explaining the Summaries

To explain what the ACSM and USSG guidelines mean we will look at the type, intensity, duration, and frequency of exercise recommended for health-related fitness.

Types of Recommended Exercise

Beneficial adaptations to physical activity are relatively independent of exercise mode. That is, physical fitness and health benefits can be achieved through a variety of physical activities, and there is no single ideal exercise mode. The USSG suggests a range of modes of physical activity at work, during leisure time, or in activities of daily living (e.g., brisk walking, cycling, swimming, home repair or yard work). Most people prefer brisk walking, and the USSG notes the importance of personal preference.

The ACSM recommends whole-body exercise or activities that use large muscle groups, such as walking, jogging, swimming, and cycling. This type of exercise promotes loss of body fat and favourable changes in **cardiovascular disease** risk factors (see chapter 13). In addition, whole-body or large muscle group exercise is less likely to cause abnormal cardiovascular responses such as excessively high blood pressure or irregular heartbeats. Intense exercise using isolated body parts, such as power weightlifting, may cause blood pressure to rise, which may be dangerous for individuals with high blood pressure (hypertension).

Intensity of Exercise

Both sets of guidelines recommend "moderate-intensity" physical activity, because this level of exercise leads to health benefits but is less likely to lead to injury; moreover, more people prefer moderate exercise to intense exercise. The USSG defines moderate-intensity activity as that which can be sustained for or accumulated over at least 30 min per day. This report notes that more vigorous exercise leads to greater health benefits and should not be discouraged in those sufficiently healthy and willing to undertake more intense activity.

The ACSM recommends exercise in the range of 50% to 85% of $\dot{V}O_2$max for healthy individuals, and starting at a lower intensity (40%) for people who are unfit or who have disease or impairment affecting exercise capacity. This wide range provides flexibility in choosing the most appropriate exercise intensity for a given individual. The higher range (above 70% $\dot{V}O_2$max) is appropriate for those who are already fit and who have performance- or sport-related goals.

During low- to moderate-intensity exercise in the range of 50% to 70% $\dot{V}O_2$max, fatty acids provide the major fuel source for ATP production; regular exercise in this range, therefore, enhances fat loss by mobilizing fatty acids from adipose tissue. Low- to moderate-intensity exercise can be maintained for prolonged periods of time (up to 60 min), increasing energy expenditure during exercise, which is important for loss of body fat and body mass. Low- to moderate-intensity exercise is also less likely to cause musculoskeletal injury. Moreover, adults are more attracted to activity in this range of intensity, which increases the probability of continued participation.

Duration of Exercise

The duration of physical activity is a function of intensity; lower-intensity exercise usually requires a longer duration to provide the same health benefits as activity at a higher intensity. At least 15 min of moderate exercise is needed to induce training adaptations. For the average adult, beneficial effects of training do not increase appreciably when exercise sessions extend beyond 60 min, but the risk of overuse injury increases exponentially with exercise durations more than 60 min. Although a training effect may occur in response to only 10 min of high-intensity exercise, above 90% $\dot{V}O_2$max, this intensity of exercise is not recommended for the average person or nonathlete. Longer-duration, low-intensity exercise is also

recommended to increase total energy expenditure and thus enhance loss of body fat.

The USSG recommends "accumulating" a minimum of 30 min of physical activity each day; this means that exercise can occur in several shorter sessions (e.g., three 10-min sessions) provided the total time is at least 30 min. Longer-duration physical activity is also recommended for those who are interested and physically capable. The ACSM recommends a range between 20 and 60 min per session, with the duration dependent on the intensity—that is, a shorter duration for more vigorous exercise and longer duration for more moderate exercise.

Frequency of Exercise

For most people, exercise frequency is a matter of personal preference, accessibility, and convenience. The health benefits are very much related to the total amount of exercise over time, as in months to years. Lower-intensity or shorter-duration activity requires more frequent sessions, whereas more intense or longer exercise can be performed somewhat less frequently.

A minimum weekly training frequency of three days is recommended to increase physical fitness. For the average adult, however, only marginal gains in fitness and health benefits occur with more than five sessions per week, while the risk of injury increases greatly with frequency. The ACSM-recommended exercise frequency of three to five sessions per week provides a good balance between optimal beneficial training effects and potential risk of injury. However, low-intensity, low-impact activities, such as walking, cycling, or swimming, are not associated with risk of injury; and these activities may be undertaken daily. The USSG recommends moderate physical activity such as brisk walking on most, preferably all, days. Exercise is more likely to become an integral part of one's lifestyle if performed on a daily basis.

Resistance Exercise

Both the ACSM and USSG recommend appropriate resistance exercise such as low- to moderate-intensity weight training. Although health benefits are most commonly associated with endurance-type exercise, resistance training also leads to health benefits such as alterations in heart disease risk; favourable changes in body composition (e.g., reduced body fat and increased muscle mass); and counteracting the age-related loss of muscular strength, bone density, and functional capacity.

As the name implies, in **circuit training** the athlete moves around a circuit of different exercise stations. Circuit training was designed for athletes as a way to develop all aspects of physical fitness, including muscular strength, power, endurance, agility, and cardiovascular endurance. Because it is time efficient and versatile, circuit training is also beneficial for developing health-related fitness.

Circuit training can be sport specific to train skills or movement patterns required in a sport. Weightlifters use circuit training to work on specific muscle groups. Circuit training can be modified for lower-intensity work to develop health-related fitness, for example, by alternating weightlifting with aerobic-type activity. The exerciser spends 30 to 90 s at each station, with shorter rest intervals between stations (figure 11.4).

A major advantage of circuit training is its versatility. Stations can be changed regularly to provide variety and to prevent overuse of a particular muscle group. For athletes, the stations may also be altered as training progresses through the season. One disadvantage of circuit training for athletes is the difficulty in designing a single circuit program to optimally train both muscular strength and cardiovascular endurance; high-performance athletes may need more than one type of circuit.

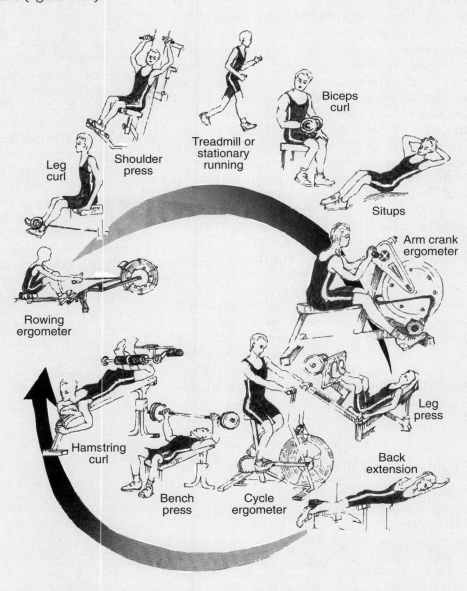

Figure 11.4 Circuit training for cardiovascular fitness and muscular endurance. Exercise includes alternating aerobic-type activities, such as running or cycling, and resistance work.

(These latter effects are discussed in more detail in the next two chapters.) Both sets of guidelines recommend 8 to 10 resistance exercises involving the major muscle groups and using lower intensity and higher repetitions (e.g., one to two sets of 8 to 15 repetitions).

Summary

Factors that limit exercise performance vary according to the duration and intensity of exercise and thus the energy systems used. Performance in power and speed activities is limited by depletion of PCr and lactic acid accumulation; performance in endurance exercise is limited by muscle glycogen depletion, dehydration, and increased body temperature. The body adapts to exercise training by delaying the onset of fatigue so that the athlete can accomplish more work or exercise at a faster pace or for a longer time before becoming fatigued. Endurance training increases $\dot{V}O_2$max, muscle glycogen stores, and the cardiovascular system's ability to deliver oxygen to and remove metabolic by-products from the working muscles. About half of endurance exercise capacity appears to be genetically determined. High-intensity sprint or strength training increases muscle size (hypertrophy), muscle glycogen and PCr stores, enzymes involved in anaerobic metabolism, and the ability to buffer and remove lactic acid. Muscle fibre type does not change after training, although oxidative capacity and fatigue resistance increase in FT fibres. Delayed-onset muscle soreness after heavy exercise is more likely to occur in muscles acting eccentrically (e.g., in downhill running); DOMS reflects minor structural damage to muscle fibres and connective tissue, which is usually repaired within one week. Regular moderate physical activity is recommended for all individuals, of all ages, whether they are healthy or have diseases or conditions that affect physical capacity.

Further Reading and References

American College of Sports Medicine. 1998. The recommended quantity and quality of exercise for developing and maintaining cardiorespiratory and muscular fitness, and flexibility in healthy adults. *Medicine and Science in Sports and Exercise* 30: 975-991.

American College of Sports Medicine. 2000. *ACSM's guidelines for exercise testing and prescription.* 6th ed. Philadelphia: Lippincott, Williams & Wilkins.

American College of Sports Medicine. 2002. ACSM's position stand: Progression models in resistance training for healthy adults. *Medicine and Science in Sports and Exercise* 34: 364-380.

Baechle, T.R., and R.W. Earle. 2000. *Essentials of strength training and conditioning.* 2nd ed. Champaign, IL: Human Kinetics.

Reaburn, P., and D. Jenkins, eds. 1996. *Training for speed and endurance.* St. Leonards NSW, Australia: Allen and Unwin.

U.S. Department of Health and Human Services. 1996. *Physical activity and health: A report of the Surgeon General* (pp. 61-260). Atlanta: U.S. Department of Health and Human Services, Centers for Disease Control and Prevention, National Center for Chronic Disease Prevention and Health Promotion.

CHAPTER 12

Physiological Capacity and Performance Across the Life Span

*M*uch of our knowledge of exercise physiology comes from research on young adult males, who are a relatively accessible group for research studies. Over the past few decades, however, work has focused on understanding the responses to exercise in a wider range of individuals—from children to persons who are elderly, and in both males and females—and identifying the mechanisms underlying any differences between ages or between sexes in the responses to exercise. Physical activity and sport can be considered a natural part of growing up in children, who are naturally active. Developing high-performance adult athletes requires that children have opportunities for skill development and competitive sport. People are continuing to compete in high-level sport as they age, retiring later than in the past. Moreover, there are many health benefits of regular exercise (discussed further in the next chapter), and virtually all people are encouraged to be physically active.

Figure 12.1 shows general trends in world record running performance by age for adult males and females. At all ages males perform better than females and, for both sexes, performance is best in the age range of approximately 18 through 30 years. This chapter examines how exercise capacity and performance change across the life span for both males and females, as well as the implications of these changes for exercise prescription in children and in persons who are elderly. The chapter also covers factors responsible for age and sex differences in exercise performance.

Responses to Exercise in Children

Children differ from adults in their metabolic, cardiovascular, respiratory, thermoregulatory, and perceptual responses to exercise, as well as in their muscular strength. Children should not be thought of simply as small adults. A different approach to exercise testing and training is needed for studying children's responses to exercise. For example, young children may lack the motivation or understanding to exercise to exhaustion, and a $\dot{V}O_2$max test may be inappropriate for assessing exercise capacity in the young.

Metabolic Response to Exercise in Children

Children's metabolic responses (both aerobic and nonoxidative) to exercise differ from adults'.

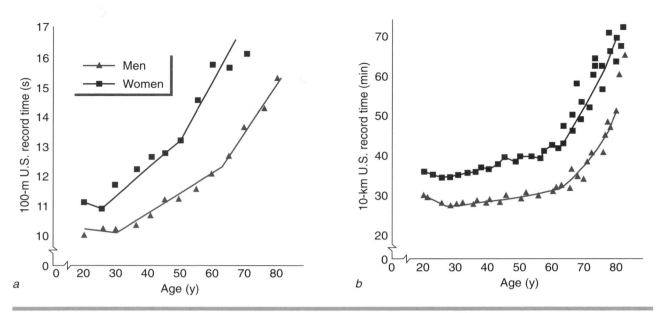

Figure 12.1 Trends for U.S. running records in the (a) 100 m and (b) 10 km by age and sex. For both events, the fastest times are recorded in the age range between 20 and 30+ years and for males compared with females.
Reprinted from Wilmore and Costill 2004.

Aerobic Exercise Capacity in Children

When expressed as an absolute rate (L/min), $\dot{V}O_2$max is much lower in children than in adults and increases proportionately with growth. In untrained males, $\dot{V}O_2$max increases until age 16 to 18 years and remains relatively constant until the mid-20s. In untrained females, $\dot{V}O_2$max increases until age 12 to 14 years, after which it may plateau or possibly decrease. In both boys and girls, much of the growth-related change in $\dot{V}O_2$max is associated with changes in body size and composition. On average, $\dot{V}O_2$max tends to be higher in boys than in girls across all ages, although there is a great deal of overlap up to age 10 years. Before puberty, absolute $\dot{V}O_2$max in girls is 85% to 90% of that in boys; after puberty, the gap widens, with $\dot{V}O_2$max expressed as L/min in girls approximately 70% that of boys.

When adjusted for growth-related changes in body size (ml $\cdot$ kg^{-1} $\cdot$ min^{-1}), $\dot{V}O_2$max is generally similar in young children compared with adolescents and young adults (figure 12.2). The sex difference in $\dot{V}O_2$max relative to body mass is smaller than that for absolute $\dot{V}O_2$max. Although $\dot{V}O_2$max relative to body mass remains constant after puberty in boys, it may actually decrease after puberty in girls. This decrease is probably due to both physiological and social factors, (e.g.,

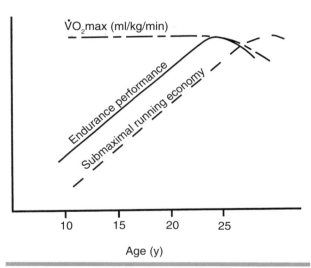

Figure 12.2 General trends of developmental changes in aerobic power, running economy, and endurance performance.
Reprinted from Rowland 1990.

the increase in proportion of body mass as fat as well as declining participation in regular exercise in girls).

$\dot{V}O_2$max relative to body mass is not as good a predictor of endurance capacity in children as it is in adults. For example, endurance training in children may improve performance without markedly improving $\dot{V}O_2$max.

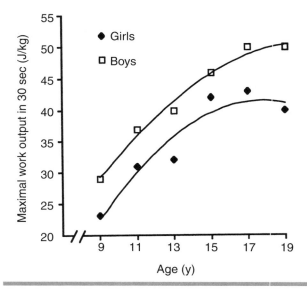

Figure 12.3 Anaerobic capacity in children measured as maximal work output, in joules/kg, during a 30-s knee extension test.
Adapted from Saavedra et al. 1991.

Although aerobic power may be similar between children and adults, there is a considerable difference in the metabolic cost of activity and the economy of movement. The oxygen cost of exercising at a set pace (e.g., running at a given speed) is higher in children than in adolescents and young adults because their economy of movement is lower than that of more mature people. This disparity, accounted for by differences in technique and body dimensions (e.g., leg length), decreases with growth.

Anaerobic Exercise Capacity in Children

Anaerobic capacity, as measured in 10- to 30-s all-out exercise, is much lower in children compared with adults (figure 12.3). Anaerobic capacity increases throughout childhood and adolescence, with the greatest changes occurring between the ages of 9 and 15. As with aerobic capacity and strength, there are few sex differences before age 9, with sex differences emerging between the ages of 10 and 13. Anaerobic capacity peaks between ages 14 and 16 in girls but continues to increase until at least age 20 in boys. Again, these differences are due to increased muscle mass associated with growth and maturation in boys.

The differences in anaerobic capacity between children and adults persist even when values are adjusted for differences in body mass or muscle size. Thus, children have a much lower capacity to generate power, and this difference is not due simply to a smaller body size or less muscle. The immediate energy system is not deficient in children, but anaerobic glycolytic capacity is lower in children than in adults. Skeletal muscle glycogen stores, glycolytic enzyme activity, and the ability to buffer acidity from lactic acid are lower in children. Although children have a lower glycolytic capacity, they recover faster than adults after brief high-intensity exercise, possibly because of the lower amount of lactic acid produced. Thus, children may be better able to tolerate interval training and stop–start activities such as soccer or basketball.

Cardiorespiratory Responses to Exercise in Children

As noted in chapter 10, maximum heart rate is a function of age, and children have very high maximal heart rates—frequently over 200 beats/min. However, both submaximal and maximal cardiac output and stroke volume are lower in children than in adults. The high maximal heart rate does not fully compensate for the lower maximal stroke volume, resulting in a lower cardiac output during exercise. Blood flow to working muscles is higher during exercise in children; this may partially compensate for the lower cardiac output. However, compared with values in adults, the oxygen-carrying capacity of blood is lower in children due to lower levels of the oxygen-carrying protein haemoglobin. In contrast, during exercise, skeletal muscle oxygen extraction and muscle oxidative capacity are similar in children and adults.

The respiratory system is less efficient in children than in adults. Children require a higher minute ventilation at any given oxygen consumption; that is, in children, more air must be inspired by the lungs to provide the same amount of oxygen to the muscles. The reason is the higher respiratory rate and shallower breathing in children. The respiratory muscles of children must work harder during exercise; and respiratory muscle fatigue contributes to the higher metabolic cost, feelings of discomfort, and early fatigue during intense exercise. These differences in the cardiovascular and respiratory system responses to exercise limit oxygen delivery to working muscles, resulting in a lower endurance exercise capacity in children. This does not mean that children cannot perform endurance exercise or improve endurance exercise capacity; rather, it means that children cannot be expected to perform endurance exercise, or train for endurance events, at the level expected of adults.

Thermoregulatory Response to Exercise in Children

Children produce more metabolic heat during exercise than adults and must lose more body heat to avoid an increase in core temperature during exercise. During exercise, the body loses heat mainly by sweating. Compared with adults, children begin to sweat at a higher relative work rate and also sweat less during exercise. The lower rate of heat loss via sweating is partially compensated for by a higher rate of heat loss via circulatory adjustments and conduction and convection from the skin's surface. This is so because children have a higher surface area-to-mass ratio. However, the combination of higher metabolic heat production and poorer ability to remove heat via sweating means that children are less tolerant of prolonged exercise, especially in a warm environment. Children also require a longer time to acclimatize (adjust) to exercise in warm environments. In addition, children have a less responsive thirst mechanism; that is, they are less likely to voluntarily replace water lost during exercise.

Coaches and teachers who supervise children exercising in the heat should be aware of the lower heat tolerance in children and should plan sport training and classes accordingly. For example, sport training or physical education classes should schedule different sports according to the seasons and should include interval-type exercise with longer rest intervals and frequent water breaks on hot days.

Muscular Strength in Children

As described in earlier chapters, muscular strength is related to muscle size. As expected, then, muscular strength increases with growth in children and adolescents. Muscular strength is similar between boys and girls up to age 8 to 9 years. In boys, muscular strength increases linearly until age 13 to 14 years; then an accelerated increase occurs during the adolescent growth spurt. In girls, strength increases linearly until age 14 to 16 years, after which it remains relatively unchanged.

These differences in muscular strength are related to differences in body composition, especially muscle mass. During childhood, body size and body composition are only modestly related to muscular strength; after puberty, strength is more closely related to body size and muscle mass. Body size, somatotype, and muscular strength are more closely related to one another in boys than in girls.

The large increase in strength after puberty in boys is due to a higher level of the male hormone testosterone, which increases muscle mass. In both boys and girls, gains in strength during and after puberty are also due to simultaneous maturation of the neural pathways that control movement (see also chapter 16). The rate of strength gain during maturation varies widely between individual children, and muscular strength measured in childhood is not a good indicator of strength in adulthood.

Adaptations to Exercise Training in Children

In the past, people questioned whether children can increase physical capacity with training. Research over the past 20 years, however, shows that children are trainable and that children demonstrate significant improvements in muscular strength and aerobic and anaerobic exercise capacity with training.

Aerobic and Anaerobic Training in Children

$\dot{V}O_2$max and endurance exercise performance improve in children when training is based on sound physiological principles as described in chapter 11. Children, however, show less relative improvements than adults—whereas $\dot{V}O_2$max may increase 20% to 40% in adults, it increases by only 5% to 25% in children. The reasons for this limited response of aerobic power are not completely known, but may relate to the lower haemoglobin and thus lower oxygen-carrying capacity of blood in children. Other cardiorespiratory adaptations include

- decreased resting heart rate and heart rate at a given submaximal work rate,
- increased maximal cardiac output and stroke volume,
- increased work rate (as a percentage of $\dot{V}O_2$max) at the lactate threshold, and
- increased maximum minute ventilation.

Aerobic performance often increases more than $\dot{V}O_2$max, suggesting that economy of movement (technique, skill) improves with endurance training in children (figure 12.2).

Anaerobic training has also been associated with improved performance. In brief high-intensity exercise, as measured in the 10-s and 30-s all-out cycling tests, biochemical adaptations include higher blood and muscle lactic acid concentrations during and after maximal exercise, increased skeletal muscle stores of glycogen and ATP-PCr, and enhanced glycolytic enzyme activity in skeletal muscle.

Strength Training in Children

Until the mid-1980s, children and adolescents were not advised to perform strength training. This view was based on the belief that children have a limited capacity to increase strength beyond gains due to normal growth and that strength training may be harmful to the growing musculoskeletal system. However, scientific research since then confirms that suitably prescribed programs can improve muscular strength without increasing the risk of injury in boys and girls.

Musculoskeletal injuries arising from strength training in youth usually occur because of poor technique and during maximal lifts above the head or body in unsupervised programs. Recommendations for strength training in children and adolescents are given in "Summary of Guidelines for Resistance Training in Youth", and include the following precautions:

- Programs should be supervised and should include spotting for all lifts above the head or body using free weights.
- Programs should be noncompetitive and should emphasize proper form and technique rather than the amount lifted.
- Training should focus on developing muscular endurance; that is, it should involve high-repetition, low-resistance work with at least 7 to 10 repetitions per set.
- Persons under age 16 years should not perform maximal lifts.

Exercise Prescription for Children

It is clear, then, that because children can experience training effects through physical exercise, current recommendations that everyone should be physically active on a regular basis also apply to children. The recent increased incidence of childhood obesity (now at 20-25%) has been attributed partly to less physical activity.

SUMMARY OF GUIDELINES FOR RESISTANCE TRAINING IN YOUTH

Mode

- Training should be sport specific to the extent possible.
- Resistance provided by body weight (e.g., callisthenics) or pin-weight machines is preferable to free weights.

Intensity

- Training should be of low intensity and higher repetitions, focusing on muscular endurance rather than strength (e.g., 8-15 repetitions per set).

Volume

- Volume and frequency should be low (one or two sets, one or two days per week).
- The program should incorporate 1- to 2-min rest between sets.

Safety Issues

- People under the age of 16 years should not perform maximal lifts.
- Focus should be on proper technique and breathing.
- Young exercisers should perform complete warm-up and cool-down, including stretching of muscles used in training.
- Children who are performing resistance training, especially novices, should have adult supervision.

Compiled from American College of Sports Medicine 2000; Sports Medicine Australia Safety Guidelines for Children 1997.

Exercise prescription in children should roughly follow the general recommendations for enhancing fitness in adults described in chapter 11. At minimum, children should participate in 30 min of moderate-intensity physical activity on most, preferably all, days. For young children (under 6 years), this should occur through enjoyable creative play. In addition to 30 min/day of moderate exercise, older children and adolescents should perform 20 to 30 min of vigorous exercise at least three times per week.

Children have very different reasons and motivation for exercising than adults. Children exercise for fun, peer recognition, and social interaction and prefer interval-type exercise in the context of recreation and games. Self-selected activities that are fun and that can be incorporated into the daily lifestyle (e.g., cycling to school) or social environment (e.g., in-line skating) are more likely than others to attract children and adolescents to habitual physical activity.

Exercise Capacity During Aging

It is well known that physical work capacity and sport performance decrease during aging. As seen in figure 12.1, world record performance declines after about age 30 in both male and female athletes.

Aging is associated with changes in many physiological variables that influence physical work capacity and sport performance. Extreme loss of physical work capacity may also limit the ability of older persons to maintain an independent lifestyle (e.g., walk, climb stairs), resulting in institutionalised care of persons who are elderly. In this discussion, the elderly group consists of persons over age 65 years.

Until recently, a decline in functional capacity was regarded as simply an inevitable consequence of growing older. However, it has recently been questioned whether the dramatic changes in work capacity are due solely to the aging process itself or also to the increasingly sedentary lifestyle that usually accompanies aging; older individuals tend to be less active during both leisure and work. Many of the age-related changes in functional capacity resemble those seen with detraining (cessation of exercise training). It is important to distinguish the inevitable effects of aging from those due to inactivity. If age-related changes are due primarily to lifestyle, then maintaining physical activity patterns throughout adulthood and into old age may prevent or delay many of the adverse physical consequences of aging.

Reasons for Decreases in Exercise Capacity During Aging

Both aerobic and anaerobic exercise capacity decline with age, as a consequence of changes in many physiological variables such as aerobic power ($\dot{V}O_2$max), maximum heart rate (HRmax), cardiac output, blood composition, and decreased muscle mass and strength. Although some degree of change is inevitable with aging, continued

exercise training can slow the rate of decline and help maintain exercise capacity in people who are elderly.

Variability in Rates of Physiological Aging

Individual variability—that is, the differences between individuals—increases with aging. In other words, people age at different rates with regard to physical capacity. Spirduso (1995) classifies older individuals into five categories based on physical function:

- *Physically dependent*—those who are debilitated by disease
- *Physically frail*—those with conditions affecting functional capacity in tasks of daily living
- *Physically independent*—the majority of elderly people who are free of disease but do not exercise regularly
- *Physically fit*—those who are physically active to achieve health benefits at least twice per week
- *Physically elite*—a small group who train intensely, sometimes for competitive sport or adventure recreation

We need to consider this diversity of physical activity patterns and physical capacity when discussing physiological responses to exercise in older people. The following sections on physiological responses and adaptations to exercise provide a general discussion in relation to the "average" older individual—that is, a person who would be considered physically independent or fit according to Spirduso's classification.

Aerobic Capacity During Aging

Both cross-sectional and longitudinal studies indicate that aerobic power, as measured by $\dot{V}O_2$max, declines after about age 30 years. The rate of decline averages about 0.5% to 1% per year but may vary depending on the group studied (figure 12.4). Between ages 40 and 50 years, $\dot{V}O_2$max in the average sedentary person may decline by 5 ml · kg^{-1} · min^{-1} (up to 10%). In sedentary people, the rate of decline accelerates after age 75 years.

The decrease in $\dot{V}O_2$max during aging in a sedentary individual relates to several factors, including loss of muscle mass, decreases in maximum heart rate and cardiac output, and a decline in the oxidative capacity of muscle.

In sedentary persons, the age-related decline in $\dot{V}O_2$max is generally correlated with changes in body composition. Loss of muscle mass and

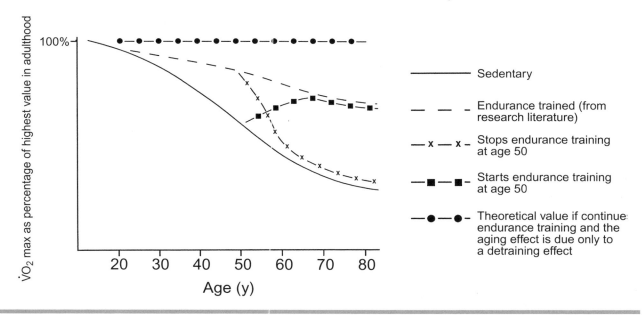

Figure 12.4 Trends in $\dot{V}O_2$max with age. This graph was drawn from research data and theoretical trends hypothesized if age-related changes were due solely to effects of detraining.

an increase in fat mass collectively decrease the amount of metabolically active muscle tissue that consumes oxygen to produce energy for exercise.

About 50% of the decrease in $\dot{V}O_2$max is due to a decrease in maximum heart rate, resulting from changes in neural input to the heart; this is not altered by training at any age. Since cardiac output is a function of heart rate and stroke volume, the decrease in maximum heart rate results in a decrease in maximum cardiac output. As discussed in chapters 10 and 11, cardiac output is an important determinant of $\dot{V}O_2$max. In contrast to cardiac output and heart rate, maximum stroke volume may not change much during aging.

The ability of skeletal muscle to extract and use oxygen during exercise also decreases with age in a sedentary population. This decrease is due to a reduction in the number of capillaries within skeletal muscle and in the capacity to redirect blood flow to muscle, which compromises oxygen delivery to working muscles during exercise. The oxidative capacity of skeletal muscle also declines during aging, possibly due to decreased mitochondrial function and protein synthesis in skeletal muscle. Thus, there is both less delivery of oxygen to working muscles and lower capacity of the muscles to use oxygen to produce ATP.

Anaerobic Capacity During Aging
Compared with aerobic exercise capacity, there has been far less research on anaerobic capacity

in older sedentary individuals. As described in chapter 10, tests of anaerobic capacity involve 30-s sprints, often on a cycle ergometer. It is difficult and possibly unsafe for older individuals unaccustomed to intense exercise to exert a maximal effort in such a test; it is also unlikely that older sedentary people would be motivated to perform maximally on such a test.

Anaerobic capacity peaks at about age 20 years but may be maintained with continued high-intensity training until the late 30s to early 40s. In older sedentary people, anaerobic capacity declines by about 6% per decade. This decline is closely related to loss of muscle mass, especially of the thigh muscles, the main source of power in cycling tests of anaerobic capacity. The decreases in anaerobic capacity and muscle mass are due to decreases in muscle fibre size (especially in the larger, stronger FT fibres), loss of FT motor units, and changes in coordination. Anaerobic capacity and muscle size decrease with age more in women than in men, although this may occur because older women are less likely to perform intense physical activity. These sex differences are discussed in more detail starting on page 175.

Muscular Strength
Muscular strength is important to sport performance as well as to basic functions such as posture, balance, and coordination, and to simple activities such as walking and climbing stairs. Dramatic

changes in muscular strength accompany aging and may limit an older individual's ability to be physically active and to maintain an independent lifestyle.

For untrained persons, maximum muscular strength is achieved in the early 20s and may be maintained with regular strength training through the late 30s to early 40s. Strength decreases by about 2% to 4% per year in aging sedentary individuals. Declines in strength are attributable to several factors. Lean body mass decreases gradually between ages 30 and 50, after which the loss of muscle mass accelerates. Between 20 and 80 years of age, people typically lose 40% of their muscle size. As explained in chapters 5 and 11 muscular strength is generally related to muscle mass or cross-sectional area (see "Strength Training in People Who Are Elderly: How Adaptable Are Aging Muscles?" on p. 173). During aging, muscle mass decreases due to atrophy of the larger, stronger FT muscle fibres and loss of FT motor units. Slow-twitch (ST) fibres do not atrophy as much as FT fibres because ST fibres are recruited during normal activities such as standing and walking. As they age, people are less likely to perform high-intensity exercise, the type needed to recruit FT fibres.

During aging, the amount of connective tissue within skeletal muscle may increase while the size of muscle fibres may decrease. Because connective tissue does not contribute to active force generation, a muscle's overall size may not decrease as much as its force-generating capacity. That is, the muscle may appear to be the same size, but a relative decrease in force-producing muscle mass and relative increase in nonforce producing connective tissue mean the muscle is weaker. In addition, age-related changes in neural input, particularly loss of FT motor units in skeletal muscle, may influence recruitment patterns and reduce force generation. Finally, older untrained persons may be reluctant to voluntarily exert maximal force, and this may result in lower strength readings during testing.

Preventing or Reversing Age-Related Changes in Work Capacity

Because people tend to become more sedentary with age, it is unclear how much the age-related changes in exercise capacity are due to a reduced level of physical activity and how much to the aging process itself. In sedentary persons, many of the age-related changes in exercise capacity resemble the effects of detraining—for example, decreases in $\dot{V}O_2$max and muscular strength.

Comparisons between masters (veteran) athletes and age-matched sedentary individuals indicate a much slower rate of decline with age in $\dot{V}O_2$max and muscular strength for the athletes. Moreover, training studies generally indicate that the responses to exercise training in older individuals are relatively similar to those seen in younger people, provided that similar training programs are used.

Aerobic Capacity in Masters (Veteran) Athletes

In masters athletes, $\dot{V}O_2$max decreases with age at half the rate observed in an aging sedentary population. For example, while $\dot{V}O_2$max may decline with age by 0.5 ml $\cdot$ kg^{-1} $\cdot$ min^{-1} per year in sedentary people, the rate is about 0.25 ml $\cdot$ kg^{-1} $\cdot$ min^{-1} per year in masters endurance athletes who continue to train (see figure 12.4). Thus, about half of the decrease in aerobic power appears to be inevitable with aging, and half seems to be preventable with continued training. However, it is not clear whether masters athletes continue to train at the same (high) intensity as when they were younger.

In the veteran athlete, the decrease in $\dot{V}O_2$max results from the decrease in maximum heart rate, which cannot be changed with training. Relatively smaller decreases in maximum stroke volume and skeletal muscle oxygen extraction and oxidative capacity may also contribute to the slower rate of decline in $\dot{V}O_2$max in the athlete. In contrast, other factors, such as skeletal muscle capillary number, muscle fibre size (especially in ST fibres), and skeletal muscle oxidative activity may be maintained with continued training. The higher the volume and intensity of training, the more these physiological functions are retained during aging.

Anaerobic Capacity and Muscular Strength in Masters Athletes

Few studies have addressed anaerobic capacity and muscular strength in masters athletes. As with endurance capacity, continued training halves the rate of decline of anaerobic capacity and muscular strength with age. Much of the age-related decline in muscular strength can be accounted for by loss of muscle mass, even in masters athletes. However, continued training partially prevents the skeletal muscle atrophy, particularly of FT fibres, that normally occurs in sedentary individuals. Recent research suggests that strength training induces muscle hypertrophy and improves strength in older people (see "Strength Training in People

IN FOCUS: STRENGTH TRAINING IN PEOPLE WHO ARE ELDERLY: HOW ADAPTABLE ARE AGING MUSCLES?

During aging, loss of muscle mass and strength can significantly impair functional capacity and possibly limit the ability to be physically active and live an independent lifestyle. Poor muscular strength, loss of bone mineral density, and poor balance due to loss of muscular strength increase the risk of falling and fracturing a bone. Several research groups have studied the relationship between muscular strength, bone health, and balance in elderly persons to determine whether resistance training can counteract some of the adverse changes.

In postmenopausal women (those most at risk of bone fracture), one year of high-intensity strength training (80% of 1-repetition maximum or 1RM) two days per week resulted in significant increases in bone mineral density (BMD) in the spine and hip as well as total body bone mineral content (BMC) (Nelson et al. 1994). Muscle strength and mass, BMD, and BMC increased over the year in the exercise group, whereas these variables declined over the same period in a nonexercise control group. The beneficial changes in muscular fitness were associated with enhanced dynamic balance in the exercisers.

It is important to know how much resistance training is needed to maintain muscular fitness in persons who are elderly, since some may not be able or willing to train frequently. One recent study compared the responses to two types of resistance training in elderly persons (Hunter et al. 2001). Older men and women (ages 61-77 years) performed resistance training for 45 min three days per week for 25 weeks. One group always trained at a constant high intensity (80% of 1RM), while another group varied their training intensity (from 50% to 80% of 1RM). The two groups showed similar increases in muscular strength and muscle mass after the 25 weeks. However, the group that varied training resistance showed more improvement in the ability to perform a functional task (walking while carrying a weight). The authors concluded that variable resistance training leads to improvements in muscular fitness similar to those obtained with high-intensity training but may offer additional benefits in performance of tasks of daily living.

In another study in the elderly group (Taaffe et al. 1999), men and women aged 65 to 79 years performed high-intensity resistance training (80% of 1RM) on one, two, or three days per week for 24 weeks. All exercisers increased muscular strength of the back and the upper and lower body, and there were no differences in strength gains between the three exercise groups. Training only once per week was as effective as training two or three times per week. Moreover, these improvements in muscular strength were correlated with beneficial changes on tasks such as rising from a chair or backward walking, measures of coordination and balance.

Together, these studies suggest that, in the elderly population, appropriate resistance training can enhance muscle mass, fitness, and dynamic balance and thus help reduce the risk of falling and fracturing a bone. One session per week of high-intensity training, or two to three sessions of variable resistance training, may be just as effective as three sessions of high-intensity training.

SOURCES

- Hunter, G.R., C.J. Wetzstein, C.L. McClafferty, P.A. Zuckerman, K.A. Landers, and M.M. Bamman. 2001. High-resistance versus variable-resistance training in older adults. *Medicine and Science in Sports and Exercise* 33: 1759-1764.

- Nelson, M.E., M.A. Fiatarone, C.M. Morganti, I. Trice, R.A. Greenberg, and W.J. Evans. 1994. Effects of high-intensity strength training on multiple risk factors for osteoporotic fractures: A randomized controlled trial. *Journal of the American Medical Association* 272: 1909-1914.

- Taaffe, D.R., C. Curet, S. Wheeler, and R. Marcus. 1999. Once-weekly resistance exercise improves muscle strength and neuromuscular performance in older adults. *Journal of the American Geriatric Society* 47: 1208-1214.

Who Are Elderly: How Adaptable Are Aging Muscles?" on p. 173).

Despite inevitable changes with age, compared with the average untrained young individual, the veteran athlete exhibits higher levels of many variables such as $\dot{V}O_2$max, muscular strength, and anaerobic capacity.

Training Effects in Previously Sedentary Individuals

Older people are capable of increasing exercise capacity and performance in both endurance and brief, high-intensity exercise. In general, however, responses may be less and may take longer than in younger individuals.

- **Aerobic training effects in previously sedentary persons.** $\dot{V}O_2$max increases with endurance training in older individuals, although it may take longer to see these increases than in younger adults. As in younger individuals, changes in $\dot{V}O_2$max are closely related to both the total amount of exercise and training intensity. Increases in $\dot{V}O_2$max in the range of 20% to 40% may occur in older people following endurance training. Increases in maximum cardiac output (due to increases in stroke volume) and increases in skeletal muscle oxygen extraction and oxidative capacity accompany increases in $\dot{V}O_2$max.

- **Anaerobic training in previously sedentary persons.** Few studies have examined the effects of anaerobic training in older untrained people. The limited data suggest that high-intensity aerobic training up to 85% $\dot{V}O_2$max also improves anaerobic capacity as measured in the 30-s sprint cycle ergometer test. The mechanisms are unclear at present but may be related to increases in muscular strength and size, especially of FTa fibres, as well as enhanced oxidative capacity in muscle.

- **Strength training in previously sedentary persons.** Increases in isometric and isokinetic strength after resistance training have been well documented in older, previously sedentary people (see "Strength Training in People Who Are Elderly: How Adaptable Are Aging Muscles?" on p. 173). Early studies suggested that gains in muscular strength occur in the absence of muscle hypertrophy and must therefore be due to neural adaptations. However, more recent work using sensitive imaging devices to accurately measure muscle cross-sectional area, such as computer-aided tomography (CT scans) and magnetic resonance imaging (MRI), shows that muscle hypertrophy occurs with resistance training in previously sedentary older people. Muscle hypertrophy is due to hypertrophy of FT fibres, especially FTa fibres. However, the increases in strength are greater than can be accounted for by hypertrophy alone, indicating that neural adaptations also contribute to strength gains in older people. It is generally believed that the capacity for hypertrophy is somewhat limited in this population and that neural factors contribute proportionally more to strength gains in older than in younger individuals. The limited capacity for muscle hypertrophy in older compared with younger men may also be due to age-related decreases in levels of the male hormone testosterone.

Exercise Prescription for Older Adults

Exercise prescription for older people follows the general guidelines for health-related fitness discussed in chapter 11, with some modification. A conservative approach to exercise prescription is needed, given that older individuals often exhibit a low functional capacity and slower rate of adaptation to exercise training. As noted earlier, variability in physical capacity widens with age, and there can be large differences in exercise capacity between individuals.

Table 12.1 lists some of the special concerns one should take into account when prescribing exercise for elderly people. The goals of physical activity for this group are to improve or maintain functional capacity, muscular strength and endurance, and quality of life, and to slow or prevent the onset of disease. The exercise prescription should focus on developing and maintaining aerobic capacity, muscular strength and endurance, flexibility and range of motion around the joints, balance, and coordination.

Low- to moderate-intensity exercise confers many health benefits (discussed in chapters 11 and 13) while avoiding possible deleterious effects such as musculoskeletal injury or a cardiac event (e.g., heart attack). Older people are also more attracted to lower- to moderate-intensity exercise than to exercise at higher intensities, especially if they are in a group. Exercise modes such as walking, low-impact aerobics, water-based exercise, and exercise on machines (e.g., treadmill, exercise bike) are appropriate. Intensity is generally recommended

Table 12.1

Exercise Prescription for the Older Adult

Variable	Considerations for exercise prescription
Low functional capacity	Low-intensity exercise (40-50% $\dot{V}O_2$max), gradual progression, interval training to avoid early fatigue
Low muscular strength and endurance	Exercises to enhance muscular strength and endurance
Impaired coordination and slower reaction time	Simple movements, supported exercise, exercise machines
Increased risk of disease, especially heart disease, obesity, hypertension, arthritis, and osteoporosis	Health risk screening and medical examination, modification of exercise prescription according to condition
Prevalence of osteoarthritis	Low-impact, low-intensity exercise; weight-supported or water-based activities
Low heat tolerance	Frequent replacement of fluids, avoidance of exercise in the heat
Lower capacity for muscle hypertrophy and protein synthesis	Expect slower progress and improvement in exercise capacity

in the range of 50% to 70% $\dot{V}O_2$max, although lower intensities (40-50%) may be appropriate for beginners or those with low functional capacity. Self-selected pace may enhance enjoyment and compliance.

Recent research suggests that resistance training can help maintain muscle mass and prevent declines in muscular fitness in elderly exercisers. Current recommendations for healthy adults, including those who are elderly, suggest that resistance training should be performed two to three times per week and should include 8 to 10 exercises using all the major muscle groups, each performed 8 to 15 times (repetitions) per set. There is some evidence that, at least at the onset, resistance training once per week is as effective in increasing muscular fitness as two to three times per week (see "Strength Training in People Who Are Elderly: How Adaptable Are Aging Muscles?" on p. 173). Older people should aim for repetitions of eight or more per set to avoid maximal lifts and high-intensity resistance exercise, since these can cause an excessive rise in blood pressure and musculoskeletal injury.

The risk of falling increases with age; combined with age-related decreases in bone mass, a fall can lead to bone fracture (discussed further in chapters 13 and 22). In those who are elderly, regular physical activity helps to maintain bone mass and muscular strength and to prevent falls. A combination of moderate exercises, such as walking, circuit resistance training, callisthenics, and exercise to music, is effective in increasing aerobic fitness, muscle mass, muscular strength, balance, coordination, and agility.

Life Span Sex Differences in Physiological Responses and Adaptations to Exercise

With a few exceptions (e.g., equestrian events), males and females usually do not compete against each other in competitive sport; boys and girls are usually separated by about age 10 to 12. This is a consequence of obvious differences in body size, shape, and physiology, which in turn influence important factors such as muscular strength and power and endurance exercise capacity. Comparing the physiological responses to exercise in males and females helps us further understand limiting factors to performance and mechanisms underlying long-term training adaptations, as well as to develop optimal exercise programs to enhance fitness and health for both men and women.

Sex Differences in Aerobic Exercise Capacity

As mentioned earlier, $\dot{V}O_2$max tends to be higher in males than in females at all ages. This difference is rather small in young children, becoming larger with age, especially after puberty when aerobic power continues to increase in boys but plateaus in girls.

On average, boys also perform better than girls in endurance tests such as the 1-km (0.6-mile) run. The gap in endurance performance cannot be explained only by differences in $\dot{V}O_2$max, suggesting that other factors influence sex differences in endurance performance. For example, at all ages boys tend to be more active than girls. Thus, on average, boys may be considered better trained or closer to their biologically determined potential. The reasons for sex differences in activity patterns among children are complex but probably relate more to social (e.g., opportunity and support) than to biological factors, especially in young children. There is, however, great individual variability and overlap between boys' and girls' performance, and it is not unusual for some girls to be better athletes than many boys their age. The differences between prepubescent boys' and girls' performance narrow considerably when one compares child athletes who have received the same level of training and support.

Beginning during puberty, aerobic exercise capacity is higher, on average, in males than in females. The magnitude of the difference in aerobic power, however, is related to how $\dot{V}O_2$max is expressed. For example, since absolute $\dot{V}O_2$max (L/min) is related to body size, there is a large difference, approximately 40%, between sexes. The sex difference narrows to about 20% when $\dot{V}O_2$max is adjusted for body size (ml · kg^{-1} · min^{-1}) and even further, to less than 10%, when adjusted for differences in body composition (ml · kg lean body mass^{-1} · min^{-1}). This difference reflects the fact that oxygen uptake during exercise is largely a function of muscle metabolism; a larger mass of metabolically active muscle tissue leads to a higher oxygen use.

Even when $\dot{V}O_2$max values are adjusted for body composition, there is still about a 10% difference between similarly trained men and women. This residual difference is most likely due to other cardiovascular factors influencing oxygen delivery to and extraction by skeletal muscle. As shown in table 12.2, blood levels of red blood cells and haemoglobin (the oxygen-carrying protein in blood) and heart size relative to body size are lower in

Table 12.2

Physiological Sex Differences and Effects on Performance

Female compared with male	Effect on exercise performance
Smaller body, shorter limbs, narrower shoulders, wider hips	Lower muscular strength and power
Less muscle mass, proportionately more body fat	Less metabolically active tissue, lower $\dot{V}O_2$max, lower muscular strength and power, lower aerobic and anaerobic power
Smaller average muscle fibre diameter	Lower muscular strength and power
Smaller heart	Lower stroke volume, cardiac output, and $\dot{V}O_2$max
Lower blood haemoglobin level	Lower oxygen delivery to muscle and $\dot{V}O_2$max
Lower $\dot{V}O_2$max	Lower aerobic power and endurance performance
Greater reliance on circulatory adjustments than on sweating to lose heat	Lower sweat rate during exercise
Variables for which there are no sex differences	
Distribution of muscle fibre types	
Lactate threshold as a percentage of $\dot{V}O_2$max	
Tolerance of exercise in the heat	
Metabolic capacity per gram of muscle	
Relative improvement in aerobic and anaerobic capacity and muscular strength after training	
Psychological factors influencing sport performance (e.g., motivation, competitiveness, pain tolerance)	

Compiled from Wells 1992; McArdle, Katch, and Katch 1996.

women. The combination of lower red blood cell and haemoglobin concentrations and cardiac output means that, in women, less oxygen is delivered to working muscles during maximal exercise.

It should be noted that these sex differences in aerobic power and related cardiovascular system measures become very small, less than 10%, when similarly trained female and male endurance athletes are compared. In addition, males and females show similar improvements in $\dot{V}O_2$max and endurance performance relative to initial levels after endurance training.

Sex Differences in Anaerobic Capacity

When one is examining sex differences in anaerobic exercise capacity, it is important to include similarly trained individuals. In comparisons of similarly trained males and females, total work output and peak power in a brief all-out exercise test are lower in females. These differences are mainly due to the smaller body size and amount of muscle in females. As with $\dot{V}O_2$max, sex differences in anaerobic exercise capacity become smaller when adjusted for body mass but do not disappear completely, mainly due to differences in body composition and other factors such as relative limb length (i.e., men generally have longer limbs than women, conferring mechanical advantage). Peak muscle and blood lactic acid levels may be lower after maximal exercise in females, again possibly due to the smaller muscle mass. However, during submaximal exercise at the same percentage of $\dot{V}O_2$max, blood lactic acid levels and the lactate threshold are similar in males and females. Males and females generally show similar training responses provided that training programs are similar, except that females generally do not show the same magnitude of muscle hypertrophy as men (discussed in the next section).

The sex differences in anaerobic power and capacity are reflected in performance differences in power events such as jumping, throwing, and weightlifting. There are greater sex differences in performance in these power events, which are more related to muscle mass, than in other events such as sprinting or longer-duration activities.

Sex Differences in Muscular Strength

As discussed earlier in this chapter, strength does not differ between girls and boys until about age 11, after which muscular strength continues to increase in boys but plateaus in girls. In general, muscular strength in adult women is about two-thirds that in

men, although there is large individual variability depending on body size and muscle mass as well as the muscle group measured, fibre type, and individual fitness level. As with other measures, sex differences in strength become smaller when adjusted for body size and limb size, indicating that much of the sex difference in strength is due to differences in muscle mass. When expressed relative to lean body mass or muscle cross-sectional area, strength is similar between males and females; that is, strength per gram of muscle is similar between men and women. Sex differences are also smaller when strength is compared in similarly trained male and female athletes. For example, sex differences in strength are smaller among powerlifters than among average sedentary persons.

Sex differences in muscular strength also depend on the body segment and muscle groups compared. Muscular strength in females is closer to that of males in the lower body than in the upper body. For example, chest and arm strength relative to body weight in females is about 50% to 60% of that in males; in contrast, strength of the hip flexors and extensors relative to body weight is similar in males and females. These differences by body segment are probably due to two factors: first, anatomical differences such as wider shoulders and longer arms, and proportionately more total muscle mass in the upper body, in men, and second, the traditional tendency on the part of men to participate in sport and activities of daily living that train the upper body.

As with endurance training, males and females show similar improvements in strength relative to initial levels after resistance training, provided that the training programs are similar. Recent research using sensitive imaging techniques to visualize muscle cross-sectional area shows that muscle hypertrophy occurs in females, at least in short-term studies lasting up to several months. Muscle hypertrophy may not be as obvious because of the thicker subcutaneous fat layer in the female. There have not been enough long-term studies to demonstrate whether or not muscle size continues to increase with strength training in females. However, comparison of elite male and female bodybuilders or powerlifters shows greater muscle mass and definition in males, suggesting a greater capacity for hypertrophy with long-term training in males.

Sex Differences in Sport Performance

Although there are sex differences in world record performance in most sports, the gap between males and females has been narrowing over the past few decades. For example, the difference between the men's and women's 100-m (109-yd) world records decreased from 11.40% to 7.74% between the years 1956 and 1986. Over the past few decades, women's sport performance has improved at a much faster rate than men's, most likely due to social rather than biological factors. It has been suggested that if this trend continues, there might, in time, be no sex differences in world records in some sports. It is likely that swimming would be the first sport in which this might occur. Differences in world record performance between men and women are smaller in swimming than for other sports because the higher body fat levels in women provide an advantage in extra buoyancy. However, given the physiological, anatomical, and biomechanical differences between sexes, as described previously, it is unlikely that sex differences will completely disappear in all sports, especially those involving primarily pure speed, strength, power, and cardiorespiratory endurance.

Summary

To effectively prescribe exercise programs, whether for elite athletes or for general fitness and health, it is important to understand the effects of growth, development, aging, and sex differences on the physiological responses and adaptations to exercise. In most sports, best performances occur between the ages of 20 and the mid-30s, and in males compared with females at any age.

Children are not just scaled-down adults, and the physiological and psychological changes that accompany growth and development greatly influence exercise capacity. Whereas $\dot{V}O_2$max achieves adult values relatively early in youth, endurance performance continues to improve through young adulthood because of changes in body dimensions and economy of movement. Anaerobic capacity also improves with growth because of increased muscle mass and enhanced muscle metabolic capacity. Children show improvements in both aerobic and anaerobic exercise capacity with training, although the magnitude of change may be somewhat lower in children than in adults.

Physical work capacity and sport performance begin to decline at about age 30 in both men and women. The rate of decline is slower, by about half, in those who continue exercise training throughout adulthood and old age. Some changes are inevitable, such as the age-related decline in maximum heart rate; other changes may be slowed or prevented, such as loss of muscle mass. Older people are capable of improving both aerobic capacity and muscular fitness with appropriate exercise training that takes into account a slower, more gradual rate of improvement. Importantly, resistance training not only improves muscular fitness but also prevents falls by enhancing balance and coordination.

Physiological sex differences relating to body size, body composition, and blood chemistry account for most of the differences between men's and women's sport performance. Males and females show similar physiological changes with equivalent exercise training, except that men exhibit greater muscle hypertrophy after many months of resistance training.

Further Reading and References

American College of Sports Medicine. 1998. ACSM's position stand: Exercise and physical activity for older adults. *Medicine and Science in Sports and Exercise* 30(6): 992-1008.

Cotton, R.T., C.J. Ekeroth, and H. Yancy, eds. 1998. *Exercise for older adults.* San Diego: American Council on Exercise; Champaign, IL: Human Kinetics.

Jenkins, D., and P. Reaburn. 2000. *Guiding the young athlete.* St. Leonards, NSW, Australia: Allen and Unwin.

Rowland, T.W. 1996. *Developmental exercise physiology.* Champaign, IL: Human Kinetics.

Spirduso, W.W. 1995. *Physical dimensions of aging.* Champaign, IL: Human Kinetics.

Wells, C.L. 1992. *Women, sport and performance: A physiological perspective.* 2nd ed. Champaign, IL: Human Kinetics.

CHAPTER 13

Applications of Exercise Physiology to Health

The major learning concepts in this chapter relate to
- the major causes of disease, death, and disability in developed countries;
- the financial cost of physical inactivity;
- the fact that few adults are sufficiently physically active to derive health benefits;
- evidence that physical activity helps prevent and manage many diseases and impairments;
- physical activity, cardiovascular disease, and metabolic syndrome; and
- physical activity and other diseases or conditions, such as asthma, cancer, osteoporosis and bone health, and arthritis.

*R*egular physical activity has many health benefits. Moderate exercise can prevent obesity, reduce the risk of heart disease and hypertension (high blood pressure), guard against or reverse adult-onset or **type 2 diabetes,** and possibly prevent osteoporosis (loss of bone mass in older people). Exercise is considered part of a healthy lifestyle. Governments, doctors, and health promoters now recommend regular moderate physical activity for virtually all individuals, including children, persons who are elderly, and those with disease or disability.

Physical Activity and Public Health

Regular exercise is encouraged among the general population for reasons other than merely enhancing lifestyle; society as a whole benefits from reduced health care costs, increased productivity at work, and improvements in quality of life that accompany increased physical fitness. This chapter examines the major causes of disease, death, and disability in developed countries; the extent to which the disease and disability can be attributed to physical inactivity; and the role of a physically active lifestyle in preventing and managing major diseases and impairments.

The Public Cost of Inactivity

The costs of inactivity to the international community are enormous. We first examine the major causes of disease and death in developed countries and the costs associated with physical inactivity. Then we discuss how changing physical activity habits worldwide might reduce these costs.

Major Causes of Disease and Death in Developed Countries

The causes of disease and death have changed dramatically over the past 100 years. Whereas infectious diseases such as influenza ("flu") and tuberculosis were the major killers in the early

Table 13.1

Top Four Leading Causes of Death in Some Countries, 1997-2001

Country	Leading cause	Second leading cause	Third leading cause	Fourth leading cause
Australia	Cancer	Heart disease	Stroke	Respiratory diseases
Canada	Cancer	Heart disease	Stroke	Respiratory diseases
Denmark	Heart disease	Cancer	Respiratory diseases	Digestive system diseases
Japan	Cancer	Stroke	Heart disease	Respiratory diseases
United Kingdom	Cancer	Heart disease	Respiratory diseases	Stroke
United States	Heart disease	Cancer	Stroke	Respiratory diseases
Germany	Diseases of the circulation (including heart disease)	Cancer	Stroke	Respiratory diseases

Compiled from UK data: www.dphpc.ox.ac.uk/bhfhprg/stats/2000/2000/keyfacts/index.html; USA data: www.cdc.gov/nchs/fastats/deaths.htm; Canadian data: www.statcan.ca/English/Pgdb/People/Health/health36.htm; Danish data: www.dst.dk/665; German data: www.destatis.de/basis/e/gesu/healtab3.htm; Japanese data: www.stat.go.jp/English/data/nenkan/1431-25.html; Australian data: www.abs.gov.au/ausstats/abs@nsf/Lookup/.

part of the last century, the major causes of illness and death in most developed countries are now heart disease, stroke, and cancer, which collectively account for approximately two-thirds of all deaths each year. Table 13.1 presents the four leading causes of death in several developed countries: cancer, heart disease, respiratory diseases, and stroke. Physical activity has potential to prevent three of these diseases—heart disease, cancer and stroke—and can help manage all four.

There are many reasons for the change in causes of disease and death over the past century, including advances in hygiene (such as water and waste treatment and food processing), in medicine (such as antibiotics and other drugs), and in public health (such as mass immunization). Some diseases, such as heart disease and cancer, have become more prevalent because of our increasingly affluent and sedentary lifestyle.

Financial Burden of Physical Inactivity

Physical inactivity is more than just a personal lifestyle issue; physical inactivity contributes to many diseases and high health care costs. In many developed countries, health care costs have increased dramatically and now consume up to 10% of national income, in the hundreds of millions to billions of dollars each year for many countries. In most developed countries, the cost of health care is shared among the general population, either via government-supported medical

care, as in Europe, Canada, and Australia, or via private and employer-supported health insurance funds, as in the United States. Thus, society as a whole must pay the high cost due to lifestyle factors such as physical inactivity. While it is difficult to estimate additional social costs associated with death and disability arising from these diseases, these costs also contribute to the overall burden of disease on society.

It has been estimated that more than half of the incidence of the major killers such as heart disease, cancer, stroke, obesity, and diabetes may be prevented, particularly through alterations in lifestyle factors such as diet, physical activity, and smoking. Encouraging regular physical activity among the general population is an important, relatively low-cost way to reduce the incidence and costs of disease. For example, it has been estimated that about 12% of deaths each year (about 300,000 deaths) in the United States are attributable to physical inactivity. Australian data indicate that 18% of deaths due to heart disease and 16% of those due to stroke each year can be attributed to physical inactivity. Another way of looking at these data is to suggest that we might prevent 12% to 18% of heart disease deaths each year by becoming more physically active. Because physical inactivity contributes to many diseases, simply increasing the number of physically active people could provide savings in the tens to hundreds of millions of dollars.

Exercise Participation and Health Benefits in Adults

Although most adults agree that regular physical activity is important for good health, few exercise sufficiently to derive protective health benefits. For example, the World Health Organization estimates that about 60% of the world's population does not perform sufficient physical activity to derive health benefits; participation is especially low among women and those living in urban areas, in both developed and developing countries. In most developed countries, between 30% and 60% of adults do not perform sufficient regular physical activity (as described in chapter 11) to prevent disease. For example, about 50% to 60% of American adults are not sufficiently active to derive health benefits, and up to 30% may be sedentary (i.e., perform no regular physical activity). In Australia, about 44% of adults are considered to be at risk of premature death due to physical inactivity, with 20% considered sedentary. In the European Union, an average 32% of adults are considered sedentary.

Patterns of participation in physical activity are not uniform across a given population. In most developed countries, participation in regular physical activity decreases with increasing age. For example, more than 35% of Americans over age 55 years are considered sedentary. Participation also declines dramatically during adolescence, particularly among girls; nearly half of American and European young people (12-21 years) are not vigorously active on a regular basis. At all ages, men are more physically active than women. Participation also varies by education level, socioeconomic status, and racial or ethnic group. In general, the socioeconomically disadvantaged and people of racial or ethnic minorities and migrant groups tend to be the least active. In many developed and developing countries, participation in physical activity has declined over the past 20 years, leaving an ever increasing number of people at risk of diseases associated with physical inactivity.

The 1996 landmark U.S. Surgeon General's report, *Physical Activity and Health,* clearly stated that health and quality of life can be dramatically improved through inclusion of moderate physical activity in the daily lifestyle (the types and amount of physical activity recommended to enhance health were discussed in chapter 11). Moreover, the health benefits are proportional to the amount of physical activity; that is, the more physical activity performed, the greater the ben-

efits. The greatest impact on health is associated with moving from a sedentary to moderately active lifestyle, suggesting that even moderate physical activity has great potential to enhance health and prevent disease (see "Quantifying the Health Risk of Physical Inactivity: How Much Can Physical Activity Prevent Disease?" on p. 182).

Physical activity prevents disease both directly and indirectly. For example, physically active people have less than half the heart disease risk of sedentary people. This reduced risk arises from both a direct effect of exercise (strengthening the heart) and several indirect effects, such as reduced body fat, blood cholesterol, blood pressure, and blood glucose levels, each of which lessens the risk of disease. The diverse biological mechanisms by which physical activity helps prevent disease include positive effects on metabolism (e.g., increased tissue uptake of glucose from the blood), function (e.g., stronger heart pumping action), structure (e.g., less fat, more muscle), body chemistry (e.g., lower blood cholesterol levels), and psychological well-being (e.g., reduced anxiety and depression). The effects of exercise on various health risk factors are discussed in further detail later in this chapter.

Promoting a Physically Active Lifestyle

For most people, simply recognizing that exercise is important to good health is not enough to maintain the habit of regular physical activity. While many people enthusiastically begin exercise programs, fewer than 50% are still exercising regularly after six months.

Reasons for Differences in Activity Levels

As might be expected, people who are frequently active tend to be young, to be nonsmokers, to be fairly lean and relatively fit, and to have previously participated in physical activity or sport (table 13.2). Regular exercisers tend to exhibit a higher degree of self-motivation especially toward health-related behaviour and to have supportive friends and family members. Regular exercisers also tend to be more highly educated and of a higher socioeconomic level than those who do not exercise regularly. Physically active people commonly exercise because they wish to derive health benefits and improve physical appearance, and because they enjoy physical activity.

There are many reasons why people do not exercise regularly or drop out of exercise programs soon after beginning. Two commonly cited reasons

IN FOCUS: QUANTIFYING THE HEALTH RISK OF PHYSICAL INACTIVITY: HOW MUCH CAN PHYSICAL ACTIVITY PREVENT DISEASE?

As discussed throughout this chapter, physical activity helps prevent several diseases such as heart disease, stroke, hypertension, obesity, osteoporosis, cancer, and type 2 diabetes. Because heart disease and stroke are major killers in most developed countries, much of the research on physical activity and health has focused on preventing cardiovascular disease (CVD).

One of the first studies to document a relationship between physical activity and mortality involved 16,000 graduates of Harvard University who were studied over many years as they aged (Paffenbarger et al. 1986). Death rates were closely related to weekly energy expenditure. Compared with inactive men ("500 kcal/week), active men (>2,000 kcal/week) had a 25% to 33% lower death rate. The protective effect of physical activity was most pronounced in the older men.

In 1995 and 1996, Blair et al. published two important studies showing a strong relationship between poor physical fitness and increased risk of early death due to CVD. The largest difference in CVD death rate was between those in the lowest 20th percentile of physical fitness and the middle fitness level (20th to 60th percentile). Men who were initially in the low-fit category but improved their fitness level over the five-year study showed a 50% reduction in death rate.

A more recent study also shows a strong relationship between physical fitness and death due to CVD among high-risk men (i.e., those with risk factors for CVD) (Myers et al. 2002). Physical work capacity in metabolic equivalents (METs) was a strong predictor of death over a subsequent six-year period. Each 1-MET increase in exercise capacity was associated with a 12% decrease in risk of death.

These studies show that low physical fitness is strongly associated with risk of dying from CVD among both healthy and high-risk individuals; the risk of death can be reduced significantly with only moderate increases in fitness. Fortunately, not much exercise is actually needed to improve physical fitness. Exercising at an intensity as low as 30% of $\dot{V}O_2$max can increase physical fitness in people who are unfit to start with. This exercise intensity is easily achievable by almost everyone. In other words, for unfit, older persons, even low- to moderate-intensity exercise will improve physical fitness and enhance health.

SOURCES

- Blair, S.N., J.B. Kampert, H.W. Kohl, C.E. Barlow, C.A. Macera, R.S. Paffenbarger, and L.W. Gibbons. 1996. Influences of cardiorespiratory fitness and other precursors on cardiovascular disease and all-cause mortality in men and women. *Journal of the American Medical Association* 276: 205-210.

- Blair, S.N., H.W. Kohl III, C.E. Barlow, R.S. Paffenbarger Jr., L.W. Gibbons, and C.A. Macera. 1995. Changes in physical fitness and all-cause mortality: A prospective study of healthy and unhealthy men. *Journal of the American Medical Association* 273: 1093-1098.

- Myers, J., M. Prakash, V. Froelicher, D. Do, S. Partington, and J.E. Atwood. 2002. Exercise capacity and mortality among men referred for exercise testing. *New England Journal of Medicine* 346: 793-801.

- Paffenbarger, R.S. Jr., R.T. Hyde, M.A. Alvin, A.L. Wing, and C.-C. Hsieh. 1986. Physical activity, all-cause mortality, and longevity in college alumni. *New England Journal of Medicine* 314: 605-613.

are perceived lack of time and the inconvenience of getting to exercise facilities. Many people are poorly informed about the level of exercise needed to derive health benefits, believing they must exercise far more than is actually necessary (see chapter 11 for levels of exercise recommended for health benefits). People often drop out of an exercise routine because of injury or illness, boredom, lack of enjoyment, or failure to meet unrealistic expectations because of inappropriate exercise programs. Some people stop exercising because of lack of family support or feelings of self-consciousness. In addition, social and cultural norms may discourage physical activity in many groups. (Chapter 19 presents more detailed discussion of motives for exercising and adhering to exercise programs.)

Table 13.2

Factors Influencing Exercise Participation and Adherence

Factors promoting participation and adherence	Factors reducing participation and adherence
Younger age	Older age
Higher socioeconomic status	Illness, injury, disability
Higher educational level	Perceived lack of time
Spousal and peer support	Cigarette smoking
Convenient and safe facilities	Obesity
Previous participation in exercise and sport	Inappropriate type of exercise
Exercise in groups or with a partner	Boredom, lack of variety

Ways to Improve Exercise Participation

The 1996 U.S. Surgeon General's report, *Physical Activity and Health,* has prompted many national and international efforts to increase the number of physically active people. Because so much of the population is sedentary, we need to intervene at the local community and population level. Although whether a given individual exercises regularly is a personal decision, the local community and governments can increase the likelihood that people will be physically active. To be effective, efforts must involve coordinated action between government (local, regional, national), nongovernment organizations, employers, community groups, and the health professions. The current trend is toward a "partnership" approach involving all these groups.

The physical environment is important in determining whether people within a given community are physically active. For example, lack of places to walk or cycle safely is a major barrier to exercise for many, especially the elderly. Governments can enhance participation by providing walking and cycling paths with adequate lighting and also through policy decisions, for example, by requiring housing and business estates to provide safe pedestrian and cycle access. Brief, simple advice from the family doctor about the amount of exercise needed to enhance health positively influences an individual's decision to become more physically active. Many work sites now provide on-site employee exercise programs or flexible work schedules to allow workers time to exercise. Work-based programs not only are convenient but also benefit companies by reducing absenteeism and increasing productivity. School physical education and sport programs can influence the health of entire generations by helping young people establish a physically active lifestyle early on. Various forms of media are also useful in promoting the benefits of physical activity among the wider community. Posters, brochures, television and radio advertising, and promotional community events positively influence attitudes toward physical activity.

We now consider the role of regular physical activity in preventing and managing specific diseases or conditions, such as cardiovascular disease, metabolic syndrome (obesity, hypertension, type 2 diabetes, dyslipidemia), cancer, asthma, osteoporosis, and arthritis.

Physical Activity, Cardiovascular Disease, and Metabolic Syndrome

Cardiovascular disease causes about one-third of all deaths throughout the world (see table 13.1). The World Health Organization now considers cardiovascular disease a "pandemic" (epidemic on a worldwide scale). Cardiovascular disease comprises various diseases of the heart and blood vessels, including coronary heart disease (CHD), heart failure, high blood pressure (hypertension), stroke, and peripheral vascular disease (disease of the blood vessels in the limbs). Physical inactivity is now recognized as a major contributing factor to cardiovascular disease; regular moderate physical activity reduces the risk of disease by half. Metabolic syndrome is composed of obesity, hypertension, type 2 diabetes, and dyslipidemia, which tend to occur together in some people. Each of these components by itself increases risk of cardiovascular disease, and together they increase the risk even more. This section addresses the risk factors for, and the role of physical activity in preventing, cardiovascular disease. We also present recommendations for exercise prescription to lower cardiovascular disease risk and beneficially alter each component of metabolic syndrome.

Cardiovascular Disease

The death rate due to cardiovascular diseases has been declining since the late 1960s in some regions and countries, such as North America, much of Western Europe, the United Kingdom, Australia, and New Zealand. However, the incidence remains high or is increasing in other regions, in particular in Eastern Europe and Russia. Overall, heart disease and stroke account for almost 50% of all deaths in developed countries.

Coronary heart disease, the most prevalent form of cardiovascular disease, is characterized by a narrowing of the inside (lumen) of the blood vessels supplying the heart. Cells lining the interior of these blood vessels become damaged, possibly in response to consistently elevated blood pressure, high levels of circulating cholesterol (a type of lipid or fat, discussed in more detail further on), or both. Smooth muscle cells proliferate and move into the inside of the blood vessel, narrowing the vessel's inside diameter and restricting blood flow and delivery of oxygen and nutrients to various areas of the heart muscle. This process is called atherosclerosis.

Although genetic predisposition plays an important role, much of the incidence of heart disease can be prevented by lifestyle modification, including regular physical activity. Risk factors for heart disease have been identified from decades of research. Primary or major risk factors—factors most closely related to the incidence of cardiovascular disease—are listed in table 13.3. It is important to note that these risk factors are based on population studies and thus denote the statistical probability of developing disease when risk factors are present; they cannot be used to predict with any certainty the risk of developing disease for a particular individual.

Cardiovascular and heart disease risk increases with age in adults. Before menopause, women have a lower risk than men, but the difference becomes much smaller after menopause. Family history of early heart disease in a close relative (e.g., sibling, parent, or grandparent) also increases disease risk. Cardiovascular disease risk is elevated in individuals with impaired glucose tolerance (the ability of the body to regulate blood glucose levels) or diabetes (disease involving impaired glucose tolerance), because persistently high blood glucose levels cause damage to blood vessels throughout the body.

Elevated blood pressure and cholesterol are associated with development of atherosclerosis and heart disease. Cholesterol is carried in the blood in several different molecules, usually combined with special proteins. Low-density lipoproteins (LDL) carry most of the choles-

Table 13.3

Major Cardiovascular Disease Risk Factors

Risk factor	
Family history	Heart attack or heart disease before age 55 years in male close relative or 65 years in female close relative (parent, sibling)
Cigarette smoking	Current smoker or having quit less than six months ago
Hypertension	SBP ≥140 and/or DBP ≥90 mmHg, or on medication for blood pressure
Dyslipidemia	TC >5.2 mmol/L or LDL-C >3.4 mmol/L or HDL-C <0.9 mmol/L, or on medication for blood lipids; HDL-C >1.6 mmol/L is a negative (beneficial) risk factor
Impaired glucose tolerance	Fasting blood glucose ≥6.1 mmol/L or on medication to control blood glucose level
Obesity	BMI ≥30 kg/m² or waist circumference >100 cm
Sedentary lifestyle	Not participating in regular physical activity that meets the minimum recommendations of the U.S. Surgeon General

SBP = systolic blood pressure, DBP = diastolic blood pressure, TC = total cholesterol, LDL-C = low-density lipoprotein cholesterol, HDL-C = high-density lipoprotein cholesterol, BMI = body mass index.

Compiled from American College of Sports Medicine 2000.

terol; high levels of LDL increase the risk of heart disease. High-density lipoproteins (HDL) are considered protective against heart disease in that they remove cholesterol from the body. Thus, a favourable lipid profile for prevention of heart disease includes low blood levels of total cholesterol (TC) and LDL along with high levels of HDL. The ratio of TC to HDL-cholesterol is an important measure. Cigarette smoking also contributes to the process of atherosclerosis, although the exact mechanisms are unclear at present. Smoking may elevate blood pressure, and certain chemicals contained in cigarette smoke may contribute to blood vessel damage. A sedentary lifestyle (i.e., no physical activity at work or in leisure time) increases risk by about twofold.

Physical Activity and Cardiovascular Disease

It is estimated that 12% to 38% of all cardiovascular disease may be attributed to physical inactivity, depending on the country from which data are obtained. Regular moderate exercise is now widely recognized as an effective means of prevention. On average, the risk of cardiovascular disease is halved in those who participate in regular moderate physical activity compared with sedentary individuals. Regular exercise has both direct effects, by strengthening the heart and making it a more efficient pump, and indirect effects, by altering many other risk factors. Table 13.4 summarizes the effects of regular physical activity on other risk factors.

Regular moderate exercise lowers blood pressure in those with mildly to moderately elevated blood pressure (see later). Regular moderate physical activity increases the protective HDL fraction and may lower TC level in blood. Exercise helps prevent obesity by controlling body weight and fat levels and maintaining lean body mass. Physical activity also increases glucose tolerance, helping to prevent or treat non-insulin-dependent diabetes mellitus (see later), or both. Finally, smokers who exercise regularly may be more likely to stop smoking once an exercise program is established.

Exercise Prescription to Prevent and Treat Cardiovascular Disease

The general recommendations for exercise for good health, as described in chapter 11, apply to prevention of cardiovascular disease. The risk of disease decreases proportionally with increasing amount of exercise performed. It is the total amount of exercise, rather than exercise intensity, that is most important in reducing risk. Moderate exercise such as brisk walking (in the range of 50-70% $\dot{V}O_2max$) appears to be nearly as effective as more intense exercise such as jogging or running; moderate exercise is certainly more attractive to, and safer for, most people. Brisk walking for 30 to 45 min, four to five times per week, markedly reduces the incidence of cardiovascular disease.

Table 13.4

Effects of Regular Physical Activity on Cardiovascular Disease Risk Factors

Risk factor	Effect of exercise and implications for disease prevention
High blood pressure	Lowers blood pressure in those with mild to moderate hypertension
	In conjunction with medication, lowers blood pressure in those with severe hypertension
Blood lipids	May lower blood cholesterol level
	Increases blood high-density lipoprotein levels
	May decrease blood low-density lipoprotein-cholesterol level if accompanied by weight loss
Obesity	Reduces body weight and body fat
	Increases lean body mass
Type 2 diabetes	Prevents obesity
	Improves glucose tolerance and insulin sensitivity
	Reduces body fat levels
Cigarette smoking	Smokers who exercise may smoke less or stop smoking

Regular moderate exercise is also advocated in the rehabilitation of patients with, or at high risk for, cardiovascular disease. Exercise rehabilitation increases functional capacity and slows further progression of the disease in cardiac patients. Exercise rehabilitation programs must be individually prescribed based on the extent of disease and symptoms. Medical supervision is generally recommended during the early stages of an exercise program. Patients typically have a low exercise capacity and need moderate and gradual programs.

Metabolic Syndrome: Obesity, Hypertension, Type 2 Diabetes, and Dyslipidemia

Certain diseases often occur together. One common example in developed countries is the **metabolic syndrome** or **syndrome X,** composed of obesity, hypertension, diabetes (impaired glucose tolerance), and dyslipidemia (blood lipid levels outside of the recommended range). Each of these conditions is an independent risk factor for cardiovascular disease, but there is a further additive effect on disease risk when all four are present. Although each has a strong genetic component, lifestyle factors such as diet, physical inactivity, and smoking contribute to each component of the syndrome.

In the past 20 years, the proportion of adults exhibiting metabolic syndrome has increased alarmingly, and the syndrome has been appearing at earlier ages—even in adolescents. This parallels a trend of increasing prevalence of obesity while participation in regular physical activity has remained relatively low. Through its many effects on each component, regular physical activity is important to preventing and treating metabolic syndrome. This section addresses each component of metabolic syndrome and the role of physical activity in prevention and treatment. Recommendations for exercise prescription in prevention and treatment of metabolic syndrome are summarized in table 13.5.

Obesity

Although often considered an aesthetic rather than a health issue, obesity is a major health concern because it contributes to hypertension, cardiovascular disease, certain forms of cancer (breast and bowel), osteoarthritis, and type 2 diabetes. The incidence of overweight and obesity has increased over the past 20 years in many developed countries. More than half of adults in the United States, the United Kingdom, and Australia and between 30% and 50% of Europeans are overweight or obese. The increase in the number of overweight and obese children and adolescents is particularly alarming since obesity at an early age greatly increases the risk of adult obesity. In many developed countries, 20% to 25% of children and adolescents may be overweight or obese.

Table 13.5

Exercise Prescription for Metabolic Syndrome

Exercise variable	Recommendation	Reason for recommendation
Mode	Endurance-type + circuit resistance training	Expend sufficient energy to lose body weight and fat
	Increasing incidental physical activity	Increase lean body mass
Intensity	Low to moderate	Possibly low fitness to start
	Gradual progression	Less likely to cause injury
		More attractive to people who are unfit
		High intensity has minimal further benefits
Frequency	Daily if possible, but at least three to five times per week for endurance exercise	Expend sufficient energy to lose weight and alter risk factors
	Resistance exercise two to three times per week	Part of active lifestyle
Duration	30-60 min per session	Expend sufficient energy to lose weight and alter risk factors
	>900 kcal per week	

Table 13.6

Definitions of Overweight and Obesity

Measure	Overweight	Obese
Quetelet index (QI) or body mass index (BMI)	25-30	>30
Height–weight tables	10-20% over recommended weight for height	>20% over recommended weight for height
Body composition	Adult male 18-20% fat	Adult male >20% fat
	Adult female 27-30% fat	Adult female >30% fat
Waist-to-hip ratio	Adult male >1.0	
	Adult female >0.85	
Waist measurement	Adult male >100 cm (39 in.)	
	Adult female >90 cm (35 in.)	

Overweight and obesity can be defined in several ways (table 13.6). With use of the **Quetelet index (QI)** or body mass index (BMI) (see chapter 3), overweight is defined as a QI between 25 and 30 kg/m² and obesity as a QI over 30. The recommended QI range for good health is 20 to 25. If standard height–weight charts are used, overweight is usually defined as 10%, and obesity as 20%, over recommended weight for height. Because of the simplicity of measurement and calculation, these two definitions are appropriate for the general population. However, athletes often carry additional body mass as muscle and many athletes, especially those in strength and power sports, would be classified as overweight using these simple definitions. For physically active people, including athletes, body composition—although it is a more complex measure—is a better method to define obesity because it distinguishes fat from lean body mass. After all, it is the excess fat, not total body mass, that is associated with disease. For men, overweight is defined as body fat between 18% and 20%, and obesity is above 20% fat. For women, the values are 27% to 30% fat for overweight and above 30% fat for obesity.

Obesity is usually caused by an imbalance between energy intake and expenditure, or a higher amount of energy consumed as food than expended. Obesity is rarely caused by hormonal factors, but it is now widely believed that there is a strong genetic influence.

The pattern of fat distribution around the body is important. Central or abdominal obesity is defined as an excess of fat in the abdominal region. This pattern of fat deposition primarily in the upper body (chest and waist) occurs more often in men and carries a higher risk of heart disease, stroke, and diabetes than the female pattern of fat deposition primarily around the hips and thighs. This relationship holds true even when one is comparing fat deposition within a given sex. For example, women with fat deposition primarily in the upper body are at higher risk of cardiovascular disease than women with fat deposition primarily in the lower body. The two patterns of fat deposition and obesity have been termed the "apple" and "pear," or android and gynoid, shapes for male and female patterns, respectively (figure 13.1). The pattern of fat distribution may be estimated using the **waist-to-hip ratio (WHR),** calculated by dividing the circumference of the waist by that of the hips. The recommended WHR is less than 0.85 for women and less than 1.0 for men. A simple measure of only the waist circumference can also be used to indicate excess fat; recommended values are less than 90 cm (35 in.) for women and less than 100 cm (39 in.) for men.

Exercise and Obesity

Obesity is caused by an imbalance between energy intake and expenditure, so body weight should be controlled through decreasing energy intake, increasing energy expenditure, or some combination of the two. Weight reduction programs that rely solely on dieting are generally ineffective for permanent weight loss. Regular physical activity is important to long-lasting control of body weight for many reasons (see "Benefits of Exercise in Weight Loss," p. 188).

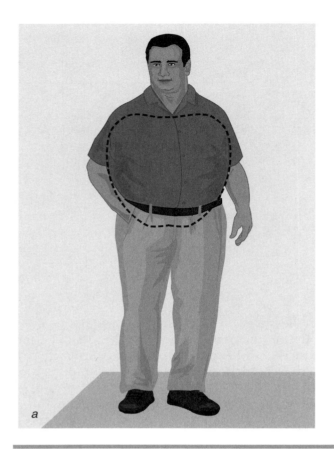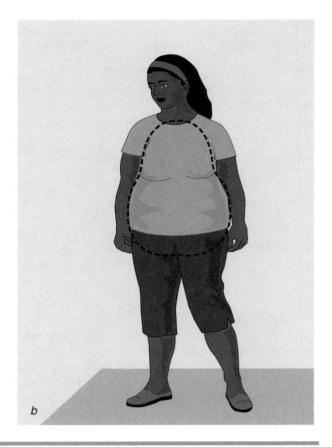

Figure 13.1   Pattern of fat distribution in the body. *(a)* Android or male pattern ("apple" shape); *(b)* gynoid or female pattern ("pear" shape).
Reprinted from Wilmore and Costill 2004.

BENEFITS OF EXERCISE IN WEIGHT LOSS

- Faster rate of weight loss when combined with dietary restriction
- Less dietary restriction required for moderate weight loss
- Greater fat loss than that achieved through dietary restriction alone
- Maintenance of and possible increase in lean body mass and resting metabolic rate
- Favourable changes in body composition
- Higher probability of maintaining new (lower) body weight once achieved

Permanent weight loss requires either exercise alone (for the moderately overweight) or a combination of exercise and diet (for the seriously over-

weight and obese). Exercise to reduce body weight helps maintain lean body mass, whereas diet without exercise may result in loss of muscle mass. It is important to maintain muscle mass because skeletal muscle is a major site of fat metabolism; losing muscle decreases the body's capacity to burn fat. Exercise is also important to maintaining the body's resting metabolic rate (RMR), the minimum amount of energy needed to maintain bodily functions. In most people except athletes in heavy training, RMR accounts for between 60% and 75% of total daily energy expenditure. Resting metabolic rate is strongly related to age, gender, body size, and lean body mass. RMR may decrease by 10% to 20% after significant weight loss through diet alone. This reduces the amount of energy needed by the body (i.e., the amount that can be consumed without gaining weight) and is one reason people find it so difficult to maintain weight loss after dieting. In contrast, weight loss through exercise helps prevent the decline in RMR, thus ensuring permanent loss of body mass and fat. Moreover, exercise increases muscle mass, enhancing lean body mass and leading to a more favourable body composition. In addition, in moderately obese individuals, low- to moderate-intensity exercise helps correct components of the metabolic syndrome, such as abdominal obesity, glucose tolerance, and blood lipid levels, even with only modest changes in body weight.

Exercise Prescription for Obesity

Exercise programs must consider that people who are obese need to increase energy expenditure but may be unfit. Exercise should be low impact to prevent injury to load-bearing joints. Good examples include walking, low-impact or water-based aerobic activities, swimming, cycling, and circuit-type resistance exercise using weight and stationary exercise machines. Each session should be long enough to require the exerciser to expend at least 300 kcal (1,260 kJ) of energy, equivalent to at least 30 min of exercise per session. Since energy expenditure increases with increasing duration of exercise, longer sessions approaching 60 min are recommended provided they do not lead to overuse injury or excessive fatigue. Exercise should be performed at least four times per week and daily if possible.

Since people who are obese may be unfit to start with, low-intensity programs that progress gradually are recommended. Exercise intensities in the range of 40% to 60% $\dot{V}O_2$max or heart rate reserve, equivalent to approximately 50% to 65% of age-predicted maximum heart rate (see chapter 11), are appropriate at the onset of an exercise program.

Hypertension

In most developed countries, hypertension or high blood pressure affects about 20% of adults; up to 60% of older adults may exhibit high blood pressure. The World Health Organization (WHO) defines hypertension as a chronically elevated blood pressure reading above 140/90 mmHg. The first number is systolic blood pressure and represents the pressure in the arteries during the heart's contraction; the second number is diastolic blood pressure and represents the pressure in the arteries during relaxation of the heart between contractions. Hypertension is associated with increased risk for stroke, heart disease, renal (kidney) failure, and peripheral vascular (blood vessel) disease. In most cases of hypertension, there are no symptoms and no single identifiable cause, and many people are unaware that they have high blood pressure.

Although there is a strong genetic predisposition, lifestyle factors such as diet, obesity, cigarette smoking, physical inactivity, and life stresses are all associated with elevated blood pressure.

Exercise and Hypertension

Individuals with mildly to moderately elevated blood pressure in the range of 140-160/90-95 mmHg benefit from regular physical activity. Moderate exercise reduces both systolic and diastolic blood pressure by about 6 to 10 mmHg, which may bring blood pressure into the normal range without the need for medication. The precise mechanisms by which exercise reduces blood pressure are not clear but may involve both direct and indirect effects. Exercise helps reduce body weight, and weight loss by itself can help lower blood pressure in individuals with mild hypertension; however, regular exercise can also reduce blood pressure without loss of body weight. Regular exercise may directly influence blood pressure by reducing heart rate at rest and during activity, by damping the body's responses to stress hormones (which act to increase heart rate and blood pressure), or both. Significantly elevated blood pressure (above 160/95 mmHg) often requires medication in addition to lifestyle modifications such as diet and exercise.

Exercise Prescription for Hypertension

Because hypertension increases the risk of cardiovascular disease, persons with hypertension should check with a doctor before beginning an exercise program. A medically supervised exercise test may be needed to determine the patient's blood pressure response to exercise. Once approved for exercise, the person with hypertension should follow the general recommendations regarding low- to moderate-intensity exercise (see chapter 11). Exercise should begin in the range of 40% to 60% $\dot{V}O_2$max and progress gradually; intensities above 70% $\dot{V}O_2$max have no additional beneficial effect in lowering blood pressure. Whole-body exercise or exercise using large muscle groups, such as walking, swimming, and cycling, are preferable, since blood pressure is less likely to increase dramatically during these types of exercise. Intense resistance (weight) training is not recommended because it may cause blood pressure to rise dangerously; however, lower-intensity resistance work that emphasizes muscular endurance is considered safe and effective for hypertensive individuals. For example, an appropriate program might include circuit training with intensity at each station in the 15- to 20RM (repetition maximum) range (see chapter 11). Since many people with high blood pressure are also overweight, overweight, hypertensive individuals should follow the general guidelines for exercise to reduce body mass (discussed previously). It may also be prudent for patients to monitor their own blood pressure response during exercise to ensure that levels stay within recommended limits.

Diabetes

Diabetes is a metabolic disease characterized by high blood glucose levels (>7.0 mmol/L during fasting). Diabetes is a major cause of death and disability in many developed countries and is a contributing factor to other illnesses such as cardiovascular, peripheral vascular, and renal diseases.

The two main types of diabetes are type 1, or insulin-dependent diabetes mellitus (IDDM), and type 2, or non-insulin-dependent diabetes mellitus (NIDDM). The onset of type 1 diabetes is not affected by physical activity or inactivity, although physical activity can in many cases help a person with type 1 diabetes manage the disease. Type 1 diabetes, which begins more often in childhood or adolescence, is an autoimmune disease; the body's immune system attacks its own cells, in this case, the insulin-producing cells of the pancreas, which then stop synthesizing insulin. Type 2 diabetes usually does not appear until later in life (after age 30) and is associated with lifestyle factors such as obesity and physical inactivity. In type 2 diabetes, the body still produces some insulin, but tissue receptors are insensitive to the insulin.

The incidence of type 2 diabetes has increased by 50% in the last 15 years; and in some countries as many as 5% to 8% of adults may exhibit impaired glucose tolerance, suggesting a predisposition toward the condition. Type 2 diabetes has become more prevalent among children, probably because of an increased incidence of obesity and physical inactivity. There is a genetic influence for both types of diabetes, although 60% to 95% of type 2 diabetes may be prevented or reversed by changes in lifestyle.

High blood glucose levels may be caused either by insufficient production of insulin by the pancreas (type 1) or by resistance to insulin in peripheral tissues (type 2). A hormone released from the pancreas, insulin is needed for glucose uptake from the blood into most tissues. Insufficient insulin (type 1), or resistance to its action (type 2), impairs the capacity for glucose uptake from the blood, thereby keeping blood glucose levels high. If left untreated, high blood glucose levels may lead to tissue damage including harm to blood vessels and nerves. Type 1 diabetes is treated with injected insulin to replace the insulin normally produced by the pancreas. Type 2 diabetes is treated with a combination of dietary and lifestyle modifications and sometimes medication to increase the pancreatic production of insulin and the body's response to the insulin that is produced (See "Can Physical Activity Prevent Type 2 Diabetes?" on p. 191).

Exercise and Diabetes

Provided that blood glucose levels can be controlled, regular moderate exercise is an important part of treatment of both types of diabetes, especially type 2. Exercise stimulates tissue uptake of glucose from the blood; a single exercise session may enhance blood glucose uptake for up to 48 h. Moreover, exercise prevents obesity, which increases the risk of type 2 diabetes. In obesity, many tissues are resistant (do not respond) to the insulin produced by the pancreas. Exercise training stimulates glucose uptake by these tissues and, as already described, helps reduce fat levels.

Exercise Prescription for Diabetes

Regular moderate exercise is recommended for improved glucose tolerance in individuals with either form of diabetes. It is best if exercise intensity and duration, as well as time of day, are relatively consistent from day to day. Individuals with either form of diabetes may need to adjust carbohydrate intake before or after exercise. The person with type 1 diabetes should avoid exercise during the peak time of insulin activity to prevent a precipitous drop in blood glucose levels. The insulin-dependent athlete should work closely with a medical practitioner to monitor possible changes in insulin and glucose requirements.

For someone who has type 2 diabetes, exercise prescription should follow the recommendations described previously for people who are obese: 30 to 60 min of low- to moderate-intensity whole-body exercise, performed at least four times per week and daily if possible. Activities such as walking, cycling, swimming, low-impact or water-based aerobics, and low- to moderate-intensity circuit exercise are recommended. Individuals with type 2 diabetes who are also obese should avoid high-impact activities such as running, at least until a significant amount of body weight is lost.

IN FOCUS: CAN PHYSICAL ACTIVITY PREVENT TYPE 2 DIABETES?

Type 2 diabetes is a major risk factor for heart disease, a major killer in most developed countries. The incidence of type 2 diabetes has increased along with the prevalence of obesity and physical inactivity. The question is, can altering lifestyle reverse the trend toward insulin resistance and development of type 2 diabetes?

In a four-year prospective study of 897 middle-aged Finnish men, the incidence of type 2 diabetes was 50% lower in men who performed moderate physical activity than in sedentary men (Lynch et al. 1996). The effect was even more pronounced in high-risk men, those with a family history, high blood pressure, and overweight; physical activity decreased risk of type 2 diabetes by 64% in these men. A fairly moderate amount of physical activity was protective—40 min of activity per week at 5.5 METs or higher, equivalent to brisk walking, moderate cycling or swimming, ball games (e.g., tennis), or gym-based exercise.

In another Finnish study, researchers asked whether lifestyle modification could influence the onset of type 2 diabetes (Tuomilehto et al. 2001). During a six-year study of 500 at-risk adults, those in an exercise intervention group who changed their diet, lost weight, and remained physically active showed improved glucose tolerance and a lower incidence of type 2 diabetes.

In a smaller study, 16 sedentary men with hyperinsulinemia (high blood insulin levels, a precursor to type 2 diabetes) were randomly assigned to two exercise groups—an endurance group and a cross-training group (Wallace, Mills, and Browning 1997). The endurance exercise group cycled and walked for 60 min three times per week, and the cross-training group performed the same endurance exercise plus circuit resistance training for an additional 60 min each session. After 14 weeks, both groups showed improvements in body fat and blood glucose and insulin levels, although the cross-training group showed greater changes in blood glucose and insulin levels.

Taken together, these studies show that moderate physical activity that reduces body fat can halve the risk of developing type 2 diabetes. Importantly, only moderate physical activity is needed to reduce risk; this is well within the capability of most people.

SOURCES

- Lynch, J., S.P. Helmrich, T.A. Lakka, G.A. Kaplan, R.D. Cohen, R. Salonen, and J.T. Salonen. 1996. Moderately intense physical activities and high levels of cardiorespiratory fitness reduce the risk of non-insulin dependent diabetes mellitus in middle-aged men. *Archives of Internal Medicine* 156: 1307-1314.

- Tuomilehto, J., J. Lindstrom, J.G. Eriksson, T.T. Valle, H. Hamalainen, P. Ilanne-Parikka, S. Keinanen-Kiukaanniemi, M. Laakso, A. Louheranta, M. Rastas, V. Salminen, and M. Uustupa. 2001. Prevention of type 2 diabetes mellitus by changes in lifestyle among subjects with impaired glucose tolerance. *New England Journal of Medicine* 344: 1343-1350.

- Wallace, J.B., B.D. Mills, and C.L. Browning. 1997. Effects of cross-training on markers of insulin resistance/hyperinsulinemia. *Medicine and Science in Sports and Exercise* 29: 1170-1175.

Dyslipidemia

Lipids or fats are needed for a variety of body functions, including synthesis of cell membranes and certain hormones. Lipids are carried in the blood attached to proteins, called lipoproteins. Although a certain level of lipoproteins in the blood is healthy, excessive lipid levels are associated with cardiovascular disease. **Dyslipidemia** is the term used to describe blood lipid levels outside the recommended healthy range. Dyslipidemia can be caused by both genetic and lifestyle factors; the most common causes are obesity, a high-fat diet, and physical inactivity. Dyslipidemia usually involves elevated blood levels of TC and low-density lipoprotein-cholesterol (LDL-C) and low levels of high-density lipoprotein-cholesterol (HDL-C). As mentioned previously, a high HDL-C level is protective since it helps remove and degrade excess lipids. Thus, low TC and LDL-C and high HDL-C levels are recommended to prevent cardiovascular disease.

Exercise and Dyslipidemia

The metabolic syndrome is associated with high TC and LDL-C levels and low HDL-C levels.

Weight loss and exercise can beneficially alter blood lipid levels. In overweight individuals, even modest weight loss, especially from the abdomen, decreases TC and LDL-C and may raise HDL-C levels. Besides medication, endurance-type exercise is most effective in raising HDL-C level. In non-obese individuals (BMI "30), regular moderate endurance exercise, as recommended for good health (see chapter 11), increases HDL-C levels even without weight loss. However, in people who are obese, weight loss through a combination of diet and exercise is required to raise the HDL-C level. The magnitude of change in blood lipids resulting from exercise depends on several factors, including initial body mass and fat, training volume, and extent of weight and fat loss. Even small changes in blood lipid levels can reduce the risk of cardiovascular disease.

Exercise Prescription for Dyslipidemia

The optimal types and amount of exercise to improve the blood lipid profile are not clearly known at present. The exercise prescription recommendations for metabolic syndrome shown in table 13.5, however, provide reasonable guidelines for this condition. A minimum weekly energy expenditure of 900 to 1,200 kcal (3,780-5,040 kJ) during exercise is needed to increase HDL-C levels. Exercise volume or energy expenditure is more important than the type or intensity of exercise. Exercise that leads to loss of body fat has an additive effect on blood lipids, especially in raising HDL-C levels. The most effective exercise includes 30 to 60 min of moderate physical activity, at intensity 50% to 80% $\dot{V}O_2$max, three to four times per week. It may be six months or more before one sees beneficial changes in blood lipids. This level of physical activity fits within the recommendations described in chapter 11.

Physical Activity and Other Major Diseases and Conditions

In addition to preventing cardiovascular disease and metabolic syndrome, physical activity also helps prevent or manage several other diseases and conditions, some of which are major contributors to death and disability in developed countries. This section deals with the role of physical activity in preventing and managing cancer, asthma, osteoporosis, and arthritis—diseases or conditions that are relatively common.

Cancer

Cancer is among the leading causes of death in many developed countries. Cancer risk is determined by genetics and some lifestyle factors. About half of all cancer deaths in developed countries may be prevented by changes in lifestyle, such as reducing cigarette smoking, alcohol consumption, and exposure to environmental and occupational carcinogens (cancer-causing substances); altering diet; and engaging in regular physical activity. Cancer is a complex and multifaceted disease that may occur in many sites in the body. There is no single cause of cancer; regardless of site, cancer is characterized by uncontrolled cell growth (malignancy), which eventually overwhelms the body's normal functioning.

Exercise and Cancer

Relatively less is known about the relationship between exercise and cancer than about exercise and other diseases. Of all the types of cancer, exercise seems to be most protective in preventing colon cancer. Physically active people are 25% to 50% less likely to develop colon cancer than inactive individuals. Regular physical activity may also protect against other types of cancer, in particular breast and reproductive system cancers in women and prostate cancer in men.

The mechanisms by which exercise protects against cancer are not known at present and probably involve different mechanisms, each specific to the site of cancer. For example, exercise may reduce the risk of colon cancer by enhancing movement through the bowel, thus limiting exposure to potential carcinogens in waste material as it moves through the intestines. Regular physical activity may reduce circulating levels of sex hormones, which may be causal factors in reproductive system cancer. In addition, exercise reduces body fat levels, and excess body fat is linked with increased risk of breast and colon cancer. Physically active people are likely to adopt healthier lifestyles compared to inactive people—for example, not smoking and maintaining a healthy body weight. Regular physical activity may also stimulate the body's immune system response to cancer cell growth.

Exercise Prescription in Cancer

Exercise has potential as a treatment for cancer patients in addition to standard medical treatment. Regular moderate exercise may counteract some of the debilitating effects of cancer and cancer therapy, such as loss of muscle and decreased functional capacity. Moderate exercise is also associated with enhanced psychological state, which may help some cancer patients cope with side effects of treatment such as nausea and fatigue.

Exercise prescription for cancer patients follows general recommendations for individuals with low functional capacity. Low-intensity and low-impact exercise is recommended. Patients should avoid exercise when experiencing excessive fatigue, fever, weakness, nausea, or anemia, and within 24 h of chemotherapy.

Asthma

Asthma is characterized by an increased airway responsiveness to certain stimuli that causes a narrowing of the airways (bronchoconstriction) and difficulty in breathing. Asthma is a frequently occurring respiratory disorder, especially among children, and is one of the most common reasons for seeing a doctor. The incidence of asthma varies by country but may be as high as 5% to 10% of the population and up to 20% of children; the incidence is increasing throughout the world. Asthma should not prevent participation in exercise, and many top athletes have asthma. The key is effective management through medication, regular physical activity, and other lifestyle modifications.

Exercise and Asthma

Asthma can be brought on by many stimuli, most commonly cold dry air, allergens, dust, smoke, and exercise. Most people with asthma develop **exercise-induced asthma (EIA),** also called **exercise-induced bronchospasm (EIB),** during physical activity. Exercise-induced bronchospasm is characterized by a reduction in the volume of air that can be forcefully expired (forced expiratory volume), resulting in breathlessness and inability to continue intense exercise. Severity of EIB is related to exercise intensity, and the most asthmagenic (most likely to cause EIB) form of exercise is intense continuous activity (above 75% $\dot{V}O_2$max) lasting up to 8 min.

Regular exercise is widely encouraged for people who have asthma, especially children. Regular exercise enhances respiratory and cardiovascular function, helps prevent obesity, and is important for motor development in children; participation in sport and physical activity is also an important aspect of social development. By following certain guidelines and using medication according to an asthma management plan developed with the family doctor, most children and adults with asthma can participate in the full range of physical activity and sport.

Exercise Prescription in Asthma

Each person with asthma may respond differently to exercise, and the response may also vary from time to time, depending on the type and intensity of exercise as well as environmental factors such as temperature and humidity. Appropriate use of preventive medication can significantly reduce the frequency and severity of EIB. Everyone who has asthma, especially the competitive athlete, should work closely with a doctor to best determine the optimal combination of medication and exercise.

Swimming, walking, and moderate activities are less asthmagenic than more intense activities such as jogging and running (see table 13.7). In general, the higher the demand on the respiratory system, the more likely exercise is to bring on EIB. For the nonathlete with asthma, the general guidelines for exercise training for health, as described in chapter 11, are recommended. That is, exercise should include moderate whole-body or large muscle group activity in the range of 50% to 85% $\dot{V}O_2$max or 60% to 85% of age-predicted maximum heart rate. Competitive athletes need to train according to the demands of the sport. Athletes with asthma have been successful in virtually every sport, providing evidence that few modifications need to be made for training such athletes.

The warm-up before exercise is very important, since it allows the cardiorespiratory system to adjust gradually to the demands of exercise. Warm-up should include light aerobic-type activity and muscular flexibility exercises. Many people with asthma experience a "refractory" period after vigorous exercise, during which symptoms are inhibited. For these people, a controlled 10-min warm-up of light aerobic exercise followed by five to seven 30-s sprints may prevent EIB for up to 2 h. The aerobic part of a workout should begin at low intensity and increase gradually. Interval training and games involving intermittent activity, such as soccer, tennis, or hockey, are often recommended, since EIB is less likely to occur during shorter work intervals. Avoiding exercise during the coldest and driest part of the day, or exercising indoors in a more controlled environment with fewer allergens or pollutants, may also reduce the risk of EIB. Swimming and other forms of aquatic exercise are highly recommended because the more humid environment of a pool reduces the likelihood of EIB. Respiratory infections, such as the common cold, increase the susceptibility to asthma; intense exercise should be avoided during and immediately after a respiratory infection.

Table 13.7

Exercise for People With Asthma

Recommendation	Reason
Maintain regular exercise	Onset of EIB is more predictable and controllable
Exercise intensity <80% $\dot{V}O_2$max (in noncompetitive athletes only)	Lower minute ventilation is less likely to cause EIB
Proper warm-up before all activity	Warm-up may prevent EIB or lessen its severity
Avoid cold, dry air	EIB is more likely to occur in cold dry environments
Swimming, interval exercise, or both	Higher humidity around swimming pools Interval exercise is less asthmagenic
Avoid exercise during or immediately after a respiratory infection	EIB more likely to occur after respiratory infection
Use prescribed medication before exercise	May prevent EIB

EIB = exercise-induced bronchospasm (sometimes called exercise-induced asthma, EIA).

Compiled from Tan and Spector 1998; Lemanske and Busse 1997.

Osteoporosis

Osteoporosis is a bone disorder in which bone density decreases to a critically low level, making bones susceptible to fracture. Osteoporosis is primarily a disease of older Caucasian women of European descent, although elderly men may also be at risk. It has been estimated that, in these populations, 15% of women age 70 years and more than half of women age 80 years will incur a bone fracture as a result of osteoporosis. Half of all older people experiencing a bone fracture will become disabled, and 20% of women with hip fractures die within a year. The annual cost of medical care and loss of independence due to osteoporotic bone fractures is estimated to be in the billions of dollars in North America and in European countries.

The causes of osteoporosis are multifaceted and include both genetic and lifestyle factors. As noted in chapters 2, 4, and 5, bone is a dynamic tissue that is constantly remodelled. During aging, the degradation of bone occurs faster than synthesis of new bone, gradually reducing bone density and the loads that can be tolerated. The most commonly fractured sites are the vertebrae and hip. Older postmenopausal women are most at risk for osteoporosis because low estrogen (female hormone) levels reduce bone synthesis. Although there is a strong genetic predisposition to osteoporosis, lifestyle factors such as regular exercise and adequate nutrition, especially in childhood and adolescence, are important to preventing osteoporosis later in life.

Exercise and Osteoporosis

Mechanical loading is necessary for maintaining bone density. Bone loses mineral content, mass, and density in the absence of loading. For example, bone loss occurs in astronauts who spend long periods in space and in patients after prolonged bed rest. Weight-bearing activity and muscular contractions that increase mechanical loading of bone increase bone density. Bone density is highest in athletes who participate in weight-bearing sports such as running and in activities inducing muscle hypertrophy such as weightlifting.

Because bone density can be influenced by load-bearing exercise, the question arises whether physical activity can prevent or counteract the loss of bone with aging. Achieving a high bone mass early in life (before young adulthood) is very important to preventing osteoporosis later in life. Thus, all children and adolescents are encouraged to participate in physical activity. Another recommendation is that young women should aim to achieve and maintain optimal bone mass (the "bone bank" notion introduced in chapter 4) through a combination of adequate intake of calcium, a well-balanced diet, and regular weight-bearing exercise as recommended for good health (see chapter 11).

Exercise Prescription to Optimise Bone Health

The best way to prevent osteoporosis is to achieve a high bone mass early in life through a combination of good nutrition, regular load-bearing and muscle-building physical activity, and avoidance of factors that reduce bone mass such as smoking, eating disorders, and menstrual disturbances. Children and adolescents are encouraged to perform moderate physical activity on most days, as well as vigorous activity involving weight-bearing movement and exercises using the whole body or large muscle groups at least three times per week. Adults should strive to continue regular weight-bearing (e.g., walking, jogging, aerobics) and muscle-building (e.g., resistance training) physical activity as a regular part of a healthy lifestyle (discussed in chapter 11). Moderate exercise (less than 70% maximum heart rate) seems as effective as strenuous exercise in maintaining bone density. Older individuals should continue cardiovascular and muscular conditioning to prevent or minimize bone loss in old age. Exercises to improve muscular strength and motor skills such as balance, coordination, and reaction time may also help prevent bone fractures by reducing the risk of falls in elderly persons (see "Strength Training in People Who Are Elderly: How Adaptable Are Aging Muscles" on p. 173 and also chapter 22).

Arthritis

Arthritis represents a variety of conditions characterized by inflammation, pain, and reduced mobility in the joints. More than 50% of the population may experience some type of arthritis at some time. Arthritis is among the most frequent reasons for seeking medical care, and billions of dollars are spent each year on medicine, rehabilitation, and disability costs. The two most common forms are osteoarthritis and rheumatoid arthritis.

Osteoarthritis, the most common form of arthritis, is characterized by structural changes in the joint and progressive degradation of cartilage in the joint, leading to local inflammation and pain. The incidence of osteoarthritis increases with age so that nearly 85% of individuals above the age of 70 years have some form of osteoarthritis. Rheumatoid arthritis, the second most prevalent form of arthritis, is an autoimmune disease in which the body's immune system inappropriately attacks its own cells in the joints. Immune cells migrate to the joints and produce substances that cause inflammation and pain within the joints, which may also affect nearby tendons and muscles. Rheumatoid arthritis occurs more frequently in women than in men.

Exercise and Arthritis

It was once thought that participation in sport, especially load-bearing activities such as jogging, increases the incidence of osteoarthritis. However, we know now that osteoarthritis occurs in former athletes only if they have experienced joint injury. Osteoarthritis may result from any injury to the joint. Athletes with a history of joint injury, such as football and basketball players, tend to have a higher incidence of osteoarthritis later in life. It is the joint injury itself, and not physical activity, that increases the risk of osteoarthritis.

Exercise does not influence the incidence or development of rheumatoid arthritis. However, recent research has shown beneficial effects of moderate exercise on physical work capacity and muscular strength, without increasing joint pain, in patients with rheumatoid arthritis.

Exercise Prescription in Arthritis

Regular moderate exercise is recommended for people who have either form of arthritis. Three main types of exercise are commonly prescribed for people with arthritis:

- Mobility exercises to maintain or increase range of motion around the joints
- Muscle-strengthening exercises to increase the weight-bearing capacity of joints
- General fitness exercises to increase cardiovascular endurance for prevention of conditions such as heart disease and osteoporosis

The general guidelines for exercise for good health, as outlined in chapter 11, apply to the individual with arthritis. The exercise program may need to be modified, taking into consideration the joints involved and a generally low initial fitness level. Range of motion exercises should be performed daily. Walking, cycling, and water-based activities such as "aqua-aerobics" and swimming are recommended as low-impact activities that will not stress the joints. Interval exercise training is often better tolerated than continuous training, especially during the initial stages of a program.

Summary

Regular moderate physical activity has many health benefits and is now recommended for virtually everyone, including children, persons who are elderly, and those with disease or impairment that may influence physical capacity. Regular physical activity helps prevent many diseases or conditions, including cardiovascular disease, metabolic syndrome (obesity, hypertension, type 2 diabetes, dyslipidemia), osteoporosis, stroke, and some types of cancer. Regular physical activity also assists in managing heart and blood vessel diseases, obesity, type 2 diabetes, and arthritis.

Exercise enhances health through both direct and indirect mechanisms. For those with specific health needs, the standard exercise prescription applies but with some modifications of frequency, duration, and intensity to maximize the health benefits while minimizing the health risks. Increasing participation in physical activity requires a "partnership" approach to enhance the attractiveness and convenience of access to physical activity by the wider community.

Further Reading and References

American College of Sports Medicine. 1997. *ACSM's exercise management for persons with chronic diseases and disabilities.* Champaign, IL: Human Kinetics.

American College of Sports Medicine. 1998. *ACSM's resource manual for guidelines for exercise testing and prescription.* 3rd ed. Baltimore: Williams & Wilkins.

American College of Sports Medicine. 2000. ACSM position stand. Exercise and type 2 diabetes. *Medicine and Science in Sports and Exercise* 32: 1345-1360.

American College of Sports Medicine. 2000. *ACSM's guidelines for exercise testing and prescription.* 6th ed. Philadelphia: Lippincott Williams & Wilkins.

American College of Sports Medicine. 2001. ACSM position stand. Appropriate intervention strategies for weight loss and prevention of weight regain for adults. *Medicine and Science in Sports and Exercise* 33: 2145-2156.

American College of Sports Medicine. 2002. *ACSM's resources for clinical exercise physiology: Musculoskeletal, neuromuscular, neoplastic, immunologic, and hematologic conditions.* Baltimore: Lippincott Williams & Wilkins.

Bouchard, C., ed. 2000. *Physical activity and obesity.* Champaign, IL: Human Kinetics.

Bouchard, C., ed. 2001. Physical activity and health: American College of Sports Medicine Consensus Conference. *Medicine and Science in Sports and Exercise Supplement* 33: S347-S640. (Entire issue)

Khan, K., H. McKay, P. Kannus, K. Bennell, and D. Bailey. 2001. *Physical activity and bone health.* Champaign, IL: Human Kinetics.

U.S. Department of Health and Human Services. 1996. *Physical activity and health: A report of the Surgeon General.* Atlanta: U.S. Department of Health and Human Services, Centers for Disease Control and Prevention, National Center for Chronic Disease Prevention and Health Promotion.

PART IV

NEURAL BASES OF HUMAN MOVEMENT

The Subdiscipline of Motor Control

*I*n the preceding chapters on functional anatomy, biomechanics, and exercise physiology, we gained some understanding of the material structure, design, and energetics of the human machinery for movement. To take our analogy between the human body and a motor vehicle one step further, this part of the text, in introducing the subdiscipline of motor control, examines the control mechanisms we have for movement. Understanding the control of human movement requires an understanding of the functions of the brain and nervous system in somewhat the same way that understanding a motor vehicle's control systems requires knowledge of automotive electronics. The analogy of the neural system to an electrical system is a powerful one that has, in a general way, guided much theorizing in the motor control field.

Definition of Motor Control

Motor control is that subdiscipline of human movement studies concerned with understanding the processes responsible for the acquisition, performance, and retention of motor skills. Motor development and motor learning are specialized areas of focus within the subdiscipline of motor control. Motor development (the subject of chapter 16) deals with motor control changes throughout the life span,

specifically those changes in the acquisition, performance, and retention of motor skills that occur with growth, development, maturation, and aging. Motor learning (discussed in chapter 17) deals with motor control changes that occur as a consequence of practice (or adaptation), focusing literally on how motor skills are learned and on the changes in performance, retention, and control mechanisms that accompany skill acquisition. Over the years a host of other terms such as motor behaviour, psychology of motor behaviour, motor learning and control, and perceptual-motor skill acquisition have also been used to describe the subdiscipline, or parts of it. The use of these terms is avoided here, however, as they confuse rather than clarify the subdiscipline's scope and structure.

The word "motor" in the terms "motor control," "motor development," "motor learning," and "motor skills" literally means movement. **Motor skills** are those goal-directed actions that require movement of the whole body, a limb, or a muscle in order to be successfully performed. Consequently the motor control subdiscipline has a broad focus and range of application, from the study of movements as simple as unidirectional finger or eye movements to those as complex as the ones involved in fundamental actions like walking, running, reaching, grasping, and speaking; in workplace tasks like welding, typing, and driving;

in artistic tasks like dancing and playing musical instruments; and in sporting tasks like performing a complex gymnastics routine or hitting a fast-moving tennis ball. Despite their obvious diversity, all of the motor skills used in these tasks have in common their purposefulness, their voluntary nature, their dependence to some degree or other on learning, and the direct dependence of the task outcome on the quality of the movement produced.

Typical Questions Posed and Problems Addressed

One way of gaining a feel for the breadth of the motor control subdiscipline is to note some of the many questions currently under examination by motor control researchers. A number of these questions are discussed in the four chapters within this part of the text.

- *How are skilled movements remembered?* For example:
  - Why is it we can remember how to ride a bicycle or how to swim after many years without practicing the skill?
  - What elements of movement are stored in memory?
  - How can memory for movement be enhanced?
  - Do skilled performers have better memories for movement than unskilled performers?
- *What is the most effective set of conditions for learning a new motor skill?* For example:
  - What type of feedback information is best for learning?
  - Can we learn when we are fatigued?
  - Should practice emphasize consistency or variability?
  - Does verbal instruction help or hinder learning?
- *How do skilled performers succeed where lesser-skilled performers fail?* For example:
  - Do experts have faster reaction times than lesser-skilled people?
  - Can skilled performers be identified at an early age?
  - How do expert game players manage to "read the play"?
  - What does it mean when skills become automatic?

- *How is movement control affected by fatigue, injury, and disability?* For example:
  - Why do people who have had a stroke have difficulty with speech and gait?
  - What causes clumsiness in some children?
  - What is the best way to recover movement control in an injured joint?
  - How and why does alcohol affect movement control?

Issues addressed by motor control researchers can clearly range from very basic questions (such as those related to how the nervous system implements the neuromuscular changes needed to make a movement purposefully faster, more forceful, or different in sequencing or timing) to very applied questions (such as those related to optimising performance or the rate of skill learning or relearning).

Levels of Analysis

The single greatest difficulty for scientists attempting to understand how the control of movement occurs is that the processes in the brain and central nervous system that control movement (and that are modified with maturation and adaptation) are not directly observable. Knowledge about motor control is gained indirectly from inferences about control mechanisms derived from the observation, description, and measurement of observable movement performed under a variety of carefully selected experimental conditions. The more different levels of analysis of the motor system undertaken, the greater the certainty with which inferences about motor control can be made.

Figure 1.1, which schematically represents knowledge organization within the discipline of human movement studies, indicates that the subdiscipline of motor control draws methods, theories, and paradigms from a range of cognate disciplines including physiology, mathematics, physics, computer science, psychology, and education. Of these influences, the most powerful ones both historically and today are the influences from physiology, especially neurophysiology, and cognitive science, which is a hybrid of experimental psychology and computer science. Indeed, within the modern field of motor control, it is still possible to identify separate yet complementary neurophysiological and cognitive science approaches to the examination of movement control, development, and learning.

The neurophysiological approach to motor control focuses on understanding the functioning of the components of the neuromuscular system, especially the functional properties of the movement receptors, the nerve pathways, the spinal cord, and the brain. Its aim, in a crude sense, is to describe the basic "wiring," interconnections, and organization of the neuromuscular system, in order to understand the neural architecture or (to use a computer analogy) the "hardware" of the motor system. Physiologists studying the motor system use a variety of methods. These include tracing nerve connections (using histological procedures such as staining or labeling procedures in which neurotransmitters are traced chemically); measuring metabolic activity in specific areas of the brain (by radioactively labeling a metabolic source such as glucose); examining the functional anatomy of the brain using modern imaging techniques (such as computer-assisted tomography [CAT scan], magnetic resonance imaging [MRI], and **positron emission tomography [PET]**); measuring electrical activity in the brain (via **electroencephalography [EEG]** and **magnetoencephalography [MEG]**) and in muscle (via electromography [EMG]); evaluating the behavioural effects of brain damage or selective lesions to the neural system; and observing the behavioural effects of selective electrical, magnetic, or chemical stimulation of given nerve pathways or regions of the brain. Many of these techniques can be applied only to anaesthetized animals or, in the case of examining the impact of brain damage, to patient populations—although improving technology, especially in terms of brain imaging techniques, allows increasing recordings and observations from the living, undamaged human brain. Basic information about motor control derived from neurophysiological studies is presented in chapter 14.

The cognitive science approach to motor control focuses not so much on the physical structure of the components of the neuromuscular system but rather on developing conceptual models to describe and explain the collective behaviour of the some 100 000 000 000 000 (10^{14}) neurons and neural connections that together compose the motor system. The favoured approach of psychologists examining movement control is to use experimental tasks with altered movement demands to test their conceptual models. The validity of the conceptual models of the perceiving, deciding, and acting stages involved in movement control is evaluated from measures of both movement outcome and pattern. The usual movement outcome (or product) measures are those of movement speed or accuracy, or both, while movement patterns are typically described using one or more of the various measures of kinematics, kinetics, and EMG as described earlier in the book for the subdisciplines of biomechanics and functional anatomy. In recent times, mixing the methods of experimental psychology with those of computer science has resulted in the establishment of the field of **cognitive science,** one of whose principal goals is to develop powerful computational models and simulations of the functioning of the neuromuscular system. Some of the basic findings concerning motor control derived from a cognitive science perspective are presented in chapter 15.

Historical Perspectives

The neurophysiological and psychological/cognitive science approaches to motor control have quite independent histories. A number of important findings with respect to the neurophysiological basis of movement and movement control were made throughout the 1800s. Foundations for a neurophysiological basis of movement control are seen in Charles Bell's 1830 text *The Nervous System of the Human Body,* which described the motor function of the ventral roots of the spinal cord and the sensory function of the dorsal roots, and later in the studies from 1856 to 1866 of Hermann von Helmholtz, which included estimations of nerve conduction velocity and early investigations of reaction time. Subsequently the springlike characteristics of muscle were described (by Weber in 1846); the electrical excitability of the brain was discovered (by Fritsch and Hitzig in 1870); and studies of the sensory and motor functions of the brain were commenced (by Beevar and Horsely in 1887).

Undoubtedly the major historical contributor to understanding the neurophysiology of the motor system was Sir Charles Sherrington (1857-1952), whose work on reflexes is still widely credited today. From Sherrington's work, especially his classic 1906 text titled *The Integrative Action of the Nervous System,* arose a number of key concepts such as synaptic transmission, reciprocal inhibition, final common pathway, and proprioception that are cornerstones of modern neurophysiology. Another influential figure for modern motor control theories was the Russian physiologist Nicolai Bernstein (1897-1966), whose integrative work on phase relations, functional synergies, and distributed control in natural actions such as locomotion appeared only posthumously in the English-language literature.

A wealth of knowledge on the neurophysiology of the motor system emerged in the period following World War II when improved electrophysiological recording techniques and neurophysiological mapping techniques facilitated new precision and insight into the structure and function of many levels of the neuromotor system. Of particular note in this period was the work of the Canadian neurosurgeon Wilder Penfield, who, along with Rasmussen, used electrical stimulation studies to map the topographical organization of the motor and sensory cortices of the brain.

The most noteworthy and influential early studies of movement from a psychological perspective were the basic studies on the control of arm movements by Woodworth in 1899 and the applied studies on the skill acquisition of Morse code operators by Bryan and Harter, published in 1897 and 1899. Little motor control research of any kind took place in the first part of the 20th century; and when motor control research reappeared in the 1930s and 1940s, it was very much oriented toward solving practical problems associated with specific motor tasks. Practical concerns highlighted in this period included personnel selection and training for war tasks such as flying, steering, tracking, and weapon control. The search for optimal approaches to teaching and coaching in physical education and sport also came to the foreground at this time.

An identifiable research field of movement control within psychology, aimed at understanding fundamental processes in movement control, did not emerge until the 1950s. This followed the major conceptual advance in experimental psychology in the late 1940s, triggered by Craik (1947, 1948), Wiener (1948), and Shannon and Weaver (1949), that viewed the brain and nervous system as processors of information in a manner akin to sophisticated, high-speed computers. The 1960s and 1970s in motor control were consequently dominated by information-processing models of motor control and attempts by psychologists such as Paul Fitts to quantify the information-processing capabilities of the motor system. The theoretical contributions of the psychologists Franklin Henry, Jack Adams, Steven Keele, and Richard Schmidt were particularly prominent in this period. Schmidt founded the *Journal of Motor Behavior* in 1969, giving the subdiscipline its first specialist journal. With the passage of time the historical distinctions between the neurophysiological and psychological (now more appropriately, cognitive science) approaches to movement control have become less pronounced. The two approaches are complementary—one providing knowledge of the structure of the neuromuscular system, the other theories and simulations of its collective behaviour and organization—and synthesis between the two approaches is increasingly sought.

Professional Organizations and Training

People interested in motor control come from a diversity of backgrounds. At any symposium or conference on motor control it is not unusual to find psychologists, neurophysiologists, cognitive scientists, engineers, neurologists, and therapists, in addition to researchers with backgrounds in human movement studies. It is perhaps not surprising therefore to find no single professional subgroup representing motor control internationally.

The major meetings for scientists interested in motor control take place under the umbrellas of international groups such as the International Society for Event Perception and Action (ICEPA) and the International Society for Postural and Gait Research (ISPGR) and strong national groups such as (in North America) the North American Society for the Psychology of Sport and Physical Activity (NASPSPA), the Canadian Society for Psychomotor Learning and Sport Psychology (SCAPPS), the Society for Neuroscience and the Psychonomic Society, and (in Europe) L'Association des Chercheurs en Activités Physiques and Sportives (ACAPS) in France. Increasingly, motor control scientists are also attending conferences in biomechanics, such as those hosted by the International Society of Biomechanics (listed in the introduction to part II).

In line with its eclectic nature, motor control research is published in a diversity of journals, including specialist journals such as *Human Movement Science, Journal of Motor Behavior,* and *Motor Control;* neuroscience journals such as the *Journal of Neurophysiology, Brain Research, Experimental Brain Research, Behavioral and Brain Sciences,* and *Journal of Electromyography and Kinesiology;* and experimental psychology journals such as *Journal of Experimental Psychology: Human Perception and Performance, Quarterly Journal of Experimental Psychology,* and *Acta Psychologica.* Research on motor control also appears regularly in more generalist human movement studies journals such as *Research Quarterly for Exercise and Sport* and *Journal of Sports Sciences.*

Further Reading and References

Latash, M.L., and V.M. Zatsiorsky, eds. 2001. *Classics in movement science.* Champaign, IL: Human Kinetics.

Rosenbaum, D.A. 1991. *Human motor control* (chapters 1-3). San Diego: Academic Press.

Schmidt, R.A., and T.D. Lee. 1999. *Motor control and learning: A behavioural emphasis* (chapter 1). 3rd ed. Champaign, IL: Human Kinetics.

Thomas, J.R. 1997. Motor behavior. In *The history of exercise and sport science,* ed. J.D. Massengale and R.A. Swanson (pp. 203-292). Champaign, IL: Human Kinetics.

Some Relevant Web Sites

Canadian Society for Psychomotor Learning and Sport Psychology
[www.scapps.org/]

International Society for Postural and Gait Research
[www.ispgr.org]

North American Society for the Psychology of Sport and Physical Activity
[www.naspspa.org/]

Psychonomic Society
[www.psychonomic.org/]

Society for Neuroscience
[http://web.sfn.org/]

CHAPTER 14

Basic Concepts of Motor Control: Neurophysiological Perspectives

Humans are capable of producing movements that are incredible in their diversity, complexity, precision, and adaptability. To support such an array of movement capability we need a neuromuscular system that is highly organized but also shares the flexible, adaptable, and complex properties of movement itself.

The nervous (or neural) system is designed in some ways like a modern electronic communications network. It has receivers (receptors) that pick up important signals; muscles (as effectors) that are able, when instructed, to bring about planned actions; and a vast array of electronic cables (neurons) and interconnections that allow for nearly infinite linking of receptors to effectors and therefore permit information flow from one region of the network to another. The routine operations of the communications system are achieved at a local level (the spinal level), while an overriding authority (the brain) makes executive and policy decisions regarding what tasks the communications network should attempt to achieve.

The "language" of the communications system is the coded electronic bursts (nerve impulses) that travel throughout the various links in the network. The communications system as a whole is never static, as new connections are constantly being made, damaged connections replaced

> The major learning concepts in this chapter relate to
> - the principal components of the nervous system and their functions;
> - the structure and function of nerve cells and their interconnections as the basic building blocks of the nervous system;
> - the major receptor systems that provide the sensory or perceptual information needed for the control of movement;
> - the nerve pathways within the spinal cord and the motor control functions and capabilities of the spinal cord;
> - the motor control functions of the higher centres of the central nervous system, especially the brain; and
> - some of the major disorders of movement and their neurophysiological origins.

and repaired, and frequently used connections expanded and upgraded to improve their capacity and rate of transmission. In the nervous system, as in the communications network, this capacity for constant change (termed **plasticity**) is essential for accommodating growth, development, and adaptation. Just as we are able to use communications systems such as the telephone effectively without direct knowledge, or in many cases any understanding, of the structure and function of their component parts, as humans we are able to use our neuromuscular systems to produce all manner of skilled movements with a minimum of awareness of either the neuromuscular system's component parts or its methods of operation.

Components of the Nervous System

As the analogy to a communications network suggests, the main components of the nervous system are

- the sensory receptors;
- the motor units; and
- the nerves (neurons) and their junctions (the synapses) that permit communication between the sensory receptors, the motor units, and other neurons.

There are some 10^{12} to 10^{14} neurons within the human nervous system, *each* of which may have as many as 10^4 synaptic connections with other neurons, receptors, or motor units. As a collective network the nervous system has two major subdivisions. The central nervous system (figure 14.1), consisting of the brain and the spinal cord,

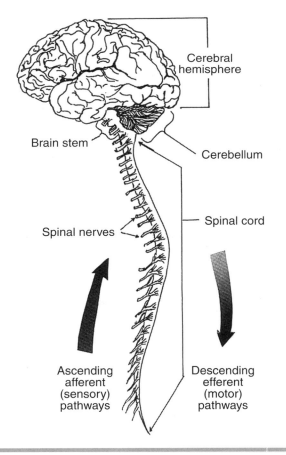

Figure 14.1 The central nervous system consisting of the brain and the spinal cord with its ascending sensory and descending motor pathways.

is responsible for overseeing and monitoring the activation of all sectors of the body, including the muscles. The **peripheral nervous system** carries information from the sensory receptors to the central nervous system and commands from the central nervous system to the muscles. The sensory information from the receptors reaches the brain via ascending **afferent** (from the Latin *afferere* meaning "to carry toward") pathways while the commands from the brain to the muscles are carried via descending **efferent** (from the Latin *efferere* meaning "to carry away") pathways.

The signals from the brain may act to either excite or inhibit those neurons (the motor neurons) that synapse directly with the fibres within muscle. Complex connections between neurons within the central nervous system provide for specialized control of movement and for the storage of information essential for memory and learning. In the sections that follow, we look in greater detail at the main components of the nervous system and their role in the control of movement.

Neurons and Synapses

The joining together of neurons through synapses—the specialized junctions between nerve cells—provides the essential foundation for the human nervous system and allows the nervous system to provide effective two-way communication between its sensory receptors and its motor units.

Structure and Function of Neurons

The **neuron** or nerve cell is the basic component of the neuromuscular system and is essential for receiving and sending messages (or information) throughout the entire system. While neurons vary substantially in both size and shape, depending on their specific function and location within the nervous system, most have a similar structure—a cell body to which are connected a single axon and (typically) many dendrites (figure 14.2). The cell body, containing the nucleus, regulates the **homeostasis** of the neuron. The **dendrites,** collectively formed into a dendritic tree, connect with and receive information from other neurons and, in some cases, sensory receptors. The **axon** is responsible for sending information away from the neuron to other neurons. Collateral branches off the axon permit communication of the nerve impulses from any particular neuron to more than one target neuron. Any single neuron can influence the activity of up to 1,000 other neurons and is itself

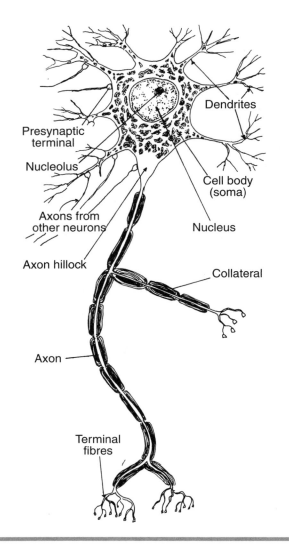

Figure 14.2 A typical neuron or nerve cell.

influenced by the excitatory and inhibitory impacts of some 1,000 to 10,000 other neurons.

There are several types of neurons; the structure of each is dictated by its function (figure 14.3). The sensory or afferent neurons are relatively linear in structure with a single axon connecting the sensory receptor ends to the cell body. The structure of the motor or efferent neurons varies according to their location. The **alpha motor neurons** of the spinal cord possess many dendritic branches and a relatively long axon, also heavily branched to innervate multiple (100-15,000) skeletal muscle fibres. The **gamma motor neurons,** as we will see in the next section, innervate contractile (intrafusal) fibres located within the muscle receptors. Consequently gamma motor neurons, which constitute about 40% of the total motor neurons within the spinal

cord, have smaller, considerably less branched, axons than do alpha motor neurons.

The **pyramidal cells,** located within the motor cortex of the brain, are so named because of their shape, derived from their branching tree of dendrites that all funnel down to a single slender axon. Pyramidal cells send motor commands over the long distances from the brain to the spinal cord and may have axons up to 1 m (3.3 ft) in length. The **Purkinje cells** within the **cerebellum** also have a single thin axon to which information is sent from an incredibly rich, systematically organized set of dendrites that provide these neurons with a characteristic treelike appearance.

Interneurons have a variety of shapes but typically have multiple dendrites and branching axons that permit the connection of multiple neurons with multiple other neurons. The structure of interneurons and their connections facilitates both the convergence of multiple input messages onto a single output cell or set of cells and the divergence of a single input message to several different motor neurons. Interneurons originate and terminate within either the brain or spinal cord. All neurons within the central nervous system are surrounded by, and outnumbered by, other cells called **glia** or **glial cells,** which provide, among other things, the metabolic and immunological support for the neurons.

Neurons carry messages from their dendrites to the terminal fibres of their axons through a series of electrical pulses, produced in the axon hillock (figure 14.2). The electrical pulse produced by the axon hillock is dependent on the spatial and temporal distribution of the pulses impinging on the cell body from its dendritic tree. Signals arriving early and originating from dendrites close to the axon hillock carry more weight than signals from distant neurons arriving late. If the summed weight of the impulses reaching the neuron exceeds its threshold voltage, the axon hillock triggers a pulse (the cell "fires") and this pulse is propagated along the axon to its terminus. The rate at which it is transmitted varies, being greatest in axons of large diameter and in those that are insulated by the fatty substance, **myelin.** Each neuron is therefore more than simply a conductor of electrical signals, constantly undertaking complex processing on the input signals it receives from other sources.

Structure and Function of Synapses

The passage of information from one neuron to another occurs via the synapses (which are in many

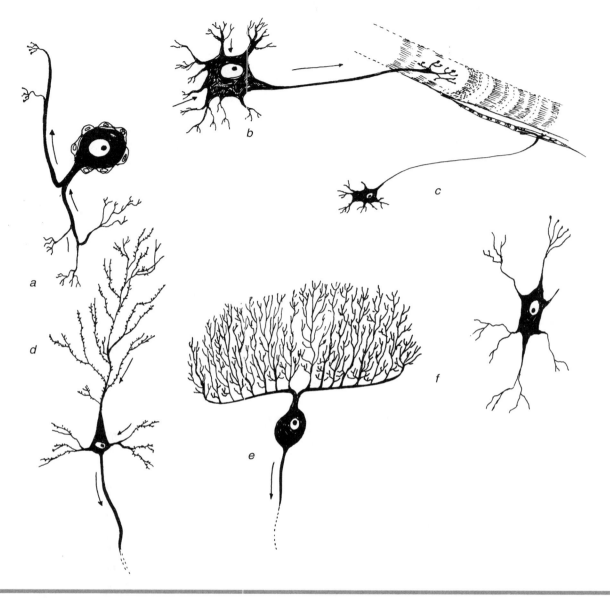

Figure 14.3 Functional types of neurons: *(a)* sensory neuron; *(b)* alpha motor neuron; *(c)* gamma motor neuron; *(d)* pyramidal cell neuron; *(e)* Purkinje cell neuron; and *(f)* interneuron.

ways the equivalent within the nervous system to the joints in the musculoskeletal system). The term **synapse,** coined by Sir Charles Sherrington, originates from a Greek word meaning "union." At the synapse, the axon of one neuron comes in close proximity to, but not direct physical contact with, the receptor surfaces of one or more other (postsynaptic) nerve cells (figure 14.4).

The electrical activity in the presynaptic neuron is transmitted across the gap (the synaptic cleft or junction) to the postsynaptic neuron either via the direct spread of electrical current or, more frequently, by the action of a chemical mediator,

called a **neurotransmitter.** In the case of chemical transmission, the nerve impulse in the axon of the presynaptic neuron triggers the release of a neurotransmitter from tiny storage sacs (vesicles) within the presynaptic membrane into the synaptic cleft. Specialized receptors on the membrane of the postsynaptic neuron detect the presence of the neurotransmitter, triggering either a heightened excitatory or inhibitory response in the postsynaptic neuron, dependent on the nature of the specific neurotransmitter. The transmission of information from one neuron to another therefore typically requires the transduction of an electrical signal

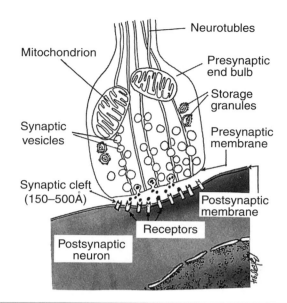

Figure 14.4 The microscopic structure of a typical synapse.

to a chemical one (at the presynaptic neuron), the diffusion of the chemical transmitter across the synaptic cleft, and then the transduction of the chemical signal back to an electrical one (at the postsynaptic neuron).

There are several different neurotransmitters; **acetylcholine (ACh)** (an excitatory neurotransmitter) is the best known. The synaptic connection can be either excitatory or inhibitory and this provides the foundation for more complex functional connections within the nervous system such as reciprocal inhibition, which, as we will see in a subsequent section, form the cornerstone of many reflex actions.

Sensory Receptor Systems for Movement

The main sensory information to guide the selection and control of movement comes from vision and proprioception. Visual information is derived from the light-sensitive sensory receptors located within the retina of the eye. **Proprioception** (from the Latin *proprius*, meaning "own") is information about the movement and orientation of the body and body parts within space and is provided via **kinaesthetic** receptors located in the muscles, tendons, joints, and skin and vestibular receptors for balance located in the inner ear (*kinesthesis* is derived from two Greek words meaning "to move" and "sensation").

Although the sensory receptors for the many facets of vision and proprioception, as well as the receptors for other senses such as hearing, taste, and smell, vary dramatically in their specific structure, all sensory receptors share the common function of transducing physical energy from either beyond the body (such as light or sound waves) or within the body (such as muscle tension) into coded nerve impulses. These nerve impulses can then be transmitted from one part of the body to another via the nervous system or integrated from one sensory system to another. In this regard the sensory receptors are very much like transducers in electronics, converting the information they receive into electrical pulses that can be transmitted along the body's many neural pathways. Humans, like all other animals, are sensitive to only a limited range of the physical signals within the environment. We know, for example, that ultraviolet and infrared wavelengths of light are present in our surrounding environment, but we do not perceive these signals without the assistance of mechanical devices because they fall beyond our usual range of visual sensitivity.

The Visual System

Our rich visual perception of the surrounding environment is achieved through the unique anatomy of the eye and a very complex set of neural processes (figure 14.5). Light reaching the retina (the light-sensitive area at the back of the eyeball) passes through a number of layers of cells to reach the photoreceptors. The photoreceptor cells (the rods and cones) contain chemicals that are sensitive to light, and they send off nerve impulses through their axons to other cells in the retina. This pattern of nerve impulses is specific to the pattern of light falling on the photoreceptors. The rods are most sensitive to light and do not respond to colour, and are therefore the primary receptors for night vision. The cones, in contrast, require high levels of illumination to function but enable us to have colour vision. The density of both types of photoreceptors is higher around the **fovea,** giving this area (corresponding to some 2° of the centre of our visual field) the highest level of sensitivity (acuity).

Nerve impulses arising from the photoreceptors are passed through a number of other layers of interneurons before being sent to the brain via the optic nerve. The arrangement of nerves in the horizontal, bipolar, amacrine, and ganglion cell layers of the retina allows for early processing of

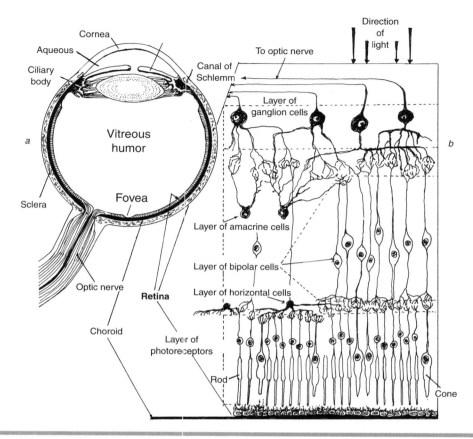

Figure 14.5   Horizontal section of the eyeball (shown on left) with the layered microstructure of the retina (shown on right).

the visual signal, especially in terms of averaging signals over a range of photoreceptors and enhancing contrast between adjacent areas of the visual field.

Visual signals from the retina are carried via the optic nerve along two major pathways, distinct in structure and function. Some 70% of the connections from the optic nerve go first to an area of the midbrain (called the lateral geniculate nucleus) and from there to the **visual cortex,** which is located toward the back of the cerebrum. This pathway, contributing to **focal vision,** is specialized for recognizing objects, distinguishing detail, and assisting in the direct visual control of fine, precise movements (such as those involved in threading a needle). Most of the remaining nerve fibres from the optic nerve terminate in another section of the midbrain called the superior colliculi. This pathway, contributing to **ambient vision,** receives information from the whole of the retina, including the peripheral retina, and is concerned with the location of moving objects within the whole visual field. This pathway is

especially implicated in providing information about our position in space and our rate of movement through the environment. Damage to the focal vision pathway results in an inability to identify objects but not the ability to locate them. The converse is true of damage to the ambient vision pathway.

The Kinaesthetic System
In addition to information provided through vision, information about the "sense of movement" is also derived from specialized receptors located within the muscles, tendons, joints, and skin.

Muscle Receptors
The principal source of sensory information from skeletal muscle is provided by the **muscle spindle.** The muscle spindle is unique as a receptor in that it also contains muscle fibres and hence also has movement capabilities. Muscle spindles are located within all skeletal muscles, although they are particularly abundant in small muscles (such as those in the hands) used to control fine voluntary

movements. Muscle spindles provide the central nervous system with information about the absolute amount of stretch plus the rate of change of stretch in a particular muscle. This, as we will see in subsequent sections, is invaluable both in the reflex control of movement and in the control and monitoring of voluntary movements.

Understanding the control capabilities of the muscle spindle requires a grasp of its unique anatomy (figure 14.6). Under normal circumstances, the contraction of any given skeletal muscle is achieved by a burst of neural activity from an alpha motor neuron that causes uniform contraction across the whole length of the large-diameter muscle fibres, called **extrafusal muscle fibres.** Lying in parallel

to the extrafusal fibres, and connected to them at their endpoints, are smaller-diameter muscle fibres called **intrafusal muscle fibres** (which form the basis for the muscle spindles). The intrafusal fibres differ from the extrafusal fibres in a number of important ways:

- They are smaller and, by themselves, are incapable of directly causing whole-muscle contraction.

- They are innervated not by alpha motor neurons from the local spinal level but independently by gamma motor neurons whose activity is controlled from descending pathways from the brain.

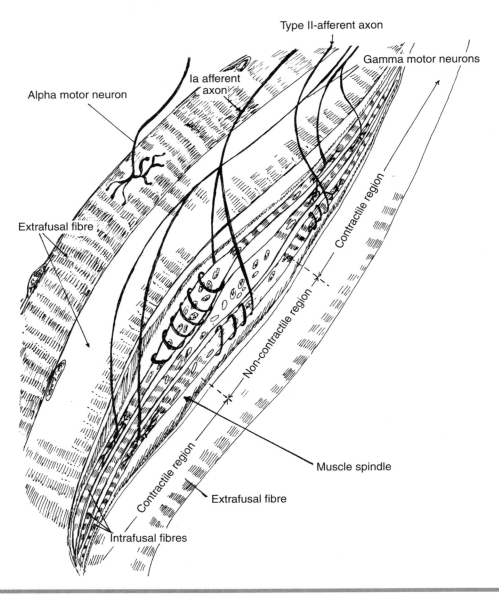

Figure 14.6   Structure of the muscle spindle.

- When stimulated, they contract only at their endpoints and not uniformly across their whole length.
- They have sensory receptors located along their length and afferent connections back to the spinal cord.

The sensory information from the muscle spindle comes from two sources: primary endings, located within the noncontractile central portion of the spindle and connected to the spinal cord by **type Ia afferent neurons;** and secondary endings, located on the contractile end portions of the spindle and connected to the central nervous system by **type II afferent neurons.** As the primary endings respond to stretch, the Ia afferent neurons send impulses back to the spinal cord either when:

- the whole muscle is stretched, or
- contraction of the ends of the intrafusal fibres by the gamma motor system is not matched by an equal shortening of the extrafusal fibres under alpha motor neuron control.

Tendon Receptors

The sensory receptors located within tendons (the attachments of muscles to bones) are known as **Golgi tendon organs.** These receptors lie close to the surface of the musculotendinous tissue and send their impulses back to the spinal cord by **type Ib afferent fibres** (figure 14.7). The Golgi tendon organs are sensitive to the amount of tension developed in the tendon. Because muscles usually shorten when they contract, tendon tension increases when a muscle contracts but decreases when a muscle is relaxed; consequently the Golgi tendon organs fire maximally when the muscle is at its shortest length. This is in direct contrast to the muscle spindles, which are most active when the muscle is stretched.

The Golgi tendon organs appear to serve two major functions with respect to movement control. The first function is a protective one to signal dangerously high tensions in muscle. The Ib afferent neurons are so connected that excessive excitation of the Golgi tendon organs acts to inhibit further muscle contraction (by inhibiting the alpha motor neuron innervating the particular muscle), thus preventing damage to the musculotendinous juncture. In this respect the Golgi tendon organ operates somewhat like a fuse within an electrical circuit. The second function of the Golgi tendon organs is to provide sensory feedback to the spinal cord, even at low levels of tension, thereby providing

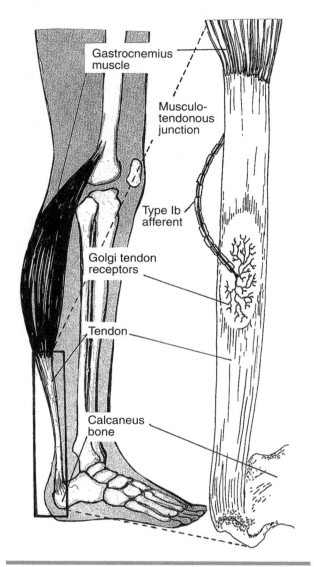

Figure 14.7 Anatomy of the Golgi tendon organs.

fine-tuned feedback information that can potentially assist in continuous ongoing control throughout a movement. The current line of thought is that the Golgi tendon organs play a particular role in controlling muscle output and tension in response to fatigue. Another theory is that the Golgi tendon organs, in conjunction with the muscle spindles, help control the relationship between a muscle's length and the force it generates—a property known as muscle stiffness.

Skin (Cutaneous) Receptors

The skin, an extremely complex and vital organ of the body, contains several types of receptors that can provide useful sensory information for the control of movement (figure 14.8). Detection of the deformation of the surface of the skin caused by movement or weight bearing, for example, may be a valuable source of information for monitoring and controlling voluntary movements. **Meissner's corpuscles** and **Merkel's discs** or **corpuscles** (on hairless parts of the skin such as the palms of the hands or the soles of the feet) and free nerve endings, **Ruffini corpuscles, Krause's end bulbs,** and nerve endings wrapped around hair follicles or other parts of the skin all provide sensory information about light touch or low-frequency vibration. Receptors such as the **Pacinian corpuscles,** located deeper in the skin, respond more to deep compression and high-frequency vibration, especially the onset and offset of such events.

Like other receptors we have considered, the **cutaneous receptors** are not uniformly distributed throughout the body but are more densely distributed in regions such as the fingertips that are used for fine, precise movements. Cutaneous sensitivity varies throughout the body in relation to the number of receptors per unit area. Cutaneous

receptors clearly have an important role in movement control: It is well known that motor performance deteriorates if these receptors are damaged. Patients with damage to the cutaneous receptors in the soles of their feet, for example, experience difficulty in maintaining balance. Likewise, engineers developing robots to perform movement tasks have discovered that robots lacking touch receptors have great difficulty in performing any tasks requiring fine precision.

Joint Receptors

Three types of receptors are located in the tissues surrounding and composing joints; each of these bears similarity to the kinaesthetic receptors located elsewhere in the body. Modified Ruffini corpuscles and modified Pacinian corpuscles, not dissimilar to those found in the skin, are located within the joint capsule itself; and Golgi organs, not dissimilar to those found in the tendons, are located within the ligaments that bind the joint together. While there is some debate over the function of the joint receptors, it appears that their main collective role

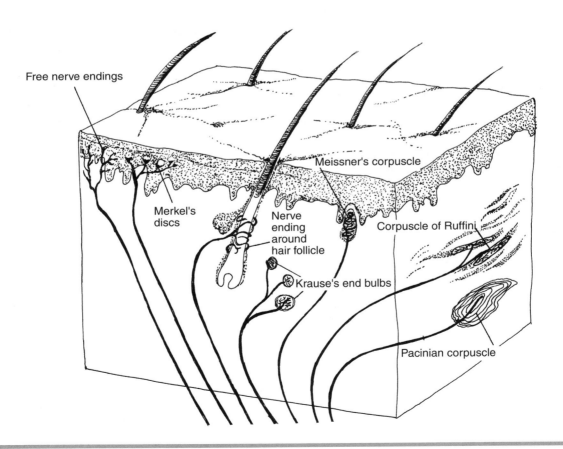

Figure 14.8 Sensory receptors for movement in the skin.

is to signal extreme ranges of motion at the joint. The joint receptors are therefore able to play a role in protecting the joint from injury by signaling to the central nervous system when the full range of motion of a joint is being reached.

The Vestibular System

While the various kinaesthetic receptors provide valuable information about the state of individual muscles, joints, and movement segments, the performance of many skilled movements also requires information about the orientation of the whole body in space. This is particularly true of movements in activities such as gymnastics, trampolining, or diving. Some of this information about whole-body orientation can be provided by the visual system, but much of it is provided by a uniquely designed receptor system (the vestibular apparatus) located adjacent to the inner ear (figure 14.9).

The **vestibular apparatus** consists of two types of receptors—the **semicircular canals** (the superior, horizontal, and posterior canals), which respond to angular acceleration in three different planes, and the **otolith organs** (the **utricle** and the **saccule**), which respond to linear acceleration. Each semicircular canal is located at right angles to the other two, allowing for separate information on the horizontal, lateral, and vertical angular acceleration of the head to be sent to the brain. In contrast, the utricle provides sensory information on the linear horizontal acceleration of the head, and the saccule provides information on the linear vertical acceleration of the head. The vestibular apparatus is centrally involved in balance such that any dysfunction of the vestibular apparatus, such as that occurring with some ear infections, can lead to loss of balance control. The vestibular apparatus is also in close reflex connection with the visual system. Discrepancies between the information provided by the visual and vestibular systems, such as those occurring in many simulated rides in amusement parks, can give rise to motion sickness.

Intersensory Integration and Sensory Dominance

Environmental events frequently are experienced by several sensory systems in the body, and the challenge for the central nervous system is to integrate these sources of information. In the maintenance of normal upright balance, for example, sensory information needs to be integrated from the visual receptors, the many different kinaesthetic

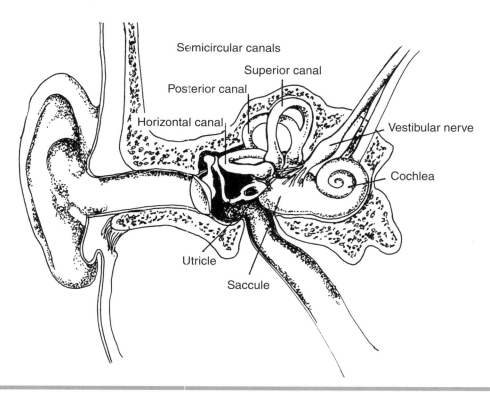

Figure 14.9 Anatomy of the vestibular apparatus.

receptors, and the vestibular system in order to determine whether balance is being maintained correctly or being lost. That integration is possible at all is a consequence of the sensory receptors transducing their very different sources of physical stimulation into the common "language" of nerve impulses and the presence of a vast array of neural interconnections and pathways within the brain and central nervous system that allow signals from diverse locations to converge. While the information coming from the different sensory systems is usually in agreement, in some circumstances the information supplied to the central nervous system may be in conflict (e.g., the visual system may indicate that balance is being lost while the kinaesthetic system indicates balance is being retained). The brain and other sections of the central nervous system require some systematic means of resolving this conflict. In humans any intersensory conflict is always resolved in favour of vision, which is referred to as the dominant sensory modality. This can, on occasions, lead to misperception and in turn misguided action if the sensations provided by the visual system are inaccurate or misleading (see "Visual Dominance in Balance Control").

IN FOCUS: VISUAL DOMINANCE IN BALANCE CONTROL

An excellent demonstration of the dominance of vision over information from other sensory systems has been provided by the Edinburgh psychologist David Lee and his colleagues. Lee had participants in his study stand upright within (what appeared to be) an enclosed room and gave them the apparently simple task of maintaining their balance so as to keep their head and whole body as still as possible. Under control conditions, the information coming from the visual system as well as from the proprioceptive system would indicate reliably to the participants that they were stationary. What the participants did not know was that the "room" surrounding them (but not the surface on which they were standing) could be moved—the front wall could be subtly, but systematically, moved either toward or away from them. In this situation the visual system senses a loss of balance (overbalancing forward when the room is moved toward the participant and the converse when the room is moved away) even though the independent information from the vestibular apparatus and the various kinaesthetic receptors would indicate no loss of balance. In such instances of intersensory conflict, the participants' response was consistently to make compensatory postural adjustments. That such responses are made in the direction opposite the perceived overbalancing clearly indicates that the visual information is the information the central nervous system "believes" (figure 14.10). Even room movements as small as 6 mm can induce marked postural sway in adults and complete loss of balance in young children, demonstrating the dominance of vision (even under situations in which the information provided by vision can be shown to be incorrect).

SOURCES

- Lee, D.N., and E. Aronson. 1974. Visual proprioceptive control of standing in human infants. *Perception and Psychophysics* 15: 529-532.
- Lee, D.N., and J.R. Lishman. 1975. Visual proprioceptive control of stance. *Journal of Human Movement Studies* 1: 87-95.

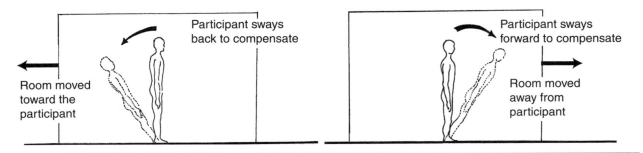

Figure 14.10   Postural adjustments induced by visual movement of room walls (*a*) toward and (*b*) away from the standing subject.
Adapted from Lee and Thomson 1982.

Effector Systems for Movement

The motor unit, as the functional unit of interaction between the nervous system and the muscular system, has been considered elsewhere in this book as the source of neural input to the muscular system (see especially figure 2.12). Consequently in this section we only briefly reiterate some of the key features of the motor unit, this time in the context of the motor unit's role as the ultimate endpoint for the output of the neural system.

A motor unit consists of a single alpha motor neuron plus all the skeletal muscle fibres (extrafusal fibres) it innervates. This may range from as few as one or two fibres for the small muscles of the eye that control precise movements to up to 1,000 in some of the larger postural muscles of the lower leg. As a general rule, the fewer the muscle fibres within a motor unit, the more precise the control that is possible. Observable contraction of a muscle or motion of a joint crossed by a muscle requires the activation of several different motor units. There are three types of motor units—corresponding to the three different types of muscle fibres it is possible to innervate (see table 10.2).

With practice and appropriate feedback it is possible to learn to selectively recruit single motor units within a given muscle, but it is not possible to voluntarily activate only some of the muscle fibres within a single motor unit. In natural movements, motor units are typically recruited in order of size, with the motor units containing the smaller, less forceful muscle fibres being recruited first. This order of activation is known as the **size principle.**

Motor Control Functions of the Spinal Cord

So far in this chapter we have examined the structures and processes by which the human body receives sensory information of relevance to movement and in turn transmits information through the motor units to produce observable movement. We have yet to examine how the central nervous system links (appropriately) this input and output information.

The central nervous system is somewhat hierarchical in structure in that the higher levels of the system, especially the brain, are responsible for higher-order creative and executive mental (cognitive) and motor control functions whereas the lower levels of the system, especially the spinal cord, are responsible for more routine, repetitive control functions. As the brain and spinal cord work together in the performance of most skilled movements, understanding the neural control of movement requires understanding of the motor control functions of each level of the central nervous system and the way in which these levels interact.

In this section we examine the basic structure of the spinal cord and its motor control capabilities. It will become clear that the spinal cord alone is responsible for the control of reflex movements (rapid movements occurring below the level of consciousness) and for the maintenance of voluntary movements, initiated by higher centres within the brain. Much of the knowledge about the motor control capabilities of the spinal cord comes from studies of reflexes in humans and other animals and from studies of the movement capabilities of spinalized animals (animals in which the nerve pathways from the spinal cord to the brain have been severed).

Structure of the Spinal Cord

As with virtually all other structures in the human body, the anatomical design of the spinal cord can be readily appreciated if its basic functions are first understood. The spinal cord serves two basic functions. Its first role is that of a dual transmission pathway to carry both input information from the sensory receptors to the brain and output information, in the form of motor commands, from the brain to the muscles. (See again figure 14.1.) Its second role is to support reflexes at the local spinal level to provide rapid, essentially automatic response to noxious (or potentially dangerous) stimuli and to ensure the successful execution of movements already under way.

The spinal cord is about as thick as an adult's little finger and runs from the base of the spine to the point where it joins the brain at the brainstem, located at the base of the skull (figure 14.1). Because of its importance to normal communications functions in the body, the cord, like a well-laid telephone cable, is protected throughout its length by the bony structures of the spine. The spinal cord runs throughout the length of the spine within the protection of a canal formed within the vertebral (spinal) column (figure 14.11).

A total of 31 pairs of **spinal nerves** are attached to the spinal cord, with each nerve attached to its side of the cord by two roots. The anterior (or **ventral**) root of each spinal nerve carries the efferent

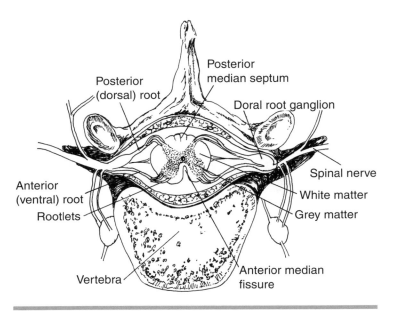

Figure 14.11 Cross section of the spinal cord and its location within the vertebral column.

or motor information away from the spinal cord to the muscles, while the posterior (or **dorsal**) root carries the afferent or sensory information from the periphery back to the spinal cord. The anterior root consists almost exclusively of the axons of alpha motor neurons, whose cell bodies are located within the anterior portion of the spinal cord. The posterior root contains both the axons of the sensory neurons and their cell bodies, the latter clustered together to form the dorsal root ganglion. The spinal cord itself, in cross section, reveals an outer covering of white matter surrounding a central mass of gray matter, roughly approximating the shape of the letter "H." The white matter of the spinal cord consists primarily of nerve fibres, and the gray matter consists of the cell bodies of neurons. Nerve fibres within the white matter are frequently bundled together to form **tracts** that carry impulses up and down the spinal cord.

Spinal Reflexes

A **reflex** (from a Latin word meaning "bending back") is the simplest functional unit of integrated nervous system behaviour. The various reflexes scattered throughout the spinal cord provide the foundation on which (directly) all involuntary and (indirectly) all voluntary movement is based.

A minimum of four basic nerve units are needed to form a reflex arc:

- A sensory receptor (to detect a pertinent stimulus)

- An afferent (or sensory) neuron (to transmit the sensory information to the central nervous system)

- An efferent (or motor) neuron (to transmit the output information from the central nervous system)

- An effector, typically a motor unit (to produce a movement response)

The simplest of all reflex systems, which has only two neurons and hence only one synapse, is called, for obvious reasons, a monosynaptic reflex. Most reflex arcs are polysynaptic, containing multiple synapses. In polysynaptic reflex systems, the sensory and motor neurons do not synapse together directly; instead the nerve impulses are passed from one to the other through a series of interneurons. The time it takes for a reflex system to work (its so-called latency or loop time) is measured from the time of stimulation to the time a response can be recorded in the muscle fibres. Not surprisingly, the more interneurons (and synapses) there are within the reflex arc, the longer the reflex loop time.

The Stretch Reflex

The best example of a monosynaptic reflex within the spinal cord is the simple stretch reflex, also known as the **myotatic reflex** (muscle-stretching reflex). The stimulus for the stretch reflex is excessive stretch on muscle as detected by the muscle spindles (see again figure 14.6). This excessive stretch may arise from a number of sources, such as postural sway (figure 14.12b) or, as is often the case in a clinical setting, unexpected stretching of the quadriceps muscle when a doctor, using a rubber mallet, delivers a sharp tap to the patellar tendon. In all cases the excessive stretch detected by the muscle receptors results in a nerve impulse being sent to the dorsal root of the spinal cord via the afferent neuron. This neuron then synapses, within the spinal cord, directly to an alpha motor neuron that transmits its nerve impulse back to the extrafusal fibres of the stretched muscle (figure 14.12a). This typically is sufficient to cause the stretched muscle to contract, thus alleviating the stretch stimulus.

In the case of the patella tap test for nerve function used by general practitioners and neurologists, it is the alpha motor neuron activation of the quadriceps muscle that causes the characteristic and forceful knee extension (kicking) response. If

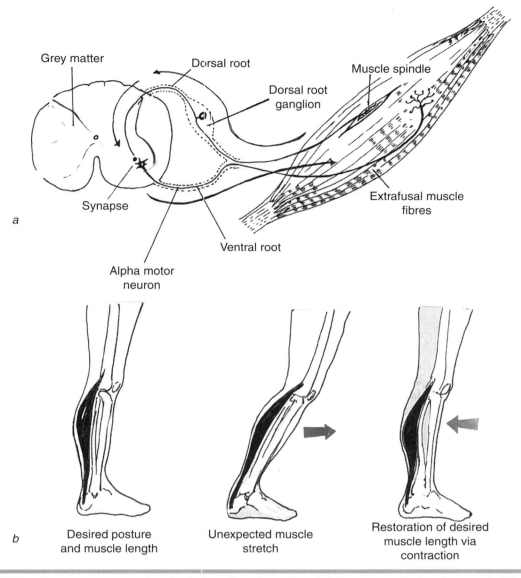

Figure 14.12 The monosynaptic stretch reflex showing (a) the neural circuitry and (b) a typical situation in which the reflex would be activated.

the initial response is insufficient to alleviate the stretch, two things will occur: (1) the same reflex arc will be activated a second time (as the stimulus still exists) and (2) commands will be sent (via interneuron connections) to other segments of the spinal cord, including higher centres. As is typical in most hierarchical organizations, the latter will produce a more powerful but slower response. Whereas the simple monosynaptic stretch reflex may have a loop time as short as 30 ms, reflex responses to alleviate stretch that involve higher segments of the spinal cord, and perhaps even regions of the brain (often referred to as **long-loop reflexes**), may have latencies two to three times this duration.

The Flexion Reflex

A common purpose of many spinal reflexes is to provide rapid protection for the body against potentially injurious stimuli while preserving, at a premium, whole-body balance. The **flexion reflex** is a polysynaptic reflex that causes withdrawal of limbs away from a potentially injurious stimulus. In this reflex, excitation of either a pain or a heat receptor in the skin activates, through interneural connections, those motor neurons innervating flexor muscles crossing joints adjacent to the stimulus, with the result that the limb is flexed away from the noxious stimulus. Equally important, this rapid limb withdrawal through flexion is facilitated by

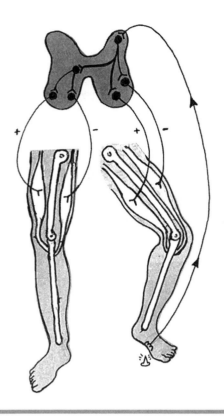

Figure 14.13 The flexion withdrawal and crossed extension reflexes. A painful stretch causes excitation (+) of the ipsilateral flexors and contralateral extensors and inhibition (–) of the ipsilateral extensors and contralateral flexors.

simultaneous "turning off" of the extensor musculature (figure 14.13). This is achieved through interneuronal inhibitory connections between the afferent neuron and the extensor motor neuron. This action provides a specific example of a general spinal reflex phenomenon called **reciprocal inhibition** (the neural control phenomenon ensuring that agonist and antagonist muscles do not typically co-contract in opposition to each other).

The Crossed Extensor Reflex

The **crossed extensor reflex** often functions in conjunction with the flexion reflex to maintain postural stability and, if necessary, to help a person push away from a painful stimulus (figure 14.13). Through interneuronal connections that again exploit the excitatory–inhibitory potential of different synaptic connections, the crossed extensor reflex ensures not only that the limb closest to a painful stimulus is flexed away

from it but also that the limb on the opposite side of the body extends (through extensor excitation and flexor inhibition). This reflex provides a good illustration of how nerve impulses pass not only to and from the spinal cord at a given segment level and up and down the spinal cord, but also across the spinal cord from one side of the body to the other.

The Extensor Thrust Reflex

The **extensor thrust reflex** is one of the more complex spinal reflexes; it aids in supporting the body's weight against gravity. Cutaneous receptors in the feet sensitive to pressure, through a vast array of interneural connections, cause reflex contraction of the extensor muscles of the leg. This reflex provides the foundation for standing balance without dependence on brain mechanisms.

Spinal Reflexes for Gait Control

All forms of human gait (such as crawling, walking, and running) are characterized by continuous patterns of limb flexion and extension, with each limb one-half cycle different from the other. The flexion and crossed extensor reflexes, with their reciprocal innervation of flexor–extension pairs across matching sets of limbs, as well as the extensor thrust reflex, with its pathways for balance preservation, provide strong building blocks for basic gait control and maintenance.

Studies conducted on animals in which spinal connections to the higher centres of the brain have been severed have demonstrated that the spinal cord has an inherent rhythmicity that plays a major role in gait control. In this respect the spinal cord is frequently described as a **central pattern generator.** Although the spinal cord per se seems incapable of initiating gait (this appears to require either motor commands from the brain or very strong sensory information from the cutaneous receptors of the feet), it seems highly capable, through its various reflex pathways, of preserving gait once initiated, even to the point of controlling a transition from one gait form (such as walking) to another (such as running). Clearly spinal reflexes play a major role, not only in involuntary protective actions, but also in the control of fundamental motor activities such as gait.

The Role of Reflexes in Voluntary Movement Control

Though it is obvious that reflexes play a major role in involuntary movement control, most neurophysiologists and motor control theorists believe that the spinal cord, through its reflex arcs, also plays a major role in ensuring that voluntary movements initiated in the brain are executed as planned. Voluntary movements must, to some degree, use reflexes as their "building blocks" because the final pathway for all motor commands, regardless of their origin, is through the alpha motor neurons at the spinal cord level to the muscle fibres. Voluntary movements largely involve modification or use of the spinal reflex pathways in ways specified by commands arising from higher centres of the central nervous system.

To see a good example of the way the reflex structure of the spinal cord can be integrated with the higher-level control provided from the brain, we can refer again to the muscle spindle (figure 14.6) and the stretch reflex (figure 14.12). We saw in the preceding section how the stretch reflex can provide a means of protecting the muscle against damage from excessive lengthening. In a more functional manner, however, the muscle spindle can act in collaboration with its sensory neurons and its alpha and gamma motor neurons to ensure that voluntary movements are executed as planned.

Any movement can be monitored and controlled through **alpha–gamma coactivation.** In this process, actual muscle length is determined by contraction of extrafusal muscle fibres controlled by the alpha motor neurons that originate at the spinal level. Intended muscle length is set by contraction of the ends of the intrafusal muscle fibres under the control of the gamma motor neurons that originate from the level of the brain. As the name implies, for any given movement, such as maintaining upright stance (figure 14.12), alpha–gamma coactivation results in the simultaneous activation of both the extrafusal fibres (by the alpha system) and the intrafusal fibres (by the gamma system).

If the movement goes as planned, the change of muscle length of both the extrafusal and intrafusal fibres will be identical, and no additional sensory impulses will be sent back from the muscle spindle to the spinal cord. If the movement does not proceed as planned (e.g., if extrafusal fibre innervation is insufficient to shorten the muscle), the sensory receptors on the intrafusal fibres will be placed on stretch and this will evoke, through the usual stretch reflex, additional alpha motor neuron activation to cause the muscle to contract. The alpha–gamma coactivation process therefore provides a good example of how movement plans from the higher centres of the central nervous system can be enacted, using spinal mechanisms to ensure that these movements are executed as planned.

Motor Control Functions of the Brain

The human brain possesses a level of complexity and organization that is perhaps without parallel. The brain serves many higher-order functions, only some of which are directly related to motor control. In this section we examine the location and function of the main areas of the brain that have been identified as having a significant role in motor control. These areas are the **motor cortex** (located immediately forward of the **central sulcus** in the frontal lobe of the **cerebrum**), the cerebellum (located off the brainstem and below the occipital lobe of the cerebrum), the **basal ganglia** (located within the inner layers of cerebrum), and the **brainstem** (located forward of the cerebellum and continuous with the spinal cord and the cerebrum). (See figure 14.14.) These areas are in constant communication through a rich, interconnecting network of nerve pathways. Some of the major pathways are shown schematically in figure 14.16 and are discussed further on.

The Motor Cortex

The cerebral cortex is the outermost layer of the cerebrum of the brain, is some 2 to 5 mm deep, has an (unfolded) surface area of some 2 to 3 m^2 (2.4 to 3.6 yd^2), and contains over half of the total neurons in the human nervous system. The cerebral cortex is divided into two halves that appear essentially symmetrical although they are somewhat different in function. These are the left and right cerebral hemispheres that join at the midline through a thick sheet of interconnecting nerve fibres called the **corpus callosum.**

Each cerebral hemisphere contains a motor cortex (lying immediately forward of the central sulcus; figure 14.14), a **premotor cortex** (lying just forward of the motor cortex), and a **supplementary motor area** (lying on the medial wall of the cerebral hemispheres and forward of the motor cortex). Each of these structures, located within the frontal lobe of the cerebrum, is intimately involved in the production and control of skilled movement. Forward of the premotor cortex are two other areas that also have important, but specialized, motor control functions. The frontal eye fields are involved in the control of voluntary eye

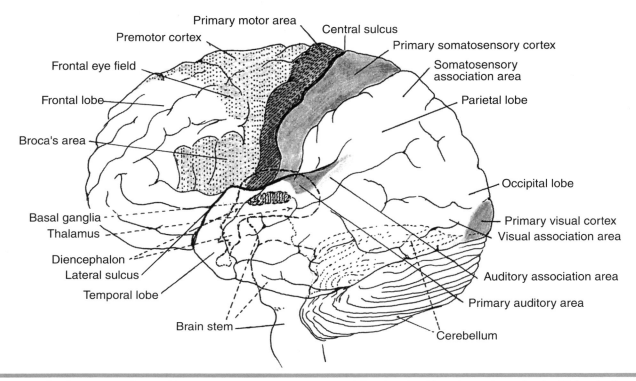

Figure 14.14   Location of the principal motor areas of the brain. Structures located within dashed lines lie underneath or within the external surface of the brain.

movements, and Broca's area (located in the left hemisphere only) has a critical role in the planning of the movements generating speech.

Each part of the motor cortex and its associated areas controls specific muscles or muscle groups within the body so that all muscles are topographically represented in the brain (figure 14.15). In pioneering studies of the human brain, two Canadian neurosurgeons, Penfield and Rasmussen, developed muscle maps of the motor cortex by applying weak electrical pulses to distinct areas of the motor cortex and observing the muscle contractions that resulted. These electrical mapping studies led to a number of important findings:

- All muscles are not represented proportionally in the motor cortex on the basis of their size; rather, representation is proportional to the precision requirements of different parts of the body, with representation of the muscles of the hand and the face occupying nearly two-thirds of the total area of the motor cortex.

- Muscles located more proximally are represented in the motor cortex of the cerebral hemisphere on both the same **(ipsilateral)** and the opposite **(contralateral)** side of the body.

- The representation of distal musculature, such as that crossing the joints in the hands and feet,

is entirely contralateral; consequently the left motor cortex controls the right hand and vice versa.

- As stimulation is moved forward into the premotor cortex, gross movements of muscle groups rather than fine movements of a discrete muscle group are observed.

- Perhaps most importantly, these electrical mapping studies suggest that the motor cortex acts as something of a relay station, being the final neural station for the organization and release of the coordinated motor commands to be sent out to the specific muscles or, more correctly, muscle fibres.

The motor cortex has two principal means of relaying its commands to the muscles. The most direct route is via the **pyramidal tract**—that is, the pathway through which neurons from the motor cortex synapse directly in some cases (and through a minimum of interneurons in most cases) with the alpha motor neurons at the spinal level. This tract carries impulses that are primarily excitatory in character. Damage to this tract can result in either partial **(hemiparesis)** or complete paralysis **(hemiplegia)** of contralateral movements. Alternative routes, known collectively as the **extrapyramidal tract,** allow nerve impulses from the motor cortex

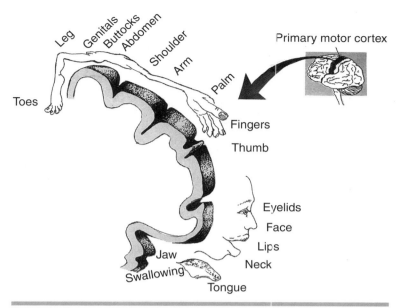

Figure 14.15 Schematic view of the primary motor cortex and the mapping between electrical stimulation of the motor cortex and regions of the body in which movement consequentially occurs. The inset shows the location of the primary motor cortex within the brain.

stem, from the vestibular apparatus, and, via the spinal cord, from the kinaesthetic receptors located on the same (ipsilateral) side of the body.

The cerebellum itself has a very regular anatomical structure based primarily on two different types of afferent fibres (called **climbing fibres** and **mossy fibres**) and one output fibre (the Purkinje cell; see again figure 14.3). This structure enables the cerebellum to perform a number of very complex signal-processing operations fundamental to many aspects of motor coordination. Indeed the cerebellum has frequently been referred to as "the seat of motor coordination." The cerebellum has two major outputs—one to the thalamus and one to the brainstem (figure 14.16).

A number of major motor control functions have been attributed to the

to reach the spinal level through a range of pathways via the cerebellum, basal ganglia, **thalamus,** and brainstem (figure 14.16). Outputs from these pathways are primarily inhibitory. Damage to the extrapyramidal tract can result in **spasticity**.

Damage to the cerebral cortex can result in **apraxia**. Specific damage to the motor cortex results in a loss of fine movement control, especially in the fingers and toes. Damage to the premotor cortex results in a disruption to movement planning and selection, especially for gross movements involving a number of muscle groups. Damage to the supplementary motor area disrupts the planning of sequential movements and performance of many tasks requiring bimanual coordination. Collectively, the premotor cortex and the supplementary motor area appear to have a particular role to play in movement planning and generation.

The Cerebellum

As seen in figure 14.14, the cerebellum attaches to the brainstem and is located behind and below the cerebral hemispheres. Like the cerebrum, the cerebellum has an outer cortex that is divided into two distinct but interconnected hemispheres. Beneath the cortex are four deep cerebellar nuclei. The cerebellum receives input information from a vast array of areas in the cerebral cortex (including the motor areas), from various areas in the brain-

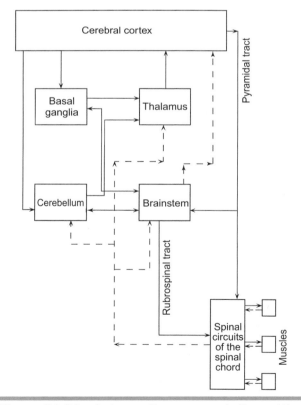

Figure 14.16 Schematic representation of the major motor pathways. Direction of information flow is shown by arrows, with afferent input denoted by dashed lines. Two-way information flow is indicated by double-headed arrows.
Adapted from Greer 1984.

cerebellum, all broadly related to the translation of abstract movement plans into specific spatial and temporal patterns that can be relayed to the muscles via the motor cortex. Principal cerebellar functions appear to be the regulation of muscle tone, the coordinated "smoothing" of movement, timing, and learning. Patients with cerebellar damage demonstrate one or more of the symptoms of low muscle tone, incoordination or **ataxia** (especially in standing, walking, speaking, or performing precise aiming movements), poor temporal control of muscle recruitment, and difficulty in learning new movements or adapting old ones. Fast, ballistic types of movement appear to be particularly affected.

The Basal Ganglia

The basal ganglia are a group of five pairs of interconnected nuclei (the globus pallidus, the caudate nucleus, the putamen, the subthalamic nucleus, and the substantia nigra) located deep within each of the cerebral hemispheres and close to the thalamus. The basal ganglia receive input from two major sources (the motor areas of the cerebral cortex and the brainstem) and, similarly, send their output to two locations (the thalamus and the brainstem). Therefore, like the cerebellum, the basal ganglia, while not synapsing directly with spinal neurons, are able to influence alpha motor neuron activity through both the pyramidal tract (via connecting to the cortex through the thalamus) and the rubrospinal tract (via the brainstem) (figure 14.16). The basal ganglia work together as a loosely connected unit, although each of the component nuclei is quite different and all are generally connected in an inhibitory fashion with one another.

Insights into the function of the basal ganglia in motor control have come primarily from studies of patients with one of two identifiable diseases of the basal ganglia. **Parkinson's disease** is a degenerative disease resulting from deficiency in the natural neurotransmitter substance **dopamine** that assists in carrying nerve impulses from one nucleus within the basal ganglia to another. Parkinsonian patients typically demonstrate a range of motor symptoms including shuffling, uncertain gait, limb tremor, difficulty in initiating movement, and high degrees of muscle stiffness. **Huntington's disease** is a hereditary degenerative disease resulting from damage to the dendrites that produce one of the neurotransmitters used to communicate between selected nuclei within the basal ganglia. Patients with this disease experience uncontrollable, invol-

untary rapid flicking movements of the limbs, facial muscles, or both. Damage to any structure within the basal ganglia may cause slowness of voluntary movement and involuntary postures and movements.

Despite knowledge of the obvious movement problems caused by basal ganglia dysfunction, the precise function of the basal ganglia in movement control remains elusive. Some favoured suggestions include the control of slow movements, the retrieval and initiation of movement plans, and the scaling of movement amplitudes as required in daily tasks such as handwriting. At a general level, the basal ganglia appear to permit selected movements to proceed while inhibiting unwanted movements.

The Brainstem

The brainstem contains three major areas that have significant involvement in motor control. These are the **pons**, the **medulla**, and the **reticular formation**. The brainstem's principal function, as revealed by figure 14.16, is to act as a relay centre, especially for the transmission of information to and from the cerebral cortex.

The pons and medulla, as the main structures within the brainstem, receive input from the cerebral cortex, cerebellum, and basal ganglia as well as all the sensory systems. These structures then integrate this information for output to the spinal cord for use in the control of many involuntary movements, such as those related to posture and cardiorespiratory activity. The brainstem not only functions in the control of muscle tone and posture but also is fundamental to the operation of a number of supraspinal reflexes. Prominent among these are the **righting reflexes**, which are designed to maintain the orientation of the body with respect to gravity, and the **tonic reflexes**, such as the tonic neck reflex, which are involved in maintenance of the position of one body part (such as the neck) in relation to other body parts (such as the arms and legs). Damage to the pons or medulla disrupts the control of involuntary movements and key orienting reflexes and endangers the control of vital physiological systems.

The reticular formation is a network of neurons that extends throughout the brainstem; through its ascending connections to the cerebral cortex, it has a major role in regulating the activity of the cortex. In this way the ascending reticular formation controls the activation of the cortex and therefore, in turn, the state of arousal experienced by the person.

(The issue of arousal is examined in more detail in chapter 18.) The descending fibres of the reticular formation input directly to the spinal reflexes and may modify reflex activity at this level as necessary to ensure that basic postural needs are met.

Integrative Brain Mechanisms for Movement

Given the complexity of both the human nervous system and the movement it produces, it is perhaps not surprising how little is yet known about the neural mechanisms underlying movement control in the brain and how much remains to be discovered. At this point it is possible only to advance some speculative ideas on the likely flow of neural information through various brain structures and the functional consequences of such information flow. The prefrontal cortex appears to be central to overall movement planning, the basal ganglia and

cerebellum to the programming of specific motor commands, and the motor cortex to the release of organized commands to the muscles, via the spinal pathways (figure 14.17). Readers should be aware, however, that neurophysiologists interested in motor control are still many years away from a complete integrative model of the brain mechanisms for motor control. One approach to hasten understanding may be to look alternatively, or better still, simultaneously at motor control from a conceptual perspective. The next chapter does this by examining basic cognitive science perspectives on motor control.

Summary

The neural control processes underlying skilled human movement are extremely complex. The foundations of the neural control system are built on nerve cells (neurons), sensory receptors, motor units, and the intricate synaptic interconnecting

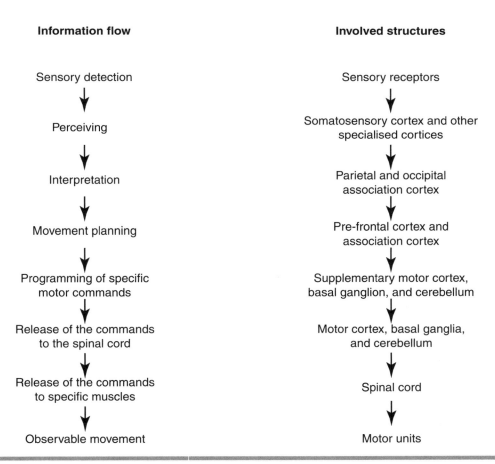

Information flow

Sensory detection
↓
Perceiving
↓
Interpretation
↓
Movement planning
↓
Programming of specific motor commands
↓
Release of the commands to the spinal cord
↓
Release of the commands to specific muscles
↓
Observable movement

Involved structures

Sensory receptors
↓
Somatosensory cortex and other specialised cortices
↓
Parietal and occipital association cortex
↓
Pre-frontal cortex and association cortex
↓
Supplementary motor cortex, basal ganglion, and cerebellum
↓
Motor cortex, basal ganglia, and cerebellum
↓
Spinal cord
↓
Motor units

Figure 14.17 Speculation on some of the major functional roles of brain structures in movement control.

of each of these components. Sensory information for movement comes primarily from vision and proprioception, with proprioceptive information being provided by a host of specialized kinaesthetic receptors located in muscles, tendons, joints, and skin, as well as by the vestibular system. Motor units, composed of single motor neurons and the muscle fibres they innervate, provide the working interface between the nervous and muscular systems. Control of reflex movements and control of the basic flexion–extension pattern generation necessary for locomotion are possible through pathways and connections present at the level of the spinal cord. More sophisticated, voluntary movement requires the involvement of the brain and its specialized pathways and regions responsible for movement planning and initiation. Deficiencies in movement control arising from localized neural damage help provide some insight into the neural basis for movement and postural control.

Further Reading and References

Kandel, E.R., J.H. Schwartz, and T.M. Jessell. 2000. *Principles of neural science.* 3rd ed. New York: McGraw-Hill.

Latash, M.L. 1998. *Neurophysiological basis of movement.* Champaign, IL: Human Kinetics.

Rothwell, J.C. 1994. *Control of voluntary human movement.* 2nd ed. London: Chapman and Hall.

Shumway-Cook, A., and M.H. Woollacott. 2001. *Motor control: Theory and practical applications* (chapter 3). 2nd ed. Philadelphia: Lippincott, Williams & Wilkins.

CHAPTER 15

Basic Concepts of Motor Control: Cognitive Science Perspectives

*I*n the previous chapter we examined in some detail the structure and function of the main components of the neuromuscular system. While this neurophysiological approach to understanding motor control provides valuable information about the receptors and effectors for movement and about the major pathways that connect the two, the sheer size and complexity of the nervous system (with its some 10^{14} neurons, each having up to 10^4 synaptic interconnections) mean that it is impossible to fully appreciate how movement is controlled by studying nerve pathways only. What is needed to complement the neurophysiological knowledge are conceptual theories and models that describe and explain the overall control logic used by the nervous system to acquire, perform, and retain motor skills. Such theories and models have typically originated from the work of experimental psychologists who have focused on the broad overall functioning of the motor system rather than the more specific detailed anatomy and physiology of its individual physical components. In recent years the methods of experimental psychology have been merged with aspects of computer science as scientists (now referred to as cognitive scientists) have attempted to understand the computational capabilities and methods of the human neuromotor system.

In this chapter we examine the role of conceptual models from cognitive science in understanding motor control; outline the key properties of skilled human movement that must be explained

The major learning concepts in this chapter relate to

- the importance of models of motor control;
- key properties that must be explained by models of motor control;
- information-processing models of motor control;
- processes and limitations related to perceiving, deciding, and acting; and
- the possibilities for alternative models of motor control.

by such conceptual models; and then describe in some detail one popular model (an information-processing model) of motor control, at the same time examining some of its basic assumptions and practical implications.

Using Models to Study Motor Control

In its structure and function, the human neuromuscular system is clearly incredibly complex. Simplified models of the system provide a valuable means of helping scientists begin to understand how movement skills might be controlled and how this control might change with practice and with age. Effective models of movement control will help explain the many unique and essential properties of movement control.

The Role of Models in Scientific Study

In all branches of science, models serve the important purpose of aiding in the understanding and advancement of theory. Models are aids to theory, enabling a theory to be visualized and understood, frequently by drawing comparison to the operation of simpler, everyday systems. Systems of infinite complexity, such as the physical system of electricity or the biological system of the heart and lungs, can be more easily understood through the use of simplifying models such as those of water flow or the action of a pump. The value of models therefore is in their potential to simplify a complex system to a level at which understanding can be achieved and experiments to further understanding can be formulated. In a system as complex as the human motor system there would appear to be great value in developing conceptual models as a means of aiding and advancing our understanding of how the system works.

It needs to be noted, however, that a model is not a theory and therefore should not be taken too literally. A model is also worthwhile only insofar as it accurately captures the key characteristics of the system we are trying to understand. Just as good models can aid understanding, poor models can hamper understanding. As more becomes known about a particular system, the shortcomings of old models are frequently realized and new models proposed in their place. As we will see in this chapter, one particular model (the information-processing model) has dominated most thinking to date about how movement is controlled, but some of the limitations in this model have become more apparent recently and alternatives have been suggested.

Key Properties to Be Explained by Models of Motor Control

A starting point for the development of a model of any system is consideration of the key features of the system that the model must be able to encapsulate. These key features, in a sense, form the constraints for the model. Within human motor control is an impressive array of unique properties that any worthwhile model or theory must be able to adequately explain. Some of the principal motor control properties that require explanation are the following:

- **Degrees of freedom** refers to the capability of the brain and nervous system to simultaneously and continuously consider, and somehow control, the enormous number of independent variables (motor units, joints, limb angles) that contribute to skilled movement. In theory there is an overwhelmingly enormous computational challenge to the capacity of the brain and ner-

vous system—one that far exceeds the capability of sophisticated computers used to control even simple movements by robots. The nervous system must use some clever solutions to overcome this problem and reduce control demands to manageable levels.

- **Motor equivalence** is the capability of the motor system to perform a particular task, and produce the same movement outcome, in a variety of ways. Even actions as apparently simple and repetitive, and as consistent in outcome, as writing one's signature can be achieved through recruitment of different motor units or even different muscle groups. Motor equivalence is a consequence of the many degrees of freedom (joints, muscles, motor units) we are able to independently control. Any plausible model of motor control must be able to account for the way in which the nervous system rapidly and apparently effortlessly selects just one combination of joints, muscles, and motor units from all the options available to it in order to perform a particular task effectively and efficiently.

- **Serial order** is the capability of the motor system to structure movement commands so as to reliably produce movement elements in their desired sequence. Correct sequencing of movement components is fundamental to the performance of virtually all skilled actions, and errors in sequencing inevitably result in errors in performance. Serial order errors in speech give rise to spoonerisms ("mumam hovement" instead of "human movement") and, in typing, to transposition errors (for instance, "cat" typed as "cta"). In gross motor skills, such as throwing, misordering of the recruitment of large proximal muscles (such as those crossing the trunk) and smaller distal muscles (such as those crossing the wrist joint) undermines the effective summation of forces and thereby limits performance. A useful model of motor control must therefore be able to account for how serial order is generated in movement sequences (and hence also how errors may sometimes arise).

- **Perceptual-motor integration** is the capability of the motor system to produce movements closely matched to the current environmental demands (perceived by the performer). Skilled movement is always subtly adjusted to meet changing environmental situations. For example, the skilled tennis player is able to adjust his or her racket swing if the ball deviates unexpectedly in flight, and all of us adjust our gait patterns if the surface we are walking over becomes irregular. Such adjustment can be achieved only if perception and action are tightly coupled and integrated. Such coupling must therefore be a key element of any satisfactory model of motor control. As the role of particular muscles in either producing or opposing limb movement will frequently vary according to contextual factors such as joint position or orientation of the limb with respect to gravity, the central nervous system, in issuing motor commands, must be continuously and accurately informed about the body's posture and position in space by its many perceptual systems.

• **Skill acquisition,** as we shall see in chapter 17, is the capability of the motor system to learn and improve, given appropriate conditions of practice. Explaining skill acquisition (or motor learning) requires a motor control theory that can, in turn, explain how experience is stored and how, once acquired, movements can be modified to meet task conditions never previously encountered. A viable model of motor control must therefore be able to adequately account for the paradoxical capabilities of skilled performers to produce movements that are both *adaptable* and *consistent.*

Several models of motor control have been proposed over the years, varying in the extent to which they attempt to explain the key properties of movement as well as incorporate what is known from the neurosciences about the structure of the motor system. The models of motor control that have been most widely developed and used are information-processing models.

Information-Processing Models of Motor Control

Experimental psychologists have, for a long time now, used the computer as an analogy for the nervous system in order to simplify thinking about the complex neural processes underpinning motor control. By first considering the (relatively) simple operations of a computer, we can perhaps attempt to better understand the much more complex operations of the human motor system.

The Human Motor System As Computer-Like

Computers are elaborate, engineer-designed devices for the processing of information.

Information-processing models of movement control are based on the notion that the nervous system is computer-like in that, although it is biological in nature, it is also capable of highly sophisticated information processing.

Hardware and Software

A computer is basically a dedicated electronic device that, through its stored programs, is able to convert input information of one type or another into output of a specific, desired form. The input information may come from data stored on a disc or may be input directly from a keyboard (as occurs when we type in letters or numbers) or from some other information acquisition system. The output information may also take various forms. It may take the form of text (letters and numbers) or graphics appearing on the computer screen or to be sent to a printer (the "hard" copy), or it may be in the form of electronic commands sent on to other devices controlled by the computer.

The conversion of input information to output information is not a passive process but rather an active reorganization of information controlled by the commands within the computer's program(s). The type of processing the computer is capable of doing, the speed with which it can complete its operations, its capacity to store information in memory, and ultimately the quality and diversity of the output it can produce are limited by two interacting factors. These are the physical construction of the computer's electrical circuits (or its *hardware*) and the computer programs that have been written specifically for the computer and that reside in the computer's memory (its *software*).

How then does the motor control system act like a computer system? The input information for movement control is the sensory information sent to the central nervous system from (primarily) the visual, kinaesthetic, and vestibular receptors. The output is the patterns of movement we can observe and describe, which arise as a consequence of the coordinated set of motor commands sent from the central nervous system to selected muscle groups. Input information is converted to output information through a number of information-processing (or computational) stages that take place within the central nervous system. The success of the resulting movement (the output) depends primarily on the computational "programs" within the nervous system that are responsible for selecting and then controlling the movement (figure 15.1). Some of

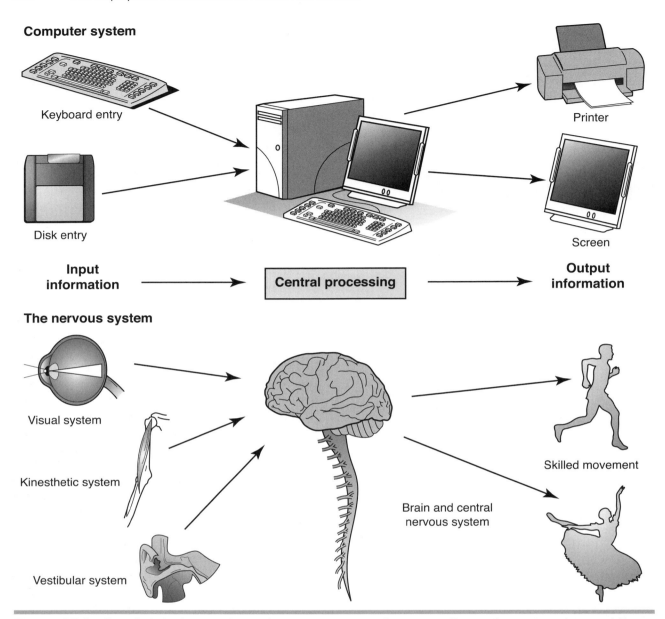

Figure 15.1   Parallels between the information-processing operations of a computer and those of the central nervous system.

the programs, such as those controlling balance and gait, may be "hard-wired" into the central nervous system (especially the spinal cord) from birth and therefore may be considered part of the hardware of the central nervous system. Other skills require programs that are not innate but that rather must be developed and constantly modified and improved through repeated use and practice. These programs are therefore analogous to specialized software written for a particular computer's hardware to fulfil a specific task and stored within the system's memory. In the nervous system this storage may require some changes to "hardware" as new neural connections are formed.

Just as one cannot understand how a computer works and controls its output simply by inspecting what it produces as output, one cannot expect to understand movement control by simply describing the observable movement patterns produced as output by the motor system. It is important to recognize that movement does not simply occur spontaneously as a consequence of unplanned muscular activity; rather movement is the end product of a long series of computations that take place largely beyond observation and within the confines of the central nervous system. To understand movement one must therefore attempt to understand these hidden processes and computations that occur in the central nervous system and that form the link between sensory input and observable movement output. Cognitive scientists have directed much of their energy toward uncovering the computational code and programs the central nervous system has inherited and developed for movement; elaborating on the processing stages used to link input and output information; and determining the capacities and limitations of the various processing stages.

Processing Stages

Most information-processing models of movement control assume that there are distinct and sequential stages through which information must pass (or, more correctly, be processed) from input to output. Figure 15.2 presents a typical information-processing model. It shows environmental and internal information (physically present in such forms as light and sound waves and muscle lengths and tensions) being picked up (transduced) through the various sensory receptors described in chapter 14 and then being transmitted along the afferent pathways to the central nervous system. This information then provides the input for central nervous system processes that ultimately produce, as output, motor commands that are transmitted along the efferent pathways out to the muscle fibres. Here the commands individually cause muscular contraction and collectively generate observable movement patterns. Feedback from the movement itself is monitored via the afferent pathways and can be used to either correct errors in the movement (if the movement is sufficiently slow) or else make improvements to the commands the next time the same or a similar movement is to be produced.

The stages proposed in figure 15.2 are themselves unremarkable and are entirely consistent with the structure and function of the receptors and effectors for movement described in the previous chapter. What remains unclear is the nature of the central processing stages, and it is in their conceptualisation of these central processing stages that most information-processing models differ. Most

models, however, accept that at least three sequential processing stages must occur in the brain and other areas of the central nervous system prior to the initiation of any movement. For simplicity, we refer to these stages as

- perceiving,
- deciding, and
- acting.

Figure 15.3 illustrates these stages. Consider in this example the perspective of player #40—the player with the ball. The player is surrounded by a nearly infinite array of physical signals (e.g., light and sound waves, pressure signals, vibrational signals, and chemical signals), only a very limited range of which can be detected by his eyes, ears, or vestibular, kinaesthetic, and other receptors. Despite this limited range of sensitivity, the brain and spinal cord of the player are bombarded every millisecond by an enormous array of input information from the billions of sensory receptors in the body. There is, for example, visual information about the player's own location on the court relative to the basket, plus information about the location, velocity, and direction of motion of all four teammates and all five opponents; auditory information from the bouncing ball and the calls of teammates, opponents, coaches, and spectators; kinaesthetic information from stretches on muscles, strains on joints, tension on tendons, and pressure on the soles of the feet; tactile information from the hand bouncing the ball; proprioceptive information from the vestibular apparatus; and smell **(olfactory information)**, taste **(gustatory information)**, and other information. Some of this information is highly relevant to the task at hand whereas much of it is quite irrelevant.

It is from this enormous array of sensory information that skilled perceiving, deciding, and acting must emanate. In the perceiving stage, the player's focus is on determining what is currently happening and what is about to happen in both the external and internal environments; in the deciding

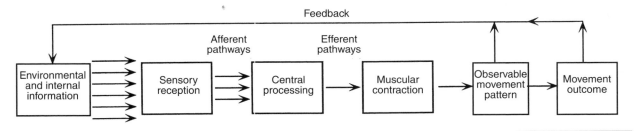

Figure 15.2 A typical information-processing model of motor control.

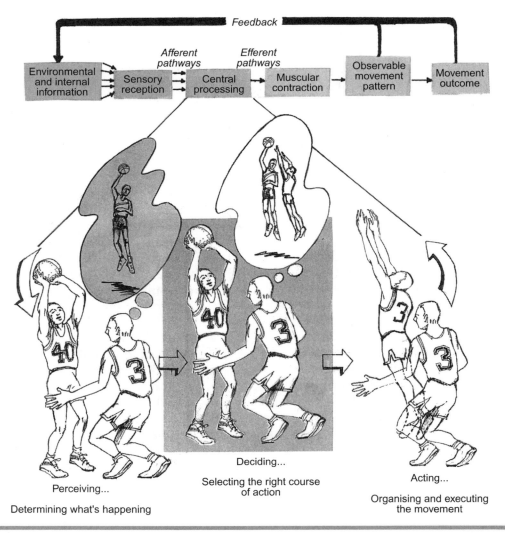

Figure 15.3 Central processing stages within a typical information-processing model of motor control.

stage, the focus is on deciding what action, if any, is needed in response to current and future events; and in the acting stage the focus is on organizing and executing the required movement response in terms of the sequence and timing of the motor commands that have to be sent to the muscles. In the sections that follow we look further at the example of the basketball player and use it to help examine each of the key central processes in more detail.

Perceiving: Determining What Is Happening

The first important process the brain and central nervous system must perform on the available sensory information is that of **perception**. This involves selecting out only the most relevant information for further processing and then

using this information to determine both what is currently happening and, implicitly, what is about to happen in the near future. In the case of our basketball player this will involve making important judgments about external events as well as more internal ones. Externally the player will need to determine, among other things and primarily through vision, the location and posture of defending players, the structure of the offence as well as the defence, the position and direction of movement of any unguarded teammates, their position on the court and especially their proximity to the basket, and the relative heights and agilities of teammates and opponents. Internally the player will need to be aware of his own body posture and balance plus monitor the tactile and auditory sensations from the ball to ensure that it is well in control.

Underlying Processes

Perceiving involves determining what is occurring in the outside world (e.g., where is each of my teammates located?), what is occurring within our own bodies (e.g., am I in or out of balance? what position are my arms in?), and the current and ongoing relationship between these internal and external "worlds" (e.g., where is my foot in relation to the sideline?). Perception is more than simply the passive reception of sensory information by our various receptors. It is rather an active process through which we interpret and apply meaning to the sensory information we receive. As our prior experiences, accumulated knowledge, expectations, biases, and beliefs all contribute to perception, it is not surprising that two people presented with the same pattern of stimulation (e.g., looking at the same picture or experiencing the same kinaesthetic sensations) often perceive and report different things. This is true not only for simple visual images (figure 15.4) but also, as we will see in chapter 17, for the more complex images typical of natural movement tasks and, as we will see in chapters 18 and 19, for perception of social environments such as sport and exercise settings.

Perception involves many subprocesses. Central ones include

- **detection** (determining whether a particular signal is present or not),
- **comparison** (determining whether two stimuli are the same or different),

Figure 15.4 A typical simple but ambiguous figure used by psychologists to demonstrate the subjective nature of perception. The pattern may be differentially perceived as either a vase or two faces looking at each other.

- **recognition** (identifying stimuli, objects, or patterns), and
- **selective attention** (attending to one signal or event in preference to others).

Processing Limitations

All of the subprocesses of perception are limited in their capacity to process information and can therefore potentially act to limit performance on any motor task. In detecting stimuli, humans are limited not only by the range of physical stimuli to which their sensory receptors can respond but also by the capability to distinguish the firing of one or more sensory neurons from a background of general neuronal activity. Human ability to detect stimuli consequently varies from situation to situation, from one sensory system to another, and with factors such as arousal (chapter 18). Humans are also limited in their ability to compare and detect differences between two or more stimuli. For example, in judging the approach velocity of objects (as when balls are thrown toward a person to catch), the object must increase in velocity from, on average, 4.4 m/s (16 km/h or 9.9 miles/h) to 5 m/s (18 km/h or 11 miles/h) before the change in speed can be reliably detected. The sensitivity for detecting differences varies from one sensory system to the next, with the visual and auditory systems being able to reliably detect the smallest changes in stimulation.

Regarding the attempt to recognize stimuli, objects, or events, laboratory experiments have shown that humans can only store some seven items before they start to make errors of identification. In natural settings, the number of patterns we are able to recognize (such as friends' faces or offensive patterns in basketball) is enhanced through the use of multiple attributes (such as hair colour, eye shape, nose size, and ear type in human faces; or player location, posture, and size in basketball offensive patterns).

Selective attention is both a limitation and an advantage to human performance. We all know from personal experience that our processing capacity is limited in that we cannot (typically) listen to two separate conversations or attend to two separate visual signals (such as events in the left and right extremes of our field of view) simultaneously. If a movement task requires processing information from two or more separate locations at the same time, performance of the task will generally be difficult. Being able to selectively attend to typically only one thing at a time can be an advantage, however, in that it provides a means of preventing irrelevant or potentially distracting stimuli from using up some of our valuable processing capacity. A golfer focusing on a putt, or a microsurgeon focusing on a suture, generally benefits from being able to apply all his or her attention to the specific movement, thus effectively blocking out surrounding noise and other potentially distracting events.

Deciding: Determining What Needs to Be Done

Once the person has determined what is currently happening and has predicted future events, the second important stage of information processing involves the process of decision making, that is, determining what, if any, new action or response is required. In many movement tasks, such as sport tasks, the decision-making process equates to "picking the correct option" from a range of possible response options. Clearly the quality of the decision will be determined, in part, by the accuracy of the preceding perceptual judgment. In our basketball example, the player with the ball has at least six broad response options to choose among on any occasion: continuing to dribble the ball, shooting the ball at the basket, or passing to one of four possible teammates. Furthermore, each option involves a number of subdecisions. For example, if the decision is to pass or shoot, what type of pass or shot is most appropriate? Determining what option to select generally depends not only on the current perceptual information but also on other situational information such as the state of the game (the score and time remaining), the player's confidence, and knowledge about the capabilities of teammates and opponents.

Underlying Processes

Decision making is essentially the process of response selection—picking the right movement option to match the current circumstances. The quality of the decision obviously depends not only on the quality of the preceding perceptual judgments but also on knowledge of the costs and benefits associated with each option. The latter depends heavily on the extent of the individual's experience. The speed and accuracy of decision making about movement are also influenced by such things as the number of possible options (or response choices) that exist, the costs associated with making incorrect decisions, and the total time available to make decisions. Some activities, such as snooker, offer essentially unlimited periods of

time in which to select the correct action, whereas in other motor skills, such as tennis, the time constraints on decision making are severe.

Processing Limitations

The measure that is used to determine how quickly people can make decisions is called choice **reaction time** (CRT). In the laboratory, CRT is measured by presenting participants with an array of stimuli (usually lights) each of which has its own associated response (frequently a button press). The participant's task is to view the stimulus array and respond, as soon as possible after a stimulus light is illuminated, by depressing the response button corresponding to the illuminated light. Choice reaction time is then recorded as the time elapsed between the illumination of the stimulus light and the initiation of the button press. Researchers can examine the limitations on rapid decision making by recording CRT as they systematically vary the number of possible stimulus–response pairs. (A task in which there are six possible stimulus–response pairs would form a reasonable analogue of the response choice facing the basketball guard in our example.)

Studies in which the number of stimulus–response alternatives has been varied are consistent in their findings. When there is no uncertainty (there is only one possible stimulus and response), the reaction time is about 200 ms (one-fifth of a second). This is also the delay in responding observed in sprint events between the sound of the gun and the commencement of movement, and in hand–eye coordination tasks when there is unexpected movement of the object that is to be intercepted. As the amount of uncertainty is increased through the addition of more and more possible options, CRT slows substantially (figure 15.5). Each time the number of possible stimulus–response alternatives doubles, CRT slows by a constant amount, such that the increase in CRT from a two-choice situation to a four-choice situation approximates the increase in CRT from a one-choice to a two-choice situation and from a four-choice to an eight-choice situation.

This relationship between CRT and number of stimulus–response alternatives is frequently exploited in a range of movement tasks. Designers of cars and machinery attempt to minimize the number of options on machine controls to reduce the decision-making time of users. Skilled sport players attempt to familiarize themselves with the preferred options and patterns of play of their

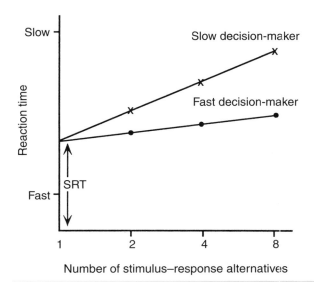

Figure 15.5 Reaction time shown as a function of the number of stimulus–response alternatives for two different people.

opponents as a means of speeding up their own rates of responding. Basketball players who recognize that an opponent can control the ball only with his or her right hand are able to respond more rapidly to the opponent's moves than players who think that both left- and right-hand alternatives may occur. Conversely, the player who is able to execute, with equal skill, a wide range of options (through having equal shooting, passing, and dribbling skills on both sides of the body) can maximize the amount of information that has to be processed by an opponent and thus substantially slow the speed of decision making of the opponent.

Close inspection of figure 15.5 reveals two independent components to decision making. One component, given by the intercept of the CRT–alternatives line with the Y-axis, corresponds to reaction time when there is only one option, and therefore no uncertainty. This component, known as simple reaction time (SRT), is not influenced by practice and reflects individual differences in the time it takes for the afferent nerve impulses to reach the brain and be registered there and for efferent commands to be sent to the muscles. The second component, given by the slope of the CRT–alternatives line, is a measure of decision-making rate; it estimates the average increase in CRT that occurs for each new additional stimulus–response option. The slope of the line is steep for people who are slow decision makers and approaches zero for individuals who are very fast decision makers. While

there is likely to be some genetic component to fast decision making, situation-specific practice and advance knowledge about the probabilities of different events occurring can reduce the amount of information to be processed and lead to faster decision making.

Acting: Organizing and Executing the Desired Movement

Having selected the desired action (e.g., a jump shot), the player must organize the movement before it can be initiated. This organization involves sending from the brain motor commands that specify the order and timing of motor unit recruitment. If these efferent commands are not appropriately structured, the resulting movement pattern may lack the force, timing, or coordination necessary to successfully fulfil the objective of the movement (in the basketball case in figure 15.3, the shooting of a goal). All three central processes of perception, response selection (decision making), and response organization and execution (acting) are completed before any observable muscular contraction takes place or whole-body movement occurs.

Feedback during the shooting movement may assist in adjusting the motor commands, although the skill of shooting is, in all probability, too short in duration for feedback-based corrections to have time to be effective. Visual feedback derived from the completed action does, however, provide a valuable source of information for the performer to assist in future repetitions of the same or a similar action. If the shot is too short, for example, this information can be used to ensure that more motor units are recruited the next time the player opts to take a shot from a similar position on the court. Comparison of visual information about the outcome of the movement with kinaesthetic information from the execution of the movement provides a valuable means of "calibrating" the force production system, enabling our basketball player to "find his or her range." The relationship between information about movement execution and that from movement outcome provides essential guidance for future attempts at movement skills of all types, as illustrated in figure 15.6.

Underlying Processes

Once a particular movement response has been selected, the central system still has significant responsibility for ensuring that the selected movement response is actually executed as desired. At least three further subprocesses are involved at this

Type of evaluation	Was the movement executed as planned?		
	Outcome	Yes	No
Was the goal accomplished?	Yes	Got the idea of the movement	Surprise!
	No	Something's wrong	Everything's wrong

Figure 15.6 The relationship between information about movement execution and information about movement outcome as a basis for guiding future attempts at a skill.
Reprinted from Gentile 1972.

stage in the processing of information for movement control:

- *Movement organization* (planning out carefully the sequencing and timing of the efferent commands to be sent out to selected motor units)
- *Movement initiation* (transmitting the required motor commands to the muscles)
- *Movement monitoring* (adjusting the movement commands on the basis of sensory information about the movement's progress)

Processing Limitations

The speed and accuracy with which movements can be executed and controlled depend on a number of factors, including the complexity of the movement (the number of joints, muscles, and motor units involved plus the difficulty that their coordination may pose for the maintenance of posture and balance), the time constraints imposed on the movement, and the acceptable margins for error in the movement. For movements of relatively long duration (greater than one-third of a second), feedback generated during the movement itself can assist with control and precision. Control based on the monitoring of feedback is known as **closed-loop control.** In contrast, very rapid movements require all the efferent commands to be structured in advance. This type of control is known as **open-loop control** and is thought to involve the use of **motor programs.**

For movements controlled in a closed-loop manner, the time taken to complete the movement **(movement time)** is directly dependent on the difficulty of the movement. Movements with high precision demands (such as movements to small targets) and those traversing a large distance take much longer than movements to large targets over a short distance. The relationship between movement time and movement difficulty is governed very systematically by the amount of information that must be processed (figure 15.7). For movements controlled in an open-loop manner, the time taken to initiate the movement is directly proportional to the amount of preplanning that must take place. This is greater for more complex movements.

Some Implications

Now that we have examined the central processing stages proposed by an information-processing model of motor control, we can consider some of the implications and insights such a model may provide.

One insight that emerges from consideration of the central processing stages relates to the sources of errors in motor skill performance (see "When Things Go Wrong," p. 234). If our hypothetical basketball guard attempted to shoot a jump shot but succeeded only in losing the ball or producing an inaccurate shot, it would be tempting to simply attribute the error to poor movement execution technique and to try to remedy this by having the player practice his shooting skills. However, it should be apparent from the preceding discussion that an error in performance could result from poor perception or poor decision making as much as from poor movement execution. Given the sequential nature of central information processing, a perceptual error (such as incorrect judgment of the distance from the basket or failure to detect the movement of one of the defensive players) or a decision-making error (such as shooting from a position where passing would have been a better option) can cause an ineffective shot as much as poor execution of a correctly selected movement. Likewise, a therapist attempting to recover the normal gait of a stroke patient or assist the skills development of a clumsy child needs to be aware that deficiencies in movement control can occur even when the ability to actually execute a selected movement is essentially normal.

A second, related implication of the information-processing model is that although the three central stages of information processing are not directly observable they may nevertheless need to be trained just as much as the more observable

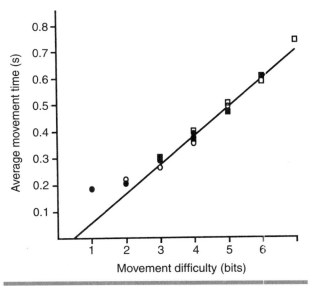

Figure 15.7 Average time to make a movement plotted as a function of the difficulty of the movement. The task involved moving as rapidly as possible for a 10-s period between two targets. Movement difficulty was manipulated by changing the distance between the targets and the size of the targets.
Data from Fitts 1954.

components of performance such as technique, strength, speed, endurance, and agility. In many motor skills, the perceptual and decision-making aspects of movement control may act as the limiting factors to performance and therefore warrant systematic training. This is especially true of activities such as the fast ball sports, in which decisions must be made in a very short time on the basis of only limited information. Conditions such as aging, injury, or disease that slow the speed with which a person can produce movements also increase the necessity for good perceptual and decision-making skills to offset delays imposed by longer movement times.

Some Alternative Models of Motor Control

Although modelling the motor control system as an information-processing system appears to be a useful way of starting to think about the neural control of movement, it is certainly not the only way. Over the past 20 years, a number of limitations and assumptions of the information-processing model have been highlighted and some alternative models suggested. Critics of the information-processing model have been concerned about the assumption that every movement is somehow represented and stored in the central nervous system. They have argued that simply assigning the responsibility of movement organization to the brain does not explain control but simply creates the (unacceptable) need for an intelligent little person (a "homunculus") somewhere else within the brain! Critics have also been concerned about the assumption that the brain and nervous system directly control the specific motor commands for all aspects of a movement, noting that many of the physical properties of the musculoskeletal system (such as the springlike characteristics of muscle and the natural oscillatory frequencies of limbs) can themselves contribute significantly to movement control without the necessity of any nervous system involvement.

Dynamical models of movement control propose that movement patterns are not represented anywhere in the nervous system by way of a plan or program but rather emerge naturally (or self-organize) out of the physical properties of the musculoskeletal and related systems. In this view, the emergence of complex patterns of organization

IN FOCUS: WHEN THINGS GO WRONG

In a system as complex as the human motor system, the chance of errors occurring is always high. Many errors in movement control (such as typing the wrong letter on a keyboard or brushing against a door frame as you walk through the door) have trivial consequences. However, in other cases, errors in movement control (such as applying too much force to a scalpel if you are a surgeon, or applying foot pedal pressure to the accelerator rather than the brake if you are a bus driver) can have catastrophic consequences. As human error rather than machine error is by far the most common cause of catastrophic accidents in the workplace, ergonomists (scientists who study people in the workplace) are constantly looking for ways to minimize the incidence of errors and to make tasks safer.

The first step toward eliminating errors is to understand how they come about. Understanding the sources of errors can also help provide insight into how movement control is normally achieved. Errors arise from multiple sources, and several classification systems for errors have been developed. Errors in the execution of movement appear to occur most frequently in situations where the ongoing control of movement is not monitored closely or continuously enough (using closed-loop control). Under these circumstances, unconscious, automatic control of movement takes over. If these automated movements are not the appropriate ones, errors inevitably occur. The driver of a car who finds him- or herself traveling on a familiar route (such as the route home) rather than on the planned route to another destination has committed an error of this type.

Ergonomists attempt to design and redesign tasks so as to minimize either the potential for processing errors or the consequences of processing errors. One important principle is to ensure that the links between task information and the required movement responses (what is known as **stimulus–response compatibility**) are made as natural as possible. Displays in which the required outcome and action are compatible (figure 15.8) allow for both faster reaction times and fewer errors than incompatible displays.

SOURCES

- Proctor, R.W., and T.V. Van Zandt. 1994. *Human factors in simple and complex systems.* Boston: Allyn & Bacon.
- Reason, J. 1990. *Human error.* Cambridge, UK: Cambridge University Press.
- Schmidt, R.A. 1989. Unintended acceleration: A review of human factors contributions. *Human Factors* 31: 345-364.

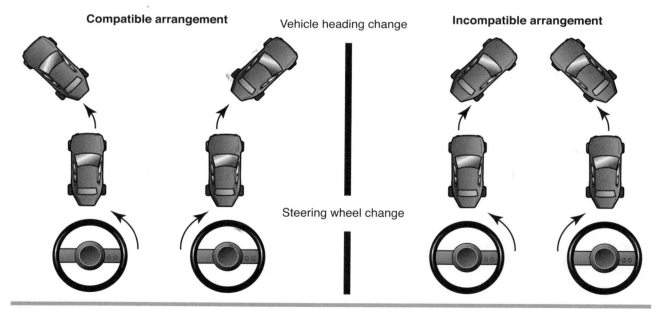

Figure 15.8 Displays of different levels of stimulus–response compatibility. In the compatible display the movement of the steering wheel and the resultant movement of the vehicle are in the same direction, whereas in the incompatible display they are reversed.

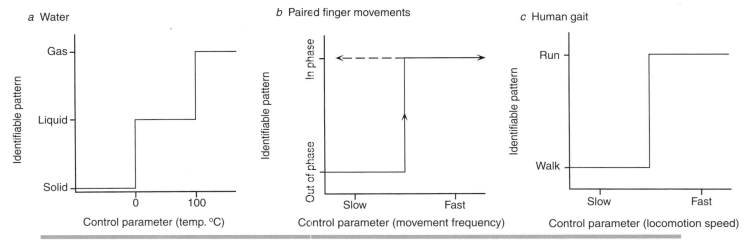

Figure 15.9 Parallels between the pattern transitions observed in nonbiological systems and those occurring in some human movements.

out of the motor system, such as the patterns that characterize gaits like walking and running, is no more in need of a pattern representation than are nonbiological systems, such as chemicals—which, under appropriate environmental conditions, are able to organize and reorganize into complex patterns without the need for a nervous system. For example, under appropriate environmental conditions (level of heat), water may undergo complete pattern reorganizations without the form of the new pattern being explicitly stored or represented anywhere. As some aspects of movement control share these same characteristics of pattern reorganization at critical levels of **control parameters** (see

figure 15.9), there is considerable current research interest in attempting to explore this model of movement control further by determining the control parameters for a range of human movements and exploring other parallels between movement pattern formation and pattern formation in physical systems (see "Pattern Transitions in Human Hand Movements" and "Pattern Transitions in Human Gait" on p. 236). As more is understood experimentally about movement control, new models that progressively encapsulate more of the many essential and unique characteristics of movement listed earlier in this chapter can and will be developed.

IN FOCUS: PATTERN TRANSITIONS IN HUMAN HAND MOVEMENTS

In purely physical systems (such as chemical and laser systems), spontaneous transitions from one form of organization or pattern to another occur when key environmental conditions (or thresholds) are reached. For example, we all know that the water molecule spontaneously reorganizes its structure as it is heated past the critical control temperatures of 0° and 100° C (32° and 212° F) (figure 15.9). The study of such transitions is called **synergetics.** Synergetic transitions share a number of common characteristics including sudden, discrete changes in organization around critical levels of the control parameter, increased variability in structure as the transition

point is approached (a property called **critical fluctuations**), and a delay in returning to stability if the system is perturbed in some way when it is near a transition point (a property called **critical slowing down**).

In an extensive series of studies, Scott Kelso from Florida Atlantic University in the United States, Herman Haken from the University of Stuttgart, Germany, and a number of coworkers set out to determine whether transitions in movement patterns also share these synergetic characteristics. They used as their task paired movements of the index fingers with the index fingers prepared either anti-phase or in-phase (figure 15.10). When

(continued)

the fingers were commenced anti-phase and the participants were required to progressively move the fingers at a faster rate, a critical frequency was reached at which the fingers moved spontaneously from anti-phase to in-phase. The phase relations between the two fingers were found to increase in variability and to be slower to respond to perturbations as the transition point (a point of pattern instability) was approached. Thus the critical fluctuations and critical slowing-down properties seen in other synergetic systems also exist in some aspects of the human motor system. The in-phase coordination was found to be a stable one that could be preserved across all finger movement frequencies.

These experimental data have made possible the formulation of suitable mathematical models to explain the synergetic phenomena in this task. This model has been the subject of extensive experimentation over the past decade across a range of multijoint coordinative tasks.

SOURCES

- Haken, M., J.A.S. Kelso, and H. Bunz. 1985. A theoretical model of phase transitions in human hand movements. *Biological Cybernetics* 51: 347-357.

- Kelso, J.A.S. 1995. *Dynamic patterns: The self-organization of brain and behavior.* Cambridge, MA: MIT Press.

- Kelso, J.A.S., and G. Schöner. 1988. Self-organization of coordinative movement patterns. *Human Movement Science* 7: 27-46.

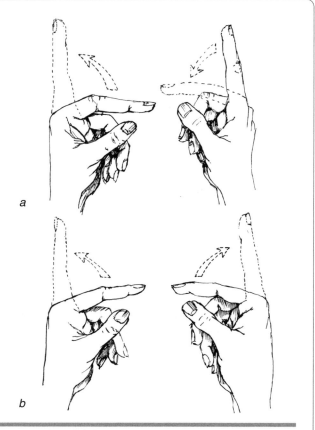

Figure 15.10 When the rate of paired finger movement is progressively increased, a spontaneous shift from *(a)* anti-phase (one finger extended while the other is flexed) to *(b)* in-phase occurs.
Reprinted from Haken, Kelso, and Bunz 1985.

IN FOCUS: PATTERN TRANSITIONS IN HUMAN GAIT

The studies of phase transitions in coupled hand movements (figure 15.10) demonstrate the potential value of examining the transitions between different stable patterns of movement as a basis for understanding the underlying control processes. One common movement pattern transition that interests many motor control researchers is the transition in human gait between walking and running that occurs when the speed of travel exceeds a threshold of some 2 m/s (~7 km/h or 4.3 miles/h). This transition occurs apparently quite automatically (without conscious attention), but the question of interest is what triggers the neuromuscular system to make this pattern reorganization.

Classical studies of walking, trotting, and galloping in horses by the biologists Hoyt and Taylor in the early 1980s suggested that physiological efficiency was the key controller, with the animals favouring speeds of travel that are optimal for minimizing their required energy expenditure

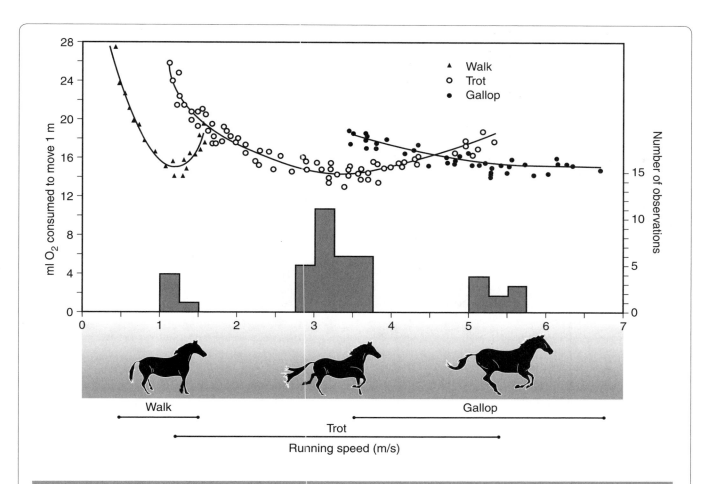

Figure 15.11 Efficiency (measured in terms of oxygen consumption for each metre of movement) for horses walking, trotting, and galloping at different speeds. The histograms at the bottom depict the frequency with which each running speed is naturally used for each gait mode. Gait modes are naturally selected that minimize oxygen consumption.

Reprinted from Hoyt and Taylor 1981.

(figure 15.11). In humans, and in other animals, the minimization of energetic costs is one of a number of potential control variables. Other possibilities include mechanical limits and the need to limit peak muscle forces and loadings on bone in order to avoid injury. Studies by Frederick Diedrich and Bill Warren from Brown University, based on a theory of dynamical systems, have demonstrated that the relative phase of the leg segments displays the same synergetic characteristics in gait as the relative phase of the fingers in bimanual tasks (see "Pattern Transitions in Human Hand Movements," p. 235). Research on movement pattern transitions is providing a valuable platform for the development of alternative theories of motor control.

SOURCES

- Dickinson, M.H., C.T. Farley, R.J. Full, M.A. Koehl, R. Kram, and S. Lehman. 2000. How animals move: An integrative view. *Science* 288(5463): 100-106.

- Diedrich, F., and W.H. Warren Jr. 1995. Why change gaits? Dynamics of walk-run transition. *Journal of Experimental Psychology: Human Perception and Performance* 21: 183-202.

- Hanna, A., B. Abernethy, R.J. Neal, and R.J. Burgess-Limerick. 2000. Triggers for the transition between human walking and running. In *Energetics of human activity,* ed. W.A. Sparrow (pp. 124-164). Champaign, IL: Human Kinetics.

Summary

Cognitive science has provided a number of conceptual models of how the brain and nervous system controls movement. The dominant model has been one that conceptualises the brain and nervous system as an information-processing/computational system. A major interest has consequently been to attempt to understand the processing stages and the processing capacities and limitations of the human computational system. The sequential processes of perceiving, deciding, and acting all have demonstrable processing limitations, whose quantification helps cognitive scientists understand some of the key characteristics of, and constraints to, human motor performance. Although there are parallels between human information processing and the functions of computers, it does not necessarily follow that the human brain and nervous system *is* a computational system. Alternative models of movement control that are more concerned with the dynamics of the human neuromuscular system are gaining increasing popularity as an explanation of how control is achieved over at least some aspects and types of movement.

Further Reading and References

Crutchfield, C.A., and M.R. Barnes. 1993. *Motor control and motor learning in rehabilitation.* Atlanta: Stokesville.

Kelso, J.A.S. 1995. *Dynamic patterns: The self-organization of brain and behavior.* Cambridge, MA: MIT Press.

Magill, R.A. 1998. *Motor learning: Concepts and applications.* 5th ed. Boston: McGraw-Hill.

Proctor, R.W., and T.V. Van Zandt. 1994. *Human factors in simple and complex systems.* Boston: Allyn & Bacon.

Schmidt, R.A., and T.D. Lee. 1999. *Motor control and learning: A behavioral emphasis.* 3rd ed. Champaign, IL: Human Kinetics.

Schmidt, R.A., and C. Wrisberg. 2000. *Motor learning and performance.* 2nd ed. Champaign, IL: Human Kinetics.

CHAPTER 16

Motor Control Changes Across the Life Span

The major learning concepts in this chapter relate to

- the development of movement in the first two years of life and the development of fundamental motor patterns in childhood;
- the motor performance of people who are elderly;
- the major physical changes in the central nervous system, changes in the sensory receptors and sensory systems, changes in the effectors (muscles), and changes in reflex systems that affect neural control of movement across the life span; and
- developmental improvements in information-processing capability and declines in information processing with aging.

*I*n chapters 14 and 15 we examined some basic concepts of motor control from neurophysiology and cognitive science. In this chapter we use these basic concepts to examine motor control changes across the life span. The field of study concerned with the description and explanation of changes in motor performance and motor control across the life span is typically called **motor development.** While the study of motor development has historically focused on the period from conception through to adolescence, and the changes and stages through which the developing human progresses in attaining adult levels of motor performance, researchers in this field are now also increasingly interested in the deterioration of motor skills apparent in the elderly population. Such a broadened focus is important given aging populations worldwide and given that many changes in the performance and control of motor skills are age related and occur throughout the entire life span.

Studies of motor development have generated knowledge about many topics, including:

- The normal rate and sequence of development of fundamental motor skills that arise out of the interaction of biological maturity and environmental stimulation
- Individual differences in the rate of development of specific skills (and motor performance comparisons of early and late developers)
- Deviations from normal development in special populations (such as those who are intellectually challenged and those with conditions such as cerebral **palsy,** Down syndrome, or clumsiness)

Such knowledge is important practically in

- assessing the normative development of children,
- screening for neurological and motor disorders as well as identifying and nurturing exceptional talent,
- informing the designers of remedial and therapeutic programs, and
- ascertaining the readiness of individual learners for new challenges.

For both young and elderly persons, knowledge from the field of motor development is valuable in highlighting the role of regular practice and physical activity in the acquisition and retention of movement skills.

Motor development knowledge is typically derived from research studies of one of two main types: **cross-sectional studies,** in which motor control and performance are compared between different people of different ages, and **longitudinal studies,** in which the motor control and performance of the same set of people are traced over a number of years as they mature and grow older. The changes in motor control and performance that occur across the life span may be described, recorded, and explained at a number of different levels of observation. In this chapter we first examine observable changes in motor skills across the life span and then, in keeping with the previous two chapters, examine underlying changes taking place at the neurophysiological level and at the level of information-processing capabilities.

Changes in Observable Motor Performance

One way—the simplest way—of beginning to understand how motor control changes across the life span is to describe and measure motor performance at different ages and to determine the specific ages at which key movement skills are first mastered.

Motor Development in the First Two Years of Life

The first 24 months of life are especially critical for maturation of the neuromuscular system and for the emergence of the basic control skills on which fundamental movement patterns are based. Motor development during this period follows some general principles; and emergence of basic control of posture, locomotion, and movements of the hands can be used as a check on normative development.

General Developmental Principles

Significant advances in voluntary motor control occur over the first two years of life; these advances are generally thought to follow two main principles. The **cephalo-caudal principle** states that development proceeds in a "head-down" manner with control developing first in the muscles of the neck, followed, in order, by control of the muscles of the arms, trunk, and legs. (Although this principle is generally accepted, some recent challenges have come from dynamical theories of movement control. Some evidence suggests that coordinated kicking movements that are dynamical precursors to walking are present before head and neck control is achieved.) The **proximo-distal principle** states that development progresses from the axis of the body outward, with control being achieved over muscles crossing the trunk (and axial skeleton) before control is achieved over the muscles controlling the limbs (and appendicular skeleton).

These cephalo-caudal and proximo-distal changes in motor control are paralleled by changes in **muscle tone.** At birth, muscles controlling the movement of the axial skeleton typically have low tone while those controlling the movements of the limbs have high tone (or stiffness). As the infant matures, this situation reverses, with the axial muscles gaining tone (to assist in maintaining posture against gravity) and the muscles of the limbs decreasing in tone (to facilitate more efficient, voluntary control of the limbs). Reflexes present at or soon after birth either disappear, are reduced in strength (attenuated), or are modified during the first 24 months of life.

Motor Milestones for Normative Development

In order to eventually achieve movement independence, it is vital in the first years of life that human infants develop control of their general body position (or posture); develop an ability to move (locomote) throughout their environment; and develop an ability to reach, grasp, and manipulate objects using their hands. Because posture, locomotion, and manual control are so vital to the infant's development, a number of researchers have studied—and documented in great detail—the ages and stages that infants pass through in the acquisition of these essential motor skills. Knowing the average and range of chronological ages at which key developmental stages are typically reached in these skills provides a set of **motor milestones** that one can use to monitor the motor development of each child and to detect, at an early age, possible movement and neurological difficulties.

The development of the control of posture occurs through three main stages (head control, sitting, and standing), each containing a number of identifiable progressions as shown in figure 16.1.

A number of the major motor milestones for the achievement of the locomotion skill of walking occur in parallel with these postural control developments (figure 16.2). Upright, unassisted walking appears, on average, at the end of the first year, although walking may first appear anywhere in the range from 9 to 17 months. A general expectation is that most babies will achieve the basics of walking during the first 6 months of their second year and will then have refined this skill considerably by the end of their second year.

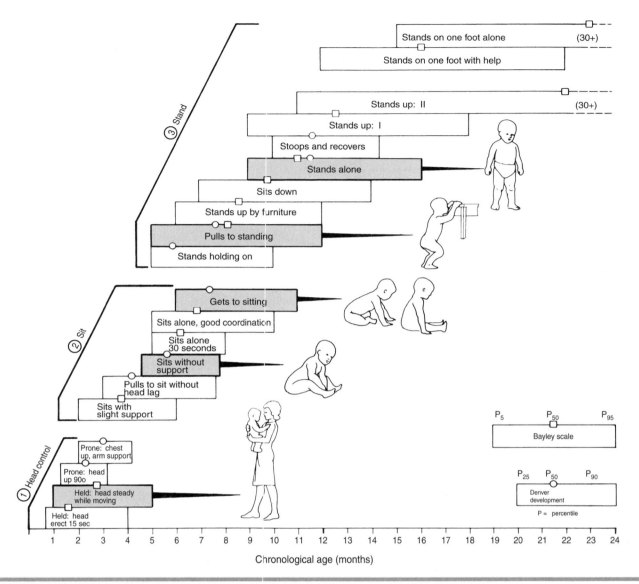

Figure 16.1 Identifiable progressions in the development of postural control.
Reprinted from Keogh and Sugden 1985.

The rudiments of mature grasping and reaching control of the hands are in place typically by the end of the first year of a child's life, although the functional use of the hands for self-help activities (such as using a spoon) and more complex manual acts (such as tying shoelaces or writing) continues to be refined for many years (figure 16.3). Key milestones in the control of grasping and object manipulation by the hands occur, on average, at 3 to 4 months (when large objects such as cubes are first picked up), at 6 to 7 months (when opposition of the thumb is achieved), and at 9 to 10 months (when the pincer grasp for picking up small objects is first mastered). Reaching skills, controlled more by the proximal muscles of the shoulder and the

arm than by the more distal muscles of the wrist and the intrinsic muscles of the hand, are achieved somewhat earlier than manipulation skills. The major motor milestone of reaching to touch a desired object is attained, on average, at some 3 to 4 months of age. Control of basic reaching and grasping lays the foundation for the acquisition of the many complex manual movements fundamental to uniquely human actions, gestures, and communications.

Motor Milestones in Special Populations
Some special populations who experience movement difficulties as adults are also typically slow in reaching a number of the major milestones for

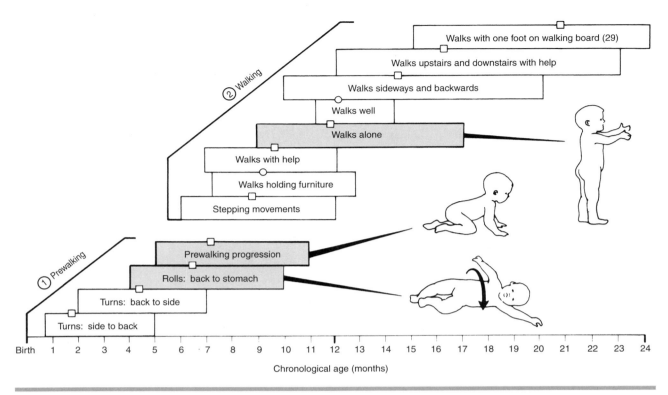

Figure 16.2 Identifiable progressions in the development of locomotion.
Reprinted from Keogh and Sugden 1985.

motor development. Children with **Down syndrome** (a chromosomal abnormality that affects many aspects of development) are systematically late in achieving virtually all major motor milestones. For example, children with Down syndrome walk unsupported on average only after 2 years of age. Mental retardation of all forms also acts to delay the attainment of motor milestones; children with the lowest intelligence quotients (IQs) are the least likely to stand or walk within the first 12 months (figure 16.4). Motor milestones may therefore act as early markers of potential adult movement problems, as well as possible neurological problems, and thus serve as valuable diagnostic and detection tools. Children described by parents and teachers as clumsy but lacking any apparent neurological problems may also be identified early through assessment of motor milestones and offered appropriate additional training and intervention.

The Notion of Critical Periods

The use and identification of motor milestones are based on the observation that the majority of children pass through the same basic stages in achieving mastery over those motor skills (such as posture, locomotion, and reaching and grasping) that are fundamental to survival. (There are, however, notable exceptions. For example, it is not uncommon for children to walk without having first learned to crawl.) The basic similarity in stages across children suggests a significant inherited or genetic influence on the emergence of the motor skills essential for survival. Obvious questions therefore are whether these normal progressions can be altered by environmental factors (e.g., can an enriched environment accelerate the rate of motor development or an impoverished one retard it?) and whether there are **critical periods** in which the child is most sensitive to learning a particular skill. Other aspects of human development show clear evidence of critical periods. Language, for example, is most easily acquired in childhood, and if it is not acquired at this time it is very difficult to pick up subsequently.

The evidence for critical periods in motor development is not clear-cut. Some studies of children from cultures that restrain their infants from movement during part of the first year of life have indicated lower than expected levels of adult motor skill. Other such studies have demonstrated no detrimental effects and have indicated that after

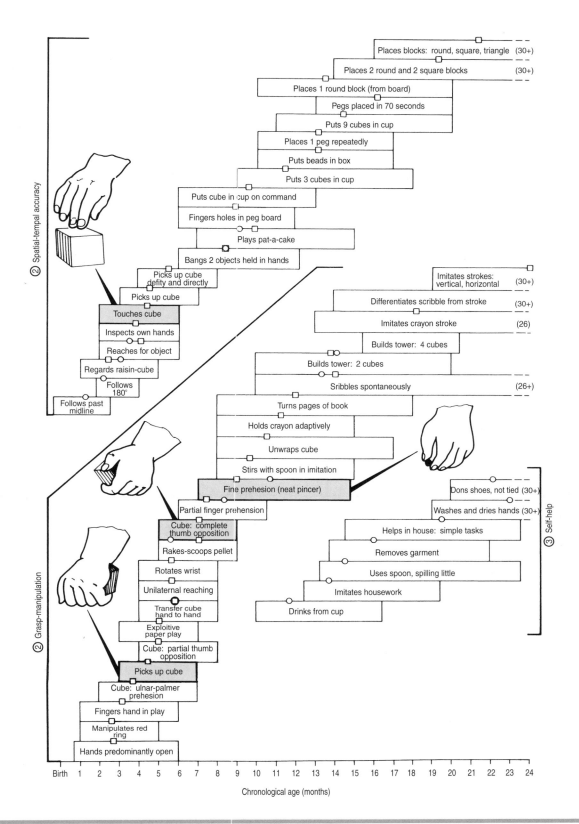

Figure 16.3 Identifiable progressions in the development of reaching (upper panel) and grasping (lower panel).

Reprinted from Keogh and Sugden 1985.

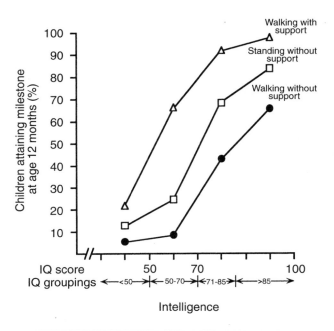

Figure 16.4 The relationship between intelligence and the attainment of specific motor milestones by the end of the first year of life. An IQ of 100 corresponds to the population mean.

Data from Von Wendt, Makinen, and Rantakallio 1984.

an early delay, motor milestone achievement is quickly recovered in later years. There is clear evidence, however, that enriched environments, in which stimulation is extensive and opportunities for motor exploration and play are numerous, can speed up the rate at which motor milestones are achieved. What is less apparent is whether or not the early achievement of milestones translates into improved adult performance levels. Children who are relatively late in achieving some milestones may attain other skills rapidly, provided that they are exposed to them during a critical period (see "Critical Periods and Enriched Environments: The Story of Johnny and Jimmy"). Programs to enhance early motor development (such as educational gymnastics, infant swimming programs, and the Suzuki method for learning musical instruments) are all based on the assumption that early attainment of milestones is beneficial to longer-term skill acquisition.

Practical Applications

Motor milestone scales (such as those presented in figures 16.1 through 16.3) provide a useful means of comparing the rate of motor development of any child against that of children of the same chronological age. As the ranges of ages

IN FOCUS: CRITICAL PERIODS AND ENRICHED ENVIRONMENTS: THE STORY OF JOHNNY AND JIMMY

In 1935 the American developmental psychologist Myrtle McGraw conducted one of the classical studies on motor development, examining the role of enriched environments and experiences on the rate of acquisition of a range of movement skills. To separate the role of environment from any hereditary factors that may influence motor development, McGraw studied intensively the early development of two male twins—Johnny and Jimmy—who were raised in different environments. Johnny was given exposure to free play with a wide variety of toys as well as extensive stimulation, practice, and experience in a wide variety of movement activities. Jimmy was given few toys and spent the majority of the time in his cot. The twins were given different movement experience conditions from 21 days to 22 months of age, and their capacity to acquire and retain

various motor skills was observed and described throughout this period and at a number of times thereafter (up to age 6 years).

McGraw found, among other things, that critical periods existed for the learning of certain skills. Johnny, for example, had considerable practice and instruction in riding a tricycle at age 11 months but showed no signs of any acquisition of the skill of tricycle riding until around 19 months of age. Jimmy, in contrast, was first exposed to a tricycle at age 22 months and learned the skill almost immediately. This suggests that a certain amount of maturational readiness is necessary before some skills can be learned.

Although in the case of the voluntary skill of tricycling, Johnny's early movement experiences were of no apparent advantage; on many other motor tasks Johnny not only acquired the skills earlier than Jimmy but also showed this superior performance for a number of years after the end of the enrichment period. McGraw reported that overall, Johnny appeared to have superior movement competence and confidence, demonstrating the important role that early movement experiences may have not only on the acquisition and retention of a number of motor skills but also on psychological attributes such as self-confidence (see chapter 18). Following 22 months of minimal motor activity Jimmy was able to "catch up" to Johnny in performance on some, but certainly not all, motor skills.

Given that enriched environments are now known to be important for motor development, studies that deliberately deprive a child of exposure to such an environment would, in current times, not be permitted to proceed on ethical grounds.

SOURCE

- McGraw, M.B. 1935. *Growth: A study of Johnny and Jimmy*. New York: Appleton-Century. (Reprinted by Arno Press, 1995)

on each milestone in the figures indicate, there is enormous normal variability on all these measures, and a child advanced on one milestone may be below the average on another. The normal range of variability also tends to increase as the child gets older and the task more difficult, probably reflecting the cumulative effect of environmental rather than genetic influences. The observation that environment can influence motor development is important as it suggests, among other things, that early intervention for children with movement difficulties offers the best option for effective development. Motor milestones may therefore, within the recognition of the scope of normal variability, be used validly by teachers, therapists, and parents as a screening tool for early detection of possible motor and neurological problems. Further, creating an environment that provides maximal opportunities for movement in the early years would appear to be a concern that should be treated seriously by all parents.

Development of Fundamental Motor Patterns in Childhood

From age 2 onward, children develop a range of fundamental motor skills that form the basis for specialized adult motor behaviour. As with the postural, locomotor, and manual control skills from the first two years of life, substantial information is available on the ages and stages at which children from 2 through to 7 years of age and older acquire basic locomotor skills (walking, running, jumping, and hopping) as well as nonlocomotor skills (such as throwing, catching, hitting, and kicking). In most cases information is available to outline developmental trends not only in performance (product measures) but also in the movement patterns used (process measures) in performance of these skills. In this section we provide two examples of developmental stages in fundamental motor patterns—one from the locomotor skill of running and the other from the nonlocomotor skill of overarm throwing.

Ages and Stages in the Development of a Locomotor Skill

A number of children begin to run at around age 18 months, and most run by the time they are 2 years old. The running pattern of a child shows a number of developmental stages before becoming mature, on average, by age 4 to 6. Initial running is characterized by short, uneven steps, a wide base of support, and no easily observable flight phase (the phase during which neither foot is in contact with the ground). This first stage of running usually occurs before adult mastery of walking is achieved. In mature running, strides are uniformly long and there is an identifiable flight phase characterized by a full recovery of the nonsupport leg (figure 16.5, and "Characteristics of Stages in the Development of Running" on p. 246). The early attainment of a mature running pattern is important given the central role that running plays in many childhood games and activities. As noted in chapter 8, running speed typically continues to improve throughout not only childhood but also adolescence, as strength is gained and subtle refinements in technique produce improvements in efficiency.

Ages and Stages in the Development of a Nonlocomotor Skill

The overarm throwing pattern is a discrete action used to propel objects and implements, such as balls and javelins. It is a basic movement pattern

Initial

Elementary

Mature

Figure 16.5 Stages in the development of a mature running pattern. (See "Characteristics of Stages in the Development of Running" below for greater detail on the characteristics of each stage.) Reprinted from Gallahue 1989.

fundamental to many other more specific sport skills such as throwing a baseball or cricket ball, serving a tennis ball, spiking a volleyball, or passing a water polo ball. Overarm throwing performance (regardless of whether the throwing is for distance, accuracy, or form) continues to improve throughout childhood into adolescence. Even though rudimentary throwing patterns are in place certainly by the second year of life, throwing velocity continues to improve steadily throughout the primary school years (figure 16.6).

As with running and all other fundamental motor skills, the transition from initial throwing form to mature throwing patterns involves a number of stages. Initial attempts at throwing are characterized by an absence of backswing, weight

CHARACTERISTICS OF STAGES IN THE DEVELOPMENT OF RUNNING

Initial Stage

1. Short, limited leg swing.
2. Stiff, uneven stride.
3. No observable flight phase.
4. Incomplete extension of support leg.
5. Stiff, short swing with varying degrees of elbow flexion.
6. Arms tend to swing outward horizontally.
7. Swinging leg rotates outward from hip.

8. Swinging foot toes outward.

9. Wide base of support.

Elementary Stage

1. Increase in length of stride, arm swing, and speed.

2. Limited but observable flight phase.

3. More complete extension of support leg at takeoff.

4. Arm swing increases.

5. Horizontal arm swing reduced on backswing.

6. Swinging foot crosses midline at height of recovery to rear.

Mature Stage

1. Stride length at maximum; stride speed fast.

2. Definite flight phase.

3. Complete extension of support leg.

4. Recovery thigh parallel to ground.

5. Arms swing vertically in opposition to legs.

6. Arms bent at approximate right angles.

7. Minimal rotary action of recovery leg and foot.

Common Problems

1. Inhibited or exaggerated arm swing.

2. Arms crossing the midline of the body.

3. Improper foot placement.

4. Exaggerated forward trunk lean.

5. Arms flopping at the sides or held out for balance.

6. Twisting of the trunk.

7. Poor rhythmical action.

8. Landing flat-footed.

9. Flipping the foot or lower leg in or out.

Reprinted from Gallahue 1989.

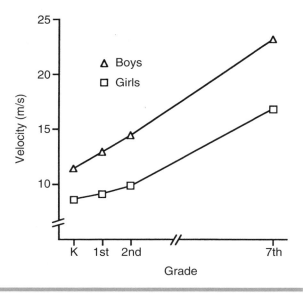

Figure 16.6 Improvements in overarm throwing velocity with age.

Data from Halverson, Roberton, and Langdorfer 1982.

transference, and upper and especially lower body involvement. Intermediate levels show increased weight transference, trunk rotation, and the involvement of a forward step, albeit from the leg ipsilateral (on the same side) to the throwing arm. Only at the mature stage does the nonthrowing hand lead toward the target, the trunk rotate fully, and the lower body contribute to the action through a forward step of the leg contralateral (opposite) to the throwing arm (figure 16.7 and "Characteristics of Stages in the Development of Overarm Throwing," p. 249). By age 12 most boys have acquired the mature throwing pattern whereas only a minority of girls have. In overarm throwing, in contrast to any of the other fundamental motor skills that have been examined, there appear to be differences in skill performance between boys and girls that cannot be accounted for simply by higher amounts of practice among boys.

Practical Applications

The information available on the ages and stages in the development of fundamental motor skills has potentially great value to teachers of motor skills, especially at the primary school level, for at least three reasons:

• First, such information may provide a reasonably objective method for monitoring the motor development of individual children and of detecting any potential movement problems (e.g., most test batteries for movement clumsiness either are based on or include fundamental motor skill items).

• Second, the information may provide the teacher with a guide as to forthcoming progressions in movement sequence development and therefore may provide a basis for accelerating

Initial

Elementary

Mature

Figure 16.7 Stages in the development of a mature overarm throwing pattern. (See "Characteristics of Stages in the Development of Overarm Throwing" on p. 249 for greater detail on the characteristics of each stage.)
Reprinted from Gallahue 1989.

acquisition of specific motor skills. For example, given the information provided in figure 16.7 and "Characteristics of Stages in the Development of Overarm Throwing" (p. 249), it may be wise to emphasize an instruction such as "point at the target" to facilitate the transition from the intermediate to the mature stage of throwing. Equally, an instruction such as "step toward the target with your left leg" may be inappropriate for someone at the initial stage of the throwing skill, since an intermediate stage involving stepping with the other (right) leg is typically involved as

a transition from the initial to the mature pattern of throwing.

• Third, assessment of fundamental motor skills can provide an indication of the readiness of children for involvement in more structured activities such as sports in which performance is based around proficiency in one or more fundamental motor skills. There is growing concern in many countries about declining fundamental motor skill proficiency among school-aged children (see "Fundamental Motor Skill Proficiency in Australian Children," p. 250).

CHARACTERISTICS OF STAGES IN THE DEVELOPMENT OF OVERARM THROWING

Initial Stage

1. Action is mainly from elbow.

2. Elbow of throwing arm remains in front of body; action resembles a push.

3. Fingers spread at release.

4. Follow-through is forward and downward.

5. Trunk remains perpendicular to target.

6. Little rotary action during throw.

7. Body weight shifts slightly rearward to maintain balance.

8. Feet remain stationary.

9. There is often purposeless shifting of feet during preparation for throw.

Elementary Stage

1. In preparation, arm is swung upward, sideward, and backward to a position of elbow flexion.

2. Ball is held behind head.

3. Arm is swung forward, high over shoulder.

4. Trunk rotates toward throwing side during preparatory action.

5. Shoulders rotate toward throwing side.

6. Trunk flexes forward with forward motion of arm.

7. Definite forward shift of body weight.

8. Steps forward with leg on same side as throwing arm.

Mature Stage

1. Arm is swung backward in preparation.

2. Opposite elbow is raised for balance as a preparatory action in the throwing arm.

3. Throwing elbow moves forward horizontally as it extends.

4. Forearm rotates and thumb points downward.

5. Trunk markedly rotates to throwing side during preparatory action.

6. Throwing shoulder drops slightly.

7. Definite rotation through hips, legs, spine, and shoulders during throw.

8. Weight during preparatory movement is on rear foot.

9. As weight is shifted, there is a step with opposite foot.

Common Problems

1. Forward movement of foot on same side as throwing arm.

2. Inhibited backswing.

3. Failure to rotate hips as throwing arm is brought forward.

4. Failure to step out on leg opposite the throwing arm.

5. Poor rhythmical coordination of arm movement with body movement.

6. Inability to release ball at desired trajectory.

7. Loss of balance while throwing.

8. Upward rotation of arm.

Reprinted from Gallahue 1989.

IN FOCUS: FUNDAMENTAL MOTOR SKILL PROFICIENCY IN AUSTRALIAN CHILDREN

A recent study (Booth et al. 1999) assessed the fundamental motor skills of a sample of 5,518 schoolchildren in the Australian state of New South Wales. The children were drawn from four grades (grades 4, 6, 8, and 10) and from some 90 schools varying in location, size, and structure. The average ages of the children tested in each grade were 9.3, 11.3, 13.3, and 15.3 years, respectively. Each child performed six motor skills (run, vertical jump, catch, overarm throw, forehand strike, and instep kick) that are fundamental to performance in a wide range of physical and sport activities. Each child was assessed on each skill to determine whether a mature motor pattern had been achieved and whether the child had mastery or near mastery of the skill.

The pattern of findings was similar for all six skills. As illustrated for the run and the overarm throw (figure 16.8), the percentage of children who had achieved mastery of the fundamental movements by each grade level was surprisingly low. At the age when involvement in organized sports like baseball, softball, cricket, and volleyball is likely to first take place, less than half of the boys and less than a quarter of the girls had mastery over the fundamental overarm throwing skill upon which these sports are based. It is interesting in this context to note the high numbers of young children who drop out of organized sport because they do not perform basic skills well enough to experience success (see also chapter 19). Proficiency in fundamental motor skills is significantly related to, though it predicts only a small portion of, participation of adolescents in organized physical activity (Okely, Booth, and Patterson 2001). Other evidence suggesting a link between involvement in daily physical education in primary school and fundamental movement skill proficiency (Cooley et al. 1997) has been used to argue the case for increased physical education time within the school curriculum.

SOURCES

- Booth, M.L., T. Okely, L. McLellan, P. Phongsavan, P. Macaskill, J. Patterson, J. Wright, and B. Holland. 1999. Mastery of fundamental motor skills among New South Wales school students: Prevalence and sociodemographic distribution. *Journal of Science and Medicine in Sport* 2: 93-105.

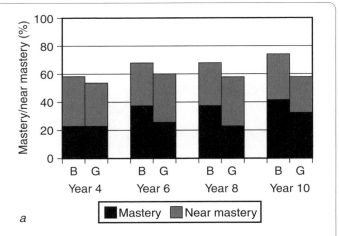

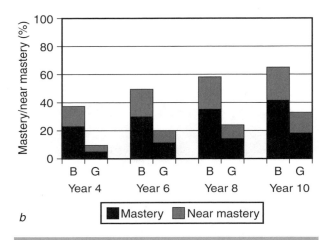

Figure 16.8 Performance of a sample of 5,518 Australian schoolchildren of different ages on the fundamental skills of (a) running and (b) overarm throwing. The dependent measure is the percentage of boys (B) and girls (G) at each age level who have attained mastery or near mastery of the skill.

Reprinted from Booth et al. 1999.

- Cooley, D., R. Oakman, L. McNaughton, and T. Ryska. 1997. Fundamental movement patterns in Tasmanian primary school children. *Perceptual and Motor Skills* 84: 307-316.

- Okely, A.D., M.L. Booth, and J.W. Patterson. 2001. Relationship of physical activity to fundamental movement skills in adolescents. *Medicine and Science in Sports and Exercise* 33: 1899-1904.

Motor Performance in Older Adults

Although, as noted in chapter 12, the peak performance of motor skills (at least as evidenced by sport performance such as world records) appears to be achieved typically from the late teens through to the mid-30s, most movement skills, with regular practice, can be retained at a high level throughout most of adulthood. Older adults, however, show some declines in motor performance in skills as fundamental as balance and locomotion that can seriously impair their movement capabilities. Some motor skill deterioration with age appears inevitable, although the extent of the decline can clearly be limited by regular practice.

Changes in Balance and Posture

Falls are a major cause of injury and loss of independent living capability in the elderly population. Falls may be particularly incapacitating for older women, as discussed in chapters 4 and 13, because of their susceptibility to osteoporosis and hip fractures. As a major cause of falls is loss of balance, scientists who study aging (gerontology) have been particularly interested in investigating balance and posture control in persons who are elderly. Such studies have revealed an increase in postural sway as people get older (with this effect being most pronounced in women) and slower and less expert recovery of unexpected losses of balance. Poorer vision, increased use of medication, and changes in gait pattern that result in less lifting of the foot off the ground also contribute significantly to increased losses of balance and falls among older people.

Changes in Walking Patterns

As noted in chapter 8, a number of discernible changes in walking patterns occur with aging. In comparison to younger adults, older people typically lift their feet less, walk more slowly, have shorter stride lengths, have a reduced range of motion at the ankle, and have greater "out-toeing" in their foot placement. The net purpose of all these changes is to attempt to increase stability, although this is inevitably to the detriment of mobility. Interestingly, with age, the motor patterns of people who are elderly regress in many respects to characteristics seen in the immature gait of children. This is especially true of the out-toeing of the foot, a movement pattern used by both young children and the elderly to improve lateral stability.

Changes in More Complex Motor Skills

As with fundamental motor skills, the performance of acquired specialized motor skills also declines with age, although the extent of deterioration does not appear to be uniform across all types of skills. Motor tasks that have to be performed under time stress, require complex decision making, or involve high anxiety (see chapter 18) typically show the most marked deteriorations with aging. Importantly, even in late adulthood, people can still learn and improve motor skills, so extensive practice on new motor skills can more than offset many of the systematic effects of the aging process.

Practical Applications

Although it is clear that some decline in motor performance is inevitable with aging, it is also clear that a number of very positive things can be done to offset these effects:

- Regular exercise (to offset losses in strength, flexibility, and endurance)
- Regular practice (to consolidate and improve existing movement patterns)
- Developing and practicing "smart" strategies (such as pacing effort to conserve energy, anticipating to avoid reaction time delays, and using encoding strategies to improve memory)

All these activities can assist in the retention of a high level of motor performance through into senescence.

Changes at the Neurophysiological Level

Examination of the neurophysiological changes in the motor system that accompany development and aging can offer some useful insight into the mechanisms underpinning the observable motor performance changes across the life span. Knowing how the nervous system changes with age also provides a foundation for understanding age-related changes in the "hardware" available for processing the information necessary for movement control.

Major Physical Changes in the Central Nervous System

The growth and development of the nervous system are controlled by a complex interaction between genetic and environmental factors. Early nervous system development, primarily genetically regulated, involves in the **prenatal** (before birth) period the formation of nerve cells and in the **postnatal** (after birth) period the branching and insulation (myelination) of the dendrites and axons of these nerve cells. The critical period for the development of the nervous system is from conception through to the end of the first year; and during this period in particular, the structural development of the nervous system is vulnerable to environmental influences. For example, sedative drugs (barbiturates) in the bloodstream of the mother prenatally, or malnutrition postnatally, can dramatically impair the normal development of the nervous system.

At birth the brain weighs some 300 (10.6 oz) to 350 g (12.3 oz) or about 25% of its adult weight. It reaches half its adult weight by around 6 months, 75% by 2 1/2 years, and near 100% by 6 years. The increased brain mass in the early years of life results from increases in size and branching within the neurons, myelination, and growth of the supporting glial cells. Rates of growth vary dramatically in different regions of the central nervous system. Within the brain the cerebral cortex is identifiable from about eight weeks following conception; and the two cerebral hemispheres are formed, but not functional, at birth. The spinal cord, while present, is small and quite short at birth; it matures, through the growth of myelin, in a top-down (cephalo-caudal) manner. The pyramidal tract is myelinated and functional by about 4 to 5 months, coinciding with the first appearance of voluntary movement control in the infant.

The process of myelination involves surrounding the axons of the nerve cells with a fatty sheath (myelin) that acts to insulate the nerve and substantially increase its speed of impulse conductance, its capability for repetitive firing, and its resistance to fatigue. Myelination of the sensory and motor neurons begins five to six months before birth and is completed within the first six months postnatally. Higher centres of the central nervous system, especially the cerebellum and cortex, begin and complete myelination much later than subcortical regions of the system. Damage to the myelin sheath, such as occurs in the disease **multiple sclerosis,** causes tremor, loss of coordination, and, on occasion, paralysis.

IN FOCUS: BRAIN MECHANISMS FOR MOVEMENT IN OLDER PEOPLE

In the introduction to part IV of this book we note that neurophysiologists now have available a number of powerful techniques for examining brain mechanisms of movement control. In recent years a number of research groups have used these techniques to examine whether the brain mechanisms for movement control are any different for older adults as compared to young adults.

Alexandra Sailer, Johannes Dichgans, and Christian Gerloff from Tuebingen University in Germany used electroencephalography (EEG) to record the electrical activity of the brain while older (55-76 years) and younger (18-27 years) people performed a simple motor task. They found that the electrical activity in different frequency bands in the EEG differed markedly between the two age groups, with the older participants recruiting greater activation of the primary sensorimotor and premotor regions of both cerebral hemispheres (see again figure 14.14). A group of researchers from the National Institute of Mental Health in the United States used functional magnetic resonance imaging (MRI) to examine the activation of different regions of the brains of young ("35 years) and older (>50 years) people as they performed a reaction time task. Consistent with the EEG findings of the Tuebingen group, the U.S. team discovered greater activation of a number of motor areas of the brain (contralateral sensorimotor cortex, premotor and supplementary motor areas, and ipsilateral cerebellum) in the older participants. Further, some areas of the brain that were not activated for the younger participants in completing this task (ipsilateral sensorimotor cortex, basal ganglia, and contralateral cerebellum) were activated by the older participants. Collectively, these findings suggest that even for the performance of simple movement tasks, additional areas of the brain of older participants are recruited. This is possibly an adaptive mechanism to compensate for some of the inevitable neural function loss that occurs with aging.

SOURCES

- Mattay, V.S., F. Fera, A. Tessitore, A.R. Hariri, S. Das, J.H. Callicott, and D.R. Weinberger. 2002. Neurophysiological correlates of age-related changes in human motor function. *Neurology* 58: 630-635.
- Sailer, A., J. Dichgans, and C. Gerloff. 2000. The influence of normal aging on the cortical processing of a simple motor task. *Neurology* 55: 979-985.

A progressive loss of nerve cells occurs throughout life (at a rate of some 10,000 per day) such that by age 65 to 70, some 20% of the total neurons present at birth are lost. Glial cells, on the other hand, increase with age. The net effect is a decrease in brain weight in people who are elderly. Brain activity during the performance of simple movements (as measured using techniques such as electroencephalography and functional magnetic resonance imaging) appears to be fundamentally different for older people (see "Brain Mechanisms for Movement in Older People," p. 252). There is also a general slowing of sensory and motor function with aging, resulting from a reduced impulse (signal) strength relative to background neural activity (noise).

Changes in the Sensory Receptors and Sensory Systems

Maturation of the key sensory systems for movement (i.e., the visual, kinaesthetic, and vestibular systems) occurs at different rates, each limiting the time at which mature movement control can first occur. The changes in movement "hardware" affect both information-processing capability and observable motor performance.

The Visual System

The eye, like the brain, undergoes most of its growth prior to birth, even though the size of the eye at birth is only about half its final size at maturity. The retina is fairly well developed at birth; the myelination of the optic nerve has commenced at this time and is complete some one to four months after birth. The neural pathways to the visual cortex are functional at birth. **Visual acuity** (sharpness of vision) is poor at birth, and at the first month of life the acuity of the human infant for stationary objects is about 5% that in mature adults. (The infant of one month sees the same degree of visual detail from a distance of 6 m [6.6 yd] that an adult would see from approximately 250 m [273 yd]!) Acuity improves rapidly during the first years of life; adult levels are attained by about age 10 for the viewing of stationary objects and age 12 for the viewing of moving objects (figure 16.9). This improved acuity results, to some degree, from an improved capability to adjust the shape of the eye's lens (called **accommodation**) but primarily from improved neuronal differentiation in the retina and connections in the cortex. The visual system matures at a rate sufficient to provide the visual information needed to guide movement at each stage of development.

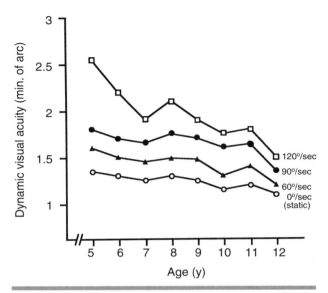

Figure 16.9 Changes in dynamic visual acuity as a function of age and the angular velocity of the target being viewed. The lower the score, the better the visual resolution of the moving target.

Data from Cratty, Bergel, and Apitzsch 1974.

With aging come a number of declines in visual function that can, in turn, adversely affect movement capability. By about age 40 there are clinically significant losses in the ability to accommodate to near objects. Material that can be read at 10 cm (4 in.) at age 20 must be progressively moved away to distances of 18 cm (7 in.) at 40, 50 cm (20 in.) at 50, and 100 cm (39 in.) at 70 years of age to retain sharp focus. In addition to acuity losses with aging, there is reduced sensitivity to glare, declining sensitivity to contrast, narrowing of the visual field size, and increased visual difficulty at low light levels. All of these changes make it increasingly difficult for people to perform movement tasks (such as driving or catching) that require precise visual information and accurate judgment of the location and speed of moving objects.

The Kinaesthetic and Vestibular System

Because of their central role in many reflex systems essential for the survival of the newborn, the kinaesthetic receptors develop early and are functional essentially from birth. Cutaneous receptors in the mouth are functional from as early as 7 to 8 weeks, and those in the hand from 12 to 13 weeks after conception. Muscle spindles are evident in muscles of the upper arm from 12 to 13 weeks postconception, although the main development of the muscle spindles, as well as the Golgi tendon organs, joint receptors, and cutaneous receptors, occurs during the period of four to six months after conception (three to five months before birth). The vestibular apparatus is completely formed two to three months after conception and may function reflexively from this point onward. The kinaesthetic and vestibular systems are thus prepared early in life to support infant activity before the visual system matures.

Although relatively little is known about the effect of aging on the kinaesthetic and vestibular receptors, there are clear functional losses in elderly persons with respect to balance, as detected through the vestibular system, and sensitivity to touch, vibration, temperature, and pain as detected through the cutaneous receptors. We know that some 40% of the vestibular receptors and nerve cells are lost by age 70. We also know that the number of Meissner's corpuscles in the skin (figure 14.8) decreases with age and that those remaining undergo changes in size and shape. In addition to a loss of receptors is a decrease of as much as 30% in the number of sensory neurons innervating the peripheral receptors (a condition known as **peripheral neuropathy**).

Changes in the Effectors (Muscles)

Some aspects of the growth and aging of muscle were considered in chapters 4 and 12. The number of muscle fibres increases prenatally and also for a short period postnatally, approximately doubling between the last trimester of gestation and four months after birth. Most fibres do not differentiate until the foetus is around seven months old. Slow-twitch (or type I) fibres begin to appear five to seven months after conception, and these fibres constitute about 40% of all muscle fibres present at birth. Fast-twitch (or type II) fibres make up about 45% of all fibres present at term. Slow- and fast-twitch muscle fibres continue to increase in the first year of life, and the relative distribution of fibre type appears to reach steady state by the age of 3 years.

As children mature, their muscle fibres become wider (figure 16.10) and longer, and this comes about through an increase in both the number and length of the contractile units (the **myofibrils**) within each muscle fibre. Adult-sized muscle fibres are attained in adolescence. The reductions of muscle size that occur in the elderly appear to be a consequence of reduced effectiveness of the nerves that activate the muscles and the selective loss of fast-twitch muscle fibres and motor units

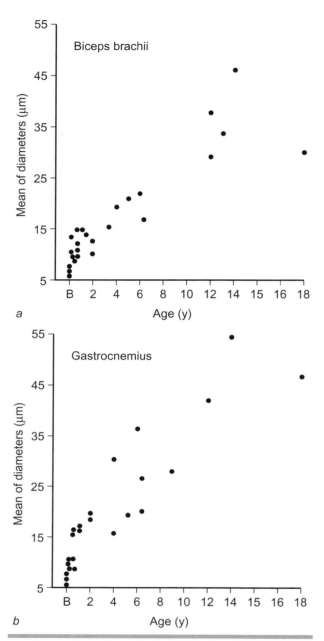

Figure 16.10 Changes in fibre diameter of major muscles of the upper arm (biceps brachii) and lower leg (gastrocnemius) from birth (B) to 18 years of age.
Reprinted from Malina and Bouchard 1991.

primarily. Growing evidence indicates significant changes in the control properties of the motor units with aging. For instance, discharge rates of motor units become more variable in older people, and this contributes to a reduced capability to perform steady, submaximal muscle contractions.

Changes in Reflex Systems

Receptors and effectors, as we saw in chapter 14, are the key elements of reflex systems. The development, modification, and frequently the extinction of reflexes have a complexity greater than that of the maturational changes of the sensory and effector components of the various reflex systems. Reflexes present at birth or soon thereafter can be broadly distinguished with respect to whether their principal function is to assist survival of the newborn or to lay the foundations for the development of voluntary movement control.

Primitive Reflexes

The human newborn is extremely vulnerable because of its limited mobility and capacity for voluntary movement. Consequently, in the early stages of life, infants must depend heavily on adult caretakers and some reflexes for survival and protection. Those reflexes present at birth that function predominantly for protection and survival are referred to collectively as **primitive reflexes.** Examples are the sucking reflex, which enables the newborn to instinctively gain nutrition from the mother's breast; the searching or rooting reflex, which helps the newborn locate the nipple; and the Moro reflex, which assists with initial respiration. The primitive reflexes, which dominate movement control at birth, typically weaken with advancing maturity to the point that they are either completely inhibited or at least highly localized by three to four months after birth. This coincides with the increased maturity of the cortex, suggesting a transition from involuntary control toward greater voluntary control. Persistence of reflexes for extended periods after their expected time of disappearance is used by paediatricians (doctors who specialize in the care of children) as a sign of neurological problems.

Postural and Locomotor Reflexes

Postural reflexes, such as body righting, neck righting, and the parachute reflex, serve the function of keeping the head upright and the body correctly oriented with respect to gravity. These reflexes are not present at birth but generally appear after two months, around the time at which early stages of postural control are being established (figure 16.1). They disappear after the first year of life and are progressively replaced by more voluntary movement control. The **locomotor reflexes,** such as the walking and swimming reflexes, are present from birth or soon thereafter. These reflexes disappear after some four to five months and before voluntary walking or swimming is attempted (figure 16.2). While the exact role of the postural and locomotor reflexes in the development of future voluntary movement control is not entirely clear, it appears that these reflexes collectively play a role in preparing the nervous system and its pathways for the emergence of the voluntary fundamental motor skills discussed earlier in this chapter. Localized reflexes such as the stretch reflex described in chapter 14 persist throughout the life span and appear to alter relatively little in old age.

Changes in Information-Processing Capabilities

Young people and elderly people must process information through the same basic central processing stages of perception, decision making, and movement organization and execution used by others and as outlined in chapter 15. What changes with development, and with aging, are the speed, efficiency, and sophistication with which information can be processed.

Developmental Improvements in Information-Processing Capability

The processing of information for perceiving, deciding, and acting improves with maturation and then appears to undergo some decline in people who are elderly. These alterations in information processing are directly responsible for the parallel changes in observable motor performance described in the previous section.

Perception

At least three general principles can be observed in the development of perceptual skills. One principle is that *the maturation of perceptual skills continues well after the sensory system and receptors have matured structurally.* We noted earlier that in the visual system, the optic nerve is fully myelinated and pathways from the eye to the visual cortex are functional by a few months after birth, yet visual acuity does not reach adult levels until around age 10. Similarly the kinaesthetic "hardware" for life is essentially complete at birth, yet improvements in the ability to make precise kinaesthetic judgments continue at least until age 8 and typically beyond (figure 16.11a).

A second principle is that *the more complex the perceptual judgment that has to be made, the longer it takes the developing child to reach adult levels of performance.* In the visual system, acuity for moving objects matures after acuity for stationary objects. Similarly, relatively simple visual-perceptual judgments, such as those involved in comparing object size or depth, differentiating an object from its background, or distinguishing a whole image from its component parts, reach maturity within the first decade of life. In contrast, more complex judgments, such as those involved in anticipating the direction of an opponent's stroke in racket sports, may continue to improve well into the third decade of life. As with vision, kinaesthetic

judgments of complex movement patterns mature much later than simple acuity judgments. (Compare figures 16.11*a* and 16.11*b*.)

A third principle relates to the integration of information between the different sensory systems. Perhaps surprisingly, *the integration of visual and kinaesthetic information does not follow the individual maturation of the visual and kinaesthetic systems but rather occurs simultaneously with it.* For simple tasks, such as shape or pattern recognition, in which the child is permitted active kinaesthetic exploration of the shape, the integration of visual and kinaesthetic is mature by around age 8. By this age, children exploring an object through touch alone are then able to also identify the object visually from among other possible shapes or objects. More complex visual-kinaesthetic integrations (such as those that occur in hitting and catching tasks, in which movements must be initiated and guided kinaesthetically to coincide with the arrival of a ball perceived visually) take longer to mature, although the attainment of adult levels of performance does not lag behind the separate maturation of the visual and kinaesthetic systems.

Decision Making

Reaction time to a single, unanticipated stimulus (simple reaction time) decreases rapidly (that is, gets faster) until the mid- to late teens when adult levels are attained (figure 16.12). Being

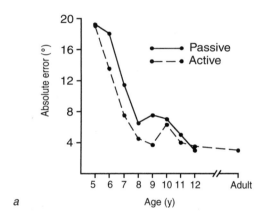
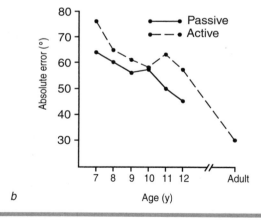

Figure 16.11 Changes in (a) kinaesthetic acuity and (b) kinaesthetic memory for complex movement patterns as a function of age. Active movements are those controlled by the participant, whereas in passive movements the limb is moved to different positions by the experimenter.
Reprinted from Bairstow and Laszlo 1981.

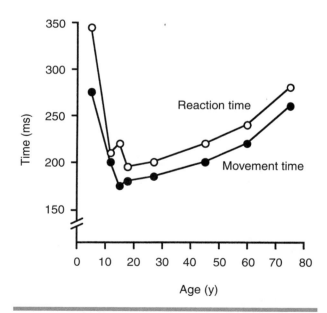

Figure 16.12 Changes in simple reaction time and movement time across the life span.
Reprinted from Hodgkins 1962.

more complex, reaction time for tasks involving decision making (choice reaction time) takes somewhat longer to mature. Children are slower than adults in making decisions for a number of reasons in addition to their slower simple reaction times. Compared to both older children and adults, younger children have a slower information-processing rate (as shown by a steeper slope in figure 16.13) and also frequently process more information in selecting any particular response. This occurs because of their relatively poor ability to separate out relevant from irrelevant information. It is not surprising, therefore, that children's performance in tasks that require rapid decision making, such as a number of the team sports, shows rapid and continued improvements with practice into the mid-20s and beyond.

Organizing and Executing Movement

Like simple reaction time, the time taken to make simple movements varies across the life span (figure 16.12). Movement time for simple actions reaches its minimum around the midteens and remains at that level, typically, until the mid-30s. More complex movements (movements with greater programming or control requirements, or both) are made more slowly by young children than by adults, although it is not clear whether this is due to a slower information-processing rate for movement control for children. Compared to adults, children show a reduced ability to perform one or more tasks concurrently with a movement task, suggesting that movement is controlled more automatically by adults. A large part (perhaps all) of the information-processing capacity of children is needed for the control of even apparently simple movements, leaving little free capacity to allocate to the performance of other simultaneous tasks.

Declines in Information Processing With Aging

Much remains to be learned about the changes in information-processing capability that accompany aging. What is most apparent is the systematic decline in both the speed of reacting and the speed of moving that appears to occur from about the 30s onward (figure 16.12). This reduced speed of responding becomes pronounced in people who are elderly and is apparent in a wide range of tasks, from walking to driving a car or typing. The loss of speed in responding is more pronounced in complex than in simple tasks. For example, longitudinal studies reveal that slowing with age is much more pronounced for choice reaction time than for simple reaction time, and this appears to be especially true for women (figure 16.14).

The phenomenon of slowing with age follows a *last-in-first-out rule* in that simple actions (such as reflexive movements) acquired early in life are more resistant to slowing and loss than are complex, coordinated skills learned voluntarily later in life. Although the declines in reaction and movement time with aging appear somewhat inevitable, the rate of decline may be slowed considerably by regular physical activity.

The slowing of simple reaction time with age appears to reflect primarily the neurophysiological changes in the central nervous system, especially nerve cell loss and changes in some of the sensory receptors. In choice reaction time tasks, slowing reflects a lowered signal-to-noise ratio in the central nervous system plus a cautious strategic change toward emphasizing accuracy to the detriment of speed. The slowing of movement time with aging reflects similar central rather than peripheral factors. With aging comes a reduction in the number of functional neuromuscular units and, from age 20 to 60, a 15% to 35% increase in the amount of neural stimulation needed to excite a muscle to contraction. These changes, along with a loss of nerve cells, effectively lower the signal-to-noise ratio, slowing the rate at which movements can be initiated and can be completed. As new movement

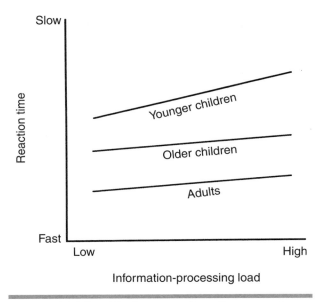

Figure 16.13 Choice reaction time as a function of information-processing load and age.
Adapted from Keogh and Sugden 1985.

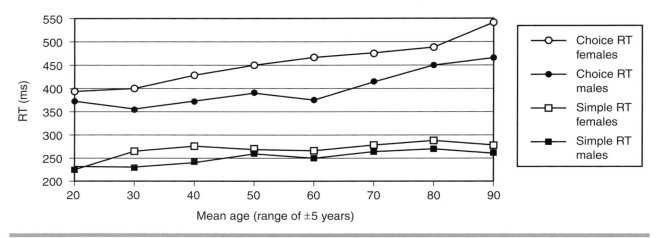

Figure 16.14   Longitudinal changes in simple and choice reaction time as a function of aging. Reprinted from Fozard et al. 1994.

skills may possibly be more difficult to learn for older adults, an important preparation for later adulthood is for people to learn, early in life, skills they can maintain with relative ease by means of ongoing practice throughout the life span. Chapter 17 examines how skills are learned and acquired through practice.

Summary

Throughout the life span there are continual changes in the capacity to perform various motor skills. In the first two years of life, development occurs in a head-down (cephalo-caudal) and centre-out (proximo-distal) manner, with the emergence of essential postural, locomotor, and manual manipulative skills following a similar sequence and general time scale for most children. Fundamental movement patterns (for running, jumping, throwing, catching, etc.) develop throughout childhood, again in a predictable sequential manner, as voluntary control of movement progressively overrides the reflex control dominant in the early years of life. Motor performance declines in older people due to a combination of aging effects and, in many cases, a reduction in activity or training levels. These changes in motor performance that

occur across the life span are a consequence of age-related changes in both the underlying functional anatomy and information-processing capacity of the neuromuscular system.

Further Reading and References

Burton, A.W., and D.E. Miller. 1998. *Movement skill assessment*. Champaign, IL: Human Kinetics.

Grabiner, M.D., and R.M. Enoka. 1995. Changes in movement capabilities with aging. *Exercise and Sport Science Reviews* 23: 65-104.

Haywood, K.M., and N. Getchell. 2001. *Life span motor development*. 3rd ed. Champaign, IL: Human Kinetics.

Keogh, J.F., and D.A. Sugden. 1985. *Movement skill development*. New York: Macmillan.

Shumway-Cook, A., and M.H. Woollacott. 2001. *Motor control: Theory and practical applications* (chapters 8 and 9). 2nd ed. Baltimore: Lippincott, Williams & Wilkins.

Spirduso, W.W. 1995. *Physical dimensions of aging*. Champaign, IL: Human Kinetics.

Sugden, D.A., and J.F. Keogh. 1990. *Problems in movement skill development*. Columbia, SC: University of South Carolina Press.

CHAPTER 17

Motor Control Adaptations to Training

*I*n this chapter we use the basic neurophysiological and cognitive science concepts of motor control introduced in chapters 14 and 15 to examine the changes in motor control that occur as a consequence of practice or training. The specialized field of study concerned with the description and explanation of changes in motor performance and motor control with practice is typically referred to as **motor learning** (or occasionally perceptual-motor skill acquisition).

Learning is the change in the underlying control processes that is responsible for the relatively permanent improvements in performance that accompany practice. Like many aspects of the motor control field, learning is difficult to study because it is a process that cannot be directly observed or measured. Scientists attempting to understand motor learning depend on making accurate inferences about learning from observable (and measurable) changes in performance. However, because it is possible to acquire significant skill without necessarily improving performance, changes in the underlying control processes are not always directly or faithfully reflected in observable performance. The difficulty scientists have in accurately measuring learning is shared with practitioners, such as physical educators and coaches, who are charged with the responsibility of attempting to objectively measure skill learning.

> The major learning concepts in this chapter relate to
> - training-related changes in observable motor performance, including stages in the acquisition of motor skills and the characteristics of skilled performers;
> - training-related changes in motor control at the neural level;
> - neurophysiological accounts of learning and plasticity;
> - training-related changes in information-processing capabilities, including changes in perception, decision making, and movement organization and execution; and
> - factors affecting the learning of motor skills, including the quantity and type of practice, instruction, and feedback.

Studies of motor learning have generated knowledge about many topics, including

- the characteristics of expert performers;
- the effectiveness of different types and schedules of feedback for skill learning;
- the relative merits of different types of practice;
- the transfer of skill from one practice setting to another;
- the retention of skills over time; and
- the relearning of skills following traumas, such as joint injury, or central nervous system damage such as occurs with stroke.

This knowledge is clearly important practically for any profession involved with the assessment, measurement, and improvement of motor skills.

As is the case with motor development, examined in the previous chapter, knowledge about motor learning comes from both cross-sectional and longitudinal studies. In cross-sectional studies the motor control and performance of people with different levels or types of practice are compared, whereas in longitudinal studies the motor control and performance of the same set of people are examined on a number of occasions throughout the learning or retention of a particular skill. Comparisons of skilled (expert) and lesser-skilled (novice) performers are examples of the first type of study, whereas training studies are typical of the second type of study. Motor learning can be studied at many levels and, again in keeping with the previous chapter, this chapter addresses the changes that take place with practice at the observable motor performance level, the neurophysiological level, and the level of information-processing capabilities.

Changes in Observable Motor Performance

Describing the observable characteristics of expert performance, and the observable changes in motor performance that occur as a new skill is acquired, provides a useful starting point for the systematic study of motor learning.

Characteristics of Skilled Performers

Skilled performance is the learned ability to achieve a desired outcome with maximum certainty and efficiency—that is, the ability to attain a desired result on a particular task with a minimum outlay of time, energy, or both. Comparisons of the observable characteristics of expert performers with those of people less skilled on the same task typically reveal a number of differences. Experts, in contrast to the lesser skilled, are frequently characterized as

- having all the time in the world,
- picking the right options,
- reading the situation well,
- being adaptable,
- moving in a smooth and easy manner, and
- doing things automatically.

Skilled performers are apparently able to balance well the conflicting needs to be

- fast yet accurate,
- consistent yet adaptable, and
- maximally effective yet with a minimum of attention and effort.

Knowing how skilled performers are able to achieve this balance requires an understanding of the changes that occur with practice in the underlying neurophysiology and information processing.

Stages in the Acquisition of Motor Skills

To proceed from being a novice performer to being an expert requires extensive practice over a long time. As a general rule, the acquisition of expertise in motor skills, like expertise in cognitive skills such as playing chess, appears to take some 10 years, 10,000 hours, or literally millions of trials of practice. Although skill acquisition is a continuous process, learners pass through at least three identifiable stages during the long transition from novice to expert.

Stage 1: The Verbal-Cognitive Phase

In the **verbal–cognitive phase,** the movement task to be learned is completely new to the person. Consequently the learner is preoccupied at this stage with trying to understand the requirements of the task, especially what needs to be done in order to perform the skill successfully. All of the learner's limited information-processing capacity is directed to such issues as where to position the whole body and limbs, where best to gain ongoing feedback about performance, or how the correct movement feels. Consequently the major activity at this stage of performance is thinking and planning of movement strategies (that is, cognition), and significant benefits can be gained from good (verbal) instruction and especially demonstrations.

The movements used to make initial attempts at the new task are typically pieces of movement patterns from existing skills that are joined together to meet the challenges of the new task. In other words, "old habits" are reshaped into new patterns. Instruction that highlights the similarities (and also the differences) between the new skill and skills already learned may therefore be beneficial in speeding up the rate at which the new skill is learned. Performance fluctuates dramatically in the early stage of learning new skills as a wide range of movement strategies are tried and many discarded.

Stage 2: The Associative Phase

In the **associative phase** of learning, performance is much more consistent as the learner settles on a single strategy or approach to the task. Learners consequently spend the majority of their time and effort in "fine-tuning" the selected movement pattern rather than constantly switching from one movement pattern to another. With the basic knowledge of the requirements of the task established, learners become better able in the associative phase to both produce the movement pattern they had planned and adjust to changes in the conditions in which the movement is to be performed. These developments ensure increased levels of task success.

In contrast to the verbal–cognitive phase, in which the instruction provided by others may be the single most beneficial thing for skill acquisition, in the associative phase there is no substitute for specific practice on the task itself. Progressive increases in task complexity (e.g., through adding more complex rhythmic patterns for a pianist, more difficult terrain or traffic conditions for a car driver, or more opponents or time constraints for a basketball player) provide a valuable means of fostering the systematic continuation of skill development.

Stage 3: The Autonomous Phase

With sufficient practice on a particular skill, some learners reach the third stage of learning—the **autonomous phase.** The autonomous phase is so named because at this stage the performance of the skill appears largely automatic. The movement apparently can be controlled without the person having to pay attention to it. At this stage of learning, movements are performed consistently with such precision and accuracy that there is no longer a need to constantly monitor feedback to ensure that the movement is correctly performed. Open-loop control therefore largely replaces closed-loop control (see p. 232), and skilled performers have spare attention that can be allocated to other tasks. Consequently, one sign of the expert performer, as we will see later, is the ability to do two or more things at once. For example, typists in the autonomous phase of learning are able to conduct sensible telephone conversations with minimal interference to their concurrent typing speed or accuracy.

The only major drawback to reaching the autonomous stage of learning is that performers at this stage find it difficult, if not impossible, to change their movement pattern if, for example, an error in technique becomes ingrained. Experts also often cannot verbally describe how they perform the skilled movements that they execute; in the autonomous phase, movement control operates below the level of consciousness. Even at the autonomous phase of learning, there is room for improvement in both motor performance and control. As we will see later, there is no reason that learning of movement skills cannot occur continuously even after the learner has reached the autonomous phase.

The Specificity of Motor Skills

Skill acquisition is highly specific, and there is generally little or no transfer of training from one motor skill to another. Level of performance on any one particular motor skill is typically of limited use in predicting the rate of learning or the ultimate level of performance of an individual on any other motor skill. Despite the intuitive appeal of the idea, there is little or no evidence that motor skill is generalizable. It appears that someone who is expert at one activity has the same probability of being a novice on a different skill as someone unskilled in the expert's activity of specialization. The person therefore needs to pass through the verbal–cognitive, associative, and autonomous stages of learning for *each* skill in which he or she wishes to become highly proficient.

Changes at the Neurophysiological Level

Learning, as we noted in our earlier definition, is reflected in a relatively permanent change in performance. This relatively permanent change in observable performance must be underpinned by biological changes at the level of the nervous system. Consequently, neuroscientists have made extensive efforts in searching for neural correlates of learning and memory.

Challenges for a Neurophysiological Account of Learning

Understanding the physiology of learning and memory represents perhaps the ultimate challenge to researchers in the neurosciences. The task is difficult for a number of reasons. First and foremost, both learning (and memory) and the brain (and the rest of the central nervous system) are incredibly complex, so matching the complex phenomena of learning and remembering with the complexity of the structure of the brain and central nervous system is always going to be an extremely difficult, if not impossible, task. Any attempt to locate neural changes underpinning learning is complicated by the possibility that learning a particular motor skill may be associated with structural and functional changes in only a few of the brain's many billions of neurons or, more likely, may be associated with relatively subtle changes in the relationships between a large number of neurons distributed at diverse locations throughout the brain. The neural changes that underpin learning may take place over very short or quite lengthy time periods, and techniques are only now being developed that may be sufficiently sensitive to record some of the subtle neural changes that accompany the acquisition of skill.

Given the complexities within both the learning phenomena and the structure of the brain, and the technical difficulties in observing and recording possible neural events associated with learning, most neurophysiological studies of learning have, by necessity, used animals (other than humans) and very simple learning tasks (rather than the complex ones typical of human movement). Learning can be of many different types, yet to date, the majority of learning studies from a neurophysiological perspective have focused on simple stimulus–response learning or conditioning (figure 17.1). Such studies have typically sought to find a neural mechanism for memory, as memory for past events and experiences is seen as fundamental for learning. Without changes in the organism through past experience, so that future behaviour can be influenced by experience, learning would be impossible. Although the inferences that one can make about the neurophysiological bases of human motor learning at this stage are only very preliminary and necessarily somewhat speculative, it does appear that the search for a single neural location for memory, in particular, is likely to be fruitless. Increasingly, it appears that memory is a phenomenon that is distributed throughout networks of neurons and not confined to a single site.

Plasticity As the Basis for Learning

Although psychologists frequently conceptualise memories as records stored in filing cabinets or other similar archives, this is not the way experience is retained within the nervous system. Expe-rience is not stored literally; rather it influences the way we perceive, decide, and organize and execute movement by physically modifying the neural pathways and circuits responsible for perceiving, deciding, and acting. Remembering and learning are therefore possible only because the nervous system is highly plastic; that is, it is able to dynamically alter its structure to accommodate the new functions it is required to perform. Plasticity exists on a continuum from short-term changes in synaptic function to long-term structural changes in nerve connections and the organization of neural networks. These changes in the time scale of neural modifiability are matched by parallel changes in learning state (figure 17.2).

While the nervous system retains some degree of plasticity throughout life, the brain is especially plastic during the early years and during the critical periods for skill acquisition described in the previous chapter. If, for any reason, a region of the brain is injured or damaged during the first few years of life, other regions of the brain are often able to take over the functions normally performed by the injured region. The level of plasticity is reduced later in life, making full recovery of function following traumas, such as stroke (which damages the brain, commonly resulting in deficits in motor skills such as speech and gait), difficult in persons who are elderly. Significant neurological reorganization in adults is nevertheless possible.

Injury to the central nervous system can cause complete loss of some neurons, disruption to some of the axonal projections of others, and denervation of still other neurons innervated by the injured neuron. Full or partial recovery of function is achieved through a number of means, both local

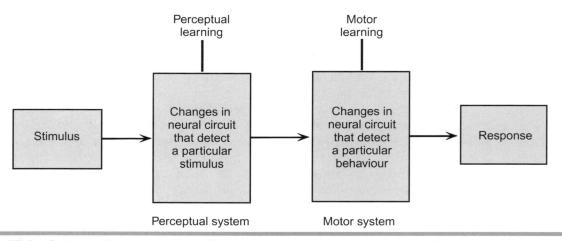

Figure 17.1 Schematic view of the different types of learning examined by neurophysiologists. Reprinted from Carlson 1994.

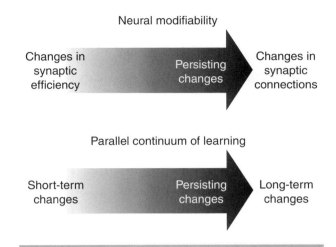

Neural modifiability

Changes in synaptic efficiency → Persisting changes → Changes in synaptic connections

Parallel continuum of learning

Short-term changes → Persisting changes → Long-term changes

Figure 17.2 Short- to long-term changes in neural modifiability (plasticity) are matched by parallel short- to long-term changes in learning.
Reprinted from Shumway-Cook and Woollacott 2001.

and global. Locally, postsynaptic neurons connected to the damaged neuron(s) may become hypersensitive (and hence easier to activate); previously silent synapses may become functional; injured axons may regenerate; and collateral axons may sprout from neighboring nerve cells to create alternatives to the damaged connections. More global means include the reorganization of the connections between the motor cortex and the affected limb or region and increased contributions of parallel nerve pathways on the other side of the body.

As with the response to injury, the nervous system may undergo a number of structural and functional changes either as a cause or as a consequence of learning. Learning may result in a change in the response characteristics of neurons, the establishment of new synaptic connections (as well as the atrophy and eventual disappearance of old, unused connections) (figure 17.3), and changes in the response characteristics of synapses. All of

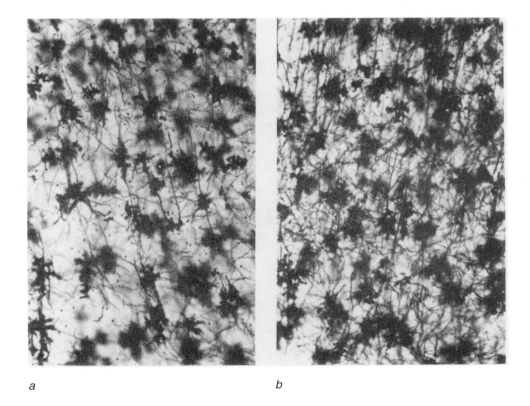

a   b

Figure 17.3 Golgi-Cox-stained microscopic sections of the somatosensory cortex of (a) untrained and (b) trained rats. A greater density of synaptic connections is apparent in the trained rats.
Reprinted from Spinelli, Jensen, and Di Prisco 1980.

these changes are interrelated and contribute to synaptic plasticity. Collectively, changes in the nature of synaptic structure and function across one or more synapses modify neural circuitry that in turn can modify perceiving, deciding, and acting.

These neural changes are underpinned by biochemical and molecular changes. For example, the reinforcement or repetition of successful responses is associated with increased receptivity to the chemical neurotransmitter dopamine. (We saw in chapter 14 that this chemical is important in transmitting messages within the basal ganglia—among other places—and that deficits in this chemical within specific regions of the basal ganglia appear to be responsible for Parkinson's disease.) Recent evidence also suggests that long-term memory may require the synthesis and growth of new proteins whereas short-term memory may not.

Synaptic Changes and Long-Term Potentiation

It has been recognized for a long time that synaptic changes must be fundamental to learning and memory. In the 1940s, the Canadian neuropsychologist Donald Hebb proposed that the strengthening of synapses provides the neural foundation for learning and that a synapse is strengthened by simultaneous activity in both pre- and postsynaptic neurons (figure 14.4). More specifically, Hebb hypothesized that if a synapse repeatedly becomes active at around the same time that the postsynaptic neuron fires, structural or chemical changes, or both, will take place in the synapse to strengthen it. At the time Hebb proposed these effects, techniques were not available to test them.

More recent neurophysiological studies have demonstrated one mechanism through which the type of synaptic strengthening hypothesized by Hebb could be achieved. When intense, high-frequency electrical stimulation is applied to afferent neurons in some regions of the brain, a subsequent increase in the magnitude of the neural response is frequently seen to standard (test) stimuli of known electrical strength. This increased response may last for up to several months and is referred to as **long-term potentiation.**

Long-term potentiation provides a measure of increased synaptic efficiency in that a test stimulus of the same magnitude elicits an increased response. This suggests that the effectiveness of the involved synapse(s) has been improved.

The exact cellular and biochemical mechanisms underpinning long-term potentiation are still not clear but appear to involve the combined effects of depolarisation of the postsynaptic membrane and the activation of unique receptors sensitive to a neurotransmitter called glutamate. Long-term potentiation has been observed primarily in one specific area of the brain (the **hippocampus**) but also at other sites. As disruption to long-term potentiation impairs learning of the type usually attributed to the hippocampus, it seems reasonable to conclude that improved synaptic efficiency through long-term potentiation is an important neural foundation for skill acquisition.

Learning appears to be associated with long-term changes not only in synaptic efficiency but also in synaptic connections. Recent studies, using a procedure known as **transcranial magnetic stimulation (TMS),** have demonstrated significant changes in the topographical organization of the motor cortex and surrounding regions (see again figure 14.15) as the learning of skills progresses from a cognitive stage, in which conscious control is exerted over movement, to a more automatic stage, in which movement is controlled without explicit knowledge.

Changes in Information-Processing Capabilities

Comparative studies of expert and novice performers on different motor skills reveal a number of changes in information-processing capability with the acquisition of skill. Changes are apparent in each of the central processing stages of perception, decision making, and movement organization and execution described in chapter 15, as well as in the observable movement patterns and outcomes produced (figure 15.3). In all cases expert–novice differences are systematically present only for the processing of information specific to the motor task for which expertise has been developed.

Sensory Reception

The sensitivity of the key sensory systems for movement (the visual, kinaesthetic, and vestibular systems) appears to change relatively little as a consequence of learning, practicing, and improving specific motor skills. Although it may be possible to improve characteristics like visual or kinaesthetic acuity by specifically training these attributes, as a general rule sensitivity to the sensory information

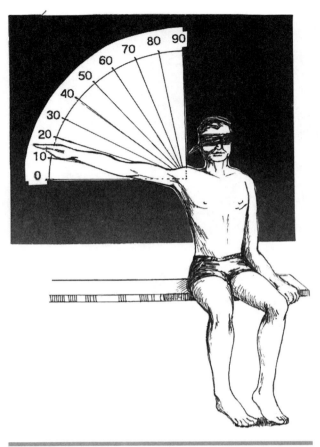

Figure 17.4 A standardized test of kinaesthetic acuity. In this test, acuity is determined by the person's accuracy in reproducing the joint angle or end location of a previously experienced movement.

needed to control movement is not the limiting factor to motor performance. Perhaps surprisingly, expert performers, from a range of motor skills, are not characterized by above-average levels of visual acuity or kinaesthetic sensitivity, at least when these characteristics are measured using standardized tests (figure 17.4). What appears to be more strongly related to skilled performance is the way in which the basic information provided by the various sensory receptors is processed, interpreted, and used by the central nervous system.

Perception

With practice and the acquisition of expertise, a number of systematic changes take place in the way performers perceive environmental and internal events. Among other things, experts are superior to novices in recognizing patterns and predicting (anticipating) forthcoming events. The

experts' superiority is not general, however, and holds only for patterns and events drawn from the experts' domain of expertise. Expert pianists, for example, are better at recognizing patterns within pieces of music than novice pianists but are typically no better than novices in recognizing patterns present in skills other than music.

The improved pattern recognition that accompanies skill acquisition is especially evident in team sports, such as hockey, basketball, volleyball, and football, in which selecting the correct movement response depends on the ability to quickly and accurately recognize the defensive or offensive patterns of play (or both) of the opposing team. In a number of studies, experimenters have taken full-colour slides of an opposing team during a game and then shown these slides to the study participants for brief periods (usually 5 s). Participants are then asked to recall the position of all the players (both attacking and defensive) shown on the slide. When the slides depict structured patterns of play (figure 17.5), the recall of expert team sport players is superior to that of lesser-skilled players. This perceptual advantage disappears when the players shown on the slides are in random position (in other words when no familiar pattern is present). This finding, illustrated in figure 17.6, is a systematic one and demonstrates that the experts' superior perception is due to skill-specific experience and not to generically superior perceptual skills that hold across all situations. To use the computer analogy introduced in chapter 15, the main change with the acquisition of expertise on a task is improvement in the specific programs (or "software") used to process the input information provided by the sensory receptors.

Many motor skills, especially skills such as those performed in fast ball sports, involve substantial time constraints that pose major demands on human information-processing capabilities. However, as noted earlier, one of the distinguishing characteristics of expert performers is their ability to give the impression of having "all the time in the world." One important means by which expert performers overcome time constraints is to anticipate likely events from any advance information available to them. Experts are better able than lesser-skilled players to predict actions in advance on the basis of information available from the actions and postures of their opponents. For example, expert badminton players are able to make more accurate predictions of where an opponent's stroke will land from information available before the opponent

Figure 17.5 A typical structured pattern recognition situation from the sport of field hockey.

Photo copyright by the National Sports Information Centre, Australian Sports Commission. Reprinted with permission.

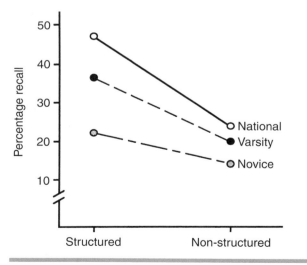

Figure 17.6 Pattern recall accuracy of field hockey players of different skill levels for displays with and without structure.

Reprinted from Starkes 1987.

strikes the shuttlecock than can novices (figure 17.7). Experts are also able to pick up information from cues in addition to those used by novices. In the badminton example, experts are able to pick up advance information from the motion of the racket and the arm holding it whereas novices can only use the racket as a cue (figure 17.8). Expert–novice differences in the ability to anticipate occur even when the two skill groups may be looking at the same features of their opponent. This demonstrates that the limiting factor in the perceptual performance of untrained individuals is not the ability to pick up the necessary sensory information but rather the ability to interpret, understand, and use it to guide decision making and movement execution.

Decision Making

A capability to make decisions both quickly and accurately would be clearly beneficial for the

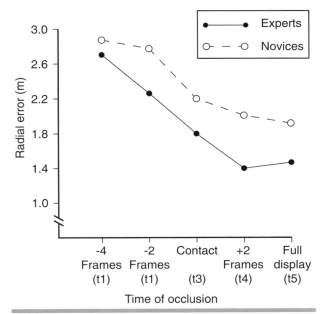

Figure 17.7 Error in predicting the landing position of a badminton stroke made from watching a film of an opposing player. Error is shown as a function of both skill level of the participants and occlusion of the display. Time of occlusion is expressed in relation to the contact of the opponent's racket with the shuttle (each frame is approximately 40 ms).

Reprinted from Abernethy and Russell 1987.

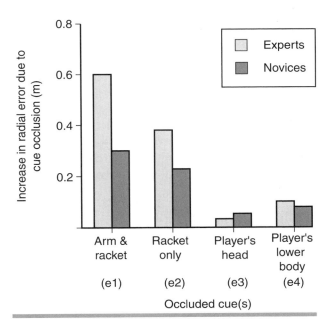

Figure 17.8 Error in predicting the landing position of a badminton stroke when visibility is occluded to different regions of the opposing player's body. An increase in error indicates that the occluded region contains cues that are of use in anticipating the stroke direction and force.

Reprinted from Abernethy and Russell 1987.

performance of many motor skills. As we saw in chapter 15, a faster choice reaction time on any particular occasion may be achieved with practice in one of two ways—either through facilitation of the overall decision-making rate (decreasing the slope of figure 15.5) or through a decrease in the total amount of information that has to be processed before making a decision. Choice reaction time tasks, which use stimuli and responses that are skill specific, provide some support for the proposition that expert performers have faster decision-making rates than novice performers. Nevertheless, the more potent strategy used by skilled performers to decrease their decision-making time is to reduce the absolute amount of information that they have to process. Expert performers do this by using knowledge, acquired through experience, of options and event sequences that either are not possible or at least are improbable. Experts are more accurate than novices in predicting the probability of particular events occurring; this knowledge is invaluable in reducing the amount of information to be processed and, in turn, allowing decisions to be made more rapidly ("Response Speed in Soccer Players," p. 268).

Movement Organization and Execution

A number of changes in the way movement is organized and executed are apparent with training. The rate of processing movement information improves with practice; and this is due, at least in part, to transition away from an exclusively feedback-dependent closed-loop type of motor control early in learning toward greater use of open-loop control. Open-loop control is via motor programs that are prestructured sets of commands, allowing a movement sequence to be executed without reliance on feedback. As skills are learned, it appears that bigger and better motor programs are formed, bringing progressively more movement elements under the control of a single program. The collective effect of a reduced need to constantly monitor feedback, as well as a reduction in the number of separate programs needed to produce a particular action, is a reduction of the demands that the organization and execution of movement place on the information-processing resources available within the central nervous system. Consequently, highly skilled performers have significant amounts of spare information-processing capacity that they can allocate to the performance of other tasks. Expert performers are therefore markedly superior to lesser-skilled performers in the performance of a second task concurrent with their usual movement

IN FOCUS: RESPONSE SPEED IN SOCCER PLAYERS

In order to examine the decision-making ability of soccer players of different skill levels, Werner Helsen and J.M. Pauwels of Leuven University in Belgium developed an experimental setting in which players were required to respond to life-size dynamic simulations of situations drawn from typical soccer games. The player would watch a soccer match scenario unfold on a large screen; then, at a critical moment, one of the attackers depicted would play the ball in the direction of the viewer. The viewer's task was to move as quickly and accurately as possible to execute the movement most appropriate for the situation. The options available were shooting at goal, dribbling around the goalkeeper or defender, and passing to a free teammate. Time taken to respond and the accuracy of the responses were recorded. Where the viewer was looking was also measured using an eye movement recorder.

According to the results, not only did the expert soccer players have total **response times** that were, on average, nearly 260 ms faster than those of nonexperts; they were also more accurate in their responses. Experts chose the correct option 92% of the time compared to 82% for the nonexperts. These differences occurred in the absence of any major differences in the eye movement patterns of the different skill groups. These general observations have since been replicated by Mark Williams and colleagues from Liverpool John Moores University on a number of subgame situations (e.g., 3 vs. 3; 1 vs. 1).

SOURCES

- Helsen, W., and J.M. Pauwels. 1993. The relationship between expertise and visual information processing in sport. In *Cognitive issues in motor expertise,* ed. J.L. Starkes and F. Allard (pp. 109-134). Amsterdam: North-Holland.

- Williams, A.M., and K. Davids. 1998. Visual search strategy, selective attention, and expertise in soccer. *Research Quarterly for Exercise and Sport* 69: 111-128.

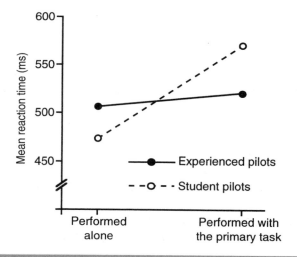

Figure 17.9 Mean reaction time on the secondary (memory) task for experienced and student aircraft pilots when the task is performed alone and in conjunction with a primary task.
Reprinted from Crosby and Parkinson 1979.

control skills. Figure 17.9 illustrates the superior performance of experienced pilots undertaking a reaction time task while performing a landing procedure on a flight simulator.

Observable Movement Pattern and Movement Outcome

By definition, the movements produced by skilled performers more consistently and precisely match the requirements of the movement task than those of lesser-skilled performers. In other words, expert performers develop, with practice, a greater capability to produce exactly the movement outcomes needed for successful performance. They achieve these successful movement outcomes through movement patterns that also show some reliable differences from those of novice performers. Movement patterns, as discussed in chapter 6, can be best described in terms of their kinematics, kinetics, and underlying neuromuscular patterns using the techniques of biomechanics.

Studies drawing on both biomechanics and motor control have revealed a number of important changes in observable movement patterns with growing expertise. In terms of kinematics, the movement patterns of expert performers are characterized by greater consistency with respect to overall movement duration as well as movement trajectories and displacement-time characteristics. This greater consistency in the observable movement pattern is necessarily also reproduced in the underlying movement kinetics. Typically the force–time curves not only are more consistent on a trial-to-trial basis for expert performers but also show a clearer, more distinct pattern of force pulses. Expert performers make greater use of the external forces (such as gravity and reactional forces) available within movements and restrict the injection of muscular force generated by the body to only those points in the movement where it is needed and can act most effectively. The time course of power generated in movements consequently varies significantly between experts and novices. Novices tend to supply muscular force more frequently throughout a movement, often either inefficiently in opposition to external body forces or as an unnecessary supplement to external forces. It is therefore not surprising to observe that neuromuscular recruitment patterns (as revealed from electromyography) become more discrete with practice and that there is a general reduction in recruitment as muscular contraction extraneous to the movement of interest is eliminated (figure 17.10).

Implications for Training

The value of knowledge about expert–novice differences in information-processing capability is that it provides a guide to where energy and attention should be directed in practice and training. Training based on improving the information-processing factors known to be related to the expert's advantage on a task would appear to be more sensible than training focusing on factors providing little or no discrimination between experts and novices. This logic is unfortunately not always fully appreciated or considered in the design of practice and recommendations for practice. For example, the generalized visual and kinaesthetic training programs that become popular from time to time, and that are designed to improve motor skills through improving the general sensitivity of the sensory systems for movement, are most unlikely to be beneficial for skill learning. The reason is that they do not, under most circumstances, train any of the limiting information-processing factors for skill performance. The available evidence on motor expertise also clearly suggests that the training of perceptual and decision-making skills is, in many cases, just as important as the training of movement execution skills, if not more so—yet this is also not frequently reflected in current training practices.

Given that good practice is clearly fundamental to motor skill learning, the next section focuses on some of the major factors known to affect the learning of motor skills.

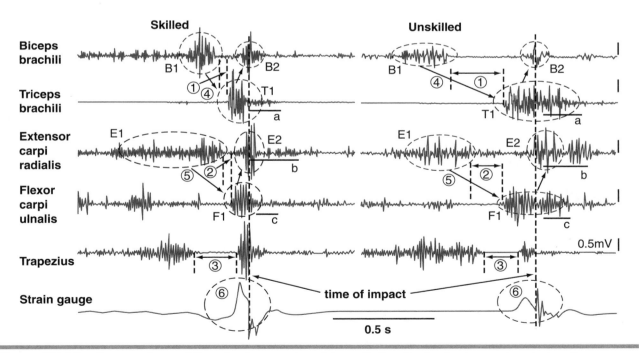

Figure 17.10 Comparison of the electromyographic activity of selected muscles during the execution of an overarm hitting action by skilled and unskilled performers. Note that the bursts of muscle activity for the skilled performers are more discrete than those for the unskilled.
Reprinted from Sakurai and Ohtsuki 2000.

Factors Affecting the Learning of Motor Skills

An age-old adage about skill learning is that "practice makes perfect." Studies by experts from a range of motor domains demonstrate the extraordinary amounts of deliberate, effortful practice (typically >10,000 h) undertaken in the acquisition of expertise and show that the sheer volume of practice is one of the key discriminators between experts and lesser-skilled performers (figure 17.11).

Although it is certainly true that extensive amounts of practice are necessary for high levels of skill to be developed, practice is a necessary but not a sufficient condition for learning. The "practice makes perfect" adage, therefore, while essentially true, needs to be qualified in a number of ways.

Imperfectability of Skills

One important qualification to the "practice makes perfect" adage involves the recognition that although motor skills improve with practice, there is no reason to suggest that they ever become perfect—in other words, that there is no room for further improvement through learning. In even extremely simple tasks, like hand rolling cigars in a factory, improvements in performance are still apparent after as many as 100 million trials of practice (figure 17.12)! In more complex motor skills, which involve many more components that can be potentially improved with practice, improvements are likely to extend over an even greater time scale. There is no evidence that skill learning ever ceases, provided that practice is ongoing. The leveling out of performance observed after a number of years of practice or performance in various motor skills is more likely attributable to either psychological or physiological factors or to measurement difficulties. In light of this, a more appropriate adage may be "practice makes better."

Necessity of Feedback for Learning

Although practice is necessary for learning, practice alone does not guarantee learning. In particular, learners must be able to regularly derive feedback about their performance in order for practice to be effective in improving learning. If learners are not able to gain information about the success or lack of success of each attempt they make at a

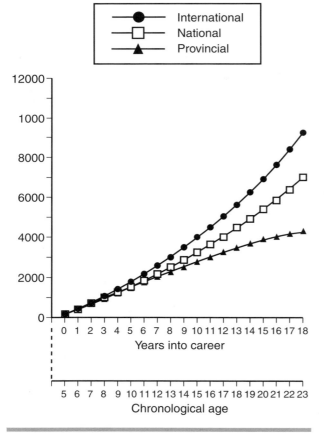

Figure 17.11 Accumulated hours of practice as a function of years of playing and chronological age for soccer players of different skill levels.

Reprinted from Helsen, Starkes, and Hodges 1998.

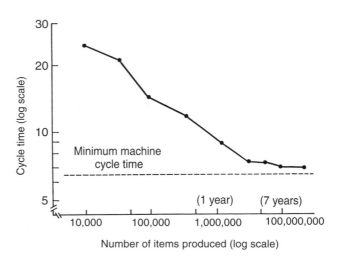

Figure 17.12 Time taken to hand roll a cigar as a function of amount of practice for factory workers.

Reprinted from Crossman 1959.

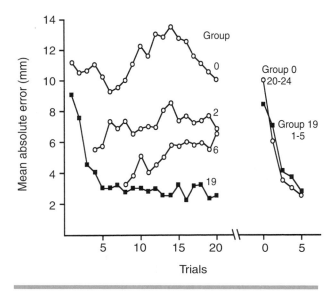

Figure 17.13 Error in reproducing a simple movement as a function of the amount of feedback received.

Reprinted from Bilodeau, Bilodeau, and Schumsky 1959.

new task, learning will be impaired and indeed may not occur at all.

In the study shown in figure 17.13, participants had to learn to perform an accurate arm-positioning movement on the basis of kinaesthetic information alone. The participants were divided into four groups who differed in the number of successive trials on which they were given feedback information (Group 0 had no feedback; Group 2 had feedback for the first 2 trials only; Group 6 for the first 6 trials; and Group 19 for the first 19 trials). Performance was directly related to the amount of feedback given: Group 19 performed the best, and Group 0 performed the worst. Significantly, Group 0 apparently learned nothing at all over the duration of the training period. This can be ascertained from the observation that this group, when eventually given feedback, improved their performance at a rate essentially identical to that seen over the initial trials of practice for the group given feedback on every trial. All 20 trials of practice done without feedback were therefore apparently of no benefit for learning. This suggests another modification of the "practice makes perfect" adage. "Practice, the results of which are known, makes better" may more accurately encapsulate the nature of the relationship between practice, feedback, and learning.

Not all feedback is, of course, equally effective. As a general rule, feedback information is more effective to the extent that it is more specific and also provides information from the learner's perspective rather than from an external, third-person perspective. This is true whether the feedback information is derived by the learners themselves, gained from media such as video, or provided by a teacher, coach, or instructor. Feedback information must be limited to key features to avoid overloading the information-processing capacity of the learner. For this reason, summary feedback presented at the end of a block of trials rather than after each individual trial may be advantageous. This approach also encourages learners to develop the ability to extract their own feedback information (rather than relying on feedback from external sources), and this ability is essential for learners' progression toward the autonomous phase of learning.

Importance of the Type of Practice

Just as different types of feedback vary in their effectiveness for learning, so too do different types of practice. The teacher or coach of motor skills charged with the responsibility of designing training or practice regimes for maximal effectiveness must consider a range of issues such as the length of the rest intervals between practice attempts, the extent to which fatigue should be included or avoided in the practice sessions, and the degree to which the practice should be repetitive as opposed to variable.

A common guiding principle, which holds across all motor skills, is one of specificity. This is the notion that skills should be practiced under conditions that most closely replicate the information-processing demands of the situation in which the skills must ultimately be performed. Consequently the best type of practice differs for different motor skills. Importantly, in some cases, the best type of practice may differ from commonly and traditionally accepted methods of practice for a particular skill or set of skills. Consideration of practice for the motor skills involved in golf may well illustrate this point.

To play golf successfully, a person must be able not only to drive the ball a long way with wooden and long iron clubs but also to accurately pitch and chip the ball, play from sand bunkers, and putt with precision. The golfer must also be able to adapt these skills on a shot-by-shot basis to accommodate such factors as the lie of the ball, the force and direction of the wind, and the position of the hazards and obstacles. A key issue for the player and coach is how these skills might best be practiced.

The traditional form of structured practice for golfers is to take a bucket of balls to the practice range and hit the same club over and over again until the skill is executed effectively and ingrained. This form of practice, in which each component skill is performed repetitively, is referred to as **blocked practice.** Most practice that involves drills is of this type. This type of practice can be contrasted with **random practice,** in which clubs and shots are practiced in essentially random order, in a manner not dissimilar to that which occurs in actually playing a round of golf. A random practice schedule might involve, for example, hitting in order a driver, a 7 iron, a sand wedge, and a putter. With such a practice method the same specific task is never repeated on two successive practice trials. Normally different clubs are used on successive trials, but if the same club (e.g., the putter) were to be used twice in a row it would be from a different position or distance. The critical question is which type of practice is most effective for learning and performing golf skills.

We may gain some insight into this question from the results of a study on blocked and random practice as shown in figure 17.14. In this study, participants learned three different rapid hand and arm movement tasks. One group of participants learned the tasks through blocked practice and the other through random practice. The total amount of practice was the same for the two groups. During practice, the blocked group performed best. This is not surprising given that they were constantly exposed to the same task requirements trial after trial.

What is important, however, is how well the two groups were able to retain what they had practiced (recall that learning is a *relatively permanent* change in the ability to perform skills). The skills were retained best (and hence, by inference, learned best) by the group who had experienced random practice. Blocked practice was particularly ineffective when subsequent performance of the practiced skill was under random conditions. This is a crucial observation, as this is precisely the situation that exists for the golfer who does repetitive practice at the driving range and then attempts to use these skills under playing conditions. Thus, while blocked practice may be valuable in the very early stages of learning, it is important to recognize that blocked practice alone may give an inflated view of actual skill level and may be a less than optimal preparation for a task that ultimately requires execution of component skills in essentially random order. This example illustrates the importance of designing practice in a way that best simulates the demands of the actual performance setting.

Limitations of Verbal Instruction and Conscious Attention

The conventional approach to skill acquisition relies heavily on the premise that detailed verbal instruction about how to perform the skill and conscious attention to this information by the learner are beneficial to learning. Learning achieved under these circumstances is referred to as **explicit learning,** with one of its defining characteristics being the concurrent acquisition of both the movement skill and knowledge about the performance of the skill.

Although explicit learning may be beneficial in the early stages of skill acquisition, there is a danger in becoming "bogged down" with too much knowledge. An excessive focus on explicit knowledge can prevent

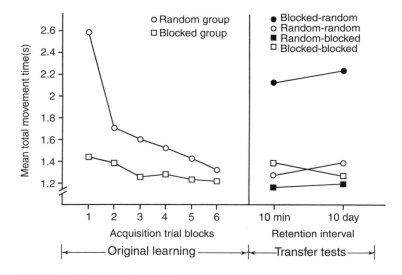

Figure 17.14 Performance during the acquisition and retention of a complex movement task for groups practicing under either random or blocked conditions.
Reprinted from Shea and Morgan 1979.

desirable progression to the autonomous stage of learning, and too much conscious attention to performance can interfere with the automatic production of movement and lead to errors of "paralysis by analysis." Making movement control conscious can leave skill performance open to disruption from other conscious processes, such as those that will be discussed in chapter 18 in relation to anxiety. It is for this reason that interest in **implicit learning** of motor skills is growing. In implicit learning, conscious attention is directed away from the task at hand (typically through the use of a concurrent secondary task); verbal instruction is minimal or absent; and the participants simply practice and acquire the skill without the concurrent acquisition of explicit knowledge about the performance of the skill. In cognitive tasks, such as pattern recognition, skills learned implicitly have

been shown to be both more resistant to forgetting and more resistant to stress and anxiety than the comparable skills learned explicitly. Preliminary evidence suggests that this is also likely true for the learning of motor skills (see "Implicit Learning of Motor Skills," below).

Dependency of Learning on Readiness

In chapter 16, in the context of the notion of critical periods, we saw that if the learner is not developmentally ready, practice may be spectacularly ineffective in improving skill; yet when the learner reaches an appropriate developmental stage, skills may be acquired with surprisingly few trials of practice (see "Critical Periods and Enriched Environments: The Story of Johnny and Jimmy" on p. 244). While developmental factors have the most pronounced effects in moderating

---

IN FOCUS: IMPLICIT LEARNING OF MOTOR SKILLS

Rich Masters (at the time from the University of Birmingham, in the United Kingdom, and now at the University of Hong Kong) was interested in the question of the relative merits of implicit and explicit approaches for learning golf putting. He randomly allocated novice golfers to one of five experimental groups. Participants in the implicit learning (IL) group and the implicit learning control (ILC) group received no instruction on how to putt and were required to do a concurrent random letter generation task at the same time they practiced putting in order to divert their conscious attention away from the putting skill. Participants in the explicit learning (EL) group were given very specific instructions on how to putt and were required to follow these instructions throughout practice. Participants in the stressed control (SC) and the nonstressed control (N-SC) groups were given no instructions or any secondary task but were simply instructed to improve as much as possible. After four practice sessions of 100 putts, the groups were tested at a final session, again of 100 putts. In this test condition, three of the groups (IL, EL, and SC) were subjected to performance stress (through monetary rewards for performance and evaluation by an "expert" assessor) whereas the others (ILC and N-SC) were not. Analysis of putting performance showed that those who learned implicitly were

less likely to suffer performance decrements under stress.

Sian Beilock from Michigan State University and colleagues examined the performance of experienced golfers under two conditions. In one condition the golfers undertook a dual task so that their attention would be distracted from putting, and in the other they gave conscious attention to step-by-step monitoring of movement execution. In both this experiment and a second experiment on soccer dribbling, the findings indicated that deliberate attention to the conscious step-by-step monitoring of performance can actually impede the performance of experts whereas it benefits the performance of novices and less skilled performers.

SOURCES

- Beilock, S.L., T.H. Carr, C. MacMahon, and J.L. Starkes. 2002. When paying attention becomes counterproductive: Impact of divided versus skill-focussed attention on novice and experienced performance of sensorimotor skills. *Journal of Experimental Psychology: Applied* 8: 6-16.
- Masters, R.S.W. 1992. Knowledge, knerves and know-how: The role of explicit versus implicit knowledge in the breakdown of complex motor skill under pressure. *British Journal of Psychology* 83: 343-358.

the effectiveness of practice, other psychological factors present throughout the life span are also important in determining the extent to which practice translates into actual learning. Foremost among these are the motivation of the learner and his or her level of arousal and anxiety. These concepts are considered in the next part of the book, which introduces the subdiscipline of sport and exercise psychology.

Summary

Practice (training) results in changes in information-processing capabilities and in the underlying structure and function of the brain and neuromuscular system, and these changes collectively produce significant observable changes in motor performance. With practice, skills become less consciously controlled and more automatic—this provides highly skilled performers with both the capacity to perform multiple tasks concurrently and improvements in efficiency that help delay the onset of fatigue. Clear expert–novice differences are evident in all three aspects of central information processing (perceiving, deciding, and acting), indicating that the nervous system responds to training through functional adaptations in much the same manner as do other key systems for human movement. Although the precise neural mechanisms for learning are not yet well understood, it is apparent that the nervous system possesses considerable plasticity. Short-term changes in synaptic efficiency and long-term changes in synaptic connectivity appear to be fundamental neural foundations for learning. Under the appropriate set of practice conditions, and in the presence of suitable feedback, continuous refinement and improvement of all motor skills seem possible (even simple tasks performed by acknowledged experts). The challenge for researchers and practitioners alike is to understand more fully the optimal practice conditions for the continuous learning of different types of motor skills.

Further Reading and References

Abernethy, B., J. Wann, and S. Parks. 1998. Training perceptual-motor skills for sport. In *Training in sport: Applying sport science*, ed. B.C. Elliott (pp. 1-68). Chichester: Wiley.

Christina, R.W., and D.M. Corcoz. 1988. *Coaches guide to teaching sport skills*. Champaign, IL: Human Kinetics.

Ericsson, K.A., ed. 1996. *The road to excellence: The acquisition of expert performance in the arts and sciences, sports and games*. Mahwah, NJ: Erlbaum.

Magill, R.A. 1998. *Motor learning: Concepts and applications*. 5th ed. Boston: McGraw-Hill.

Schmidt, R.A., and C. Wrisberg. 2000. *Motor learning and performance*. 2nd ed. Champaign, IL: Human Kinetics.

Shumway-Cook, A., and M.H. Woollacott. 2001. *Motor control: Theory and practical applications* (chapter 4). 2nd ed. Baltimore: Lippincott Williams & Wilkins.

Starkes, J.L., and K.A. Ericsson, eds. 2003. *Expert performance in sports: Advances in research on sport expertise*. Champaign, IL: Human Kinetics.

PART V

PSYCHOLOGICAL BASES OF HUMAN MOVEMENT

The Subdiscipline of Sport and Exercise Psychology

Sport and exercise psychology is the subdiscipline of human movement studies involving the scientific examination of human behaviour and cognition (thought processes) in the context of physical activity. The distinction between sport psychology and exercise psychology is not always clear. Sport psychology obviously is concerned with human behaviour in the sport environment, whereas exercise psychology focuses on the exercise environment. Issues relating to competition and sporting performance traditionally fall under the jurisdiction of sport psychology. Exercise psychology typically involves the study of why people do or do not exercise and the psychological effects of exercise on the person. Some overlap between the two fields is inevitable. Many competitive athletes participate in exercise that is not sport specific to improve various aspects of fitness. Similarly, some outdoor recreation activities could be classified as either sport or exercise.

Typical Questions Posed and Problems Addressed

Sport and exercise psychologists address two major questions:

- What effect does participation in physical activity have on the psychological makeup of the participant?
- What effect do psychological factors have on physical activity participation and performance?

Areas of interest related to the first question include the effects of exercise on psychological well-being and the effect of participation in youth sport on the development of character. Examples of themes related to the second main question are the effects of anxiety on competitive performance and the effects of self-confidence on participation in physical activity.

Additional issues in the field include the following:

- The effect of personality on participation or performance
- Reinforcement, feedback, and performance
- Leadership effectiveness
- The enhancement of sporting performance or exercise adherence through the development of psychological skills
- Techniques for developing self-confidence

- The effects of external rewards on motivation
- The influence of an audience on performance
- Determinants and consequences of team cohesion
- Psychological predictors of athletic injuries
- Goal setting
- Concentration and attention
- Drug abuse in sport

An introductory text does not allow for the coverage of all of these issues; however, some of them are explored in the following four chapters. Chapter 18 addresses some of the basic concepts in sport psychology including personality, competitive anxiety, and the use of imagery. Chapter 19 introduces exercise psychology. Chapters 20 and 21 follow the structure of the previous chapters by focusing on psychological factors across the life span (e.g., psychosocial development of children, life satisfaction in the aged, and termination of athletic careers) and psychological adaptations to training (e.g., psychological stress, burnout, and the transfer of mental skills).

Levels of Analysis

Because sport and exercise psychologists do not have the technology to directly record individuals' thoughts or feelings at a specific moment, they must rely on the use of multiple levels of analysis. The most commonly used levels of analysis are the behavioural or observational level, the cognitive level, and the physiological level. The behavioural or observational level involves watching individuals and recording what they do. For example, when studying anxiety during competition, psychologists might use the behavioural or observational level of analysis for indications of anxiety such as irritability, yawning, the trembling of muscles, the inability to make decisions, or the inability to concentrate. The study of competitive anxiety using the cognitive level of analysis typically entails questionnaires or inventories. The physiological level of analysis involves the direct measurement of physiological variables. In studying competitive anxiety this might consist of measurement of heart rate, respiration rate, or muscle tension. There is frequently a lack of coherence across different levels of analysis within sport and exercise psychology. Therefore, making generalizations on the basis of

a single level of analysis may be misleading. For this reason it is important to use multiple levels of analysis within this subdiscipline.

Historical Perspectives

Sport and exercise psychology is a much newer area of study than some of the other biophysical subdisciplines of human movement studies. The first recognized studies in sport psychology appeared in the late 1890s. These were isolated studies that included the investigation of topics such as reaction time, audience effects, mental practice, and the personality of athletes. In North America the first systematic research in sport psychology is attributed to Coleman Griffith, who wrote books on the topic, worked as a practitioner with athletes, and established a sport psychology laboratory at the University of Illinois in the 1920s. At the same time, organized sport psychology was beginning in Eastern Europe with the establishment of the Institutes for Physical Culture in Moscow and Leningrad. Extensive research began in the Soviet Union in the 1950s to help cosmonauts control bodily functions and emotional reactions while in space. Later these methods were applied to Soviet and East German elite athletes. Although very little of the sport psychology research and practice from within Eastern Europe was available outside the region, it is apparent that there was a strong regime of applied research on the enhancement of the performance of elite athletes.

In North America, little was accomplished in the area of sport psychology after Griffith until the 1960s, when numerous researchers became actively involved in the area and professional organizations began to form. Many of these research workers had been trained in motor control but saw the need to develop sport psychology independently of the study of motor control. In 1965 the International Society of Sport Psychology was formed in Rome, with professional societies for sport psychology and motor control in the United States and Canada forming soon after (1967 and 1969, respectively). Although the emphasis in Eastern Europe remained on field research, North America emphasized laboratory research. It was not until the 1980s that greater importance was placed on applied and field-based research outside Eastern Europe. Active research on, and practice of, sport psychology are now evident in most countries throughout the world.

Although studies on exercise and psychological factors have been intermingled within sport psychology studies during the past 30 or 35 years, and the International Society of Sport Psychology helped to identify exercise psychology as a separate area of study in 1968, exercise psychology has only recently emerged as a specialist area. In 1988 the *Journal of Sport Psychology* became the *Journal of Sport and Exercise Psychology*. The first textbooks focusing solely on exercise psychology were not published until the 1990s. In most professional organizations, exercise psychology is seen as a subdivision of sport psychology. However, just as sport psychology and motor control have split into two distinctive fields, the future might also see the professional division of sport psychology and exercise psychology.

Professional Organizations

The International Society of Sport Psychology (ISSP) was organized in 1965 with the stated purpose of promoting and disseminating information throughout the world. In addition to publishing the *International Journal of Sport Psychology*, the ISSP holds a World Congress in sport psychology once every four years. The location of the Congress varies. For example, the 8th, 9th, and 10th World Congresses were held in Portugal, Israel, and Greece, respectively. The 11th World Congress in 2005 will be held in Sydney, Australia.

Numerous regional organizations of sport psychology have also been formed. For example, in Europe there is the European Federation of Sport Psychology, which is officially called the Fédération Européenne de Psychologie du Sport et des Activités Corporelles (FEPSAC). There are also several national sport psychology organizations in Europe, such as the Associazione Italiana Psicologia dello Sport (Italy), the Sociedade Portuguesa de Psicologia Desportiva (Portugal), and the Société Francaise de Psychologie du Sport (France).

In North America, two major associations have emerged. The primary goal of the North American Society for the Psychology of Sport and Physical Activity (NASPSPA) is advancement of the knowledge base of sport psychology through experimental research. Perhaps because of its relatively early founding in 1967, NASPSPA has a comparatively heavy emphasis on motor learning and control and motor development. Founded in 1985, the Association for the Advancement of Applied Sport Psychology (AAASP) is concerned with ethical and professional issues related to the development of sport psychology and the provision of psychological services in sport and exercise settings. AAASP promotes the development of research and theory, but also focuses on intervention strategies in sport psychology.

The organizations just listed are not intended to be representative of all sport psychology associations, societies, and organizations. Sport psychology is growing rapidly in many countries in addition to those in Europe and North America—particularly in Australia, Brazil, China, India, Korea, and Nigeria.

Further Reading and Reference

Weinberg, R.S., and D. Gould. 2003. *Foundations of sport and exercise psychology* (chapter 1). 3rd ed. Champaign, IL: Human Kinetics.

Some Relevant Web Sites

Active Australia Web site
[www.activeaustralia.org]

Association for the Advancement of Applied Sport Psychology
[www.aaasponline.org/cc/index.php]

College of Sport Psychologists (Australian Psychological Society)
[www.psychsociety.com.au/units/colleges/sport/default.asp]

Commuter Challenge (Canada) Web site
[www.commuterchallenge.net]

Exercise and Sport Psychology (Division 47 of the American Psychological Association)
[www.psyc.unt.edu/apadiv47/]

Health Canada SummerActive Program Web site
[www.summeractive.canoe.ca]

Healthy People 2010 (U.S.) Web site
[www.healthypeople.gov]

International Society of Sport Psychology
[www.issponline.org]

Journal of Sport & Exercise Psychology
[www.humankinetics.com/products/journals/journal.cfm?id=JSEP]

National Center for Chronic Disease Prevention and Health Promotion: Physical Activity and Health
[http://www.cdc.gov/nccdphp/dnpa/physical/index.htm]

North American Society for the Psychology of Sport and Physical Activity
[www.naspspa.org/]

Sport and Recreation New Zealand (SPARC)
[www.hillarysport.org.nz]

The Sport Psychologist
[www.humankinetics.com/products/journals/journal.cfm?id=TSP]

West Australia Transport TravelSmart Web site
[www.dpi.wa.gov.au/travelsmart/]

World Health Organization (WHO) World Health Day Web site
[www.who.int/world-health-day]

CHAPTER 18

Basic Concepts in Sport Psychology

Sport psychology covers many topics, and an introductory text such as this one obviously cannot provide comprehensive coverage of the field. Instead, this chapter introduces four of the major domains of the field. In terms of the analogy between the human and the car, first introduced in the preface, sport and exercise psychology—with its focus on mental processes and behaviour—is about investigating the driver of the car. What assumptions might be made about an individual who owns a brand-new Jaguar versus someone who drives an old, beat-up Ford? One might presume that these individuals have different personalities. In addition to presenting information about personality and sport, this chapter addresses a few of the factors that might influence the performance of the driver. Specifically, we introduce and briefly consider motivation, anxiety, and imagery. We also deal briefly with the practice of applied sport psychology—what sport psychologists actually do.

In keeping with the strong tradition in psychology of focusing not only on average (group) behaviour but also on individual differences, this chapter and subsequent chapters in this part of the book make liberal use of applications of key concepts to hypothetical individuals. The approach to examples is therefore more personalized than that used in the preceding chapters.

Personality

Personality has been defined in various ways, but certain elements are common to all definitions. In

> The major learning concepts in this chapter relate to
> - the difference between the trait and interaction frameworks of personality;
> - limitations of personality research in sport;
> - the three components of motivation—direction, intensity, and persistence;
> - different definitions of success and their influence on motivation;
> - the roles coaches, parents, and teachers can play in establishing the motivational climate;
> - the difference between arousal and anxiety and how they relate to performance; and
> - what imagery is, how it works, and why it should be used.

its simplest form, personality is the composite of the characteristic individual differences that make each of us unique. Think about two people you know who act very differently. What is it about them that makes them different? For example, is one outgoing and the other shy? These differences are what make each of us unique. In portraying the differences between the two people you have considered, you have depicted aspects of their personalities.

Trait Framework of Personality

One framework for studying personality, the trait framework, suggests that everything we do is the result of our personalities. In other words, our behaviour is determined completely by our personalities. This idea can be formalized by the equation:

$$B = f(P)$$

where B stands for behaviour, f stands for function of, and P stands for personality. According to the trait framework, each individual has stable and enduring predispositions to act in a certain way across many situations. These predispositions, or **traits,** predict how we will respond. For example, if Peter had the trait of shyness he would be expected to be reserved or timid when joining a new team. If Sue, on the other hand, had the trait of being outgoing, she would be expected to be extroverted and sociable when first meeting new teammates.

The trait framework of personality suggests that personality traits can be objectively measured. Since traits are considered to be enduring and stable, inventories are used in an attempt to measure personality characteristics.

Traditional sport personality research has administered personality inventories to groups of athletes and nonathletes and then examined the results for any differences. Some studies have shown differences indicating that nonathletes are more anxious than athletes, and that athletes are more independent and extroverted than nonathletes.

There have, however, been problems with some of the traditional sport personality research. Most personality inventories were designed for a specific purpose with a specific population in mind. Several of these inventories were created for use with clinical populations and therefore are inappropriate for use with nonclinical populations such as athletes. Additionally, some of the questionnaires used in the research have not been shown to be valid and reliable. In other words, it has not been demonstrated that the questionnaires actually measure what they purport to measure, or that if the same questionnaire is given to a person more than once the same results will be obtained.

Aside from possible difficulties with the inventories used in the research, complications arise regarding the definitions of athletes and nonathletes. Trying to generalize findings across studies is difficult because the terms have not been defined in the same manner in the different studies. How does one define an athlete? In some studies, athletes have been defined as those competing in intercollegiate sport. Does this mean that individuals participating in club sports are not athletes? Other studies consider individuals to be athletes only if they have achieved a particular level of performance—not if they train and compete regularly at a lower level of proficiency. Would people who run on their own many times per week be considered athletes if they do not participate in competitions? What about someone who competes in social tennis matches

only once or twice per year? As you can see, even if valid, reliable, and relevant personality inventories can be obtained, the problem of determining who is an athlete and who is not makes comparing the personalities of athletes and nonathletes difficult.

Interaction Framework of Personality

Another model for studying personality is the **interaction framework.** The interaction framework recognizes that personality traits influence behaviour; however, it also acknowledges the situation or the environment as influencing how we act. This framework can also be expressed as an equation:

$$B = f (P \times E)$$

where E stands for environment. Within the interaction framework, the environment refers to all aspects of the situation that are external to the individual, including the physical surroundings as well as the social milieu (other people). It is important to note that the interaction framework reflects that personality is just that, an interaction. It is not only that personality traits and the environment both influence behaviour but also that the two interact and affect each other. For example, if Peter, the shy person, was attending the first training session of a new team, aspects of the environment could influence his behaviour. If the other team members were inseparable and suspicious of newcomers, Peter's behaviour would probably be different than if the team members went out of their way to welcome newcomers. Similarly, if Sue, the extroverted athlete, joined a new team, her outgoing personality would influence the behaviour of the people around her, thus influencing the team environment.

The interaction framework considers not only traits but also states. Traits, as previously described, are stable, enduring personality predispositions. **States,** on the other hand, refer to how someone feels at a particular point in time. Traits may influence states, but they do not directly determine states. For example, Bruce may have a low level of trait anxiety. Generally speaking he is a relaxed and calm person who does not get anxious easily. In most situations he is tranquil and unruffled. Bruce could be in the final of the local basketball competition and still remain calm. Certain situations, however, may cause Bruce to react with high levels of state anxiety—for example, if he found himself hurtling down the ramp of a ski jump or, more commonly, shooting a free throw with his team down by one point and with one second left in the game.

The main difference between the interaction framework and the trait framework is that the interaction framework acknowledges the ability of the situation or the environment to influence how we react. Behaviour is not determined solely by personality traits. Because the situation can influence behaviour, the use of personality inventories as a method of establishing how individuals will perform in all situations is not viable. Although these inventories may give some indication of the personality traits of individuals, measures that take into account specific situational factors are needed to determine states.

Practical Implications of Personality in Sport

If it could be determined that people with certain characteristics or traits perform better in some sports than in others, then personality inventories could be given to people to determine which sports suit them best. This process should also incorporate anthropometric, biomechanical, physiological, and motor control facets of the individuals. The challenge, of course, is to base any decisions of this type on factors that will continue to be predictive of performance over time.

Motivation in Sport

Many coaches complain that certain people would be great athletes if only they were motivated. The athletes are seen to have all of the anthropometrical, biomechanical, physiological, and skill components necessary for performing at a high level except that they just do not seem to care. They might arrive late for practice, fail to try hard during drills, or just not show up at all. People interested in community health may not be so concerned about the performance of specific individuals but are instead troubled by the large number of people who no longer participate in sport at all. Both concerns relate to the concept of motivation.

Definition of Motivation

Motivation is made up of three components: direction, intensity, and persistence.

- *Direction* refers to where people choose to invest their energy. There are few unmotivated people in the world; it may be instead that the direction of

a person's motivation is not what others would prefer. For example, a roommate may be motivated to go to the movies but not motivated to clean the house. Similarly, an athlete with great performance potential may be motivated to go to a party instead of to practice. Neither of these people lacks motivation; they have just chosen a different direction in which to invest their energy.

- *Intensity* refers to how much energy is invested in a particular task once the direction has been chosen. Two athletes may both choose the direction of practice, but where one invests very little effort, the other tries hard and works at a high level of intensity. It is worth noting, however, that a given exercise or drill requires different amounts of intensity from different individuals depending on their levels of fitness and skill.

- The third component of motivation is persistence. *Persistence* refers to the long-term component of motivation. It is not enough to have an athlete choose the direction of practice and work out at a high level of intensity during one or two sessions. Athletes need to continue to practice and participate over time. It is preferable to have athletes work out at a moderate level of intensity over an entire season than to have them work out at an extreme level of intensity at the beginning of the season and then drop out.

People interested in enhancing the motivation of sport participants should keep in mind all three components. One can influence direction by making the path of participation enjoyable. One can increase intensity by stipulating a reason for doing any particular activity. For example, how much effort are workers likely to put into a job if they think there is no reason for doing it? The same principle holds true for sport. Athletes will be more likely to put effort into a sporting drill if they perceive there is a reason for doing so. One can improve persistence by providing positive feedback to participants. Athletes are more likely to continue participating if they are convinced of the long-term benefits. Would students continue to study if they honestly felt that studying did not accomplish anything? If, however, they felt that studying gave them greater understanding of the subject, better prepared them for more advanced courses, led to higher grades, or caused them to get positive recognition from someone they cared about, they would be more likely to persist. Believing that continued participation leads to greater success increases persistence.

Definition of Success

Traditionally many have considered success in sport in terms of winning and losing; winning equals success and losing equals failure. If outcome is the only criterion for success, anyone who does not win fails. Using this definition, only one team in a given league can be considered successful at the end of the season; only one swimmer in a particular event is successful; and only one runner in a marathon is successful. Everyone else is a failure. Given that experiencing success enhances motivation, the outcome definition of success leaves the majority of participants lacking in motivation as they experience only failure.

Success, however, can mean different things to different people. Some people do define success in terms of outcome. They consider a performance successful only if the athletes were able to demonstrate that they were better than everyone else. Nevertheless, many athletes feel successful if they have improved their own performance, regardless of outcome. In other words, if technique improved, times decreased, or some aspect of performance was enhanced, the performance was successful. In this case, the athlete is making comparisons with his or her own previous performances rather than with those of others.

Some people do not use performance as the basis of their definition of success. Success for these people is achieved when they receive social approval or recognition from others. For example, a field hockey player may consider the season successful because she made many friends or obtained media recognition. Similarly, a basketball player may consider a game a success because he was complimented by the coach for his efforts.

These different definitions of success are called **achievement goal orientations.** The two most frequently studied achievement goal orientations are the task and ego orientations. A person with a strong **task orientation** has a self-referenced definition of success, whereas someone with a strong **ego orientation** defines success as being better than others. Athletes often have a combination of task and ego achievement goal orientations. They may want to improve their skills as well as win. Also, a single individual may have a stronger ego orientation for one sport but a stronger task orientation for another. For example, Victor may consider winning to be the primary definition of success in tennis but improvement in specific skills to be the principal basis for determining success in soccer.

Achievement Goal Orientations and Motivation

These different definitions of success influence motivation. For example, if a swimmer is in a race and is predominantly ego oriented, effort may decrease if he or she is well ahead of the others or is well behind and perceives that there is no chance of winning. If, however, the swimmer has a strong task orientation, he or she will maintain effort regardless of placement in the race, as for this person success is related to improving one's own time or technique.

Coaches should be aware that athletes often vary in their achievement goal orientations. Unfortunately, many coaches assume that all their athletes define success in the same way they do and that athletes' reasons for participating are also identical. This can be problematic if the goals of the coach and the athletes are in fact different. Athletes are more likely to drop out of a situation if they perceive that their needs are not being met. Therefore, if a coach is interested only in winning but the athletes are interested in making friends or learning new skills, the athletes may feel that they are not getting what they want and may abandon the sport altogether. Similarly, if a coach stresses individual growth and improvement by ensuring that every athlete gets similar playing time, and some of the athletes are solely interested in winning, friction and unhappiness may result.

Motivational Climate

Although dispositional tendencies predispose people to be high or low in task and ego orientations, the environment also has an impact. Perceived motivational climate is determined by the emphasis of situational goal perspectives. Coaches, parents, and physical education teachers create environments that differ in the amount of emphasis on task or ego. The majority of research on motivational climate in physical activity has taken place in physical education classes. A task or mastery climate is linked to enjoyment, perceived ability, and effort. An ego environment increases the tension and pressure felt by participants and negatively affects their interest and enjoyment. Goal orientations of children in physical education are related to the perceived motivational climate emphasized by parents, teachers, peers, and sporting heroes. See "The Influence of Parents, Teachers, Peers, and Sporting Heroes on Achievement Goal Orientations" on page 283.

IN FOCUS: THE INFLUENCE OF PARENTS, TEACHERS, PEERS, AND SPORTING HEROES ON ACHIEVEMENT GOAL ORIENTATIONS

Research has shown that physical education students' perceptions of motivational climate predict their achievement goal orientations. Literature also suggests that through the media, children feel they know their sporting heroes and can predict their behaviours, values, and beliefs. Two researchers from England decided to extend this area of research and explore the relationship between children's achievement goal orientations for physical education; perceptions of motivational climate emphasized by parents, peers, and teachers; and perceptions of sport heroes' achievement goal orientations.

English schoolchildren (145 males and 121 females) aged 11 to 15 years anonymously completed questionnaires designed to measure their achievement goal orientations and their perceptions of the motivational climates emphasized by parents, peers, teachers, and sporting heroes. Task orientation was positively related to perceptions of a task orientation in sport heroes and perceptions of a learning climate emphasized by parents, peers, and teachers. Ego orientation was positively related to perceptions of an ego orientation in sport heroes and perceptions of a comparison-oriented climate from parents, peers, and teachers.

These results demonstrate a link between children's achievement goal orientations and their perceptions of their sporting heroes' achievement orientations and the motivational climate emphasized by significant others. The results do not, however, indicate whether motivational climate and the perceived goal orientations of sporting heroes influence the children's goal orientations, or if the children's goal orientations affect their perceptions of motivational climate and their heroes' goal orientations. Future research needs to engage in longitudinal studies (studies across time) that contain interventions designed to manipulate the motivational climate and the perceptions of sporting heroes' achievement orientations. This latter intervention might involve producing videos of contemporary sporting heroes in task-involving situations. In the meantime, parents and teachers should be aware that the motivational climate they create may influence the achievement goal orientations of children. With this in mind, they should consciously emphasize the importance of effort and learning.

SOURCE

- Carr, S., and D.A. Weigand. 2001. Parental, peer, teacher and sporting hero influence on the goal orientations of children in physical education. *European Physical Education Review* 7(3): 305-328.

Arousal, Anxiety, and Sport Performance

Motivation is not the only psychological factor that influences participation and performance in sport. Arousal and anxiety levels have a profound effect on the performance of athletes.

Arousal

Arousal is traditionally considered to refer to physiological activation. When individuals are highly aroused or activated, they often have tense muscles and higher blood pressures, heart rates, and respiration rates than when they are calm. Arousal can also involve mental activation in addition to physiological activation. Therefore, one can consider arousal to be the degree of mental and physical activation or intensity. An example of a low level of arousal is the grogginess people experience when first awakening in the morning. An example of high arousal is the heightened mental alertness and physiological activation experienced before a major competition.

Anxiety

Confusion often results when people use the terms arousal and anxiety interchangeably. **Anxiety** is the subjective feeling of apprehension and is usually accompanied by increased arousal levels. High levels of arousal, however, are not always accompanied by anxiety. If a team just beat the defending league champions, they would probably be fairly aroused, yet they would almost certainly interpret this arousal as excitement rather than anxiety.

People usually experience anxiety when they perceive a situation as threatening. This commonly occurs when there is a perceived imbalance between the demands of the situation and the individual's ability to meet those demands. For example, if Diane believes that she must make the final two free throws in a basketball game yet does not believe that she is very good at making free throws, she is likely to experience anxiety. This anxiety will increase the more Diane believes that there may be negative repercussions if she misses. If, for example, she did not care about the outcome of the game or her own performance, and was playing only to have fun with her friends, she would probably experience only mild anxiety, if any. On the other hand, if she felt the outcome of the free throws to be extremely important, she would likely experience a relatively high level of anxiety. Anxiety, therefore, is determined by the perceptions of the individual. If two individuals are in the same situation, one may perceive the situation as slightly challenging and the other perceive it as extremely threatening.

Trait and State Anxiety

This sensation of anxiety is a good example of the interaction framework mentioned earlier in the discussion of personality. The level of anxiety experienced is a result of the interaction of personal factors (e.g., personality, needs, capabilities) and situation factors (e.g., opponent, task difficulty, and the presence of other people). An athlete with a predisposition to perceive situations as stressful or threatening would be considered as having a high level of **trait anxiety. State anxiety,** on the other hand, is the experience of apprehension at a particular point in time.

Cognitive Anxiety and Somatic Anxiety

State anxiety is a multidimensional concept. **Cognitive anxiety** is the mental facet of state anxiety. Worry, perceived threat, and self-defeating thoughts are possible aspects of cognitive state anxiety. **Somatic anxiety** refers to physiological anxiety responses such as butterflies in the stomach and sweaty palms. Although somatic and cognitive state anxiety are related, it is possible to experience a high degree of one without experiencing a high degree of the other. For example, students experience the anxiety of examinations differently. Some get very tense physically, whereas others drive themselves (and everyone else) crazy with their

worrying. The same is true for the experience of anxiety in sport. Some athletes have more signs of somatic anxiety, while others experience greater cognitive anxiety.

The Arousal–Performance Relationship

If Wendy had just awakened and was feeling sluggish and tired, she probably would not perform very well if asked right away to engage in competitive sport. Her performance would probably improve as she became more alert and awake. In other words, her performance would improve with increases in arousal. This proposed linear relationship between arousal and performance is called the **drive theory** and is illustrated in figure 18.1. According to the drive theory, performance continues to improve with further increases in arousal or activation. Therefore, if the drive theory is to be believed, one way to ensure optimum performance is to arouse athletes as much as possible.

Some coaches have done strange things because of their belief in drive theory. Coaches have had their players bite the heads off of live chickens, castrate a bull, stage a mock gun battle in a school cafeteria (complete with fake blood), and watch films of prisoners being murdered in concentration camps. Fear, anger, and horror were seen as emotions that could raise arousal levels. The point of the activities was to arouse the athletes as part of precompetition preparation.

Unfortunately, there are still some coaches who continue to base precompetition preparation on the drive theory. The **inverted U hypothesis** provides a more useful method of considering the relationship between arousal and performance. As seen in figure 18.2, when the relationship between arousal and performance is plotted according to this hypothesis, the graph forms an upside-down "U." As with the drive theory, performance increases with increases in arousal. However, these increases in performance occur only up to a certain point of arousal. If arousal continues to increase past that point, performance begins to deteriorate. The point at the top of the inverted U is called the point of optimal arousal. It is at this point that performance is best. If arousal is below that point (i.e., underarousal), performance will be less than optimal. If, however, arousal is past the point of optimal arousal (i.e., overarousal), performance will also be less than optimal. Therefore, continually increasing arousal could be problematic, causing overarousal and thus impairing the athlete's performance.

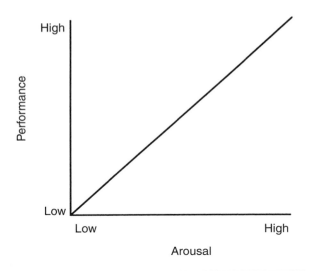

Figure 18.1   Drive theory.

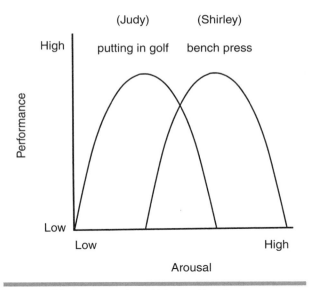

Figure 18.2   Inverted U hypothesis.

The challenging aspect of this relationship for coaches is that the **optimal level of arousal** varies across individuals even within the same sport (figure 18.3). Judy may perform her best when feeling relaxed. Shirley may perform her best when feeling pumped up and activated. One can also see differences in optimal arousal when considering different sports. If Judy is putting in golf and Shirley is attempting to lift a personal best in bench press, Judy's optimal level of arousal will be lower than Shirley's.

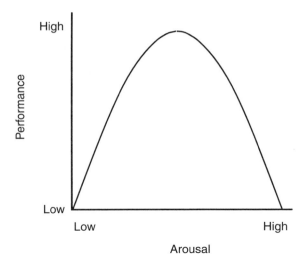

Figure 18.3   Variance in optimal levels of arousal.

In practical terms, these individual differences in optimal arousal mean that the identical precompetition buildup will not be equally effective for everyone. Athletes who arrive at the competition feeling sluggish and underaroused may benefit from a big motivational speech from the coach. However, if other athletes on the team are already at their optimal levels of arousal, or already overaroused, a motivational pep talk will probably have a negative impact on their performances. Some athletes may need to focus on relaxing and lowering their levels of arousal before competing.

The Anxiety–Performance Relationship

As mentioned previously, state anxiety is considered multidimensional. Somatic and cognitive state anxiety have different effects on performance. The relationship between somatic state anxiety and performance is virtually identical to that between arousal and performance. That is, the relationship is curvilinear. As state somatic anxiety increases, there are initial increases in athletic performance. If somatic anxiety continues to increase, then performance gradually decreases.

Cognitive state anxiety, however, is believed to have a negative linear relationship with performance. As cognitive anxiety increases, performance decreases (figure 18.4). In simple terms, as people begin to worry, their athletic performance begins to suffer. The more worry and distress, the worse the performance.

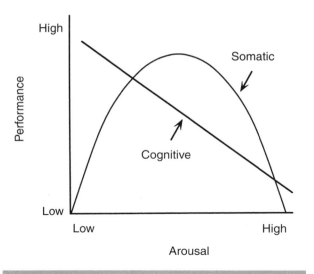

Figure 18.4 Relationship between multidimensional state anxiety and performance.

Things become more complicated when we try to consider the simultaneous impact and interaction of cognitive anxiety and somatic anxiety (or arousal) on performance. When cognitive anxiety is low, the relationship between arousal and performance takes the form of the inverted U. However, when cognitive anxiety is high, the relationship between arousal and performance takes the form of the **catastrophe model**. According to the catastrophe model, if arousal continues to increase beyond the optimal point, performance shows a sharp and rapid decline rather than a gradual decrease. Basically, a catastrophe occurs.

If cognitive anxiety is high, an increase in arousal can cause a terrible performance. For example, if Rob is playing basketball and has a high level of cognitive state anxiety, an increase in arousal past the optimal point could cause him not only to miss easy shots but also to make multiple errors in shot selection, defence, and passing. According to the catastrophe model, for Rob to return to his optimal level of performance while experiencing high cognitive anxiety, he would need to allow his physiological arousal to return to a relatively low level before gradually building his arousal and performance back up to optimal levels. That is, although a relatively small increase in arousal can cause a catastrophe in performance, a subsequent minor decrease in arousal is not sufficient to allow performance to return to its previous level. Instead, a significant decrease in arousal is needed before the previous standard of performance can be obtained.

In the catastrophe model, which explains the relationship between arousal and performance under high levels of cognitive anxiety, arousal is something like a wave crashing on the beach. Once the wave reaches its highest point, it does not just gradually decrease in height but instead decreases suddenly and dramatically. The water is unable to wash down the beach and instantly create a new wave. Instead, it washes back into the ocean before once again gradually building up to its optimal height.

Measuring Anxiety

Most researchers and practitioners use self-report measures of anxiety. Although physiological measures such as respiration or heart rates, biochemical changes, and electrophysical factors can provide information about anxiety, they are problematic because the results do not tend to agree from one measure to another and because individuals differ in how they physiologically respond to anxiety. In addition, most sports involve physical activation that affects the measures. Generally these measures can be used only in stationary sports such as rifle shooting without confounding anxiety and physical activation.

Self-report measures of anxiety use questionnaires designed to measure either trait or state anxiety. The trait measures indicate the individual's typical level of anxiety. State measures determine the level of anxiety experienced in a particular situation. Most questionnaires contain separate scales for cognitive and somatic anxiety. Traditionally the scales have measured the intensity of anxiety experienced. In the 1990s, some researchers added a directional component to scales, as two people may experience the same intensity of a symptom of anxiety but interpret its impact differently. For example, Julie might interpret butterflies in the stomach as a sign that the adrenalin is pumping and she is ready to perform (facilitative), but Marg might interpret the same butterflies as a sign that she is getting nauseated and will perform poorly (debilitative).

The Practice of Applied Sport Psychology

What sport psychologists do varies across practitioners. Those who are trained through the sport and exercise sciences tend to focus on performance enhancement, helping athletes improve the quality and consistency of their performances by teaching them psychological skills such as goal setting, imagery, self-talk, and relaxation. This approach to applied practice is often termed educational sport psychology. Sport psychologists who are educated in the areas of clinical or counseling psychology tend to have a broader view of applied practice and to treat and care for athletes both in and out of sport. Not surprisingly, this approach to applied practice is usually termed clinical sport psychology. Some organizations also have created a third category, research sport psychologists, for those who conduct investigations to increase knowledge in the field but rarely work directly with athletes. Ideally, people entering the field will obtain training in both psychology and the sport and exercise sciences. In addition, people who will be working with athletes need to complete supervised practica. The specific courses and the number of hours of supervised experience required by prospective sport psychologists vary across countries.

Imagery: An Example of a Psychological Skill

Both educational sport psychologists and clinical sport psychologists use cognitive–behavioural interventions. Performance enhancement service delivery is based on techniques that allow athletes to learn about the relationships between cognitions, feelings, and behaviours and to control factors that prompt or reinforce behaviour. As there are entire textbooks devoted to these psychological skills, what follows is only a brief introduction to one of the skills, namely imagery.

Imagery involves using all the senses to create or recreate an experience in the mind. Terms such as mental practice, mental rehearsal, and visualization have been used interchangeably with the term imagery. Mental practice and mental rehearsal could be considered specific forms of imagery; the person uses imagery to mentally practice or rehearse particular skills or techniques. Visualization and imagery are often considered synonymous. However, the term imagery is preferable in that visualization implies restriction to the sense of vision, whereas imagery more obviously includes sound, smell, taste, and feel, as well as sight.

How Imagery Works

Although sport psychologists, athletes, and coaches generally agree that imagery helps individuals learn new skills and improve performance, there is less agreement about how it works. Here we briefly describe three of the existing theories of how imagery works; however, no single theory has proven to be the definitive answer. Nevertheless, each of the following theories should help to create an understanding of imagery.

- **Psychoneuromuscular theory.** The psychoneuromuscular theory states that when we image ourselves moving, our brain is sending subliminal electrical signals to our muscles in the same order as when we physically move. In other words, all of the coordination and organization of movement take place in our brain. When we image movement, all of that organization and coordination takes place. Our brain sends electrical signals to our muscles about which muscles should contract when and with what intensity. Therefore, whether an athlete is actually performing a movement or vividly imaging the performance of the same movement, similar neural pathways to the muscles are used. The psychoneuromuscular theory claims that it is through this mechanism that imagery may aid skill learning.

- **Symbolic learning theory.** The symbolic learning theory is similar to the psychoneuromuscular theory except that electrical activity in the muscles is not required. According to the symbolic learning theory, imagery helps the brain to make a blueprint (or plan) of the movement sequence without actually sending any messages to the muscles. Imagery helps to develop a blueprint that can be followed when action is required.

- **Attention–arousal set theory.** According to the attention–arousal set theory, imagery works because it helps the athlete to reach the optimal level of arousal and focus attention on what is relevant. According to this theory, imagery does not send messages to the muscles or help develop a blueprint but instead primes the athlete for performance. Physiologically, imagery helps the athlete raise or lower arousal levels to the appropriate point. Cognitively, imagery helps the athlete attend to what is important for good performance, decreasing the chance of distraction.

Reasons to Use Imagery

Imagery has many advantages. It is not physically fatiguing, so practicing with imagery just before competition does not cause a decrease in the energy available. Imagery can be practiced anywhere and anytime—when one is sitting on a bus, waiting in the marshalling area, or taking a shower. Imagery can also be a welcome change of pace during a practice session. When an athlete is physically working hard, it can be useful to take a time-out and image specific aspects of technique. Imagery also uses a language that is understood by the body. Sometimes an athlete understands technical corrections of a skill but this understanding does not translate to the actual movement. Using imagery can enhance the translation. Additionally, in these times of expensive equipment, shoes, and gadgets, imagery is free!

Imagery can be used in different ways. Athletes can use imagery to help to learn to control emotions. For example, Jay is a rugby player who contributes much to the team except when she feels that an official has made a poor decision. She has a tendency to lose her cool, which often results in penalties. Through imagery, she can imagine herself shrugging off a poor call and focusing on what she needs to do to perform well. If she practices an appropriate reaction through imagery, the desired reaction will become natural when the situation occurs in real life.

Imagery can also be used to improve confidence. If athletes can see and feel themselves performing the way they want to immediately before they go out to perform, they will feel better about themselves. Another way of thinking about this notion is to consider how previous performances relate to confidence. If Cliff was about to shoot a free throw in the final seconds of a basketball game, he would probably feel more confident if he had just made his last 20 attempts than if he had just missed his last 20 attempts. Basically, the more times people do something, the easier it is to do. When athletes image themselves performing the way they want to immediately before the physical performance, they can create that feeling of confidence.

Imagery can be used in combination with physical practice to enhance the learning of new skills and the performance of known skills. Although there is some disagreement in the literature, most of the research supports the idea that combining physical practice and mental practice is more effective than doing either alone. For example, if Javier is learning how to serve in tennis, he will learn the skill more quickly if he both images himself serving correctly and physically practices serving.

There are, however, examples of individuals achieving a great deal with the sole use of imagery. Athletes from many sports have been able to

maintain their technique through the use of regular imagery sessions while injured and unable to physically practice. For example, a golfer who broke his wrist in a non-golfing-related incident was not able to play for almost two months. Although unable to physically practice, he imaged himself playing every day. He imaged successful tee shots, approach shots, chips, and putts. Shortly after the removal of his cast, he competed in his club's championship tournament. Although he did not win, his score was slightly better than his scores in the three tournaments prior to his injury.

Although there are many other uses of imagery, we consider only one more here. Imagery has been shown to help with the control of pain. The most common form of pain control imagery is probably removing oneself mentally from the situation. By conjuring up scenes that are unrelated to or incompatible with pain, people can learn to distract themselves from pain. For example, instead of thinking about how much her reconstructed knee hurts, Terry can image herself relaxing at the beach with the waves gently breaking at her feet, the sun warming her shoulders, and the breeze wafting across her face.

Developing Vividness and Control

If imagery is to be effective, two factors need to be developed—vividness and control.

The more vivid an image, the more likely the brain will be convinced that the image is real. When physically participating in sport, most athletes are aware of the feel of the movement and of sounds and sights in the environment. Some sport settings, such as chlorinated pools for swimming, are also associated with smells and tastes. The more senses that can be included in an image, the more vivid and therefore the more effective the image will be. Athletes get more out of imagery if they image in Technicolor with surround sound than if they image in fuzzy, silent, black and white. See "Imaging With All the Senses" (p. 289) for an example of guided practice in these techniques.

Vividness alone, however, is not enough. The athlete also needs to be able to control imagery in order for it to be effective. If we try to image a good performance, and instead image only mistakes and errors, anxiety will increase and self-confidence waiver. Following prewritten or prerecorded imagery scripts can often help the athlete image the desired performance. Following

IMAGING WITH ALL THE SENSES

Try to image the following sensory experiences. Are some images more vivid than others?

See

- A colourful sunset
- The face of a friend

Hear

- The roar of a crowd after the home team has scored
- A door slamming shut

Feel

- Stepping into a cold shower
- A bear hug from a close friend or relative

Smell

- Freshly mown grass
- Cigarette smoke

Taste

- Sour milk
- Chocolate

a guided imagery script is often easier than trying to develop an image from scratch. If vividness and control can both be developed, then imagery can be a productive technique for athletes.

Summary

Personality, or the composite of the characteristics that make each of us unique, has been studied using trait and interaction frameworks. Because of problems with some of the questionnaires used, as well as inconsistent operational definitions, traditional sport personality research has been inconclusive regarding whether athletes and nonathletes have different personalities. Motivation is made up of direction, intensity, and persistence. Making the direction of participation enjoyable, providing reasons for particular activities, and providing positive feedback can enhance the motivation of athletes.

Success in sport is not always equivalent to winning. Individual definitions of success are called

achievement goal orientations. The two most widely studied achievement goal orientations are task and ego. People with a high ego orientation define success as the ability to demonstrate that they are better than everyone else. People with a strong task orientation define success as improving their own performance. Coaches, parents, and physical education teachers create motivational climates that differ in the amount of emphasis placed on task or ego. The majority of research indicates that creating a task environment is preferable, as it is linked to enjoyment, perceived ability, and effort.

Arousal is traditionally considered to be physiological activation. Anxiety is the subjective feeling of apprehension usually experienced when a situation is perceived to be threatening or when there is a perceived imbalance between the demands of the situation and the ability to meet those demands. Cognitive anxiety, or worry, is the mental side of anxiety. Somatic anxiety, similar to arousal, is the physical facet of anxiety. Although some coaches continue to subscribe to the drive theory, most people involved in sport now recognize that the inverted U hypothesis more accurately reflects the relationship between arousal and performance. Athletes can have poor performances due to either over- or underarousal. Good coaches realize that the optimal level of arousal varies across sports and individual athletes. The relationship between somatic anxiety and performance is virtually identical to the relationship between arousal and performance. Cognitive anxiety, however, has a negative linear relationship with performance. When cognitive anxiety is low, the relationship between somatic anxiety and performance takes the form of the inverted U. However, when cognitive anxiety is high, the relationship between somatic anxiety (or arousal) and performance takes the form described by the catastrophe model.

Educational sport psychologists and clinical sport psychologists have different training and different views regarding what is involved in the practice of applied sport psychology. Both types of sport psychologists, however, help athletes develop and use psychological skills. Imagery is one example of a commonly taught psychological skill. Imagery involves using all the senses to create or recreate an experience in the mind. The psychoneuromuscular theory, the symbolic learning theory, and the attention–arousal set theory attempt to explain how imagery works. Although people do not agree on how imagery works, most in the field agree that imagery can be used to learn to control emotions, improve confidence, enhance the learning of new skills and the performance of known skills, and control pain. For imagery to be effective, however, individuals need to develop vividness and control of their images.

Further Reading and References

Andersen, M.B. 2000. *Doing sport psychology* (Introduction, pp. xiii-xvii). Champaign, IL: Human Kinetics.

Cox, R.H. 2002. *Sport psychology concepts and applications* (chapter 13, "Alternatives to Inverted-U Theory," pp. 211-228). 5th ed. Boston: McGraw-Hill.

Vanden Auweele, Y., K. Nys, R. Rzewnicki, and V. Van Mele. 2001. Personality and the athlete. In *Handbook of sport psychology*, ed. R.N. Singer, H.A. Hausenblas, and C.M. Janelle (pp. 239-268). 2nd ed. New York: Wiley.

Vealey, R.S., and C.A. Greenleaf. 2001. Seeing is believing: Understanding and using imagery in sport. In *Applied sport psychology: Personal growth to peak performance*, ed. J.M. Williams (pp. 247-283). 4th ed. Mountain View, CA: Mayfield.

CHAPTER 19

Basic Concepts in Exercise Psychology

The major learning concepts in this chapter relate to

- the difference between exercise participation motivation and exercise adherence motivation;
- the biological, psychological, sensory, and situational factors that interact to influence exercise adherence motivation;
- the role of goal setting in enhancing motivation;
- setting goals that are challenging but realistic, positive, controllable, specific, and measurable;
- the various stages of the transtheoretical model;
- possible explanations of why exercise enhances psychological well-being; and
- the influence of exercise on negative mood states and cognitive performance.

*T*he purpose of this chapter is to examine the reciprocal links between psychology and exercise, specifically the effects of psychological functions, such as motivation, on exercise and the effects of exercise on psychological factors such as feelings of well-being, mood states, and mental performance.

Effects of Psychological Factors on Exercise

The three components of motivation, mentioned in the previous chapter in relation to sport, apply to exercise motivation too. The direction facet clearly relates to whether an individual chooses the direction of the gym or the couch, the stairs or the elevator, the pool or the bath. Once the direction of exercise has been chosen, the intensity and persistence aspects become important. For example, John and Ian both decided as a New Year's resolution to join a gym. They have both chosen the direction of exercise. During their first day at the gym, John rides the bicycle for 30 min, tries the stepper and the rowing machine, completes multiple sets on the different weights available, and finishes off with an aerobics class. Ian, on the other hand, begins with just a few minutes on the bicycle and then completes just one set of each of the various weight exercises using light weights. John has exhibited high intensity and Ian low intensity.

If we stopped at this point in the example, we might conclude that John is more motivated than Ian. Our opinion might change when we consider the third component of motivation, persistence. Two days later Ian is once again at the gym, adding a couple of minutes to his time on the bicycle and sticking to his workout with light weights. John, on the other hand, does not make it to the gym. He is home in bed, so sore he has trouble sitting up. Six months down the road, Ian is still regularly attending the gym. John, on the other hand, believes that exercise is painful and avoids it whenever possible.

Exercise Participation Motivation

Exercise participation motivation refers primarily to the direction component of motivation. Exercise participation motivation deals with the initiation of exercise. A variety of factors influence whether or not people initiate an exercise program.

Knowledge, attitudes, and beliefs about exercise influence motivation toward exercise participation. People who understand the importance and value of regular exercise are more likely to initiate an exercise program than those who do not. Similarly, if people have positive attitudes about the value and importance of regular exercise they will have greater motivation to participate in exercise than do people with negative attitudes.

Valuing the importance of exercise, however, is not the only determinant of exercise participation. Beliefs about ourselves influence motivation as well. Even if Jenny understands that exercise is important, she is unlikely to begin an exercise program if she believes that she cannot succeed at it. If she believes that the exercise program is too difficult or that it requires more fitness, strength, coordination, or time than she has, then it is doubtful that she will join the program. This confidence in one's ability to succeed at an exercise program is called **exercise self-efficacy.** It is logical that self-efficacy would influence behaviour. How likely would people be to do something they were convinced they could not do, particularly if they had to pay to do it? How likely would they be to invest any energy in pursuing that activity? Most people in this situation would not attempt the activity. People who have such feelings are described as having low self-efficacy. As **self-efficacy,** or one's belief in one's ability to succeed at a particular task, increases, so does the likelihood of undertaking that task.

What does this mean in terms of enhancing exercise participation motivation? Educating people about the importance and value of exercise can be a valuable first step, as persons who understand the merit of exercise are more likely to adopt an exercise program. Unfortunately, imparting knowledge is not enough. Enhancing the exercise self-efficacy of individuals will increase their motivation. Some people have low self-efficacy about exercise because they believe that they are too unfit to begin exercising. They may equate exercise with young, thin, Lycra-clad gym enthusiasts whom they see in the media. Programs that emphasize choice of activities, portray exercisers who are similar to the potential exercisers in age and fitness level, and communicate that exercising can be enjoyable may increase exercise participation motivation. Exercise programs that begin with activities people already know they are capable of doing, such as walking and climbing stairs, may also increase self-efficacy and thus increase motivation. To be effective, exercise programs need to recognize how the motivating factors for exercise differ between younger and older adults ("One Size Does Not Fit All: Younger and Older Adults Differ in Motivating Factors for Exercise," p. 293).

Exercise Adherence Motivation

Although many people get motivated to begin an exercise program, many who begin fail to continue. As we saw in chapter 13, approximately 50% of those who begin a regular physical activity program drop out within the first six months. These people had exercise participation motivation but lacked **exercise adherence motivation,** the persistence angle of motivation.

Biological, psychological, sensory, and situational factors all interact to influence exercise adherence. Biologically, body composition, aerobic fitness, and the presence of disease influence adherence. Unfortunately, it is usually the people who could gain the most from exercise who are the least likely to adhere. People who are overweight or obese, low in fitness, or chronically ill are less likely than thinner, fitter, and healthy people to adhere to an exercise program.

Just as with exercise participation motivation, attitudes and beliefs influence exercise adherence motivation. Attitudes and beliefs about the importance of exercise play a role. However, expectations about the effect of exercise, and beliefs about the effect that exercise is having on one personally, also influence adherence. For example, if David believes that major changes in fitness and body composition should occur after six weeks of regular exercise and does not perceive major improvement in his own body after six weeks, he may believe that exercise does not do what it should and therefore quit. Even though it is unrealistic to believe that six weeks of exercise can make up for many years of inactivity, it is David's beliefs that influence his behaviour. Therefore, newcomers being introduced to exercise need to acquire realistic expectations about the time and effort required and the anticipated effect of the proposed program.

In addition to attitudes and beliefs, other psychological factors influence exercise adherence motivation. **Extroverts** (people who are social and outgoing) tend to adhere to exercise programs better than do **introverts.** Exercise programs that are executed in the presence of other people are probably more comfortable for extroverts than for introverts. Extroverts tend to enjoy the interaction with class members and exercise partners, and this may encourage their adherence. Introverts, on the

IN FOCUS: ONE SIZE DOES NOT FIT ALL: YOUNGER AND OLDER ADULTS DIFFER IN MOTIVATING FACTORS FOR EXERCISE

As people get older they tend to exercise less. Exercise promotion programs that broadly target the entire community may not be effective for all age groups. As most physical activity intervention studies have focused on younger adults, there has been a need to investigate the factors that are likely to motivate older adults to exercise.

To address this issue, researchers used computer-assisted interviewing to interview 916 adults, who proportionally represented the different regions of Northern Ireland, about 13 motivating factors for exercise. Respondents were divided into two age groups, those 16 to 44 years of age (*n* = 520) and those 45 to 74 years of age (*n* = 396). Results related to how important the factors were perceived to be, the ability of exercise to achieve the various outcomes, and the motivational power of the various factors.

Younger adults rated "to improve or maintain your health" and "to have fun" as the 2nd and 4th most important personal goals out of 13, whereas the older adults rated the same goals 9th and 11th. Younger adults also rated "to learn new things" and "to seek adventure and excite-

ment" as significantly more important personal goals than did older adults. Significantly fewer older adults compared to younger adults felt that exercise could help them have fun, feel in good shape physically, improve or maintain their health, or feel a sense of achievement. Younger adults rated "to feel in good shape physically," "to improve or maintain your health," "to have fun," "to feel a sense of achievement," "to learn new things," and "to seek adventure and excitement" as having significantly more motivational power than did older adults.

According to these results, specific age groups should be targeted through emphasis on the benefits of exercise related to variables that are likely to be powerful motivators for those particular age groups. To summarize, one size does not fit all. The same exercise intervention will not be equally effective with all age groups.

SOURCE

- Campbell, P.G., D. MacAuley, E. McCrum, and A. Evans. 2001. Age differences in the motivating factors for exercise. *Journal of Sport and Exercise Psychology* 23(3): 191-199.

other hand, may adhere better to individual, home-based exercise programs. The bulk of the research on exercise adherence has involved programs that take place on-site at a fitness facility with other people. This setting may have led to the conclusion that extroverts are better adherers than introverts. If the research had been done on independent home-based exercise programs, the results might have shown that introverts were better adherers. Clearly, efforts need be made to match the social environment of the exercise program to the personality of the exerciser.

People with high levels of self-motivation are more likely to adhere to exercise programs than those with low levels of self-motivation. It is logical that highly self-motivated people have better adherence rates. The challenge is to help persons with low levels of self-motivation. One of the most effective methods of helping these people is to encourage their involvement in the goal-setting process.

Goal Setting

Setting goals can help to enhance motivation for a number of reasons. Goal setting addresses all three components of motivation. Goals give direction by providing a target. Goals are also a way of enhancing intensity, strengthening effort in that they provide reasons for the activity. We are all more likely to put in effort when we feel there is a reason for doing so. If you are given two jobs to do at work, one that has a particular target and objective and one that seems vague and purposeless, which are you more likely to put your effort into? Having goals helps to focus attention and effort. In addition, goals can augment persistence by fostering new strategies. If people have a goal that they are committed to and initial tactics appear unsuccessful, they will search for alternative strategies to achieve their aim. If the goal has not been set in the first place, giving up is more likely than persisting with different plans of action.

Goals are also beneficial because they reflect improvement. Too often people make short-term comparisons regarding their strength, fitness, flexibility, or weight. Because the positive effects of exercise take time to emerge, people often do not notice improvements if they are using a short time frame for comparison. If a goal is achieved, evidence of improvement exists.

Goal setting involves a number of steps:

- Setting the goal
- Setting a target date by which the goal is to be achieved
- Determining strategies to achieve the goal
- Evaluating the goal on a regular basis

If a target date is not set, then the goal is really just a dream: "One of these days I'll ride the exercise bike continuously for one hour." For this idea to be a goal, one needs to set a specific date by which the behaviour is to be achieved. Goal setting usually involves both long-term and short-term goals. The long-term goal provides direction. The short-term goals provide the increase in intensity and effort. People often err by only setting long-term goals. They begin to work toward achieving the goal, but success seems so far away that they give up before they get there. If you have decided to ride the bike for 1 h and are currently having trouble lasting 10 min, 1 h will seem virtually impossible. Achieving short-term goals along the way to the long-term goal boosts confidence and motivation because it is obvious that the effort is worthwhile as improvement is being made. Target dates are set for each short-term goal in a progressive order until the long-term goal is achieved. This pattern of goal setting can be likened to a staircase, where each short-term goal is a step on the way to the long-term goal (figure 19.1).

However, for goals to be effective, they need to be properly set. Goals can be considered to be good goals if they meet certain criteria (table 19.1). Goals need to be challenging but realistic. Goals that are not challenging are probably not requiring any real change in behaviour and therefore will have little impact. Nevertheless, there needs to be a balance between challenge and realism. If goals are so challenging that they are unrealistic, then people are setting themselves up for failure. Continued failure leads to lowered confidence and less motivation.

Goals also need to be specific and measurable. Saying "I want to be fitter" or "My goal is to be stronger" does not provide any way of knowing

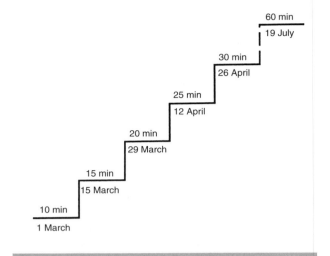

Figure 19.1   The goal-setting staircase.

when success has been achieved. What is "fitter"? How strong is "stronger"? One needs some way of knowing whether or not the goal has been achieved when the target date arrives. The easiest way of doing this is to make goals numerical. It is easy to use numbers to represent time spent exercising, distance traveled, repetitions accomplished, weight lifted, or exercise sessions attended.

In addition to being specific and measurable, goals need to be positive. If Diane makes it her goal not to recline and rest during the abdominal exercise section of her aerobics class, she will be thinking about reclining and resting. If instead she made it her goal to complete first 1 min and then 2 min of the abdominal exercises she would be thinking about doing the exercises, increasing her chances of doing them. If I tell you not to think about pink elephants, what is the first thing you think about? Similarly, if you set a goal of not letting your back arch off the bench when you do bench press, you will be thinking about your back arching. When you think about your back arching, your brain may be sending messages to the muscles that make your back arch (as in the psychoneuromuscular theory of imagery discussed in the previous chapter). So by having "not arching" as your goal, you may actually be increasing the likelihood of arching your back. Much more effective would be the goal of keeping your back pushed flat against the bench. Goals should stipulate the desired behaviour. This helps you think about, plan for, and prepare to do what it is you want to do.

For goals to be effective, it is also important that the person setting the goal has control over the

Table 19.1

Goals Need to Be . . .

Principle	Question to determine if the principle has been met
Challenging	Will the goal require effort?
Realistic	Is the goal reasonable?
Specific	Is it obvious what the precise objective is?
Measurable	Will there be an exact method of determining whether the goal has been achieved?
Positive	Does the goal stipulate the desired behaviour?
Controllable	Does the goal relate to the performance, technique, or behaviour of the goal setter?

activity. Goals should be related to the performance, technique, or behaviour of the goal setter. Goals are ineffectual when they rely on the behaviour of others. Ultimately, people have control only over what they themselves do. Goals such as being the strongest person in the gym or getting everyone in one's family involved in regular exercise are not completely under personal control. For this reason, best efforts may result in failure. Andrew may want to be the strongest person in the local gym; but for all he knows, the defending world powerlifting champion may suddenly move to the neighborhood and join the gym. So Andrew's goals should refer to how much weight he will lift, how many sets or repetitions of what specific exercises he will do, or how many times each week he will train. Similarly, if Rosie really wants everyone in her family involved in regular exercise, she should set goals related to what she will do to try to achieve that. For example, she might set a goal of organizing weekly family fun days that involve physical activity, creating a package of exercise options that fit other family members' schedules, or babysitting so her brother and his wife can go to the gym together. The focus is on what individuals can do themselves, rather than on what other people may or may not do.

In summary, goal setting can be an effective method of improving exercise adherence motivation. The goals need to be challenging but realistic, specific and measurable, positive, and controllable. If the goals require a lot of work or a major change in behaviour, the person should set a series of short-term goals that lead to the long-term goals. For each goal there should be a target date and a list of strategies that can be used to achieve the goal. Finally, the goals need to be evaluated on a regular basis. If a goal has been achieved, great! People can reward themselves: give themselves a pat on the back, buy that CD they have been want-

ing, or just feel good because of their achievement. If, however, the goal has not been achieved, they need to think about what may have happened. Possibly the goal was too big a step on the staircase; maybe a smaller step would be more appropriate. Maybe sickness or injury prevented work toward achieving the goal. In that case, a new target date should be set. Perhaps the goal was realistic, but the strategies selected to achieve it were not suitable. In that case a new target date should be set and different strategies implemented. Goals can strengthen self-motivation, which in turn can enhance exercise adherence motivation.

Sensory Factors

Different people perceive a given actual workload differently. If someone perceives an exercise program as excessively stressful, that person is apt to drop out. Therefore, when one is prescribing exercise, it is important to take into account the individuals' perceptions of the difficulty of the exercise and how much distress they experience when participating.

Problems sometimes arise when fit individuals who are used to training with their peers invite less fit (or unfit) friends to join them while exercising. Although the intention may have been to help their friends by encouraging them to exercise, the opposite actually occurs. The usually sedentary friends perceive the exercise sessions as anything but enjoyable. Because of the negative experience, they are even less likely to exercise in the future.

Exercise professionals who are qualified to prescribe exercise take into account the initial fitness levels of clients. This process decreases the chances that novice exercisers will experience excessive stress when exercising. However, when working with a large number of clients, exercise management professionals sometimes are tempted to generalize. People with certain fitness levels tend to be given particular exercise programs. Although individual variations in fitness are accounted for, individual differences in perceived exertion and perceived effort are sometimes ignored. People exercising at the same relative work intensity may have diverse experiences of that exercise. It is not only the fitness levels that need to be considered but also the individual perceptions of the exercise. One of the challenges of the exercise management profession is to cater to a large number of people while taking into account not only individual differences in physiological fitness but also subjective individual differences in perceptions of the exercise experience.

One determinant of how exercise is perceived is whether we dissociate or associate while exercising. **Associating** is attending to the body while exercising, being aware of what our muscles are feeling, how we are breathing, and even our heart rates. **Dissociating** is using attentional strategies designed to distract ourselves from the fatigue-producing effects of exercise. Dissociation can involve listening to music, daydreaming, planning our day, checking out the bodies of the other exercisers, or focusing on anything else that keeps our minds off the actual exertion of our own bodies. People who dissociate while exercising tend to have better rates of adherence than do those who associate. This phenomenon may be one reason why aerobics have become so popular. It is doubtful that many of the people currently participating in aerobics classes would continue to do the same exercises on their own in silence. Focusing on moving the right way at the right time and attending to the music, the instructor, or the other people in the class keep the mind off how the heart is beating, the muscles are straining, or the breath is gasping.

Situational Factors

A number of situational factors also influence exercise adherence motivation. The size of the exercise group has been shown to influence adherence. Although the ideal group size has not been determined precisely, if the group is too large, adherence may decrease—the individual may feel lost and unimportant. The larger the group, the less individual feedback each participant receives from the exercise leader. Additionally, many find it difficult to get to know people in a large group. With a great number of people, there is less chance that anyone will notice the absence of a single individual. Some exercisers are motivated to adhere

to a program because they do not want others to think they could not cope. Smaller groups allow for the development of social relationships and more attention and recognition from the instructor. It should be noted, however, that some exercisers, particularly those with low self-confidence, prefer the anonymity of a large class.

Ease of access to the exercise venue also affects adherence. Convenience of the program obviously influences how often people are likely to attend. Convenience can involve a number of components. If the program is in a location close to home or to work, there is greater chance of attendance than if a long trip is required. Similarly, even if we drive straight past the venue on the way home from work, difficulty finding a parking place could make us think that stopping is not worth the hassle. Exercise programs are more accessible if child care facilities are available to parents of young children. Additionally, how pleasant the exercise environment is may sway individuals to attend more or less frequently. Although Mike may think that a smelly, confined space in need of a coat of paint is perfectly fine for lifting weights, others may be immediately dissuaded by these conditions.

Having the social support of others also can influence adherence. If a spouse, friend, or significant other is supportive of our exercise behaviour, many advantages may ensue. First, it is likely that we will receive positive reinforcement from the person, which will increase our self-worth and feelings of competence. In addition to encouraging our exercise behaviour, others may provide informational support by giving advice or suggestions that may decrease the chance of injury or increase the benefits gained through exercise. Others can also provide tangible or instrumental support. For example, they may lend exercise equipment, provide transportation, or share child care. Social support can be a definite advantage.

A lack of social support would be better, however, than social disapproval. Ridicule of exercise attempts indisputably can have a negative effect on adherence. Outright resistance to exercise involvement can create additional hurdles. For example, a person may try to make a positive health change by stopping by the local gym for a 1-h exercise class after work three times per week. A spouse who condemns this activity can become an insurmountable barrier to adherence. This disapproval may be exhibited by the silent treatment, sarcasm, or outright rage. When one person makes a behaviour change, this may have an impact on others. Ride-sharing arrangements may have to be changed, dinner may be an hour later, or child care responsibilities may vary. If the partner is opposed to the lifestyle alterations, the exerciser may experience the exact opposite of social support—social disapproval.

In summary, many factors influence exercise adherence motivation. Understanding the benefits of exercise is only a preliminary step to encouraging exercise behaviour. Biological factors, attitudes, beliefs, personality characteristics, goal setting, sensory perceptions, group size, program convenience, and social support all influence exercise adherence. If people want to enhance the exercise adherence of themselves or others, they should consider all these factors.

The Transtheoretical Model

When one is developing interventions to increase the exercise behaviour of individuals, it is important to note that no single intervention will be effective for everyone. Some of these individual differences are the result of people being at different stages of change. Changing behaviour is a process that occurs over time. People do not instantaneously change from smoker to nonsmoker, from junk food junkie to health nut, or from coach potato to exerciser. The **transtheoretical model** suggests that people progress through a series of stages of change: precontemplation, contemplation, preparation, action, and maintenance.

In the **precontemplation stage,** people are not even thinking about changing their unhealthy behaviour. In terms of exercise, precontemplators are not exercising and have no intention of beginning an exercise program in the next six months. Individuals at this stage gain more from increasing their awareness of the positive effects of exercise, and recognizing the barriers that may be preventing them from exercising, than from specific information about how to start an exercise program.

In the **contemplation stage,** people still are not ready for structured exercise programs, but they are thinking about exercising within the next six months. They are probably aware of the benefits of exercise but even more so of the barriers. At this stage it may be beneficial to encourage individuals to try a variety of activities to get a taste of what might be involved in regular exercise.

People in the **preparation stage** have taken some steps toward engaging in regular exercise. They may have contacted a physician, joined a gym, or bought a new pair of running shoes or the latest exercise gadget advertised on television. They are probably exercising irregularly but have plans of exercising three or more times per week beginning sometime in the next month. In this stage, people can benefit from information about goal setting, suggestions for safe and enjoyable activities, and the recognition of obstacles to regular exercise.

Those in the **action stage** have modified their behaviour and have begun to exercise regularly, but have been doing so for less than six months. At this stage information about techniques for staying motivated, overcoming obstacles, and enhancing confidence can be useful.

The **maintenance stage** is achieved when there is little risk of returning to sedentary behaviours, usually after a period of six months of regular exercise. Efforts, however, still need to be made to avoid relapse—returning to an earlier stage of change. To prevent relapse, people can focus on refining specific types of exercise behaviour, injury avoidance, rewarding themselves for the attainment of goals, and methods of reducing boredom.

When interventions are matched to the relevant stages of change, intervention programs have a much higher chance of success. This means that "success" in some programs may be moving people from the precontemplation stage to the contemplation stage. Research has demonstrated the effectiveness of basing exercise interventions on the transtheoretical model with adolescents, college students, sedentary employees, and adults 65 years and older.

Exercise Addiction

Some people have no trouble adhering to exercise (achieving the maintenance stage). In fact, some exercisers are actually addicted to exercise. One can say that a person is addicted to exercise when the person experiences physical or psychological withdrawal symptoms after 24 to 36 h without exercise. An **addiction** occurs when dependence on or commitment to a habit, practice, or habit-forming substance is present to the extent that its cessation causes trauma.

Some believe that exercise can be a positive "addiction," as there are many physical and psychological benefits of exercise. If discomfort or other negative effects are present in a person who lets more than one to two days pass without exercising, that individual may be more likely to exercise on a regular basis. In this situation, when the individual has control over the exercise, exercise should be considered a positive health habit, not an addiction.

In contrast, addicts do not have control over their exercise. Although exercise is usually a positive and healthy habit, some people think that more is always better. Such persons can end up with overuse injuries and psychosocial problems. Although minimum exercise guidelines have been suggested for fitness benefits, we have sparse information about how much is too much. For those addicted to exercise, exercise becomes a detriment. They may exercise despite pain and injury. Similar to the situation with addiction to gambling or drugs, the substance of the addiction, in this case exercise, becomes more important than anything else.

In some cases, the lives of exercise addicts are determined by when and where exercise is available. Minor injuries can develop into serious conditions. But in addition to the probable physical damage that excessive exercise can cause, psychosocial problems can turn into major predicaments. Negative exercise addicts have lost jobs and families because exercise became more important than work or relationships. In addiction, exercise controls the individual.

Encouraging participation in a wide range of exercise and recreational activities can be the first step in helping individuals who are addicted to exercise. If one part of the body is injured, a person can engage in some other form of exercise or recreational activity that can allow the affected body part to recover. In addition, alternative forms of recreational activity can replace compulsive exercise behaviour. Providing a range of activities increases the chance that the individual will interact with others, possibly becoming less self-absorbed. Also, making decisions about the activity in which to participate begins to give the individual power and control over exercise.

Effects of Exercise on Psychological Factors

So far this chapter has concentrated on how psychological factors influence exercise participation and adherence. The remainder of the chapter focuses on how exercise can influence psychological factors. Traditionally, discussions of the positive effects of exercise portray the majority of the benefits as physiological. For example, as discussed in chapter 13, regular physical exercise is associated with lower cholesterol, lower blood pressure, reduced weight, and a decreased percentage of body fat. But exercise has many psychological benefits as well. Exercise is associated with enhanced psychological well-being, decreased depression, reduced state anxiety, and improved cognitive performance.

Exercise and Psychological Well-Being

Exercise has been shown to influence how people feel about themselves. When people exercise they tend to feel more positive and self-confident. This relationship between exercise and psychological well-being has been demonstrated in terms of both long-term exercise and single bouts of exercise. Although researchers generally agree that exercise has positive psychological effects, they disagree about why this relationship exists. This section presents some of the possible explanations.

People may experience a sense of mastery or achievement through exercise. They feel they are successful when they are able to walk for a longer period of time, run faster, stretch farther, or lift more weight than they could before. It may be that it is this experience of achievement that makes people feel better about themselves. Through mastery in exercise, people may realize they have the capacity for change. They realize they were able to change their exercise habits, their fitness levels, or even their body shapes. This realization that change is possible may transfer to other areas of life, giving people a greater feeling of control.

Some argue that exercise is psychologically beneficial because it provides a distraction from problems and frustrations. Exercise does not have a positive effect in and of itself, according to this view; it really just allows people to take "time-out" from the problems in their lives. According to this viewpoint, the temporary respite from worry may be the only positive influence of exercise.

Another proposed explanation for the positive effect of exercise on psychological well-being is that exercise causes biochemical changes that in turn influence psychological factors. The most commonly suggested biochemical change is an increase in endorphin levels. Endorphins are naturally occurring substances with opiate-like qualities. Endorphins are important in regulating emotion and the perception of pain. If exercise increases the release of endorphins, then it may be the endorphins that cause the enhanced psychological well-being. Although theoretically this argument makes sense, the relationship has not been proven. Endorphin levels vary across individuals exercising at the same relative intensity, making it difficult to prove that any changes in psychological well-being are due to exercise-induced changes in endorphin levels.

Although a definitive explanation of the positive effects of exercise on psychological well-being does not currently exist, one should not overlook the psychological benefits of exercise. The fact that we do not understand why something happens does not mean we should fail to take advantage of the phenomenon. Exercise has been shown to increase the quality of life in many populations. Studies of persons with multiple sclerosis, breast cancer, kidney transplants, mild traumatic brain injury, human immunodeficiency virus (HIV), chronic respiratory diseases, end-stage renal disease, or Parkinson's disease have all demonstrated the benefits of exercise in enhancing quality of life. In addition to enhancing self-confidence, feelings of control, and general well-being, exercise also appears to reduce negative moods.

Exercise and Negative Mood States

Exercise effectively reduces depression. This effect is seen in mentally healthy individuals, but greater decreases in depression through exercise occur in those requiring psychological care. Both long-duration, lower-intensity and short-duration, higher-intensity exercise training have been found to be effective in reducing depression. Whatever the type of exercise, the more sessions per week (within reason) or the more weeks in the exercise program, the greater the decrease in depression.

Exercise also has been shown to decrease effectively both state and trait anxiety. Persons who are initially low in fitness or high in anxiety tend to achieve the greatest reductions in anxiety from exercise. However, unlike depression, anxiety appears to decrease only with endurance exercise. Short-duration, high-intensity exercise such as weight training may in fact increase anxiety. We need to keep in mind that exercise will probably not have a positive effect on mood states and quality of life for everyone. Some people find exercise extremely aversive and take part only because of pressure from medical staff, family, and friends. If coerced into exercising, they may focus on the negative aspects of the experience, find the entire experience onerous, and actually increase their stress levels.

For most people, however, exercise has been so effective in decreasing anxiety and depression that clinical psychologists have used endurance exercise training as a form of therapy. In some cases it is the sole psychotherapeutic tool, but more often it is used in conjunction with other modes of therapy. Running is the most common form of exercise chosen for therapy. Obviously, however, exercise is not a panacea for anxiety and depression. In fact, if it is difficult to get the average person to exercise regularly, we can imagine how difficult it can be to get a depressed person to exercise regularly.

Exercise and Cognitive Performance

Exercise not only influences how individuals feel; it also influences how well they think! Regular exercise is associated with improved cognitive performance. Studies have shown that regular exercisers perform better on reasoning tests, mathematics tests, memory tests, and IQ tests than individuals who do not regularly exercise. There is also some evidence that exercisers have better creativity and verbal ability than nonexercisers. Although the reasons for this relationship between exercise and cognitive functioning are not completely understood, it has been suggested that exercise in older individuals may slow neurological deterioration. For younger people, exercise may increase the vascular development of the brain as well as the number of synapses in the cerebellar cortex in the brain. More research is needed to ascertain any possible cause and effect relationship.

Summary

Motivation is made up of direction, intensity, and persistence. Exercise participation motivation is the direction component, referring to the initiation of exercise. Knowledge, attitudes, and beliefs about exercise influence motivation toward exercise. Of greater importance, however, is individuals' self-efficacy or their belief in their abilities to exercise. Exercise adherence is the persistence facet of motivation. Biological, psychological, sensory, and situational factors interact to influence exercise adherence. Setting effective goals with target dates and strategies can strengthen self-motivation that, in turn, can enhance exercise adherence.

The transtheoretical model suggests that individuals progress through five stages of change: precontemplation, contemplation, preparation, action, and maintenance. Matching intervention programs to the relevant stage of change increases the chances that interventions will succeed in changing exercise behaviour. Although much energy has been devoted to programs and research designed to increase the percentage of the population that is exercising regularly, a small percentage of people are addicted to exercise. Rather than exercise being a positive health habit for these people, it controls their lives, sometimes to the detriment of jobs and personal relationships.

Although many psychological factors influence whether and how much people exercise, the reverse is also true. In other words, exercise affects a number of psychological factors. Exercise has a positive effect on psychological well-being and quality of life. This positive effect may be due to mastery experiences, distraction, biochemical changes, or other factors. In addition to enhancing self-confidence, feelings of control, and general well-being, exercise decreases depression and anxiety. Finally, exercisers may have a cognitive advantage over nonexercisers.

Further Reading and References

Berger, B.G., D. Pargman, and R.S. Weinberg. 2002. *Foundations of exercise psychology* (chapter 13, "Motivational Strategies to Enhance Exercise Adherence," pp. 207-221). Morgantown, WV: Fitness Information Technology.

Dishman, R.K., and J. Buckworth. 2001. Exercise psychology. In *Applied sport psychology: Personal growth to peak performance*, ed. J.M. Williams (pp. 497-518). 4th ed. Mountain View, CA: Mayfield.

Hall, H.K., and A.W. Kerr. 2001. Goal setting in sport and physical activity: Tracing empirical developments and establishing conceptual direction. In *Advances in motivation in sport and exercise*, ed. G.C. Roberts (pp. 183-233). Champaign, IL: Human Kinetics.

Landers, D.M., and S.M. Arent. 2001. Physical activity and mental health. In *Handbook of sport psychology*, ed. R.N. Singer, H.A. Hausenblas, and C.M. Janelle (pp. 740-765). 2nd ed. New York: Wiley.

CHAPTER 20

Physical Activity and Psychological Factors Across the Life Span

Like the other biophysical sub-disciplines of human movement studies, sport and exercise psychology provides basic concepts that are of value in explaining changes in human movement across the life span and in response to training and practice. This chapter explores some applications of basic concepts from sport and exercise psychology to human movement at different points in the life span.

The major learning concepts in this chapter relate to

- whether participating in sport affects personalities or whether people with certain types of personalities are more likely to participate in sport,
- whether the psychosocial development of children is positively or negatively affected through sport participation,
- the influence of peers on exercise behaviour in adolescents,
- the effects of exercise on life satisfaction in people who are aged,
- how older adults may be encouraged to participate in exercise and sport,
- the termination of athletic careers as a challenging transition, and
- preparing for career termination well in advance.

Changes in Personality

In chapter 18 we noted that some personality differences have been found between athletes and nonathletes. Although some question whether these differences are real or merely artefacts arising from problems with operational definitions and the validity of questionnaires, there is also considerable debate about what these differences might mean (assuming they are real). Some conclude that participation in sport causes personality changes. That is, by participating in sport, people become more independent and extroverted. If this is the case, then if you want to help people become more independent and extroverted, having them participate in sport may be beneficial. Others, however, argue that participation in sport does not cause personalities to change. Instead, they suggest that individuals with certain types of personalities are more likely to participate in sport. In other words, people with

certain personality traits, such as extroversion and independence, gravitate toward sport.

So which side is right? Does participation in sport lead to certain personality traits, or do certain personality traits lead to participation in sport? Although there is no definitive answer to this question, most of the evidence suggests that the latter is more accurate. Individuals with certain personality traits seem to gravitate toward sport. That said, exceptions always exist. Although extroverted, independent individuals may be more likely to participate in sport, many individuals participate in sport who are either introverted or dependent on others. Additionally, some evidence indicates that participation does influence the personality development of young people. So, although involvement in sport as an adult probably has little if any effect on personality, involvement in sport when a person is young may influence personality development during maturation.

Psychosocial Development Through Sport Participation

Experiences in youth sport may have an impact on factors in addition to personality development. Values, attitudes, and beliefs can also be influenced by childhood experiences in sport. Through sport, children may learn about cooperation, respect, leadership, assertiveness, discipline, and fair play. They may learn that hard work results in positive achievement and develop self-esteem and self-confidence as a result.

Not all effects of participation in sport are positive, however. For example, if a child encounters numerous negative events in sporting situations, he or she is likely to avoid the sporting environment in adulthood. If fitness activities such as running and push-ups are used as punishment in youth sport, children may learn to associate fitness-related activities with penalties. If running is something we must do when we have made an error or misbehaved, why should we ever do it voluntarily? Also, if youth coaches believe that winning is the single most important result of participation in sport, children may learn about aggression and cheating instead of assertiveness and fair play. Similarly, just as children can develop positive self-esteem and self-confidence through participation in sport, they can also develop negative self-esteem and lose self-confidence if their experiences in sport are demeaning or humiliating. Sport itself is neither good nor bad. The psychosocial development of children through sport is largely dependent on the quality of the sport experience, and this in turn is often determined by the quality of the adult leadership provided.

Design of Youth Sport

Adults need to be aware that sport programs have the potential not only to benefit the growth and development of children, but also to be detrimental. Many adults mistakenly believe that any sport experience is better than none. Although all children should have the opportunity to participate in sport, these opportunities need to allow children to experiment, make mistakes, and succeed without fear or pressure. Many of the problems in youth sport occur because of inappropriate adult

behaviour. Adults sometimes design youth sport programs using the same structure as in professional adult sports. It is not uncommon, unfortunately, for children as young as 7 years to be cut from teams or permanently benched. Winning is frequently overemphasized; youngsters occasionally are harangued because they do not perform or think like adults; and parent behaviour on the sidelines sometimes keeps children from enjoying the sporting experience.

For sport to have a positive effect on the psychosocial development of children, adults need to remember that children are children and not miniature adults. Emphasizing fun and development instead of winning can avoid many potential problems and increase the chance that sport will be a beneficial experience. Adults need to recognize that growth and development include psychological and social considerations in addition to the traditional physical emphasis. Stressing fun and development instead of winning helps avoid the negative experiences of fear, inadequacy, anxiety, and inferiority.

Research has demonstrated that the quality or type of adult leadership in sport influences the attitudes of the children involved. Youngsters with coaches who use encouragement and praise and provide technical instruction like their coaches more and think their coaches are better teachers than children with coaches who are negative or who fail to provide information. These results are not surprising. Less obvious, however, is the fact that the children with positive coaches also like their teammates, enjoy being involved in the sport, and desire to continue participating to a much greater extent than children who play under less positive coaches. Participating in sport can be a good experience for youngsters. However, parents, teachers, and coaches need to realize that they must actively structure the sporting environment to ensure that sport is perceived as a pleasure rather than as an ordeal. See "What Children and Parents Want to See in Youth Sport Coaches" (p. 303) for a study investigating youth coaching preferences of children and their parents.

Adolescence, Peers, and Exercise

In the adolescence stage of development (approximately 13-19 years of age), peers usually are a more powerful influence than parents or coaches. Peer relationships can be thought of in terms of friendship and peer acceptance. Both of these factors influence affect and perceptions of physical self-

IN FOCUS: WHAT CHILDREN AND PARENTS WANT TO SEE IN YOUTH SPORT COACHES

Children (146 males and 93 females) aged 10 to 18 (mean age = 14 years) who were participating in summer youth sport and their parents (121 males and 118 females), aged 30 to 56 (mean age = 41.3 years), completed a modified version of the Participation Motivation Questionnaire (PMQ) aimed at eliciting their preferences for coaching qualities. None of the parents were currently coaching their child in the youth sport program. The subscales on the PMQ related to learning and improving new skills, affiliation and team orientation, being active during practice, excitement, fitness, and achievement and competitive challenge. Questions related to preferences for the coach's gender, age, athletic ability, and participation in practices also were included.

Boys, girls, mothers, and fathers all rated learning and improving new skills as the aspect they most wanted coaches to stress. Boys preferred a coach who emphasized fitness and achievement and competitive challenge more than did girls, but girls preferred a coach who emphasized being active more than did boys. More than fathers, mothers preferred their children to have a coach who practiced with the athletes, could perform the skills of the sport, and highlighted learning and improving new skills. More than their parents, the children preferred coaches who practiced with them, could perform the skills of the sport, and emphasized affiliation and team orientation, being active, and excitement.

Boys preferred male coaches more than did girls. Less than half of the boys, but 84% of the girls, indicated that the gender of the coach did not matter. Three-quarters of the parents stated that it did not matter whether the coach was male or female.

Almost half of the boys (43.1%) and girls (44.6%), and the majority of fathers (58.7%) and mothers (71.8%), stated that the age of the coach did not matter. For those who had a preference, the children preferred a coach ranging from 20 to 30 years of age, but the parents preferred a coach ranging from 31 to 40 years of age. Less than 5% of athletes and less than 9% of parents indicated a preference for a coach 41 to 50 years of age or over 50. Significant associations were found between the children's and their parents' responses regarding gender and age differences.

Although the results showed some similarities in the preferences that children and their parents had for the characteristics of youth sport coaches, the differences indicated that factors other than parents were influencing the children's opinions. It would have been interesting if the researchers had determined whether the preferences of the younger children versus the older children were more closely related to those of their parents. The researchers concluded that youth sport coaches are likely to have maximum credibility if they demonstrate the coaching characteristics that children and their parents prefer.

SOURCE

• Martin, S.B., G.A. Dale, and A.W. Jackson. 2001. Youth coaching preferences of adolescent athletes and their parents. *Journal of Sport Behavior* 24: 197-212.

worth, which in turn influence motivation and physical activity. When trying to achieve a sense of personal identity, adolescents often turn to those who are in the same developmental and chronological stage. Peers are the models that influence slang, style, values, and desires. As it is difficult and unusual for adolescents to strive to be unlike their peers, their behaviour is often dependent on what their friends are doing. This modelling of behaviour applies to sport and exercise. If their friends are into in-line skating, then there is every chance that adolescents will want to do in-line skating also. Unfortunately, if their friends do not exercise but instead spend their recreational time playing video games, then adolescents are likely to follow suit. In addition, low perceived peer acceptance predicts low levels of physical activity.

Exercise in Older Adults

Although participation in sport has a greater impact on the psychosocial development of children and adolescents than on adults, the impact of exercise on psychological health continues throughout life. Unfortunately, prejudicial and discriminatory views about aging exist. This discrimination is known as **age stratification.** Age stratification often results in lower self-expectancies of older individuals. People who think in these terms believe that older adults should "act their age," which for most people means being less competitive and poorer in physical performance. Two myths contribute to age stratification: "People need to exercise less as they age," and "Exercise is hazardous to the health of the elderly." In fact, as we saw in chapter 13, exercise becomes increasingly important for maintaining and improving the quality of life as people get older.

Exercise and Life Satisfaction in Older Adults

Research has revealed that exercise is related to life satisfaction in older people. Older adults who exercise regularly report significantly enhanced health, increased stamina, less stress, increased work performance, and more positive attitudes toward work than those who do not exercise regularly. Exercise may also increase self-efficacy in older persons. For example, a study of a five-week swimming program for adults over 60 years of age showed that the program had multiple positive effects. Not surprisingly, after the program participants had greater belief in their swimming ability and were more confident of their swimming competence. More importantly, participants also had increased generalized feelings of self-efficacy and competence. For example, they felt they could do chores more easily and were more able to use public transportation, which greatly strengthened their independence. Figure 20.1 demonstrates the

effects of a six-month exercise program for 60- to 75-year-old adults who had previously been sedentary. Happiness and life satisfaction increased and loneliness decreased; however, six months after the exercise program ended, subjective well-being had decreased.

As noted in chapter 19, exercise can improve self-concept and self-esteem. Exercise can also decrease stress, reduce muscle tension and anxiety, and lessen depression. These psychological benefits of exercise are advantageous to people of all ages, but in some respects they may be particularly useful for people who are elderly. Depression is the major mental health problem of the elderly population. Studies have shown that participation in a mild exercise program can significantly decrease depression in relatively independent adults (aged 60-80), as well as in clinically depressed residents of a nursing home (aged 50-98).

Encouraging Participation of Older Adults in Exercise and Sport

Exercise in older adults is associated with enhanced self-efficacy, life satisfaction, and happiness as well as decreased tension, anxiety, and depression. However, with age stratification, exercise participation motivation and exercise adherence motivation are even worse for older adults than for the general population. If people are concerned that exercising is undignified or inappropriate for them because they are older, or that exercising will cause a heart attack or worsen existing medical conditions, they have additional barriers to overcome. Exercise programs can help to combat age stratification by including pictures and videos of older exercisers in their advertising.

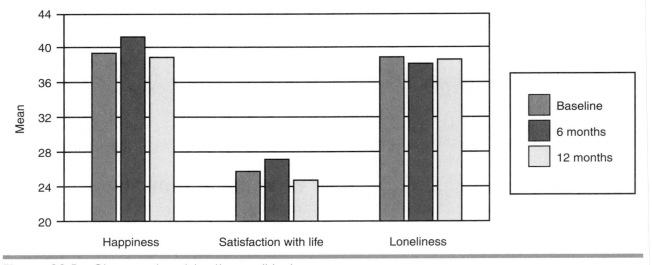

Figure 20.1 Changes in subjective well-being.

Some fitness centres are targeting the upper end of the age spectrum by having classes and social events specifically designed for older members. In addition to scheduling senior weight circuits or over-fifties aerobics classes, a few programs now offer classes for those with arthritis and exercise classes based on ballroom dancing. Additionally, hiring individuals who are expressly trained to assess older people with a variety of possible medical conditions and then prescribe suitable exercise can help overcome some of the fears and concerns of potential exercisers. (Chapter 12 provided some guidelines for exercise prescription for older adults.)

Obviously, not all older adults are hesitant about exercising. In fact, the growing popularity of masters sports should help to overcome some of the prejudicial and discriminatory views about people who are older. The Fifth World Masters Games were held in Melbourne, Australia, in 2002. Almost 25,000 competitors from 97 countries made these games the largest participatory multisport competition in the world. Some older people definitely are active! Masters competitions are held every four years at the world level and annually at the national level, and many countries hold ongoing local and regional competitions as well.

One of the more popular masters sports is swimming. Masters swim meets differ from traditional swimming competitions in that there are no heats. People are assigned to races within an event on the basis of previous best times. Each race therefore has people with similar performances, although a variety of ages may be represented. Results, however, are not on the basis of any single race. Instead results are posted on the basis of times within five-year age groups (e.g., ages 45-49 or ages 50-54). Consequently a fast 65-year-old might finish fourth in a particular race but still win the event for that age group if the first three finishers were in other age groups. One of the advantages of this system is that the emphasis is automatically placed on individual performances and participation.

Different masters sports tend to have different climates and structures regarding competition and training. A large percentage of masters swimming competitors, for example, train and compete on a regular basis but focus on participation and improving (or maintaining) their own performance. Other sports, however, organize masters competitions only at the national level. Although some of these competitors train regularly and compete locally in non-masters events, a number do not begin exercising until immediately prior to the masters competition and hope that this will get them through their events.

Greater availability of local masters events may encourage more regular involvement by more people. The growth of local and regional leagues and clubs for masters athletes in the United States suggests that this notion is correct. The largest masters sport organizations in the United States are baseball, basketball, cycling, soccer, swimming, tennis, and track and field. Once again, however, participants in different sports tend to have different attitudes toward competition as opposed to participation. Masters basketball players are known for being extremely aggressive and fiercely competitive. Masters baseball players, though, have the reputation for being relatively unfit. Although these players take pride in the fact that they are playing hardball instead of softball, they have referred to their Men's Senior Baseball League (MSBL) as the Men's Big Stomach and Butts League. Obviously, not all masters sport participants take themselves seriously. Nevertheless, organized masters sports may encourage some older people to exercise who would not otherwise do so.

Termination of Athletic Careers

For elite athletes, one of the major transitions in their lives comes when their sporting careers end. Although sporting opportunities continue to exist for many of these athletes through participation in masters sports, the structure of their lives is often altered. No longer are they representing their country at the open level or being paid by a professional sporting organization to compete. Although they still may be able to participate in other forms of competition, rarely does competitive performance continue to be the main focus of their lives.

Voluntary Versus Involuntary Termination

The effect of retirement from elite sport on the athlete depends, in part, on the circumstances surrounding career termination. If athletes voluntarily retire because they no longer enjoy training or competing, or because they have decided that other areas of their lives have priority, they may adapt more quickly to the transition because they retain a sense of control. Athletes who are forced to end their careers because of injury or deselection may find the process difficult. Research has shown that individuals who voluntarily terminate their careers feel less emptiness and disappointment and more relief and curiosity about what is to come than those who involuntarily end their careers.

Retirement from elite sport can lead to depression. Some athletes have equated the experience to the ending of a close personal relationship. For some competitors, self-esteem drops after career termination because they suddenly feel a loss of identity. (They may perceive a precipitous drop from "hero" to "zero.") Although having a strong athletic self-identity may have a positive impact on performance while one competes in sport, it may lead to greater difficulties in adjusting to career termination. The stronger the athletic self-identity, the more time is needed to adjust to career termination. The self-satisfaction and self-expression that competitors gained through their participation in sport are no longer available to them, sometimes leading them to question their own identity.

To experience depression (or a feeling of flatness) after being removed from an environment in which one has spent a lot of time is normal. Schedules change. Goals that have been a source of energy are suddenly no longer there. If athletes finish their elite careers with a poor performance, they may feel that their previous devotion to their sport was for nothing. However, finishing with a major success does not guarantee a smooth transition.

Assisting the Transition Process

When athletes retire from elite sport, either voluntarily or involuntarily, they can employ a number of strategies to ease the transition. Just as social support can help to enhance exercise adherence (as mentioned in chapter 19), it can also help in a major transition period such as career termination. Athletes can go to others for emotional support (reassurance, being loved in a time of stress), esteem support (agreement, positive comparison with others), instrumental support (direct assistance), informational support (advice, suggestions), and network support (feeling of belonging to a group of people who share a common interest). Table 20.1 provides some examples of these types of social support. Not all athletes require all types of social support, but it is worthwhile knowing that social support can be more than just a hug and an "everything is going to be OK." Making others aware of the decision to retire from elite sport is the first step toward setting up social support. Other retired athletes can also be a valuable source of social support. It can be beneficial to share experiences and discuss what has and has not been effective in dealing with the retirement process.

Retiring athletes also need to have alternative activities in which to redirect their energy. Ideally this process begins while the athletes are still participating in their sports. Some athletes may choose to become involved in other aspects of sport such as coaching, administration, or broad-

Table 20.1

Types of Social Support That May Help Athletes Retiring From Sport

Type of support	Examples
Emotional	Communicating to retiring athletes that they are loved for who they are off the field/court/track
	Supporting their decision to retire from sport
Esteem	Praising their nonsporting abilities (e.g., communication skills, organizational talents, or artistic expertise)
Instrumental	Introducing them to potential future employers
	Teaching them basic computer skills
Informational	Suggesting that the skills they used to enhance their performances in sport can also be used to enhance their performances in another career
	Informing them of the various courses they qualify to attend that are offered locally
Network	Forming a group of individuals who either are currently retiring from sport or have already successfully retired from sport
	Encouraging retiring athletes to join a social club, church group, or volunteer organization

casting. Others may prefer to invest their energy in their families, further education, or an entirely new career. When one is participating in a new activity, it is important to avoid comparisons with prior sport competence. This aspect is particularly relevant to athletes who were national or world champions, as it is improbable that they will be the best in the country or the world at their new careers. As illustrated in chapter 17, competence in any new activity is developed only over a long period of time.

Athletes may find the transition easier if they apply skills that they learned through their participation in sport. For example, goal setting can be used not only for fitness enhancement or technique improvement but also for any aspect of life that requires a change in behaviour or sustained effort over time. Many elite athletes also have become adept at controlling stress and anxiety. Skills that they employed to maintain calm and focus attention in a major championship or an international competition can serve to control stress in a new career. Similarly, if athletes have discovered that certain types of thinking are self-defeating in sporting performances, chances are good that they will find comparable thinking harmful in work performances as well. Chapter 21 provides more information on the transfer of skills from sport to work.

A percentage of retiring athletes may need to learn to take greater control of their lives. Depending on the sport, coaches, managers, and others may have made all the major decisions for the athletes in the past. Some coaches are dictatorial, and some sports traditionally regiment the lives of the athletes. People who are used to being told when to train, when to sleep, what to wear, what to eat, where to go, and how to get there may find it difficult to be suddenly on their own outside this decision-making process. Athletes may need to learn how to plan their own lives, not just in terms of a new vocation, but also in terms of day-to-day responsibilities.

Whether the person needs to gain social support from significant others, find new activities to pursue, learn to take greater control of his or her life, apply skills previously learned in sport to new situations, or change attitudes and priorities, adapting to retirement from elite sport takes time. The adaptation will be quicker for some than others, but the process is easier when the person has made preparations before retirement. Athletes who during their sporting careers enjoy partici-

pating in other activities, change their routines in the off-season, develop ideas for new careers, and create positive attitudes toward retirement are able to adjust quickly to the transition processes. Sporting organizations can help by educating athletes in advance on the importance of having a balanced lifestyle, by keeping in contact with athletes after they have left, and by providing opportunities for athletes to maintain involvement in the organization. Adaptation is quicker when athletes perceive retirement as a way of life that presents new opportunities rather than the end of life as they know it.

Summary

Involvement in sport as an adult probably has little if any impact on personality. The development of children's personalities, however, may be influenced by their involvement in youth sport. Through participation in sport, children can develop self-confidence, self-esteem, respect for others, leadership skills, and a sense of fair play. However, if the sporting experience is poorly structured, with coaches or parents who are more interested in winning than in children having fun or learning, then children can develop negative self-esteem and lose self-confidence. While coaches and parents can have a huge effect on children, adolescents tend to be influenced more by peers than by adults. The behaviour of peers largely determines whether adolescents will engage in physical activity.

Although participation in sport and exercise may not play a large role in the psychosocial development of adults, regular exercise has been shown to enhance self-efficacy and decrease depression and anxiety. Age stratification (prejudicial and discriminatory views about aging), however, means that fitness centres and exercise professionals need to develop programs and marketing campaigns that specifically target older adults. Although the number of participants in the World Masters Games is steadily increasing, more effort needs to be made at the local level to increase the percentage of older adults involved in regular physical activity.

For elite athletes who have made sport the focus of their lives for many years, the termination of their athletic careers can be a challenging transition. Difficulties in career termination arise when athletes involuntarily retire from sport, have a strong athlete self-identity, access few sources of social support, refrain from participating in other activities, and have a poor attitude toward the transition process. Preparing for career termination well in advance can help many athletes adjust more quickly to the transition.

Further Reading and References

Horn, T.S., and A. Harris. 2002. Perceived competence in youth sport athletes: Research findings and recommendations for coaches and parents. In *Children in sport: A biopsychosocial approach*, ed. F. Smoll and R. Smith (pp. 435-464). 2nd ed. Dubuque, IA: Kendall/Hunt.

Lavalee, D., and P. Wylleman, eds. 2000. *Career transitions in sport: International perspectives*. Morgantown, WV: Fitness Information Technology.

Taylor, J., and B. Ogilvie. 2001. Career transition among athletes: Is there life after sports? In *Applied sport psychology: Personal growth to peak performance*, ed. J.M. Williams (pp. 480-496). 4th ed. Mountain View, CA: Mayfield.

Weiss, M.R. 1995. Children in sport: An educational model. In *Sport psychology interventions*, ed. S.M. Murphy (pp. 39-69). Champaign, IL: Human Kinetics.

CHAPTER 21

Psychological Adaptations to Training

The major learning concepts in this chapter relate to

- the relationship between aerobic fitness and response to psychological stress,
- the limitations of research on the effects of sport participation on personality,
- the phases of training stress syndrome,
- various measures of overtraining and burnout,
- strategies to overcome and prevent training stress syndrome,
- mental skills that may be acquired through participation in sport and applied to other situations in life, and
- the importance of systematic practice of mental skills.

As mentioned in chapter 19, both single bouts of exercise and long-term involvement in exercise have been shown to have a positive effect on mood. The present chapter focuses on the psychological effects of prolonged participation in sport and exercise. We consider the role of fitness in the response to psychological stress, personality changes as a result of participation in sport, overtraining and burnout, and the development and transfer of mental skills over time.

Aerobic Fitness and the Response to Psychological Stress

As noted in chapter 19, exercise is effective in reducing depression. Depression can be related to stress. For example, failing in an achievement situation that is perceived to be important may be stressful. This failure may lead to lowered self-esteem, which in turn may lead to depression. Involvement in regular exercise has been shown both to reduce existing depression and to prevent the onset of depression. Fitter people report lower levels of depression after prolonged stress than do less fit individuals.

Depression is not the only area in which fitness level plays a role. Aerobically fit individuals are able to deal with many forms of psychological stress more effectively than less fit individuals. Probably the greatest indication of this difference is in the varied levels of cardiovascular arousal after exposure to psychological stress. Compari-

son of highly fit individuals with individuals who have low levels of fitness has shown that the fitter individuals have smaller increases in heart rate and blood pressure in response to any particular psychological stressor. Moderate endurance training, such as in aerobic dance classes, has been shown to be more effective than relaxation training in reducing heart rates before, during, and after experiences of psychological stress (see "Exercise and Firefighters' Responses to Psychological Stress" on p. 310).

Changes in Personality

Researchers have investigated changes in psychological factors in response to participation in sport as well as exercise. A number of studies have addressed the personality profiles of athletes with varying levels of skill. Studies comparing athletes within a team have shown no significant results, that is, no meaningful personality differences between successful and unsuccessful athletes or between starters and bench players

IN FOCUS: EXERCISE AND FIREFIGHTERS' RESPONSES TO PSYCHOLOGICAL STRESS

Firefighting is physically demanding, but the inherent danger to firefighters themselves and to others also means that firefighters experience a great deal of psychological stress. To determine whether exercise could decrease fire-related stress, firefighters from Texas were randomly assigned to an exercise intervention or a control group. Of the 38 males and 3 females who completed the study (mean age = 35.7 years), 21 rowed on an ergometer for 40 min, four times per week, for 16 weeks. All participants were allowed to continue their normal physical activities throughout the study. Both the rowing group and the control group completed physical tests and psychological stress tests before and after the 16-week program.

Psychological stress was induced through having firefighters complete a video-based Strategy and Tactics Drill (STD) in which they were required to make decisions based on a fire scene that increased in magnitude as the test progressed. Participants made decisions as if they were the incident commander directing personnel in fighting the fire. An official from their fire department oversaw the tests and evaluated the firefighters' decisions. Stress was measured objectively during the STD using heart rate and blood pressure and subjectively imme-diately after the STD using affect and anxiety questionnaires.

Even though the firefighters were reasonably fit before the study began, those who rowed demonstrated significant increases in $\dot{V}O_2$max and power, with no changes for the control group. Both groups showed a reduction in heart rate stress reactivity, although the reduction was larger for the intervention group than for the control group. Only the rowers showed a significant reduction in blood pressure over time. The subjective results also supported the effectiveness of rowing in decreasing fire-related stress. The intervention group showed significant reductions in negative affect and anxiety, with no changes for the control group (figure 21.1).

These results suggest that in addition to helping firefighters meet the physical demands of their profession, exercise may also help them meet the psychological demands. Exercise in the form of rowing reduced the responses to fire-related stress in firefighters.

SOURCE

- Throne, L.C., J.B. Bartholomew, J. Craig, and R.P. Farrar. 2000. Stress reactivity in fire fighters: An exercise intervention. *International Journal of Stress Management* 7(4): 235-246.

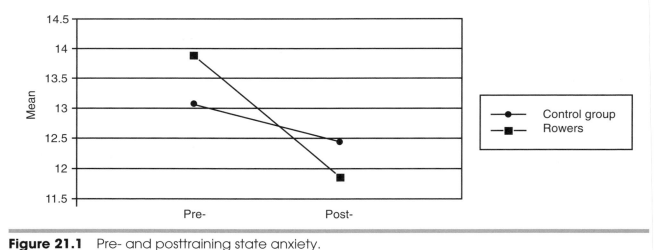

Figure 21.1   Pre- and posttraining state anxiety.

within a team. However, elite athletes (e.g., international-level competitors) can be distinguished from novice athletes on some personality characteristics. For example, there is some evidence that highly successful athletes are greater risk takers than are novices; that is, they are more likely to take chances that might result in injury or failure. Nevertheless, generalizations about expert–novice differences are difficult to draw because this research shows many of the same inconsistencies as that on personality differences between athletes and nonathletes.

First, these studies have used a variety of personality inventories and questionnaires, making interpretations across studies difficult. Also, some studies have focused on traits and some on states, and some have inappropriately merged the two. The available studies are also difficult to compare because of the different sports investigated. Generally, athletes in team sports are more extroverted and dependent than athletes from individual sports. Therefore comparing elite gymnasts to novice softball players can be puzzling. Just as defining athletes and nonathletes poses difficulties (as mentioned in chapter 18), defining elite and novice athletes can be problematic. Is someone who competes nationally in swimming and recreationally in tennis a novice or an elite athlete? Finally, even if consistent differences were found between elite athletes and novice athletes, there is no indication that training, competing, and developing physical skills over a long period of time cause personalities to change. It may be that individuals with specific personality types are more likely to develop into elite athletes than are individuals with other personality types. Not enough evidence exists, however, to support the use of personality tests for screening or selecting athletes.

Personality researchers have also tried to determine whether the type of sport in which individuals participate affects their personalities. Once again, consistent findings are few, and there are problems with the methodology frequently used in such studies. The research does not support even logical-sounding assumptions. For example, there is no evidence that athletes in body contact sports are any more aggressive than athletes in noncontact sports. One consistent finding is that participants in high-risk activities such as parachuting and auto racing usually have higher levels of sensation seeking than athletes in low-risk activities like golf or bowling. Although some studies have controlled for factors such as age and education when comparing participants in high- and low-risk sports, and although there is general agreement regarding whether the activities selected are high or low risk, the structure of the studies does not allow researchers to determine causality. In other words, descriptive studies that do not follow participants longitudinally over time cannot show whether people high in sensation seeking are attracted to high-risk sports or whether participation in high-risk sports causes individuals to be high in sensation seeking.

Changes in Motivation: Staleness, Overtraining, and Burnout

The physiological aspects of overtraining were mentioned in chapter 11 and are considered further in chapter 22. Physiologically, to achieve training gains, training stress needs to be imposed. Too much training stress, however, can lead to a negative psychophysiological reaction. Both psychological and physiological factors are involved in the negative adaptation to training. Although this section focuses on the psychological factors, it is important to emphasize that the psychological and physiological factors interact. For example, being unable to maintain training loads physically can decrease our enthusiasm for training. Similarly, having low levels of enthusiasm can decrease our ability to maintain training loads.

Definitions Related to Training Stress

Studying the area of training stress can be confusing, as people use the various terms to mean different things. Most agree that individuals progress through a series of stages when negatively adapting to training stress. Exercise physiologists often suggest that athletes first experience overtraining, then overreaching, and then staleness. In the exercise physiology literature, staleness is usually perceived as equivalent to burnout.

On the other hand, many sport psychologists refer to the negative adaptation to training as the **training stress syndrome.** This syndrome is made up of three phases: staleness, overtraining, and burnout.

- **Staleness** is the initial failure of the person to cope with the psychological and physiological stress created by training. The body and mind attempt to adapt to the training demands, but the demands exceed the person's present capabilities. Staleness is characterized by an increased susceptibility to illness, flat or poor performance, physical fatigue, and a loss of enthusiasm.

- **Overtraining,** the second phase of training stress syndrome, is the repeated failure of the person to cope with chronic training stress. Symptoms of overtraining include an increased resting heart rate, chronic illness, mental and emotional exhaustion characterized by grouchiness and anger, and being overly bothered by minor stresses.

- **Burnout,** the final phase of the syndrome, occurs when the athlete is exhausted both physiologically and psychologically from frequent but usually ineffective efforts to meet excessive training and competition demands. When experiencing burnout, athletes lose interest in the sport, are extremely exhausted, and generally do not care about their training or performance. In some cases they feel resentment toward the activity.

Monotony, repetition and boredom, too much stress, or too much training can cause staleness. Since one of the characteristics of staleness is a flat or poor performance, many coaches and athletes react to staleness by increasing the training load. This further increase in stress and training demands often leads to overtraining. In addition to too much stress or pressure, too much repetition, or too much training, overtraining can result also from lack of proper sleep, a loss of confidence, and the feeling that one is never successful. If the training load is maintained or increased at this point, burnout may result.

Burnout is the result of an ongoing process. Athletes do not just awaken one morning and suddenly experience burnout. Burnout is usually caused by excessive devotion with little positive feedback over a long time, severe training conditions, unrealistic expectations, and a lack of recovery time from competitive stress. Nontraining stressors such as having a poor relationship with one's coach or not having enough time to spend on other important relationships are also related to staleness and burnout. Burnout typically results in dropout. It should be noted, however, that although burnout often leads to dropout, many people drop out for other reasons.

Measuring Overtraining and Burnout

Traditionally sport psychologists have measured overtraining by focusing on the relationship between overtraining and mood states. The **Profile of Mood States (POMS),** used in many studies, frequently demonstrates that athletes' moods worsen during states of overtraining. As the paper-and-pencil POMS is less expensive as a means of measuring overtraining than analyses of biochemical changes that also accompany overtraining, it is sometimes used as a screening device with athletes.

Some sport psychologists have argued that the POMS is inadequate in that it does not measure the process of recovery. An alternative measure of overtraining is the **Recovery-Stress-Questionnaire for Athletes (RESTQ-Sport),** which was designed to measure the frequency of current stress and recovery-associated activities. Studies using both instruments have shown that the POMS and RESTQ-Sport scales are correlated and are therefore measuring similar constructs. The argument for using the RESTQ-Sport is that it provides useful information not available from the POMS. The POMS can reveal that mood states have worsened but cannot shed light on whether increases in stress or decreases in recovery are the reason.

Neither questionnaire attempts to measure burnout. The concept of burnout originated in the domain of human services, in people working in the helping professions. Burnout in human service providers is characterized by emotional exhaustion, a reduced sense of accomplishment, and depersonalisation. **Depersonalisation** is characterized by a detached and callous attitude toward clients. Sport psychologists have suggested that depersonalisation, although relevant to human service providers, is not a key feature of burnout for athletes. The sport equivalent of depersonalisation is **sport devaluation,** seen when athletes stop caring about their sport and their own performance. With this in mind, the Athlete Burnout Questionnaire was created to measure the three factors of emotional/physical exhaustion, reduced sense of accomplishment in sport, and sport devaluation. The validity of the Athlete Burnout Questionnaire is strong, as those who score higher on its subscales score lower on measures of social support, coping, enjoyment, commitment, and intrinsic motivation.

Strategies

A number of strategies can be implemented if signs of staleness, overtraining, or burnout emerge. The simplest course of action is to introduce a time-out. Taking a break from training (and in some cases competition) can refresh and invigorate athletes. For athletes to be able to recharge their batteries, it is important that the time-out be a break from everything, not just physical training. An emotional vacation is needed. Some coaches grudgingly allow their athletes to take a break physically, believing that a respite from physical work is all that is required. Athletes who during a physical pause in training continue to observe training sessions, analyse game film, or otherwise involve themselves mentally in their sport gain very little (if anything) from the physical respite.

Training stress syndrome is a psychophysiological process. Although reducing the intensity or volume of training may avoid the progression of the syndrome from staleness to overtraining or from overtraining to burnout, it rarely is enough to eradicate the syndrome entirely. The more significant the changes in day-to-day activities, the more effective the break will be. If Alf takes a week off from weight training, swim training, time trials, and team meetings yet spends the week living in his apartment with his three roommates who are also teammates, he is still in a swimming environment. With three roommates draping wet bathing suits around the place, leaving for and arriving back from training, complaining about the session's format, and generally talking about swimming, Alf fails to experience a real time-out. Alf's stress levels may even increase in that situation if he feels guilty or ashamed of his inactivity.

Rather than focusing on the idea of overtraining, it may be more useful to consider the problem as one of underrecovery. Recovery should be a planned component of athletes' schedules. It is fairly obvious that athletes need to have adequate nutrition and rest between training sessions, but having a mental break from the sport is just as important. Even the keenest athletes who are quite happy to eat and breathe their sport 24 h a day can benefit from a day off every once in a while. Small breaks not only allow individuals some time to develop other aspects of life (important if they are to have any semblance of balance in their lives) but also can increase the enthusiasm and energy levels of athletes once they return.

Avoiding Training Stress Syndrome

Coaches can help athletes avoid or overcome training stress syndrome by providing variety in training. Because monotony, repetition, and boredom can contribute to staleness and overtraining, activities that break the tedium and routine of training can be beneficial. Variety can be achieved in many ways. Some examples are introducing new drills; changing the training schedule; altering the training venue; encouraging fitness training through cross-training (i.e., participating in activities that are not specific to the sport); involving new faces (e.g., guest coaches, admired athletes); incorporating fun and games; and inviting and using athletes' input in the design of training.

A club swim team that was having problems with dropouts, staleness, and even behaviour problems dealt with the difficulties by introducing board games to Friday training sessions. Fridays were selected because attendance was worse at the end of the week. The process began when the coach

brought in an old game of Monopoly. Chance and Community Chest cards were replaced with Land Activity and Just for Fun cards that described silly games or activities different from those used in traditional training sessions. Buying houses and hotels earned swimmers the right to specify the next activity for their lane. The game gave the swimmers a sense of control of the session and created a sense of anticipation within the team. The game's popularity encouraged the coach and certain swimmers to adapt other board games and create entirely new board games for use at the pool. Attendance records, punctuality during the week, and personal improvements in fitness and time trials were used to determine which swimmers got to select the games for each Friday session. Attendance, training productivity, and enthusiasm all greatly increased. There were fewer signs of staleness and fewer disciplinary problems. This example suggests how effective variety in training can be.

The majority of other strategies that help to deal with staleness, overtraining, and burnout involve the development of a variety of mental skills. Goal setting can help athletes pursue more realistic expectations. Training stress syndrome is exacerbated when athletes try to achieve performances that are unrealistic and expend a lot of energy with little, if any, return. Short-term goals give meaning to training sessions, help take minds off how long the season may be, and increase positive self-concept as feelings of success accompany each achievement of a short-term goal.

Managing Stress

Because too much stress is one of the causes of training stress syndrome, skills that help athletes manage stress are extremely useful. Time-outs, greater variety in training, reduction in training, and goal setting can, as we have seen, help avoid training stress syndrome. Relaxation, positive self-talk, and learning to say no are additional strategies that can help athletes deal with negative training syndrome.

The most obvious skill in this regard is the ability to relax. The two most common physical relaxation strategies are **progressive muscular relaxation** and breathing exercises. Progressive muscular relaxation exercises involve contracting and then relaxing specific muscle groups systematically throughout the body. Often if athletes are tense, telling them to relax is ineffective because they do not know how to relax. By increasing the tension in the muscles and then relaxing, athletes become more aware of and sensitive to what the presence and absence of muscle tension feel like and learn that they can control the level of tension. Many people find it difficult to decrease the tension in a muscle voluntarily. However, when one first increases the tension and then relaxes, the level of tension in the muscle automatically drops below the initial level of tension. After regular practice of this exercise, athletes learn to relax the muscles without first having to tense them.

Breathing exercises are also useful for cultivating relaxation. When stressed, many people either hold their breath or take rapid, shallow breaths. Learning to control breathing can reduce stress. Controlled, deep, slow breathing is the aim of most breathing relaxation exercises. This type of breathing physically relaxes the body. The advantage of using breathing as a form of relaxation is that we are always breathing. With the possible exception of those who play underwater hockey, athletes can always increase their awareness of their breathing even during physical activity.

Focusing on breathing not only enables athletes to control their breathing and therefore aids in physical relaxation; it also may help them relax mentally. Stress has both physical and mental repercussions. Mentally athletes may begin to worry, create self-doubt, say negative things to themselves, and focus on factors that are irrelevant or counterproductive to performance. Concentrating on breathing is a useful way of refocusing attention and controlling self-talk. Practicing functional cue words such as "relax" or "focus" in conjunction with exhaling helps athletes stop negative self-talk. Breathing then becomes a technique for relaxing the body both physically and mentally.

Another skill that is very useful for individuals experiencing staleness, overtraining, or burnout is learning to say no. Although physical training is often a common source of stress, it is rarely the only source. Most athletes who experience training stress syndrome are hardworking, idealistic individuals who strive for high achievement. Because these people are highly motivated and often have a track record of getting things done, other people frequently ask them for favours and help. These athletes want to be successful and often agree to do too much. Any single act does not require too much effort or too much time; but when such acts are added together, the demand is greater than the individual's resources. Every time the athlete agrees to appear at a public promotion for the

sport or team, stay late at training to help another athlete, develop a textbook or training video, attend a charity fund-raiser, or complete an extra assignment or project at work or at home, the total amount of stress is increased. When the person has the time, energy, and capabilities to attend to all the demands, this is not problematic. But when the demands already exceed the person's capabilities, the inability to say no exacerbates the problem. Saying no sounds easy, but often if people know they are capable of the task, believe they may get positive recognition for doing it (remember, a lack of positive feedback is a cause of burnout), and do not want to disappoint someone, it is easier in the short term to say yes.

Severe cases of burnout may require professional counseling. However, there is no single intervention for treating athletes experiencing staleness, overtraining, or burnout. Researchers are currently seeking to determine which intervention or combination of interventions is most effective under specific circumstances.

Changes in Mental Skills

The use of mental skills to combat training stress syndrome is just one example of the way in which developing mental skills can help athletes adapt to training and competition. Participation in sport can lead to the evolution and improvement of an assortment of mental skills. These mental skills are usually viewed as serving one function—the enhancement of performance. In addition to helping athletes enhance performance, acquiring an arsenal of mental skills can increase the enjoyment of participation and positively affect the quality of life in areas outside of sport. Table 21.1 provides several examples of areas in which athletes are likely to acquire mental skills and situations in which those skills may be applicable outside of sport.

Transferring Skills From Sport to Work

Interviews with elite athletes have shown that these athletes can transfer a number of skills and abilities from sport to work. Cognitive skills (e.g., controlling attention and making decisions), self-management skills (e.g., goal setting and stress management), interpersonal skills, communication skills, leadership skills, and the ability to handle negative events are just some of the abilities that interviewees felt they had acquired through sport but then successfully applied to subsequent jobs.

Table 21.1

Examples of Mental Skills That May Be Acquired Through Participation in Sport and Applied to Other Situations in Life

Skill	Possible nonsport applications
Arousal increase	Feeling tired and needing to make a presentation
Communication	Organizing a major event
Concentration	Studying at home with loud roommates
Emotional control	Maintaining calm when the boss is rude
Goal setting	Changing health or work habits
Imagery	Rehearsing difficult confrontations
Injury rehabilitation	Dealing with work-related injuries
Preparation	Being mentally ready for a job interview
Relaxation	Getting to sleep more quickly
Self-confidence	Speaking in front of a large group
Self-talk	Remaining positive when things look bad
Team harmony	Working on a group project
Time management	Studying adequately for all exams
Travel	Avoiding jet lag

They also mentioned that personal characteristics such as assertiveness, confidence, ambitiousness, flexibility, conscientiousness, determination, persistence, and dependability, developed in sport, were beneficial in non-sport-related careers.

The ability of athletes to demonstrate these skills or personal characteristics in sport, however, does not mean that they will automatically be able to transfer the skills or characteristics to new jobs. Skills are more likely to transfer when people are interested in their jobs, are self-confident, and are aware of the skills they have. Skills are less likely to transfer from sport when individuals have no career goals, little time to devote to work, or no perceived connection in their minds between sport and work. The structure of the work environment also influences skill transfer. Job control, coworker support, recognition by supervisors, and job responsibility aid skill transfer. Low job clarity, negative reactions of others, and excessive rules and restrictions hinder the transfer of skills.

Acquiring Mental Skills in the First Place

Participation in sport does not automatically lead to the acquisition of multiple mental skills. Some successful athletes have developed mental skills through trial and error. Others have copied the skills of previously successful athletes, and some have obtained the skills through the guidance of good coaches. Unfortunately, many people never adequately develop these skills. A fairly large percentage of these people drop out from sport because they do not enjoy their participation in sport or perceive their performances to be less than satisfactory, or both. Providing a mental skills training program for participants not only enhances the formation of mental skills that individuals can use throughout life but also can increase the levels of both performance and enjoyment of participation. These two factors often go hand in hand. For example, if Laurel always gets extremely nervous before competitions to the point of being physically ill, chances are that she will not perform her best. If, on the other hand, Laurel can learn to control her anxiety and therefore avoid being sick before competition, she will not only improve her performance but also probably enjoy her participation much more.

For mental training programs to be effective, it is important that they be designed around the needs of the participants. No single technique will work for everyone. For example, although controlled deep breathing can be an effective relaxation technique, focusing attention on breathing actually increases anxiety in some people as they no longer feel that breathing is natural and develop fears of suffocation. Though this reaction is rare, it does happen. Mental training programs therefore need to be flexible and individualized. It is also imperative that people systematically practice their mental skills in the same way they practice their physical skills. Watching an instructional video about how to pole-vault and understanding the principles does not mean that you will be able to physically pole-vault with correct technique. Practice is needed. Similarly, listening to a lecture and understanding the principles of relaxation does not mean that we will be able to relax on command in an extremely anxiety-provoking situation. Extensive, systematic practice is needed to bring about this positive adaptation.

Summary

Training affects people psychologically as well as physically. Aerobically fit individuals are able to deal with psychological stress more effectively than less fit individuals. Fitter individuals have smaller increases in blood pressure and heart rate than less fit individuals in response to stress. Research also suggests that exercise results in lowered levels of negative affect and anxiety during stressful situations. Research in the area of personality is not robust enough to indicate whether or not participation in particular sports or forms of exercise affects personalities. Although some differences have been found in the personalities of team sport versus individual sport athletes, and participants in high-risk versus low-risk activities, the structure of the research does not allow us to conclude whether or not participation in the activities caused the personality differences. It may be that people with different personalities are attracted to different sports.

Although participation in sport and exercise can have many positive psychological and physiologi-cal affects, it can also result in staleness, overtraining, or burnout if participants allow themselves insufficient recovery. Burnout in sport is characterized by emotional/physical exhaustion, a reduced sense of accomplishment, and sport devaluation. Breaks from training, variety in training, goal setting, practicing relaxation skills, learning to say no, and positive self-talk can help athletes avoid and deal with negative adaptations to training.

The mental skills that athletes acquire to help them avoid burnout and enhance their performance can enhance their quality of life in areas outside of sport. Cognitive, self-management, interpersonal, and leadership skills can be transferred from sport to other careers. However, instead of assuming that athletes will automatically develop these mental skills as an inherent part of participating in sport, one should incorporate extensive and systematic practice of mental skills into the structure of training.

Further Reading and References

Berger, B.G., D. Pargman, and R.S. Weinberg. 2002. *Foundations of exercise psychology* (chapter 8, "Exercise As a Stress Management Technique: Psychological and Physiological Effects," pp. 123-135). Morgantown, WV: Fitness Information Technology.

Henschen, K.P. 2001. Athlete staleness and burnout: Diagnosis, prevention, and treatment. In *Applied sport psychology: Personal growth to peak performance*, ed. J.M. Williams (pp. 445-455). 4th ed. Mountain View, CA: Mayfield.

Weinberg, R.S., and D. Gould. 2003. *Foundations of sport and exercise psychology* (chapter 21, "Burnout and Overtraining," pp. 467-487). 3rd ed. Champaign, IL: Human Kinetics.

Weinberg, R.S., and J.M. Williams. 2001. Integrating and implementing a psychological skills training program. In *Applied sport psychology: Personal growth to peak performance*, ed. J.M. Williams (pp. 347-377). 4th ed. Mountain View, CA: Mayfield.

CHAPTER 22

AFTERWORD
Multi- and Cross-Disciplinary Approaches to Human Movement

A s the discipline of human movement studies has matured and specialized scientific research on human movement has proliferated, it has become increasingly difficult, if not impossible, for individual researchers to remain expert in more than one of the subdisciplines of the field. Indeed, knowledge is expanding at such a rate and becoming so specialized that expertise is generally confined to selected areas within a subdiscipline.

The major learning concepts in this chapter relate to

- the issue of specialization versus generalization in the discipline of human movement studies;
- the distinction between multidisciplinary and cross-disciplinary research; and
- multi- and cross-disciplinary research on injury prevention and rehabilitation, talent identification, performance optimisation, and the challenge of maximizing exercise participation among the general public.

Specialization Versus Generalization

Although the increased specialization of the discipline has allowed greatly increased understanding of many aspects of human movement, the danger of such specialization is the potential for fragmentation of the field, a concern discussed in chapter 1. With fragmentation of a discipline, researchers become so interested in their own specialist field that they fail to fully appreciate the importance and significance of knowledge from other subdisciplines and the importance of integrating information from across the various subdisciplines in order to ultimately advance understanding. Even with increasing specialization, it is important to retain a sound general knowledge and understanding of the entire field of human movement studies and an interest in, and appreciation of, how the various systems within our bodies interact.

If we are truly interested in the key disciplinary question of how and why humans move, we need to consider not just functional anatomy, biomechanics, exercise physiology, motor control, or sport and exercise psychology in isolation; rather we must integrate the various subdisciplines to produce a coherent, global view about human movement. Increasingly we should expect to see specialists from the different subdisciplines of human movement studies working together in research teams to attempt to understand complex problems (of the type to be examined in this chapter) from a number of perspectives.

Research involving more than one of the specialist subdisciplines can be classified as either multidisciplinary or cross-disciplinary, depending on the nature and extent of the integration that occurs (figure 1.2). In multidisciplinary research, specialists from the various subdisciplinary fields all investigate a common problem but do so only from their own subdisciplinary perspective. What results is a number of perspectives on a particular research issue but with no particular attempt to integrate these into a consolidated viewpoint. An example may help illustrate this approach.

Suppose a group of researchers, including a biomechanist, a sport psychologist, and an exercise physiologist, are interested in determining those features that characterize elite long distance runners. The top five male and female 10,000-m runners at a national championship are selected for study and given a battery of tests. The exercise physiologist chooses to measure $\dot{V}O_2$max and lactate threshold; the biomechanist measures leg strength and muscle activity patterns; and the psychologist conducts tests to ascertain imagery ability and pain tolerance. The researchers, then, on the basis of their data, ascertain the physiological, biomechanical, and psychological predictors, respectively, of 10,000-m running performance. No particular effort is made to ascertain the relative importance of the different measures or to determine whether a certain combination of physiological, psychological, and biomechanical factors best distinguishes elite runners. Such an approach is termed multidisciplinary because it involves a number of the subdisciplines of human movement although work in each area is essentially independent.

Cross-disciplinary research, on the other hand, relies on two or more specialists working together on a problem and bringing their own perspectives to the situation but integrating their views with those of others from different backgrounds. This approach is in many ways preferable to unidisciplinary and multidisciplinary approaches because it has the potential to create a greater understanding of human movement than would be possible if the topic were investigated in a fragmented way. Humans move as they do because of a host of interacting factors, and examination from only one perspective or level of analysis may lead to an incomplete, and perhaps flawed, understanding. Cross-disciplinary research entails a genuine attempt to cross the traditional subdisciplinary boundaries, with researchers from each subdisciplinary perspective attempting to understand the perspective of the other(s) so that a new consolidated perspective may be developed. The biomechanist and motor control specialist who work together to decide on the best type of kinematic feedback to assist learning, or the exercise physiologist and psychologist who work together to promote adherence to exercise with health benefits, are engaged in cross-disciplinary work.

Examples of Multidisciplinary and Cross-Disciplinary Approaches

The remainder of this chapter presents examples to suggest how integration of the methods and knowledge of two or more subdisciplines can contribute to understanding of some central issues in the field of human movement studies. The issues we examine, through examples of multidisciplinary and cross-disciplinary approaches to program development, are

- injury prevention and rehabilitation,
- talent identification,
- performance optimisation, and
- participation maximization.

For each example we provide references to the original sources at the end of the chapter so further details can be obtained, if required.

Injury Prevention and Rehabilitation

Prevention of, and rehabilitation from, injury are central concerns for many professions related to the discipline of human movement studies. Here we examine three examples of integrative research work on this topic. The first example relates to attempts to minimize workplace injuries arising from manual handling. The second example presents some cross-disciplinary approaches to the prevention of injuries in young people participating in the sports of baseball and cricket. The final example relates to injury prevention and osteoporosis—a chronic disease of growing public health significance.

Preventing Injuries in the Workplace During Manual Lifting

Ergonomics, as noted in chapter 15, is the field of study concerned with investigating the interface between humans and their working environment. It involves the disciplines of human movement studies and engineering, as well as other areas of biological, medical, and technological sciences. A major area of concern in ergonomics is the prevention of work-related injuries. A disproportionately large number of injuries in the workplace are caused by "overexertion" during manual lifting and carrying of objects. The back is the principal site for such injuries; over one-third of all workers' compensation claims in many countries are related to back injuries.

Guidelines related to the prevention of manual materials-handling injuries, especially those involving the back, have been developed. The National Institute for Occupational Safety and Health (NIOSH) in the United States considered information from a number of subdisciplines of human movement studies during the formulation of its guidelines. NIOSH defines limits for manual materials handling based on criteria derived from the results of biomechanical, physiological, and psychophysical studies of lifting tasks.

The compression on the base of the lumbar spine is thought to be directly and causally related to some types of back injuries, so a biomechanical criterion (based on a maximal acceptable lumbar compressive force) provides the limit to safety when lifts are performed infrequently. If only the biomechanical criterion existed, an apparently logical recommendation that might follow would be that small loads should be lifted frequently. This, however, is likely to lead to physiological fatigue, which in turn is a risk factor for back injuries. Therefore, for repetitive lifting tasks, it is more appropriate to use physiological criteria, based on cardiovascular responses, to set limits for safe lifting.

Use of the physiological approach alone could lead to a recommendation that large loads should be lifted infrequently. This would clearly be inappropriate given the knowledge from functional anatomy and biomechanics that injury may result from the excessive forces generated during heavy lifts. The third (psychophysical) criterion is based on the lifter's perception of physical work. It is believed that each person internally monitors the responses of muscles, joints, and the cardiorespiratory system to estimate an acceptable workload and this self-monitoring process appears to have a quite high level of accuracy. The psychophysical approach usually defines a limit that is a good compromise between the biomechanical and physiological results and that is applicable to most, but not all, lifting tasks (Waters, Putz-Anderson, and Garg 1994).

Preventing Injuries in Young Baseball Pitchers and Cricket Bowlers

In the 1960s, epidemiological surveys highlighted a growing problem among child athletes involved in Little League baseball (Adams 1968). These surveys revealed an alarmingly high incidence of conditions involving injury to the developing bony centres around the elbow. The term **Little League elbow** was coined to describe this general condition of the elbow because of the hypothesized relationship between the condition and the pitching action.

The first series of radiological studies revealed that the Little League players showed higher incidences of abnormal bone growth and development at the elbow than nonplayers and that there was a direct relationship between the magnitude of the changes at the joint and the amount of pitching/throwing done by the individual players. Little League elbow has its highest rate of incidence at the onset of the adolescent growth spurt (see chapter 4), in boys 13 to 14 years of age (Wells and Bell 1995). During these years of rapid growth, the bones remodel in response to the applied loads. The stresses applied during the pitching action particularly, and during throwing in general, often cause the medial epiphysis of the throwing arm to widen and develop abnormally.

Follow-up research to the initial clinical and epidemiological studies has included biomechanical investigations of the pitching technique. High-speed video analysis has helped to identify various phases of the throwing movement, including the windup, cocking, acceleration, and follow-through phases (see also figure 16.7), and to reveal that the end of the cocking phase and the start of the acceleration phase are the times when the elbow is placed under greatest stress. During these phases, the elbow is flexed and the humerus laterally rotated, with the muscles on the medial side of the joint undergoing eccentric activity prior to positively accelerating the forearm toward the target. During the follow-through, large loads are placed on the lateral side of the joint as the forearm undergoes **pronation** after ball release, leading to damage to the capitulum of the humerus and the head of the radius. Another important finding of the biomechanical research concerns the relationship between style of throwing and injury rate. Children who pitch with sidearm motions are three times more likely to develop problems than those who use a more overhand technique.

Administrators of Little League baseball were extremely concerned about the high injury rates among the athletes. In response to the medical and biomechanical research findings, various controls have been put into place to attempt to reduce the number and severity of elbow injuries to pitchers. These measures have included general advice to coaches and parents regarding the number of pitches to be thrown during training sessions, typical warm-up and cool-down activities, suitable progressions and buildup in throwing during preseason, limits on the number of innings that pitchers can throw in a game, and limits on the number of games that they can pitch within a week. Many of these rule changes were instituted in 1972, and follow-up research has been conducted to support the positive effect of these changes. While the incidence rates of Little League elbow have declined following the rule changes, the magnitude of the problem is still substantial.

The history of Little League elbow in baseball has to some degree been mirrored in cricket-playing countries, where the incidence of vertebral stress fractures in young fast bowlers is now recognized as a significant problem (Elliott 2000). A considerable understanding of back injuries in cricket fast bowlers now exists as a consequence of systematic, cross-disciplinary research. One of the first tasks undertaken by a research team at the University of Western Australia was to gather statistics on the number and types of injuries incurred by fast bowlers. This research revealed that approximately 45% to 55% of fast bowlers had abnormal radiological features in the lumbar spine (Elliott et al. 1992). This incidence of lumbar spine abnormalities was some 10 times higher than would have been expected given the players' age and sex. A subsequent series of studies including physiological, biomechanical, and anthropometric measurements has provided more detailed insight into the causes of the problem (Elliott et al. 1993).

The methods of biomechanics were used to establish a link between bowling technique and the incidence of injury. Each bowler's technique was classified as either side-on, front-on, or mixed on the basis of some key kinematic characteristics. Those bowlers who used a mixed technique requiring considerable trunk rotation about the vertical axis presented with a significantly higher incidence of injury than bowlers using the other two styles. A relationship was therefore established between biomechanics and injury. Researchers at the University of Queensland have subsequently established that asymmetry of volume of the back muscles, determined by magnetic resonance imaging techniques, is strongly correlated with the incidence of stress fractures at the level of the fourth lumbar vertebra (Engstrom et al. 2000).

Another aspect investigated was the notion of overuse. Logbook records indicated that players who bowled more had a higher incidence of stress fractures, disabling back injuries, or both. This observation has led to the formulation and implementation of recommendations as to the number of balls that fast bowlers should bowl in matches and in practice in order to minimize the likelihood of injury. It has been recognized that overuse by itself does not explain the incidence and severity of injuries, so research involving technique analysis of young fast bowlers continues in an effort to prevent the repetition of potentially injurious techniques.

Another recognition in recent years has been that injury prevention, and especially rehabilitation, has a large psychological component. Case studies describing the psychological response to injury of injured fast bowlers are a valuable adjunct to existing knowledge on injury mechanisms and their prevention (Gordon and Lindgren 1990).

Prevention of Injuries Related to Osteoporosis

Osteoporosis, as noted in a number of earlier chapters, is a major chronic disease, often linked to physical inactivity, in which bone density is reduced to levels that elevate the risk of fractures. Fractures of the hip (at the neck of the femur) and the spine are major risks. Fractures typically result from falls in which the energy of the fall cannot be dissipated by the weak bone, which then fails.

Each year in the United States alone, there are about 1.5 million osteoporosis-related fractures, including about 700,000 crush fractures of the vertebral body and about 300,000 fractures of the neck of the femur (Gatti 2002; Hellekson 2002; Messinger-Rapport and Thacker 2002). Approximately one-third of women over 65 have vertebral fractures, and one-third of men over 80 have osteoporosis. With the annual direct costs of osteoporosis estimated to be about U.S. $13 billion, there is a clear need for effective strategies to reduce the incidence of osteoporosis and the flow-on costs to society.

The understanding of osteoporosis and the development of strategies for its prevention are problems with anatomical, biomechanical, physiological, psychological, and motor control elements and implications. The maintenance of bone strength is a primary factor in the prevention of osteoporotic fractures. Bone loss can be reduced by appropriate changes in lifestyle, including increasing physical activity (especially weight-bearing exercise) and the intake of calcium and vitamin D. Smoking is strongly discouraged, as is excessive consumption of alcohol and caffeine, because of their direct effects on bone as well as their indirect effects on general health. People will benefit optimally if they adopt a healthy lifestyle when young and continue it throughout adulthood. Convincing the public about these benefits, as well as achieving a high degree of compliance so that most members of society participate in regular physical activity, is not an easy task. It requires the combined skills of specialists in each of the subdisciplines of human movement studies along with educators, psychologists, sociologists, and experts in health promotion.

For people with osteoporosis, management of the disease along with risk minimization is also a multifaceted problem. One approach is through the use of supplementation and pharmaceuticals. Bisphosphonates, which reduce the rate of bone loss by inhibiting osteoclastic activity, have been recommended for older women in particular. Hormone replacement therapy has been recommended, but recent research results have indicated unacceptable side effects of long-term therapy. A second approach is active avoidance of the risks of trips and falls. To this end, a number of interventions have proven effective in the prevention of falls in elderly patients, including strength training and balance retraining. There is also a technological solution for a person with osteoporosis who falls, which is the hip pad protector. This protects the weak bone of the hip region by absorbing the energy of impact of a fall.

Talent Identification

For many years in industry and public service, management has been concerned with selecting personnel to ensure the best possible match between the person and the task that person will perform. While person–task matching remains an important concern in ergonomics, a parallel problem in sport is talent identification—the identification of talented youngsters who have potential to become future champions in specific sports. The complex nature of elite sport performance necessitates that talent identification be broadly based on knowledge from a number of the subdisciplines of human movement studies.

Structured systems for the development and nurturing of talent in specific sports exist in many countries around the world, although the evidence base for the variables used in athlete selection is

frequently quite poor (Durand-Bush and Salmela 2001). Researchers at the Liverpool John Moores University have begun pursuing the question of the best predictors of talent in soccer (Williams and Reilly 2000). Given the multifaceted nature of soccer, there are many potentially important predictors of future success in the sport: anthropometric variables (such as height, weight, body size, bone diameter, muscle girth, body fat, and somatotype), physiological variables (such as aerobic capacity, anaerobic power, and endurance), motor control variables (such as attention, anticipation, decision making, creative thinking, and technical skills), psychological variables (such as self-confidence, anxiety control, motivation, and concentration), and sociological variables (such as parental support, education, socioeconomic background, cultural background, and hours and resources available for practice) (figure 22.1).

As a step toward determining which variables are the best predictors of talent and future success, the researchers (Reilly et al. 2000) administered a multidisciplinary battery of tests to young players (16-year-olds) classified as either elite or sub-elite. The elite players were those who had signed for a professional club and had international youth soccer playing experience. A total of 28 tests were used, including ones drawn from each of the biophysical subdisciplines of human movement studies, along with some soccer-specific practical skills tests. Of the biophysical measures used, the best discriminators between the elite and the sub-elite players were agility and sprint time (both physiological measures), the psychological measure of ego orientation, and the motor control measure of anticipation (see "Response Speed in Soccer Players," p. 268). The elite players were also significantly different from the sub-elite sample in that they were leaner, had greater aerobic power, had superior fatigue tolerance, and were better performers on the soccer dribbling test. It is noteworthy that the key discriminators were drawn from different subdisciplines.

This example clearly highlights why it is imperative that multidisciplinary rather than unidisciplinary approaches be used to examine research questions such as this. The future challenge for the researchers is to establish baseline reference data for young players.

Performance Optimisation

In contrast to the example of talent identification, which relies largely on identifying individuals with

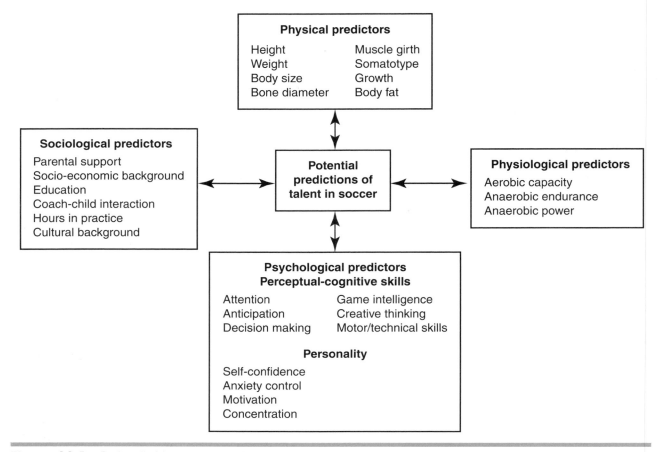

Figure 22.1   Potential predictors of talent in soccer from each sport science discipline.
Adapted from Williams and Franks 1998.

appropriate genetic predispositions for success, this section on performance optimisation illustrates how integrated knowledge from different subdisciplines of human movement studies may be used to optimise performance within the limits of each individual's physical structure. We present examples drawn from the stretch–shortening cycle of muscles, the monitoring of overtraining in athletes, and the use of psychophysiological feedback in recreational exercise and in competitive sport.

Using Stretch-Shortening Cycles in Muscle

A number of earlier chapters in this book refer to the enhanced response generated by muscle when it is prestretched. In the field of strength and conditioning, activities aimed at exploiting this effect have become known as "plyometrics." Understanding the functional basis for plyometrics is an interdisciplinary problem.

A series of experiments conducted in Finland by Paavo Komi and his colleagues in the 1970s and 1980s has led to an enhanced understanding of muscle function and techniques used in many human actions. Komi used the term **stretch–shortening cycles** to describe those activities in which the muscles are stretched prior to concentrically contracting. It necessarily follows that the muscle needs to be acting eccentrically (that is, contracting, while increasing in length) for stretching to actually occur.

Beginning in the 1960s, Cavagna among others had noted, in isolated muscle, two very important aspects about eccentric muscle contraction—that compared to purely concentric contraction, prior eccentric contraction could develop greater concentric force and result in higher efficiency (Cavagna 1977; Cavagna et al. 1965; Cavagna et al. 1968). Komi established that the same was true for intact muscles and further showed that increased or augmented work and power were available during a concentric muscle contraction if it was preceded by an eccentric phase. His work over approximately 10 years culminated in findings relating to efficiency of movement and the way in which this increased efficiency comes about. More recently, Komi has shown that enhancement of performance does not occur when the muscle is fatigued.

Komi's early work is typified by an experiment using a specially designed sled to investigate concentric and eccentric lower limb muscular work (Komi 1984). The sled was mounted on two parallel rails that were inclined approximately 40° above the horizontal. A force plate was positioned at 90° to the rails at the bottom of the sled. The force plate was used to determine the forces applied to the ground while an oxygen-measuring system was placed nearby, allowing collection of expired gases from the participants and hence calculation of the physiological cost of work. A person could sit on this sled, be positioned at a fixed point up the rails, and be released or, alternatively, be required to push him- or herself up the sled to a certain height. The work done in either the eccentric condition (being released from a fixed point) or the concentric condition (pushing up to a certain height) was constant, but the power and physiological costs were found to be different.

The stretch–shortening cycle is usually associated with the effects of gravity on the lower limb muscles during locomotion, but researchers from the Gulhane Military Medicine Academy in Turkey (Aydin et al. 2001) have recently used an isokinetic machine to produce stretch–shortening cycles of upper limb muscles. They have shown that when an eccentric action precedes a concentric action of the medial rotators at the shoulder, performance is enhanced compared with that in other conditions. This has implications for throwing training.

The explanation of the augmented work (or increased efficiency) achieved during the stretch–shortening cycle has both biomechanical and neurophysiological aspects. In terms of the biomechanics, energy is stored in elastic structures of the muscles and tendons during the eccentric phase and then used to do work during the concentric phase. Concurrently, during the stretching phase, muscle spindles and Golgi tendon organs (as described in chapter 14) are stimulated, which results in increased potentiation of the muscles. Komi and his coworkers were able to establish, through a series of cleverly designed experiments, that approximately two-thirds of the increased work in stretch–shortening cycles is due to elastic energy return and one-third is due to increased potentiation through reflex activity.

The implications of Komi's work are far-reaching since virtually all motor activities use stretch–shortening cycles. At first glance, it might appear that augmented work available through the stretch–shortening cycles would benefit only power athletes who require single, maximal efforts. In such cases, techniques reflect the fact that muscles must be prestretched prior to contracting concentrically. However, closer examination reveals that the use of stretch–shortening cycles is equally important to endurance athletes, who are clearly interested in improving efficiency. For example, in running, stretching and then shortening take place during the stance phase when the calf muscles are stretched as the body rotates over the foot and the knee flexes. This energy can then be used to do work on the body during the concentric, push-off phase. Use of this elastic energy increases the economy of gait, possibly enhancing performance.

Komi's work, along with the subsequent research by Aydin and colleagues, is an excellent example of cross-disciplinary research in which physiology, biomechanics, biochemistry, and neurophysiology have led to an improved understanding of muscle contraction dynamics (Komi 2000). By using an experimental task that required the large muscles of the lower limbs to perform sizable amounts of work, the investigators could induce sufficiently large differences between the eccentric and concentric conditions to be able to use an expired gas analysis system to estimate oxygen metabolism. If only small limb movements had been chosen, the precision of the gas analysis system would probably not have been high enough to show significant differences. Through understanding the concept of mechanical work, Komi and his colleagues recognized that the same amount of work could be done on the sled if participants accelerated the sled to a predetermined height or stopped the sled after it had been released from that height. The only difference was that the muscles had to perform concentrically in one situation and eccentrically in the other.

Knowledge of the anatomy of muscle, tendon, and neural pathways allowed these researchers to theorize on how these structures would influence the stretch–shortening cycle. If they had not known about muscle spindles, Golgi tendon organs, and the role of the Ia afferent neurons, they would not have been able to recognize that the increased work output following muscle stretch could be due to increased potentiation of the muscles through a reflex arc. Thus to obtain a complete picture of the ways in which humans move, it is important to be eclectic and to investigate problems from multiple perspectives using a variety of techniques.

Understanding Overtraining Syndrome in Athletes

Overtraining syndrome (OTS) is a condition characterized by poor performance in competition, inability to maintain high training loads, persistent fatigue, changes in mood state, disrupted sleep patterns, and frequent illness. Overtraining syndrome reflects the body's inability to adapt to the cumulative fatigue resulting from daily intense training that is not balanced with adequate rest and recovery.

The relationship between physiological and psychological causes and consequences of OTS has been the subject of much speculation and research over the past two decades. It is unclear whether one precedes and possibly contributes to the other, or whether both might result from a common cause. For example, psychological factors such as mood disturbances can lead to disrupted sleep and eating patterns that in turn can negatively affect physiological functioning. On the other hand, physiological changes such as lower lactate threshold and reduced ability to train intensely might cause athletes to become discouraged and anxious about their loss of form. Alternatively, OTS may involve changes in factors that affect both the physiological and psychological responses to intense training. Negative moods could cause poor performances, poor performances could cause negative moods, or both negative moods and poor performances could be caused by a third overriding factor. To better understand OTS, several research groups have studied simultaneous changes in physiological and psychological factors in athletes experiencing OTS.

Overtraining syndrome has been shown to reduce the lactate threshold, or exercise pace at which blood lactate starts to accumulate exponentially—a critical component of endurance exercise capacity (see chapter 11). In one study, triathletes and cyclists were followed over 19 months (Urhausen et al. 1998). In athletes showing symptoms of OTS, the ability to perform brief, maximal exercise (10-30 s) was not impaired, but endurance performance and lactate threshold declined by 27%. Athletes reported typical symptoms such as "heavy" muscles, fatigue, inability to concentrate, and disturbed sleep; mood state was also affected.

Overtraining syndrome also has been associated with changes in hormone levels. Hormones are messenger molecules that have far-reaching effects on the regulation of many physiological, psychological, and immunological functions. Overtraining seems to particularly affect the production of **norepinephrine** (NEp, also known as noradrenaline), a hormone released in response to physical stress. In a study of elite swimmers, NEp production was lower in swimmers showing symptoms of OTS than in those who were able to adapt to higher training loads (Mackinnon et al. 1997). Interestingly, low NEp levels appeared two weeks before the onset of other symptoms, such as poor performance and high fatigue, suggesting that hormonal changes may precede and possibly contribute to later development of symptoms of OTS.

Norepinephrine affects the immune system, and low NEp production might be related to the high frequency of minor illness, such as colds and the "flu," reported among top athletes, especially during overtraining. The NEp level also is associated with a number of psychological factors. The main symptoms of depression are related to a lack of NEp in the brain. Norepinephrine has also been associated with learning and memory. Blocking NEp receptors causes loss of memory. Therefore, if overtraining decreases the production of NEp, it is not surprising that overtrained athletes report an inability to concentrate and increased mood disturbance.

The consistent finding that both psychological and physiological changes occur with OTS raises the question whether there is a common cause of both types of symptoms. An interesting hypothesis recently has been put forward in an attempt to explain the physiological, psychological, hormonal, and immunological changes occurring with OTS (Smith 2000). Daily intense training without adequate recovery might cause microtrauma to the muscles, joints, and bones throughout the body, eliciting release of inflammatory molecules called **cytokines.**

Cytokines have far-reaching effects on virtually all body systems—for example, causing fever, stimulating immune cells, influencing energy metabolism, causing release of some hormones, and influencing mood state and behaviour. Of special interest is that cytokines act on the central nervous system and are associated with "sickness behaviour" such as loss of interest, sleep disturbance, reduced appetite, and depression, some of which also occur during OTS. These far-reaching effects of cytokines have probably evolved as a protective mechanism to force the body to take time to rest and recuperate from illness or injury.

Smith (2000) suggested that maintaining excessive training with inadequate recovery causes the body to adopt an abnormal and persistent state of inflammation, with chronic elevation of cytokine levels. Cytokines may be the common mediator of both the physiological and psychological symptoms of OTS. This attractive hypothesis awaits further experimental study to determine whether it can explain the myriad symptoms of OTS, and importantly, whether such a relationship can be exploited to prevent OTS among athletes.

Improving Control and Performance Through Biofeedback

Biofeedback is another performance optimisation-related area in which information from exercise physiology and sport and exercise psychology is integrated. **Biofeedback** instruments are designed to give individuals information about their biological processes. Through increasing awareness of factors such as heart rate, sweating, muscle tension, or brain activity, people can learn to modify or control their physiological activation. Biofeedback provides a means by which we can learn to control bodily functions that traditionally have been thought to be beyond our control.

A recent study in Spain addressed the effectiveness of biofeedback in controlling heart rate during low- and moderate-intensity physical exercise (Moleiro and Cid 2001). The participants were 35 females between the ages of 17 and 25 years who were randomly assigned to one of four experimental groups:

- Medium exercise intensity (50% of maximal heart rate) with heart rate biofeedback
- Low exercise intensity (30% of maximal heart rate) with biofeedback
- Medium exercise intensity with verbal instructions to try to lower their heart rates
- Low exercise intensity with verbal instructions

During the study, the participants completed a submaximal test in the first session and then had four heart rate control training sessions with either biofeedback or verbal instructions. The participants who trained with biofeedback showed a greater attenuation of the increase in heart rate produced by exercise than participants who trained with verbal instructions. The workload did not influence

the voluntary control of heart rate. These results demonstrate that people can learn to voluntarily control physiological variables through the use of biofeedback techniques (figure 22.2).

Biofeedback has also been included in intervention studies designed to enhance sporting performance. Researchers from Arizona State University (Landers et al. 1991) have directly tested the effectiveness of biofeedback in enhancing archery performance. Rather than focusing on heart rate, these researchers looked at electrocortical or brain wave biofeedback. Through use of an electroencephalogram (EEG), 24 archers were provided with correct EEG feedback, incorrect EEG feedback, or no feedback at all. The correct-feedback archers significantly improved their performance; the incorrect-feedback archers showed a significant performance decrement; and the control group showed no changes in performance. The connection of electrocortical activity to archery performance is based on research showing that shooters have reduced cortical activity in the left temporal area while shooting. In the archery study, the correct feedback was designed to reduce left temporal activation, and the incorrect feedback was designed to reduce right temporal activation. One can conclude from the study that EEG biofeedback can affect performance, with the direction of the effect being determined by whether the feedback is incorrect or correct.

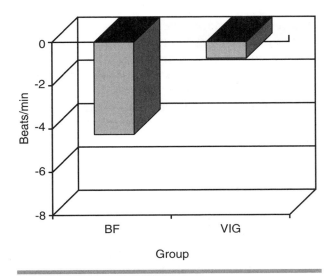

Figure 22.2 Heart rate attenuation by a group given biofeedback training (BF) and a verbal instruction group (VIG).

Reprinted from Moleiro and Cid 2001.

Together these studies on biofeedback highlight the interaction of psychological, physiological, and neural factors and further illustrate the importance of cross-disciplinary research.

Maximizing Community Participation in Physical Activity

To conclude our consideration of multidisciplinary and cross-disciplinary aspects of human movement, we examine in this section examples of programs aimed at maximizing wider participation in physical activity. As discussed in chapter 13, about half of the world's population does not perform sufficient physical activity to derive health benefits. Focusing on an entire community is an efficient means to influence the health of many people. Community interventions are more cost-effective than interventions targeted directly at individuals and can provide the necessary support and infrastructure to help people remain physically active.

To implement community strategies to enhance participation in physical activity, it is important to understand those factors influencing an individual's decision to become and remain active (as discussed in chapters 13, 19, and 20). Factors predicting adoption of activity differ from those predicting maintenance, indicating that different strategies must be used to attract individuals to initiate and then continue a physically active lifestyle (Sallis and Owen 1999). People prefer moderate physical activity, especially those over age 35 years, suggesting that effective community programs should focus on encouraging moderate activity. This strategy was incorporated in the 1996 U.S. Surgeon General's recommendations (see chapter 11) and has been part of other programs throughout the world. Knowledge about exercise is a predictor of maintenance of physical activity, suggesting that efforts to educate the community about the health benefits of physical activity may help enhance participation.

The following section is a very brief discussion of various programs from around the world designed to encourage wider participation in physical activity at the national and community levels. These are examples only and by no means constitute a comprehensive listing.

In April of each year, the World Health Organization (WHO) has a World Health Day to draw attention to a critical health issue. The 2002 World Health Day focused on physical activity. Working in partnership with many international and national organizations, WHO provided materials for local, national, and international events promoting physical activity. Events throughout the world included community walks, demonstrations, expert panel discussions, media releases, and awards for outstanding programs that encourage physical activity.

In the United States, the Healthy People 2010 objectives and strategies emphasize ways in which government, business, and individuals can encourage healthy behaviours among the population; physical activity is one of 10 leading health indicators. The program aims to increase the number of regularly active people and to decrease the number of sedentary individuals. Working with partners in government and nongovernment agencies, Healthy People 2010 provides resources to assist in marketing, media promotion, planning events, and lobbying local government for policy and infrastructure changes to encourage physical activity. The National Physical Activity Initiative provides scientific and technical assistance to promote physical activity through state education systems, health care providers, and volunteer community organizations.

Canada has for many years run national campaigns to promote physical activity. A Health Canada program, SummerActive, is a community-based campaign to increase awareness of the importance of physical activity to health, with a goal of reducing the incidence of physical inactivity. SummerActive provides community leaders and partner organizations with resources such as brochures, displays, and electronic access to information to promote events relating to physical activity. Examples of such events include public walks, opportunities for children to try different sports with encouragement by Olympic athletes, and lighthearted sport competitions. Another community event, the Commuter Challenge, was a friendly competition between Canadian cities to use active, sustainable-energy modes of transport (e.g., cycling, walking, in-line skating). The program estimated the amount of pollution saved by each city.

Active Australia is a public health initiative that combines national and local programs by departments of health, government, and sport to promote physical activity. It works through public education and media campaigns and through changes in public policy to support physically active lifestyles (e.g., providing cycle and walking paths, influencing design of residential developments).

The West Australia Department of Transport developed a program to educate and motivate people to enhance their health and the environment by using alternative forms of transport such as walking, cycling, and public transport. Evaluation of the program showed a decline in car travel, as well as increased use of active alternative transport that was sustained for two years after the program.

Sport and Recreation New Zealand (SPARC) has a variety of programs to promote physical activity. Information is available online and through the print media to help people find local physical activity programs. Individuals can consult via telephone or e-mail with an "Active Living Coordinator" to discuss personal physical activity goals. Walking maps for any location can be accessed online. General practitioners are provided with resources to encourage their patients to adopt a physically active lifestyle.

The Singapore government's Sports for Life program provides cost-effective and convenient access to facilities to encourage physical activity. One scheme teaches novices basic skills in a variety of physical activities or sports. The aim is to ensure that every housing estate in this small country has a fitness centre with qualified instructors offering beginning fitness programs to residents at low cost.

These are just a few of the many efforts throughout the world to encourage wider community participation in physical activity. Successful efforts to encourage participation in regular physical activity depend upon a cross-disciplinary and partnership approach that encompasses the physical, social, psychological, health, political, and economic factors influencing the ability of a community to support individuals in maintaining a physically active lifestyle. Working with other health professions, the field of human movement studies can play a key role in advancing community health by providing knowledge and strategies to encourage physical activity among all members of the community.

Summary

Enhanced understanding of human movement requires not only approaches that independently consider the specialist perspectives provided by multiple subdisciplines (so-called multidisciplinary approaches) but also approaches that actively seek to integrate the various perspectives. Through such integration, cross-disciplinary approaches seek to

address key challenges for the field such as those posed by the need to continuously improve injury prevention and rehabilitation, talent identification, performance optimisation, and participation maximization. The integrity and future of the discipline of human movement studies depend on the achievement of increasing specialization of knowledge without fragmentation.

Further Reading and References

Adams, J.E. 1968. Bone injuries in very young athletes. *Clinical Orthopaedics* 58: 129-140.

Aydin, T., Y. Yildiz, C. Yildiz, and T.A. Kalyon. 2001. The stretch-shortening cycle of the internal rotators muscle group measured by isokinetic dynamometry. *Journal of Sports Medicine and Physical Fitness* 41(3): 371-379.

Cavagna, G.A. 1977. Storage and utilization of elastic energy in skeletal muscle. *Exercise and Sport Sciences Reviews* 5: 9-129.

Cavagna, G.A., B. Dusman, and R. Margaria. 1968. Positive work done by a previously stretched muscle. *Journal of Applied Physiology* 24: 21-32.

Cavagna, G.A., F.P. Saibene, and R. Margaria. 1965. Effect of negative work on the amount of positive work performed by an isolated muscle. *Journal of Applied Physiology* 20: 157-158.

Durand-Bush, N., and J.H. Salmela. 2001. The development of talent in sport. In *Handbook of sport psychology*, ed. R.N. Singer, H.A. Hausenblas, and C.M. Janelle (pp. 269-289). 2nd ed. New York: Wiley.

Elliott, B. 2000. Back injuries and the fast bowler in cricket. *Journal of Sports Sciences* 18: 983-991.

Elliott, B.C., J. Davis, M. Khangure, P. Hardcastle, and D. Foster. 1993. Disc degeneration and the young fast bowler in cricket. *Clinical Biomechanics* 8(5): 227-234.

Elliott, B., P. Hardcastle, A. Burnett, and D. Foster. 1992. The influence of fast bowling and physical factors on radiologic features in high performance young fast bowlers. *Sports Medicine, Training and Rehabilitation* 3(2): 113-130.

Engstrom, C., D. Walker, V. Kippers, and R. Buckley. 2000. Quadratus lumborum asymmetry and pars interarticularis injury in cricket fast bowlers: A prospective MRI examination. In *Abstracts of the 2000 Pre-Olympic International Congress on Sport Science, Sports Medicine and Physical Education* (pp. 191-192). Brisbane: Sports Medicine Australia.

Gatti, J.C. 2002. Which interventions help to prevent falls in the elderly? *American Family Physician* 65(11): 2259-2261.

Gordon, S., and S. Lindgren. 1990. Psycho-physical rehabilitation from a serious sport injury: Case study of an elite fast bowler. *Australian Journal of Science and Medicine in Sport* 22(3): 71-76.

Hellekson, K.L. 2002. NIH releases statement on osteoporosis prevention, diagnosis, and therapy. *American Family Physician* 66(1): 161-162.

Khan, K., H. McKay, P. Kannus, K. Bennell, J. Wark, and D. Bailey. 2001. *Physical activity and bone health.* Champaign, IL: Human Kinetics.

Komi, P.V. 1984. The stretch-shortening cycle and human power output. In *Human muscle power,* ed. N.L. Jones, N. McCartney, and A.J. McComas (pp. 27-39). Champaign, IL: Human Kinetics.

Komi, P.V. 2000. Stretch-shortening cycle: A powerful model to study normal and fatigued muscle. *Journal of Biomechanics* 33(10): 1197-1206.

Landers, D.M., S.J. Petruzzello, W. Salazar, D.J. Crews, K.A. Kubitz, T.L. Gannon, and M. Han. 1991. The influence of electrocortical biofeedback on performance in pre-elite archers. *Medicine and Science in Sports and Exercise* 23(1): 123-129.

Mackinnon, L.T., S.L. Hooper, S. Jones, R.D. Gordon, and A.W. Bachmann. 1997. Hormonal, immunological, and haematological responses to intensified training in swimmers. *Medicine and Science in Sports and Exercise* 28: 741-747.

Messinger-Rapport, B.J., and H.L. Thacker. 2002. A practical guide to prevention and treatment of osteoporosis. *Geriatrics* 57(4): 16-23.

Moleiro, M.A., and F.V. Cid. 2001. Effects of biofeedback training on voluntary heart rate control during dynamic exercise. *Applied Psychophysiology and Biofeedback* 26(4): 279-292.

Reilly, T., A.M. Williams, A. Nevill, and A. Franks. 2000. A multidisciplinary approach to talent identification in soccer. *Journal of Sports Sciences* 18(9): 695-702.

Sallis, J.F., and N. Owen. 1999. *Physical activity and behavioural medicine.* Thousand Oaks, CA: Sage.

Smith, L.L. 2000. Cytokine hypothesis of overtraining: Physiological adaptation to excessive stress? *Medicine and Science in Sports and Exercise* 32: 317-331.

Urhausen, A., H.H.W. Gabriel, B. Weiler, and W. Kindermann. 1998. Ergometric and psychological findings during overtraining: A long-term follow-up study in endurance athletes. *International Journal of Sports Medicine* 19: 114-120.

Waters, T.R., V. Putz-Anderson, and A. Garg. 1994. *Applications manual for the revised NIOSH lifting equation.* Publication no. 94-110. Available via the NIOSH Web site [www.cdc.gov/niosh/pubs.html].

Waters, T.R., V. Putz-Anderson, A. Garg, and L.J. Fine. 1993. Revised NIOSH equation for the design and evaluation of manual lifting tasks. *Ergonomics* 36: 749-776.

Wells, M.J., and G.W. Bell. 1995. Concerns on Little League elbow. *Journal of Athletic Training* 30(3): 249-253.

Williams, A.M., and T. Reilly. 2000. Talent identification and development in soccer. *Journal of Sports Sciences* 18(9): 657-667.

GLOSSARY

acceleration: The time rate of change of velocity.

accommodation: The process by which the curvature of the lens of the eye is adjusted for viewing objects at different distances; achieved through control of the ciliary muscles.

acetylcholine (ACh): A common neurotransmitter in the peripheral and central nervous systems.

achievement goal orientations: Individual definitions of success.

actin: Thin protein filaments of muscle.

action stage: The fourth stage of the transtheoretical model; in this stage individuals have changed their behaviour, but have done so for less than six months.

activation heat: Heat produced in the initial stage of a muscular contraction.

adaptation: The structural or functional adjustment of an organism to its environment that improves its chances of survival.

addiction: Dependence on or commitment to a habit, practice, or substance that is present to the extent that its cessation causes trauma.

adenosine triphosphate (ATP): "High-energy" phosphate molecule produced in all cells; cleavage of the terminal phosphate bond yields energy to fuel cellular work such as development of force in skeletal muscle.

aerobic (oxidative) energy system: Energy system of oxidative metabolism requiring oxygen to produce ATP; provides the major source of ATP for endurance exercise lasting longer than 3 min.

aerobic power: See $\dot{V}O_2$max (maximal oxygen consumption).

afferent: Pertaining to sensory (or input) information transmitted from the sensory receptors into the central nervous system and brain.

age stratification: Prejudicial and discriminatory views about aging.

alpha–gamma coactivation: The process involving simultaneous activation of both alpha motor neurons (to extrafusal muscle fibres) and gamma motor neurons (to intrafusal muscle fibres), which permits comparison of actual muscle length with intended muscle length.

alpha motor neurons: Large nerve cells that innervate skeletal (extrafusal) muscle fibres.

ambient vision: That aspect of vision, deriving from the whole visual field and the nerve pathway connecting the optic nerve to the superior colliculi in the midbrain, that is responsible for the location of objects and the perception of self-motion.

anaerobic capacity: The total amount of work, in kilojoules, accomplished during brief high-intensity exercise such as the 30-s bicycle ergometer test; considered an indication of the capacity of the anaerobic glycolytic energy system.

anaerobic glycolytic energy system: Metabolic energy system that uses glycogen or glucose to produce ATP and produces lactic acid as a by-product; the major source of ATP for maximal exercise lasting between 20 s and 3 min.

anaerobic power: Peak or maximal power, in watts, achieved during brief high-intensity exercise such as the 10-s bicycle ergometer test; considered an indication of the capacity of the immediate energy system.

anaerobic threshold: See lactate threshold.

androgyny index: The use of pelvic and shoulder widths to distinguish between adult males and females.

angular acceleration: The rate of change of angular velocity. Usually expressed as two components: centripetal (or radial) and tangential acceleration.

angular displacement: The change in the angular position of a line during a time interval Δt.

angular velocity: The rate of change of angular displacement.

anterior: Front.

anthropometer: An instrument used to measure lengths of body segments.

anthropometry: Study of the size, proportions, and composition of the human body.

anxiety: The subjective feeling of apprehension or worry often experienced when a situation is perceived to be threatening.

appositional growth: Addition or erosion, or both, of bone at the outer and inner surfaces of the shaft to cause changes in shaft diameter and thickness of the compact bone.

apraxia: The inability to carry out purposeful movements in the absence of paralysis or other sensory or motor impairments, generally following damage to the cerebral cortex.

arousal: General state of physiological alertness as controlled by activation of the reticular formation.

artery: A blood vessel that carries blood away from the heart.

arthritis: Inflammation of synovial joints.

arthrology: The study of the joints of the body.

articular cartilage: Cartilage forming the smooth bearing surface of a synovial joint.

articulation: See joint.

associating: Attending to the body while exercising.

associative phase: The second phase in the learning of a new skill, in which movement patterns become more refined and consistent through practice.

ataxia: The breakdown or irregularity of muscular coordination, generally following damage to the cerebellum.

atrophic: Referring to the process whereby cells decrease in size (waste away) with disuse or disease.

attention–arousal set theory: A theory that states imagery is effective because it primes the athlete for performance by helping the athlete reach the optimal level of arousal and focus attention on what is relevant.

autonomous phase: The final phase in the learning of a new skill, in which the control of movement appears to be automatic and free of the need for constant attention.

auxology: The study of growth.

axon: A single nerve fibre extending away from the cell body of a neuron and responsible for sending nerve impulses away from the cell body.

basal ganglia: Collection of nuclei, located within the cerebral hemispheres, that are intimately involved in movement control and coordination.

bending: A combination of tensile, compressive, and shear forces within a structure.

biarticular: Referring to a skeletal muscle that crosses two synovial joints.

bicondylar calipers: An instrument used to measure widths of bones near their ends.

biofeedback: Immediate feedback of a biological phenomenon provided through use of electronic recording instruments or the use of instruments to obtain information about biological processes of which one is not normally aware. Through biofeedback training, individuals can learn to modify or control these biological processes.

biomechanics: Application of mechanics to the study of living systems.

blocked practice: A type of practice in which each component skill is practiced repetitively as an independent block; practice is fully completed on one component skill before being commenced on the next component skill.

body mass index: A simple estimate of body mass calculated by weight (kg) / height (m)^2.

brainstem: That section of the brain consisting of the medulla, pons, and midbrain, lying between the cerebrum and the spinal cord.

burnout: The third and final phase of training stress syndrome in which repeated failure to cope with training demands results in physiological and psychological exhaustion.

cadaver: A human body that has been embalmed after death for the purposes of anatomical dissection and instruction.

cadence: Frequency of a rhythmic motion, such as the gait cycle.

calcium: A mineral stored in bone and essential for muscle contraction.

callus: Unorganized meshwork of fibres or bone laid down after a fracture; eventually replaced by compact bone.

cancellous bone: See spongy bone.

cardiac muscle: The specialized striated muscle forming the walls of the heart.

cardiac output ($\dot{Q}$): The volume of blood pumped by the heart per minute.

cardiovascular disease: A general term to describe various diseases of the heart and blood vessels, including coronary heart disease, heart failure, high blood pressure (hypertension), stroke, and peripheral vascular disease (disease of the blood vessels in the limbs).

cartilage: A type of connective tissue with a high water content.

cartilaginous joint: A type of joint in which the material between the bones is mainly cartilage.

catastrophe model: When cognitive anxiety is low, the relationship between arousal and performance takes the form of the inverted U; when cognitive anxiety is high, any increase in arousal past the optimum point results in a rapid and dramatic decline in performance.

central nervous system: The nervous system consisting of the brain and the spinal cord.

central pattern generator: The capacity of the spinal cord (or networks of neurons within it) to generate rhythmic flexion–extension patterns that form the basis for locomotion.

central sulcus: A deep trench or vertical groove in the middle of each cerebral hemisphere that separates the frontal lobe from the parietal lobe; the motor cortex lies immediately forward of, and the sensory cortex immediately behind, the central sulcus.

centre of mass (or centre of gravity): The theoretical point in an object at which its entire mass appears to be concentrated.

centripetal acceleration: Component of acceleration of a point on a body given by the square of the angular velocity of the body multiplied by the radius of the circle on which the body is moving at the instant under consideration; directed toward the centre of the circle on which the body is moving at the instant under consideration.

cephalic index: Relationship between the breadth and length of the skull.

cephalo-caudal principle: Principle stating that development occurs in a "head-down" or "head-to-tail" manner.

cerebellum: A subdivision of the brain, lying below the cerebral cortex and behind the brainstem, that plays a major role in movement coordination.

cerebrum: The largest and most important region of the brain, consisting of the two cerebral hemispheres.

circuit training: Training in which the athlete moves around a circuit of different exercise stations (exercise machines or activities).

clavicle: Collarbone.

climbing fibre: A type of afferent nerve fibre within the cerebellum.

closed-chain exercises: Those exercises where an external force is applied to one or both feet during the exercise (e.g., the ground applies a force to both feet during a squatting exercise).

closed-loop control: A type of movement control in which the movement is controlled continuously on the basis of sensory feedback arising during the movement itself.

cognate discipline: A related discipline.

cognitive anxiety: The mental (cognitive) component of the situation-specific experience of apprehension.

cognitive science: The hybrid field of study between experimental psychology and computer science concerned with understanding the computational processes of the brain and nervous system.

collagen: A fibrous protein that is an important constituent of connective tissues such as bone, ligament, tendon, and cartilage.

compact bone: Dense bone that does not appear porous.

comparison: The perceptual process responsible for determining the relative strengths of two stimuli.

compass gait: The simplest form of walking in which all joints, with the exception of the hip joints, are locked in the anatomical position.

complex joint: Synovial joint containing an intra-articular structure.

compliance: The ability of a material to store energy (as strain energy) and then return it to the object that initially possessed the energy.

compound joint: Synovial joint involving more than one pair of articulating surfaces.

compression: A type of force in which the two ends of a structure are squeezed together.

compressive strength: Strength in opposing breaking when a material is compacted.

computer-aided tomography (CAT): Also referred to as computed tomography (CT); production of images of sections through the body using an X-ray source.

computer simulation: Solution of a set of mathematical (differential) equations describing the behaviour of a physical system. Complex simulations must be performed on a computer. Computer simulations of human movement are usually performed with time as the independent variable.

concentric: Referring to a dynamic muscle contraction in which the muscle shortens while developing tension.

conservative: A system in which the total mechanical energy remains unchanged.

contemplation stage: The second stage of the transtheoretical model; in this stage individuals are seriously thinking about changing their behaviour in the next six months.

contralateral: On the opposite side of the body.

control parameters: Parameters (such as movement frequency) that, when manipulated to a critical value, trigger a transition from one pattern of organization to another.

coronal plane: A plane dividing the body into front and back.

corpus callosum: The thick band of neural tissue connecting the two cerebral hemispheres.

cortical bone: See compact bone.

critical fluctuations: A term from synergetics to describe the increasing variability in a particular pattern of organization as the point of transition to a different pattern is approached.

critical period: A period during development when the organism (or a specific system or skill) is most readily influenced by both favourable and adverse environmental factors.

critical slowing down: A term from synergetics to describe the delayed response to unexpected perturbations that a particular pattern of organization shows as it approaches the point of transition to a different pattern.

cross-bridge cycling: The process by which myofilaments are pulled toward each other to produce a muscle contraction.

cross-bridge hypothesis: Proposed explanation of the mechanics of striated muscle contraction.

cross-disciplinary: Referring to an approach in which a problem is examined using the methods of more than one subdiscipline and with some integration of the information from the different subdisciplines.

crossed extensor reflex: A reflex that increases activation of the extensor muscles of the limb on the side opposite a limb undergoing flexion.

cross-sectional study: A study in which individuals of different ages are measured at about the same time.

cutaneous receptors: Receptors located in the skin.

cytokines: Molecules that play a key role in regulating the intensity and duration of the response of the immune system to stress and illness.

degrees of freedom: The number of independent variables (motor units, muscles, joint angles, etc.) that must be simultaneously controlled in order to produce purposeful movement.

delayed-onset muscle soreness (DOMS): Soreness characterized by tender and painful muscles that usually appears in the few days after unaccustomed exercise; DOMS most commonly occurs after exercise with a large eccentric component, such as downhill running.

dendrites: The branches of a neuron that synapse with, and receive nerve impulses from, other neurons.

dense bone: See compact bone.

density: Mass per unit volume of a material.

depersonalisation: A characteristic of burnout in which an individual develops a detached and disinterested attitude to others.

detection: The perceptual process responsible for determining the presence of particular stimuli.

displacement: The straight-line distance between the initial and final positions of a body.

dissection: Use of scalpels and other surgical instruments to reveal particular anatomical structures.

dissociating: Using attentional strategies designed to distract oneself from the fatigue-producing effects of exercise.

distance: Magnitude of the displacement of an object, along the path followed, between its initial and final positions.

dopamine: A neurotransmitter whose absence in parts of the basal ganglia gives rise to Parkinson's disease.

dorsal: Pertaining to the posterior side or back of the body.

dorsiflexion: Backward flexion or bending of a joint. Dorsiflexion of the foot draws the upper surface of the foot closer to the forward surface of the lower leg (the "shin").

Down syndrome: A chromosomal abnormality associated with delays in both mental and motor development.

drive theory: A proposed linear relationship between arousal and performance.

dual energy X-ray absorptiometry (DEXA): Determination of body composition using X-ray analysis.

dynamical models: Models that seek to explain the control of movement largely on the basis of the physical properties of the musculoskeletal system.

dynamics: Study of the relationships between forces and motion of (rigid) bodies.

dynamometer: An instrument or machine used to measure muscle strength.

dyslipidemia: A condition in which blood lipid (fat) levels are outside the range recommended for good health.

eccentric: Referring to a dynamic muscle contraction in which the muscle lengthens while developing tension.

ectoderm: Primary germ layer that gives rise to the outer skin and nervous system.

ectomorphy: One component of a somatotype, based on a person's height and mass.

efferent: Pertaining to motor (or output) information transmitted from the brain and central nervous system out to the muscles.

efficient: Referring to mechanical work done divided by metabolic energy consumed.

ego orientation: The tendency to define success as being better than others.

electroencephalography (EEG): The recording of electrical activity from the brain.

electromyography (EMG): The recording of the electrical signal produced by skeletal muscle as it contracts.

electron microscope: A microscope in which an electron beam passes through a very thin section of tissue.

excess postexercise oxygen consumption (EPOC) (formerly called oxygen debt): Oxygen consumption in excess of resting level during recovery after exercise.

embryo: Developing human during the first quarter of intrauterine life.

endochondral ossification: Development of bone from a cartilaginous model.

endoderm: Primary germ layer that gives rise to the organs of the body.

endomorphy: One component of a somatotype, based on the thicknesses of the skinfolds.

endurance exercise capacity: Performance measure such as the maximum time an individual can exercise at a given speed or the total amount of work that can be accomplished in a given time.

epiphyseal plate: Cartilaginous growth plate in developing long bone.

equilibrium: The state in which every point on the body has the same velocity.

excitation–contraction coupling: A series of steps between the excitation of a skeletal muscle by its nerve and the production of force.

exercise adherence motivation: The drive to maintain regular physical activity.

exercise and sport science: Term used to describe the field of study concerned with application of the methods of science to understanding of exercise and sport.

exercise-induced asthma (EIA): Bronchoconstriction (narrowing of breathing tubes) causing reduced ventilation during exercise in persons who have asthma.

exercise-induced bronchospasm (EIB): See exercise-induced asthma (EIA).

exercise participation motivation: The drive to initiate regular physical activity.

exercise physiology: The subdiscipline of human movement studies concerned with understanding physiological responses to exercise.

exercise self-efficacy: The perception of one's ability to succeed at an exercise program.

explicit learning: Learning that occurs consciously and deliberately with the concurrent acquisition of verbalizable knowledge.

extension: Joint motion in the sagittal plane that increases the angle between the limb segments.

extensor thrust reflex: A reflex extension of the legs in response to stimulation of the soles of the feet; assists in supporting the body's weight against gravity.

external force: A force that does work to move a body; only external forces are included in free-body diagrams.

extrafusal muscle fibres: The characteristic skeletal muscle fibres activated by alpha motor neurons; their contraction causes voluntary movement.

extrapyramidal tract: All the descending motor pathways from the brain to alpha motor neurons other than those direct connections contained within the pyramidal tract.

extrovert: A person whose basic orientation is toward the external world.

fibrous joint: A type of joint in which the material between the bones contains, or resembles, fibres.

First Law of Thermodynamics: Law describing the relationship between energy production, heat liberation, and the rate of work performed on an external load.

flexion: Joint motion in the sagittal plane that decreases the angle between the limb segments.

flexion reflex: A reflex that produces flexion of the joints that withdraw an injured limb away from a painful stimulus.

focal vision: That aspect of vision, deriving from the central retina (fovea) and the nerve pathway connecting the optic nerve to the visual cortex, that is responsible for object recognition and the resolution of fine detail.

foetus: Developing human during the last three-quarters of intrauterine life.

force: Measure of the effort applied.

fovea: The area near the centre of the retina in which cone cells are most concentrated and which therefore provides the most acute vision.

free-body diagram: Schematic diagram showing all the external forces acting on a system.

frontal plane: See coronal plane.

functional anatomy: The subdiscipline concerned with understanding the anatomical bases of human movement and the effects of physical activity on the musculoskeletal system.

gait: Term used interchangeably with walking in this text. One complete gait cycle is defined by the time between successive heel strikes of the same leg.

gamma motor neurons: Small nerve cells that innervate spindle (intrafusal) muscle fibres.

genotype: A genetically determined somatotype.

gerontology: The study of aging.

glia: Non-neuronal cells within the brain and spinal cord that help regulate the extracellular environment of the central nervous system through the provision of metabolic and immunological support for the nerve cells.

glial cells: See glia.

glycogen: Storage form of glucose in the cells, composed of polymers of glucose molecules.

Golgi tendon organs: Specialized receptors located within tendons that respond to tendon tension.

goniometer: An instrument used to measure range of movement at a joint.

gross anatomy: The study of structures that can be seen with the unaided eyes.

ground reaction force: The pattern of force exerted on the ground during locomotion. Usually separated into three components: fore–aft, vertical, and medial–lateral.

gustatory information: Information derived through the sense of taste.

hamstring muscles: The muscle group at the back of the thigh.

Haversian canals: The narrow system of blood vessels that supply oxygen and micronutrients to the bone cells (osteocytes).

Haversian system: The basic unit of compact bone, consisting of a canal containing blood vessels surrounded by layers of bone material.

Heath-Carter Anthropometric Somatotype: An example of a somatotyping technique based on anthropometric measurements.

hemiparesis: Muscular weakness or paralysis affecting one side of the body.

hemiplegia: Paralysis of one side of the body.

hippocampus: A portion of the brain believed to play a central role in memory and learning.

homeostasis: Maintenance of metabolic equilibrium within an organism.

human movement science: A term used to describe the field of study concerned with the application of the methods of science to the understanding of human movement.

human movement studies: The field of study concerned with understanding how and why people move and the factors that limit and enhance the capacity to move.

Huntington's disease: A degenerative, inherited disease affecting neurotransmission within the basal ganglia and characterized by rapid, involuntary limb movements.

hyperplasia: Increase in the number of cells forming a tissue.

hypertrophic phase: In weight training, the second phase of increase in muscle size.

hypertrophy: An increase in the size of each cell forming a tissue.

imagery: A mental skill involving the use of all senses to create or recreate an experience in the mind.

immediate energy system: Metabolic energy system in which phosphocreatine is split to provide energy for ATP resynthesis; provides an immediate source of energy at the onset of exercise.

implicit learning: Learning that occurs without conscious awareness and without the concurrent acquisition of verbalizable knowledge.

information-processing model: A model that considers the human nervous system as a sophisticated processor of information, like a computer.

interaction framework: An approach that views behaviour as a function of the interaction of personality and environmental factors.

interdisciplinary: Referring to an approach in which a problem is examined using the methods of more than one subdiscipline and with tight integration of the information from the different subdisciplines.

internal force: A force that does no work to move a body; internal forces are not included in free-body diagrams.

interneurons: Nerve cells that connect one nerve cell to another.

interval training: Form of exercise training using alternating work and rest intervals.

intrafusal muscle fibres: The modified skeletal muscle fibres found within the muscle spindle receptors and activated by gamma motor neurons.

introvert: A person who tends to be hesitant, reflective, withdrawn, and reserved.

inverted U hypothesis: High performance occurs with an optimal level of arousal, and lesser performances occur with either low or very high arousal.

ipsilateral: On the same side of the body.

iso-inertial: A situation involving isotonic muscle contraction.

isokinetic: Referring to movement at constant speed (e.g., of a body joint).

isometric: Referring to contraction of a muscle that is being held at constant length. At steady state, the muscle develops a constant force over time.

isotonic muscle contraction: Contraction occurring while a constant force is applied to a muscle.

joint capsule: Thick connective tissue membrane forming the boundary of a synovial joint and enclosing the joint cavity.

joint: Union of two or more bones.

joule: The Standard International unit of kinetic energy. It is equal to the work done when the point of application of a force of 1 newton is moved through a distance of 1 metre in the direction of the force.

kinaesthetic: Pertaining to sensory information provided by the receptors located in the muscles, tendons, joints, and skin.

kinanthropometry: The scientific specialization dealing with the measurement of humans in a variety of morphological perspectives, its application to movement, and those factors that influence movement.

kinematics: That branch of mechanics concerned with the description of motion.

kinesiology: The scientific study of movement; term is often used more narrowly to refer to those subdisciplines concerned with understanding the anatomical and mechanical bases of human movement.

kinetic energy: Amount of mechanical energy a body possesses due to its motion.

Krause's end bulbs: Nerve ending receptors located in the skin.

lactate threshold: Point of inflection of the curve of blood lactate concentration versus exercise intensity, above which blood lactate concentration increases disproportionately with increasing exercise intensity; considered to identify an exercise intensity that can be maintained at the upper limit of aerobic metabolic capacity without major input from the anaerobic glycolytic system.

lactic acid: By-product of anaerobic glycolytic breakdown of glucose or glycogen; hydrogen ions dissociated from lactic acid increase acidity (decrease pH), causing fatigue in skeletal muscle.

lacuna: A small space in compact bone in which an osteocyte is situated.

lamella: A layer of bone material.

lamina: Plate-like structure forming part of the arch of a vertebra.

ligament: A dense regular connective tissue joining bone to bone.

light microscope: A microscope in which a beam of light passes through a thin stained section of tissue.

line of action of the force: The direction in which a force is applied.

Little League elbow: Injury to the elbow joint in children, caused by overuse from overarm throwing; so named because of its prevalence in junior and adolescent baseball pitchers.

locomotor reflexes: Those reflexes present at birth or soon thereafter that are the primitive precursors to locomotion.

longitudinal study: A study in which the same individuals are measured over a period of years.

long-loop reflexes: A collective term for reflexes involving multiple synapses in which nerve impulses are passed from the local level of the spinal cord to higher levels of the cord and perhaps even to regions of the brain.

long-term potentiation: The prolonged increase in the efficiency of a synapse that occurs as a result of repeated stimulation; thought to be a fundamental neurophysiological mechanism for memory and learning.

macroscopic anatomy: See gross anatomy.

magnetic resonance imaging (MRI): Production of images of sections through the body using a very strong magnetic field.

magnetoencephalography (MEG): The recording of magnetic signals proportional to the electroencephalographic (EEG) waves resulting from the electrical activity of the brain.

maintenance heat: Heat produced by a muscle during a steady-state contraction.

maintenance stage: The fifth stage of the transtheoretical model; in this stage individuals have successfully changed their behaviour for at least six months.

marrow: Tissue within the shafts of long bones and in the spaces within spongy bone.

maturation: Sequence of changes occurring between birth and maturity.

maximum heart rate: Highest possible heart rate, usually achieved during maximal exercise, estimated by the equation 220 minus age.

mechanics: Study of forces and their effects; often divided into statics and dynamics.

medulla: Major anatomical component of the brainstem.

Meissner's corpuscles: Specialized receptors found primarily on the hairless surfaces of skin that respond to light pressure or touch.

menarche: Onset of menstruation.

meniscus: An intra-articular cartilaginous structure that is shaped like both a crescent and a wedge.

Merkel's discs (or corpuscles): Specialized receptors found primarily on the hairless surfaces of skin that respond to light pressure or touch.

mesoderm: Primary germ layer that gives rise to the structures of the musculoskeletal system.

mesomorphy: One component of a somatotype, based on the development of the musculoskeletal system.

metabolic syndrome: A cluster of four conditions (obesity, hypertension, type 2 diabetes, and abnormal blood lipid profile) that greatly increase an individual's risk of cardiovascular disease.

microscopic anatomy: The study of structures that can be seen in thin sections of tissue through use of a microscope.

minute ventilation: Volume of air inspired by the lungs per minute.

model: Set of mathematical equations used to describe the behaviour of a physical system. The equations are usually solved using a computer.

moment (of a force): Measure of the turning effect of the force about some point in space.

moment (of inertia): A body's resistance to rotation; mathematically, defined as the net moment applied to a body divided by the angular acceleration of the body.

monoarticular: Referring to a skeletal muscle that crosses a single synovial joint.

mossy fibre: A type of afferent nerve fibre within the cerebellum.

motivation: The drive, interest, and desire to undertake and/or achieve a particular goal. The directions and intensity of effort over time.

motor control: That subdiscipline of human movement studies concerned with understanding the processes underlying the acquisition, performance, and retention of motor skills.

motor cortex: A strip of cerebral cortex immediately anterior of the central sulcus that is responsible for the relay of many of the motor commands from the brain to the muscles.

motor development: The field of study concerned with understanding the changes in motor control that occur throughout the life span.

motor equivalence: The capability of the motor system to produce the same movement outcome in a variety of ways.

motor learning: The field of study concerned with understanding the changes in motor control that occur in response to practice (also known as skill acquisition).

motor milestones: Identifiable stages in the development of specific fundamental motor skills.

motor program: A set of motor (efferent) commands, set up in advance of commencement of a movement, that provide the potential for the movement to be completed without continuous reliance on sensory feedback (see also open-loop control).

motor skills: Skills that require movement of the whole body, a limb, or a muscle in order to be successfully performed.

motor unit: Structure consisting of a single motor nerve fibre and all the muscle fibres it innervates.

movement time: The time elapsed between the initiation and completion of a movement.

multidisciplinary: Referring to an approach in which a problem is examined using the methods of more than one subdiscipline but with little integration of the information from the different subdisciplines.

multiple sclerosis: A disease caused by damage to the myelin sheath surrounding nerves, resulting in hardening of the nerve tissue and consequent tremor, loss of coordination, and occasionally paralysis.

muscle fibre: Muscle cell; humans have three types of skeletal muscle fibres that differ in their metabolic and physiological characteristics: type I (slow oxidative), type IIA (fast oxidative glycolytic), and type IIB (fast glycolytic).

muscle moment arm: The perpendicular distance between the line of action of a muscle and the axis of rotation of the joint.

muscle spindle: Specialized receptor, located within the intrafusal fibres of skeletal muscle, that responds to stretch.

muscle tone: The tension produced in skeletal muscles under resting conditions as a result of low-frequency discharge of the alpha motor neurons.

muscular endurance: Ability of a muscle to repeatedly develop and maintain submaximal force over time.

muscular power: Muscle force multiplied by contraction velocity. A muscle develops power when it undergoes a shortening contraction, and it absorbs power during a lengthening contraction.

muscular strength: The peak isometric force that a muscle can develop. Muscles usually develop peak isometric force near their resting length. The maximum strength of a muscle is directly proportional to its cross-sectional area.

musculoskeletal system: The system consisting of bones, joints, and muscles.

myelin: The fatty insulating material covering the axons of many neurons and responsible for increasing conduction velocity within those neurons.

myoblast: Muscle-forming cell.

myofibrils: Longitudinal bundles of thick and thin contractile filaments located within muscle fibres.

myology: Study of the muscular system.

myosin: Thick protein filaments of muscle.

myotatic reflex: The simple monosynaptic "muscle-stretching" reflex in which the excitation of a muscle spindle receptor by the imposition of stretch to the muscle causes that muscle to contract.

neuromuscular training: Exercise program designed to enhance postural equilibrium and intermuscular control.

neuron: A nerve cell; the fundamental building block of the nervous system.

neurotransmitter: A chemical agent released by one nerve cell that acts upon another nerve or muscle cell by altering its electrical activity or state.

neurotrophic phase: In weight training, the initial phase of motor learning.

newton: The Standard International unit of force. It is the amount of force that imparts an acceleration of 1 metre per second on a mass of 1 kilogram.

newton·metre: The Standard International unit for measurement of the moment of a force.

norepinephrine: A hormone released in response to physical stress that causes a host of physiological responses, including increased heart rate and blood pressure; can also act as a neurotransmitter.

olfactory information: Information derived through the sense of smell.

onset of blood lactic acid accumulation (OBLA): See lactate threshold.

open-loop control: A type of movement control in which the efferent (motor) commands are preplanned before the movement is initiated and control is not dependent on any sensory feedback arising during the movement itself (see also motor program).

optimal level of arousal: The level of arousal at which performance is best; varies between sports and between athletes.

optimisation: A technique used to determine the best way of doing a task; usually used to determine optimal muscle coordination for a given task.

osteoarthritis: Degeneration of articular cartilage.

osteoblast: Bone-forming cell.

osteoclast: Bone-eroding cell.

osteocyte: Bone cell.

osteology: Study of the skeletal system.

osteon: See Haversian system.

osteopenia: Reduced bone density.

osteoporosis: Reduced bone density below a certain level, likely to be associated with fractures.

otolith organs: Set of two specialized receptors (the utricle and the saccule), located in the inner ear, in which the movement of calcium carbonate "stones" (the otoliths) against nerve endings signals horizontal and vertical linear acceleration of the head.

overtraining syndrome (OTS): See training stress syndrome.

overtraining: The second phase of training stress syndrome in which there is repeated failure to cope with chronic training stress.

ovum: Female sex cell, containing half the normal number of chromosomes.

oxygen debt: See excess postexercise oxygen consumption (EPOC).

oxygen deficit: Difference between the amount of energy required during exercise and the amount that can be supplied by oxidative metabolism; the energy requirements that cannot be met oxidatively must be supplied by the anaerobic energy systems (immediate and glycolytic systems).

Pacinian corpuscles: Specialized plate-like nerve endings located deep in the skin that respond to pressure, deep compression, or high-frequency vibration.

palpation: Using the senses of touch and pressure to identify anatomical features on a living human.

palsy: A persisting movement disorder due to brain damage acquired around the time of birth.

Parkinson's disease: A disease characterized by tremor, rigidity, and a delay in the initiation of movement caused by a deficiency in the production of the neurotransmitter dopamine within regions of the basal ganglia.

particle: An object that is infinitesimal in size but whose mass can be represented as being located at one point.

peak height velocity (PHV): Point of most rapid growth in height, a major landmark in pubertal growth.

perception: The set of processes through which a person interprets the current and future state of his or her internal and external environments.

perceptual-motor integration: The integration of perceptual and motor information in movement control so as to produce movements tightly tuned to environmental demands.

peripheral nervous system: The collection of all the nerve fibres that connect the receptors and the effectors to the brain and spinal cord.

peripheral neuropathy: Disturbances to the function or structure of the peripheral nervous system.

personality: The composite of the characteristic individual differences that make each person unique.

phenotype: A somatotype determined by both genetic and environmental factors.

phosphocreatine (PCr): "High-energy" phosphate molecule stored in muscle cells; cleavage of the terminal phosphate bond yields energy used to rephosphorylate ADP (adenosine diphosphate) to ATP at the onset of exercise.

phosphorus: A mineral stored in bone.

physical activity: According to the U.S. Surgeon General, bodily movement produced by contraction of skeletal muscle that increases energy expenditure above the basal level.

physical education: The profession concerned with education of people in, through, and about physical activity.

plasticity: Flexibility or adaptability of structure or function; often used specifically in reference to the ability of early embryonic cells to alter in structure or function to suit the surrounding environment.

polyarticular: Referring to a skeletal muscle that crosses more than two synovial joints.

pons: Major anatomical component of the brainstem.

positron emission tomography (PET): Production of images of sections through the body using positively charged electrons as the energy source.

postnatal: After birth.

postural reflexes: Those reflexes present in the first year of life responsible for keeping the head upright and the body correctly oriented with respect to gravity.

potential energy: Amount of mechanical energy a body possesses due to a change in its position; results from the gravitational acceleration experienced by the body as it moves from one position to another near the surface of the earth.

power: The time rate of doing work.

precontemplation stage: The first stage of the transtheoretical model; in this stage individuals have no intention of changing behaviour.

premotor cortex: Area of the brain situated just forward of the primary motor cortex.

prenatal: Before birth.

preparation stage: The third stage of the transtheoretical model; in this stage individuals have made some steps toward changing their behaviour.

primitive reflexes: Those reflexes present at birth that function predominantly for protection and survival.

principle of work and energy: A fundamental law of mechanics, that states that the work done on an object is equal to the change in its kinetic energy.

Profile of Mood States (POMS): A psychological inventory used to monitor individual differences in six mood state components.

progressive muscular relaxation (PMR): A relaxation technique involving the systematic contraction and relaxation of specific muscle groups throughout the body.

pronation: Movement that brings the hand or foot into the prone (palm or sole facing downward) position. Hand pronation is achieved by medial (inward) rotation of the forearm.

proprioception: The sense of the position of the body and the movement of the body and its limbs.

proximo-distal principle: Principle stating that development of control over muscles close to the body (attached to the axial skeleton) occurs before development of control over muscles of the limbs (attached to the appendicular skeleton).

psychoneuromuscular theory: A theory that states imagery aids skill learning because when we image ourselves moving our brains send electrical signals to our muscles in the same order as when we physically move.

puberty: The period of rapid physical growth between childhood and adulthood.

Purkinje cells: Large branching neurons within the cerebellum.

pyramidal cells: Nerve cells within the pyramidal tract (see pyramidal tract), so named because of their pyramid-like shape; they are broad at the top with a large branching tree of dendrites funnelling down to a single, slender axon.

pyramidal tract: The major descending motor pathway consisting of nerve cells that have their origins in the cerebral cortex and that synapse directly with motor neurons at the spinal cord level.

Quetelet index (QI): Ratio of body mass in kg to height in m^2, used for recommending appropriate body mass; also known as body mass index (BMI).

radiographic: Approaches involving the making of film records (radiographs) of internal body structures using X rays or gamma rays.

radiological: Pertaining to the use of X rays to visualize structures within the body.

random practice: A type of practice in which, in contrast to blocked practice, component skills are practiced together in an unstructured order; one trial of practice on any particular component skill can be followed, in random order, by a trial of practice on any other (or the same) skill component.

reaction time: The time elapsed between the presentation of a stimulus and the initiation of a response to that stimulus.

reciprocal inhibition: The inhibition of one motor neuron or nerve pathway by the simultaneous excitation of another having an opposing action; for example, the inhibition of extensor motor neurons during the activation of flexor motor neurons.

recognition: The perceptual process responsible for the identification of stimulus patterns.

Recovery-Stress Questionnaire for Athletes: A sport-specific questionnaire designed to measure the frequency of current stress and recovery-associated activities.

reflex: An involuntary movement elicited by a specific stimulus and typically completed rapidly and without conscious thought.

reliability: The extent to which measurements can be repeated consistently.

repetition maximum (RM): Maximum weight that can be lifted a specified number of times; for example, 3RM = maximal weight that can be lifted only three times.

resistance training: Exercise program designed to build muscular strength.

response time: The time elapsed between the presentation of a stimulus and the completion of a response to that stimulus (that is, the sum of reaction time plus movement time).

resting heat: The heat released by a muscle during its resting state.

resultant force: Vector sum of all forces acting (on a body).

reticular formation: A loosely defined network of cells extending from the upper part of the spinal cord through the medulla and pons to the brainstem, responsible for the control and regulation of attention, alertness, and arousal.

rheumatoid arthritis: Inflammation of synovial membrane.

righting reflexes: Reflexes designed to return the body to an upright position after any event causing a loss of balance.

rigid body: Idealized model of an object that does not deform or change its shape; that is, the distance between every pair of points on the body remains constant.

Ruffini corpuscles: Specialized plate-like nerve endings located in the skin that respond to pressure or touch.

saccule: The otolith organ responsible for the detection of vertical linear acceleration of the head.

sagittal plane: A plane dividing the body into left and right sides.

sarcomere: The structural and functional unit of skeletal muscle, containing thick and thin filaments.

scalar: Any physical quantity that is fully described by a number, for example, length, mass, and moment of inertia.

secular trend: Changes in physical dimensions between people of one generation and those of subsequent generations.

selective attention: The perceptual process of attending to one signal or event in preference to all others.

self-efficacy: The confidence one has in one's ability to successfully perform a particular behaviour.

semicircular canals: Set of three specialized receptors, located within the inner ear and positioned at 90° to each other, that respond to angular acceleration of the body in three different planes.

serial order: The challenge to the nervous system to structure efferent commands such that motor units are recruited, and movement elements produced, in the correct sequence.

sexual dimorphism: Structural differences between males and females.

shear: Describing a type of force in which adjacent parts of a structure move parallel to each other in opposite directions.

shortening heat: The difference between heat liberation by a muscle when it shortens and the same muscle when it contracts isometrically.

simple joint: Synovial joint involving a single pair of articulating surfaces.

size principle: The principle by which motor units are recruited in order of their size, from smallest to largest.

skeletal muscle: Striated muscle forming the major muscles of the trunk and limbs.

skill acquisition: The process through which skills are learned.

skinfold calipers: An instrument used to measure the thickness of a skinfold.

sliding filament hypothesis: Proposed explanation of the mechanics of striated muscle contraction.

smooth muscle: Muscle tissue forming the walls of organs such as blood vessels.

somatic anxiety: Physiological responses to perceived threat.

somatochart: A special diagram indicating the three components of a person's somatotype.

somatotype: A shorthand way of representing a person's shape and composition.

spasticity: An increase in the tone (resting tension) of certain muscle groups involved in the maintenance of posture; thought to arise from damage to extrapyramidal motor fibres.

specific gravity: The density of a material compared with the density of water.

speed: Magnitude of velocity.

spermatozoan: Male sex cell, containing half the normal number of chromosomes.

spinal nerve: One of the 31 pairs of nerves arising from the spinal cord.

spongy bone: Less dense bone that appears to be porous.

sport and exercise psychology: The scientific study of human behaviour and cognition in sport and physical activity.

sport devaluation: A characteristic of burnout in sport in which athletes lose interest in their sport and their performance within it.

stadiometer: Instrument used to measure height.

staleness: The initial phase of training stress syndrome in which the person fails to cope with the psychological and physiological stress of training.

state anxiety: The experience of apprehension at a particular point in time.

states: The way individuals feel at a particular point in time; transitory feelings.

static equilibrium: The state in which a body has zero velocity *and* acceleration. A body is in equilibrium if and only if the sum of all forces *and* the sum of all moments acting on the body are zero.

statics: Study of objects in equilibrium.

stature: Total body height.

step length: Distance between the left heel and the right heel when both legs are on the ground during walking.

step width: Pelvic width divided by ankle spread, where ankle spread is the distance between the two ankles when both legs are in contact with the ground.

stepwise multiple regression: A statistical procedure in which one dependent variable (e.g., body density) is determined from a number of predictor variables (e.g., skinfold thicknesses from different sites). The predictor variables are added in (or subtracted) one at a time (i.e., stepwise) to obtain the best possible prediction.

sternum: Breastbone.

stiffness: A change in force applied to a material divided by deformation of the material.

stimulus–response compatibility: A measure of the naturalness of the links between a particular signal (stimulus) and its associated (movement) response.

strain energy: Mechanical energy stored in the elastic tissues of muscle and tendon.

strength: The capacity to produce force against an external resistance.

stress: Force applied per unit area. Muscle stress, for example, is muscle force divided by muscle cross-sectional area.

stretch–shortening cycles: Cycles of activity in which muscles are placed on stretch prior to contracting concentrically (i.e., shortening).

stroke volume (SV): Volume of blood pumped by the heart with each beat.

subchondral bone: The thin layer of compact bone under the articular cartilage.

subcutaneous: Beneath the skin.

supplementary motor area: Area of the cerebrum, located forward of the motor cortex, involved in the production and control of skilled movement.

symbolic learning theory: A theory that states imagery helps to develop a blueprint of a movement sequence without actually sending any messages to the muscles.

synapse: The junction between two neurons or nerve cells.

syndrome X: See metabolic syndrome.

synergetics: The study of complex pattern formation and transitions between patterns.

synovial fluid: A type of filtrate of blood formed by the synovial membrane and contained within a synovial joint cavity.

synovial joint: A type of joint containing fluid within a cavity surrounded by a capsule.

synovial membrane: Thin membrane lining the inner surface of the joint capsule.

tangential acceleration: Component of acceleration of a point on a body given by the angular acceleration of the body multiplied by the radius of the circle on which the body is moving at the instant under consideration; directed at tangent to the circle on which the body is moving at the instant under consideration.

task orientation: The tendency to have a self-referenced definition of success.

tendon: A dense regular connective tissue joining muscle to bone.

tensile strength: Strength in opposing breaking when a material is stretched.

tension: A type of force in which the two ends of a structure are pulled apart.

thalamus: A mass of gray matter near the base of the cerebrum.

thoracic index: Relationship between breadth and front-to-back diameter of the chest.

tonic reflexes: Those reflexes that are the basis for posture, concerned with the maintenance of the position of one body part with respect to others.

torque: See moment (of a force).

torsion: A combination of compressive, tensile, and shear forces that results in twisting of a structure.

total mechanical energy: The sum total of the kinetic energy and gravitational potential energy of a rigid body.

trabeculae: Small bony rods that form the framework of spongy bone.

trabecular bone: See spongy bone.

tract: A large bundle of nerve fibres within the central nervous system.

training stress syndrome: The psychological and physiological negative adaptation to training characterized by the three phases of staleness, overtraining, and burnout.

trait anxiety: A predisposition to perceive situations as stressful or threatening.

traits: The stable and enduring predispositions each individual has to act in a certain way across a number of different situations.

transcranial magnetic stimulation (TMS): A technique for examining brain function in which a magnetic field is applied transiently to specific areas of the brain. Depending on the frequency and intensity of stimulation used, TMS can either act to stimulate the neurons in a particular area or briefly disrupt processing at the site of stimulation.

transdisciplinary: See interdisciplinary.

transtheoretical model: A model suggesting that people go through a series of five stages when changing behaviour.

transverse plane: A plane through the body parallel to the floor.

turning effect: The rotation tendency arising from the application of a force.

type 2 diabetes: Mature-onset diabetes, characterized by insulin resistance and impaired glucose tolerance and related to obesity and inactivity.

type Ia afferents: Nerves providing sensory information from the noncontractile central portion of the muscle spindle receptors back to the spinal cord.

type Ib afferents: Nerves providing sensory information from the Golgi tendon receptors back to the spinal cord.

type II afferents: Nerves providing sensory information from the contractile end portion of the muscle spindle receptors back to the spinal cord.

U.S. Surgeon General's report: A landmark report, published in 1996, that summarized research to date showing a strong relationship between moderate physical activity and health and that called for further promotion of a physically active lifestyle for all Americans.

ultrasound (US) scan: Production of images of sections through the body using an ultrasound probe; moving pictures can also be recorded in real time.

utricle: The otolith organ responsible for the detection of horizontal linear acceleration of the head.

validity: The extent to which a measurement instrument actually measures the property it sets out to measure; generally established through cor-

relation of the measure with an outside criterion or independent measure.

vector: A physical quantity that is described by its magnitude *and* direction, for example, displacement, velocity, acceleration, angular displacement, angular velocity, angular acceleration, force, and moment.

velocity: The time rate of change of displacement.

ventral: Pertaining to the anterior or abdominal side of the body.

verbal–cognitive phase: The initial phase in the learning of a new skill, in which the emphasis is on conscious understanding of the task requirements.

vertebra: One of the bones forming the spine or vertebral column.

vertebrate: Animal with a spine or "backbone."

vestibular apparatus: Sensory system in the inner ear consisting of the semicircular canals and otolith organs and responding to angular and linear accelerations of the head.

viscosity: A fluid's resistance to flow.

visual acuity: A measure of the capacity of the visual system to resolve fine detail in a viewed object.

visual cortex: Region at the back of the cerebrum, involved in the processing of visual information.

$\dot{V}O_2max$: Maximal oxygen consumption or aerobic power, usually expressed as a volume of oxygen consumed per minute as measured in a progressive exercise test to volitional fatigue; high $\dot{V}O_2max$ values are generally indicative of high endurance exercise capacity.

waist-to-hip ratio: The ratio of measurement at the waist divided by measurement at the hips; a simple measure of overweight and obesity.

watts: The Standard International unit of measurement of power. It is equal to 1 joule per second.

Z disk: Boundary of sarcomere; site of attachment of thin filaments of skeletal muscle.

zygote: Single fertilized cell that results from union of a male and a female sex cell.

CREDITS

Figure 1.2 Adapted, by permission, from E.F. Zeigler, 1990, Don't forget the profession when choosing a name! In *The evolving undergraduate major,* edited by C.B. Corbin and H.M. Eckert (Champaign, IL: Human Kinetics), 67-77.

Figure 2.1a-b Adapted, by permission, from J. Watkins, 1999, *Structure and function of the musculoskeletal system* (Champaign, IL: Human Kinetics), 112, 113.

Figure 2.8 Adapted, by permission, from W.C. Whiting and R.F. Zernicke, 1998, *Biomechanics of musculoskeletal injury* (Champaign, IL: Human Kinetics), 61.

Figure 2.11 Reprinted, by permission, from R.S. Behnke, 2001, *Kinetic anatomy* (Champaign, IL: Human Kinetics), 13.

Figure 2.13 Reprinted, by permission, from J.H. Wilmore and D.L. Costill, 2004, *Physiology of Sport and Exercise,* 3rd ed. (Champaign, IL: Human Kinetics), 42.

Figure 3.2 Adapted, by permission, from D.A. Bailey, J.E.L. Carter, and R.L. Mirwald, 1982, "Somatotypes of Canadian men and women," *Human Biology* 54(4): 813-828.

Table 3.1 Data from J.P. Clarys, A.D. Martin, and D.T. Drinkwater, 1984, "Gross tissue weights in the human body by cadaver dissection," *Human Biology* 56: 459-473.

Figure 4.3a-b Adapted, by permission, from L. Twomey, J. Taylor, and B. Furniss, 1983, "Age changes in the bone density and structure of the lumbar vertebral column," *Journal of Anatomy* 136: 15-25; and L.T. Twomey and J.R. Taylor, 1987, "Age changes in lumbar vertebrae and intervertebral discs," *Clinical Orthopaedics and Related Research* 224: 97-104.

Figure 4.5 Redrawn, by permission, from P.S. Timiras, 1972, *Developmental physiology and aging* (New York: Macmillan), 284.

Figure 4.6 Adapted, by permission, from A.E. Dugdale, V. O'Hara, and G. May, 1983, "Changes in body size and fatness of Australian children 1911-1976," *Australian Paediatric Journal* 19: 14-17.

Figure 4.8 Adapted, by permission, from J.M. Court, M. Dunlop, M. Reynolds, J. Russell, and L. Griffiths, 1976, "Growth and development of fat in adolescent school children in Victoria. Part 1. Normal growth values and prevalence of obesity," *Australian Paediatric Journal* 12: 296-304.

Figure 5.1 Adapted, by permission, from P. Kannus, L. Jozsa, P. Renstrom, M. Järvinen, M. Kvist, M. Lehto, P. Oja, and I. Vuori, 1992, "The effects of training, immobilization and remobilization on musculoskeletal tissue. 1. Training and immobilization," *Scandinavian Journal of Medicine and Science in Sports* 2: 100-118.

Figure 5.3 Reprinted, by permission, from V.L. Katch, F.L. Katch, R. Moffatt, and M. Gittleson, 1980, "Muscular development and lean body weight in body builders and weight lifters," *Medicine and Science in Sports and Exercise* 12: 340-344; and F.A. Dolgener, T.C. Spasoff, and W.E. St. John, 1980, "Body build and body composition of high ability female dancers," *Research Quarterly for Exercise and Sport* 51: 599-607.

Figure 5.4 Data from J.E.L. Carter, S.P. Aubry, and D.A. Sleet, 1982, Somatotypes of Montreal Olympic athletes. In *Physical structure of Olympic athletes. Part I. The Montreal Olympic games Anthropological Project. Medicine and sport, vol. 16,* edited by J.E.L. Carter (Basel: S Karger), 53-80; and B.D. McLean and A.W. Parker, 1989, "An anthropometric analysis of elite Australian track cyclists," *Journal of Sports Sciences* 7: 247-255.

Figure 5.5 Adapted, by permission, from R.T. Withers, N.P. Craig, and K.I. Norton, 1986, "Somatotypes of South Australian male athletes," *Human Biology* 58: 337-356.

Figure 6.6 Reprinted, by permission, from F.E. Zajac and M.E. Gordon, 1989, "Determining muscle's force and action in multi-articular movement," *Exercise and Sport Sciences Reviews* 17: 187-230.

Figure 6.7b Free-body diagram courtesy of Dr. Kevin B. Shelburne. Copyright K.B. Shelburne.

Figure 6.9 Reprinted, by permission, from K.B. Shelburne and M.G. Pandy, 1997, "A musculoskeletal model of the knee for evaluating ligament forces during isometric contractions," *Journal of Biomechanics* 30: 163-176.

Figure 6.11 Reprinted, by permission, from F. Serpas, and T. Yanagawa, and M.G. Pandy, 2002, "Forward-dynamics simulation of anterior cruciate ligament forces developed during isokinetic dynamometry," *Computer Methods in Biomechanics and Biomedical Engineering* 5(1): 33-43.

Figure 6.12 Reprinted, by permission, from K.B. Shelburne and M.B. Pandy, 1997, "A musculoskeletal model of the knee for evaluating ligament forces during isometric contractions," *Journal of Biomechanics* 30: 163-176.

Figure 6.13 Reprinted, by permission, from M.G. Pandy and R.E. Barr, 2002, Biomechanics of the human musculoskeletal system. In *Standard handbook for biomedical engineering,* edited by M. Kutz (New York: McGraw-Hill).

Table 6.1 Data from D.A. Winter, 1990, *Biomechanics and motor control of human movement* (New York: Wiley).

Figure 7.5 Reprinted, by permission, from D.A. Winter, 1979, "A new definition of mechanical work done in human movement," *Journal of Applied Physiology: Respiratory, Environmental, and Exercise Physiology* 46: 79-83.

Figure 7.6 Reprinted, by permission, from F.E. Zajac, 1989, "Muscle and tendon: Properties, models, scaling, and application to biomechanics and motor control," *Critical Reviews in Biomedical Engineering* 17(4): 359-411.

Figure 7.7 Reprinted, by permission, from M.G. Pandy, F.E. Zajac, E. Sim, and W.S. Levine, 1990, "An optimal control model for maximum-height human jumping," *Journal of Biomechanics* 23: 1185-1198.

Figures 7.8, 7.9, 7.10, and 7.11 Reprinted, by permission, from M.G. Pandy and F.E. Zajac, 1991, "Optimal muscular coordination strategies for jumping," *Journal of Biomechanics* 24: 1-10.

Figure 7.12 Reprinted, by permission, from H.J. Ralston, 1976, Energetics of human walking. In *Neural control of locomotion,* edited by R.M. Herman et al. (New York: Plenum Press), 77-98.

Figures 7.13 and 7.14 Reprinted from R. Margaria, 1976, *Biomechanics and energetics of muscular exercise* (Oxford: Clarendon).

Figure 8.1 Reprinted, by permission, from C. Vogler and K.E. Bove, 1985, "Morphology of skel-etal muscle in children," *Archives of Pathology and Laboratory Medicine* 109: 238-242.

Figure 8.2 Reprinted, by permission, from J. Rose and J.G. Gamble, 1994, *Human walking,* 2nd ed. (Baltimore: Williams and Wilkins), 5.

Figures 8.3, 8.5, and 8.9 Reprinted, by permission, from M.W. Whittle, 1996, *Gait analysis: An introduction* (Oxford: Butterworth-Heinemann), 64, 65, 103.

Figure 8.4 Reprinted, by permission, from J.B. Saunders, V.T. Inman, and H.D. Eberhart, 1953, "The major determinants in normal and pathological gait," *Journal of Bone and Joint Surgery* 35A: 543.

Figure 8.6 Reprinted, by permission, from F.C. Anderson and M.G. Pandy, 2001, "Dynamic optimization of human walking," *Journal of Biomechanical Engineering* 123: 381-390.

Figure 8.7 Reprinted, by permission, from F.C. Anderson and M.G. Pandy, 2003, "Individual muscle contributions to support in normal walking," *Gait and Posture* 17(2): 159-169.

Figure 8.8 Reprinted, by permission, from D.H. Sutherland, R. Olshen, L. Cooper, and S.L.-Y. Woo, 1980, "The development of mature gait," *Journal of Bone and Joint Surgery* 62A: 336-352.

Figure 8.10 Reprinted, by permission, from P.E. Martin, D.E. Rothstein, and D.D. Larish, 1992, "Effects of age and physical activity status on the speed-aerobic demand relationship of walking," *Journal of Applied Physiology* 73: 200-206.

Figure 8.11 Reprinted, by permission, from J.O. Judge, M. Underwood, and T. Gennosa, 1993, "Exercise to improve gait velocity in older persons," *Archives of Physical Medicine and Rehabilitation* 74: 400-406.

Figure 9.1 Reprinted, by permission, from W.J. Kraemer, S.J. Fleck, and W.J. Evans, 1996, Strength and power training: Physiological mechanisms of adaptation. In *Exercise and sports sciences reviews,* edited by J.O. Holloszy (Baltimore: Williams and Wilkins), 363-397.

Figures 9.2 and 9.3 Reprinted, by permission, from T.A. McMahon, 1984, *Muscles, reflexes, and locomotion* (Princeton: Princeton University Press).

Figures 9.5 and 9.6 Reprinted, by permission, from M.G. Pandy and K.B. Shelburne, 1997, "Dependence of cruciate-ligament loading on muscle forces and external load," *Journal of Biomechanics* 30: 1015-1024.

Figure 9.7 Reprinted, by permission, from M.G. Pandy and K.B. Shelburne, 1998, "Theoretical analysis of ligament and extensor-mechanism function in the ACL-deficient knee," *Clinical Biomechanics* 13: 98-111.

Figure 9.8 Reprinted, by permission, from T. Yanagawa, K.B. Shelburne, F. Serpas, and M.G. Pandy, 2002, "Effect of hamstrings muscle action on stability of the ACL-deficient knee in isokinetic extension exercise," *Clinical Biomechanics* 17(9-10): 705-712.

Figure 9.9 Reprinted, by permission, from T.P. Andriacchi, R.N. Natarajan, and D.E. Hurwitz, 1997, Musculoskeletal dynamics, locomotion, and clinical applications. In *Basic orthopaedic biomechanics,* edited by V.C. Mow and W.C. Hayes (Philadelphia: Lippincott-Raven).

Figure 9.10 Adapted, by permission, from M.G. Pandy, 1990, An analytical framework for quantifying muscular action during human movement. In *Multiple muscle systems: Biomechanics and movement organization,* edited by J. Winters and S. Woo (New York: Springer-Verlag), 653-662.

Figure 9.11 Reprinted, by permission, from M.F. Bobbert, and A.J. van Soest, 1994, "Effects of muscle strengthening on vertical jump height: A simulation study," *Medicine and Science in Sports and Exercise* 26: 1012-1020.

Figure 10.7b Photo courtesy of Michael Hinchey, The University of Queensland.

Figure 10.12 Reprinted, by permission, from J.H. Wilmore and D.L. Costill, 1994, *Physiology of sport and exercise* (Champaign, IL: Human Kinetics), 37.

Figure 10.13 Data from W.D. McArdle, F.R. Katch, and V.L. Katch, 1996, *Exercise physiology: Energy, nutrition, and human performance,* 4th ed. (Baltimore: Williams & Wilkins); and J.H. Wilmore and D.L. Costill, 1999, *Physiology of sport and exercise,* 2nd ed. (Champaign: Human Kinetics).

Figure 10.14 Reprinted, by permission, from J.H. Wilmore and D.L. Costill, 1994, *Physiology of sport and exercise* (Champaign, IL: Human Kinetics), 352.

Table 10.4 Compiled from L. Burke and V. Deakin, eds., 1994, *Clinical sports nutrition* (New York: McGraw-Hill); K. Inge and P. Brukner, 1986, *Food for sport* (Richmond, NSW: William Heinemann Australia); W.D. McArdle, F.I. Katch, and W.L. Katch, 1996, *Exercise physiology: Energy, nutrition and human performance,* 4th ed. (Baltimore: Williams & Wilkins); and R. Stanton, 1988, *Eating for peak performance: Gaining your full potential at work and leisure* (Sydney: Allen and Unwin).

Summary of American College of Sports Medicine Recommendations for Exercise for Health Compiled from American College of Sports Medicine, 2000, *Guidelines for exercise testing and prescription,* 6th ed. (Philadelphia: Lippincott Williams & Wilkins).

Summary of U.S. Surgeon General's Recommendations for Physical Activity Adapted from U.S. Department of Health and Human Services, Centers for Disease Control and Prevention, 1996, *Physical activity and health: A report of the Surgeon General* (Atlanta, GA: National Center for Chronic Disease Prevention and Health Promotion).

Table 11.5 Compiled from B.S. Rushall and F.S. Pyke, 1990, *Training for fitness and sport* (Sydney: Macmillan).

Table 11.6 Compiled from J.H. Wilmore and D.L. Costill, 1999, *Physiology of sport and exercise* (Champaign, IL: Human Kinetics).

Figure 12.1 Reprinted, by permission, from J.H. Wilmore and D.L. Costill, 2004, *Physiology of sport and exercise,* 3rd ed. (Champaign, IL: Human Kinetics), 557.

Figure 12.2 Reprinted, by permission, from T.W. Rowland, 1990, *Exercise and children's health* (Champaign, IL: Human Kinetics), 56.

Figure 12.3 Adapted, by permission, from C. Saavedra, P. Lagasse, C. Bouchard, and J.-A. Simoneau, 1991, "Maximal anaerobic performance of the knee extensor muscles during growth," *Medicine and Science in Sports and Exercise* 23: 1085.

Summary of Guidelines for Resistance Training in Youth Compiled from American College of Sports Medicine, 2000, *ACSM's guidelines for exercise testing and prescription,* 6th ed. (Philadelphia: Lippincott Williams & Wilkins); Sports Medicine Australia, 1997, *Safety guidelines for children* (Canberra, Australia: Sports Medicine Australia).

Table 12.2 Compiled from C.L. Wells, 1992, *Women, sport and performance: A physiological perspective,* 2nd ed. (Champaign, Illinois: Human Kinetics); and W.D. McArdle, F.I. Katch, V.L. Katch, 1996, *Exercise physiology: Energy, nutrition and human performance,* 5th ed. (Baltimore: Williams & Wilkins).

Figure 17.1 Reprinted, by permission, from Neil R. Carlson, 1994, *Physiology of behavior,* 5th ed. (Needham Heights, MA: Allyn and Bacon), 436.

Figure 17.2 Reprinted, by permission, from A. Shumway-Cook and M.H. Woollacott, 2001, *Motor control: Theory and practical applications,* 2nd ed. (Baltimore: Lippincott Williams & Wilkins), 92.

Figure 17.3 Reprinted, by permission, from D.N. Spinelli, F.E. Jensen, and G.V. Di Prisco, 1980, "Early experience effect on dendritic branching in normally reared kittens," *Experimental Neurology* 62: 1-11.

Figure 17.5 Photo copyright by the National Sports Information Centre, Australian Sports Commission. Reprinted with permission.

Figure 17.6 Reprinted, by permission, from J.L. Starkes, 1987, "Skill in field hockey: The nature of the cognitive advantage," *Journal of Sport Psychology* 9: 152.

Figures 17.7 and 17.8 Reprinted, by permission, from B. Abernethy and D.G. Russell, 1987, "Expert-novice differences in an applied selective attention task," *Journal of Sport Psychology* 9(4): 331, 338.

Figure 17.9 Data from table 1 in J.V. Crosby and S.R. Parkinson, 1979, "A dual task investigation of pilots' skill level," *Ergonomics* 22: 1301-1313.

Figure 17.10 Reprinted, by permission, from S. Sakurai and T. Ohtsuki, 2000, "Muscle activity and accuracy of performance of the smash stroke in badminton with reference to skill and practice," *Journal of Sports Sciences* 18: 910.

Figure 17.11 Reprinted, by permission, from W.F. Helsen, J.L. Starkes, and N.J. Hodges, 1998, "Team sports and the theory of deliberate practice," *Journal of Sport and Exercise Psychology* 20: 21.

Figure 17.12 Reprinted, by permission, from E.R.F.W. Crossman, 1959, "A theory of the acquisition of speed skill," *Ergonomics* 2: 157.

Figure 17.13 Reprinted, by permission, from E.A. Bilodeau, I.M. Bilodeau, and D.A. Schumsky, 1959, "Some effects of introducing and withdrawing knowledge of results early and late in practice," *Journal of Experimental Psychology* 58: 143.

Figure 17.14 Reprinted, by permission, from J.B. Shea and R.L. Morgan, 1979, "Contextual interference effects on the acquisition, retention, and transfer of a motor skill," *Journal of Experimental Psychology: Human Learning and Memory* 5: 183.

Figure 20.1 Data from E. McAuley, B. Blissmer, D.X. Marquez, G.J. Jerome, A.F. Kramer, and J. Katula, 2000, "Social relations, physical activity, and well-being in older adults," *Preventive Medicine* 31: 608-617.

Figure 22.1 Adapted, by permission, from A.M. Williams and A. Franks, 1998, "Talent identification in soccer," *Sports Exercise and Injury* 4(4): 159-165.

Figure 22.2 Reprinted, by permission, from M.A. Moleiro and F.V. Cid. 2001. "Effects of biofeedback training on voluntary heart rate control during dynamic exercise," *Applied Psychophysiology and Biofeedback* 26(4): 279-292.

INDEX

Note: The italicized *f* and t following page numbers refer to figures and tables, respectively.

ABOUT THE AUTHORS

Bruce Abernethy, PhD, is professor and director of the Institute of Human Performance at the University of Hong Kong. Previously, he was professor and head of the School of Human Movement Studies at the University of Queensland in Brisbane, Australia. He is coeditor of *Creative Side of Experimentation.*

Dr. Abernethy earned his PhD from the University of Otago. He is an international fellow of the American Academy of Kinesiology and Physical Education, a fellow of Sports Medicine Australia, and a member of the Australian Association for Exercise and Sports Science.

Stephanie J. Hanrahan, PhD, is an associate professor in sport and exercise psychology at the University of Queensland. She earned her doctorate in sport psychology from the University of Western Australia in 1989 and is a fellow of Sports Medicine Australia. Dr. Hanrahan is the 1997 recipient of the University of Queensland's Excellence in Teaching Award. She is the coauthor of *The Coaching Process: A Practical Guide to Improving Your Effectiveness* and *Game Skills: A Fun Approach to Learning Sport Skills.*

Vaughan Kippers, PhD, is a senior lecturer in the School of Biomedical Sciences at the University of Queensland.

Laurel T. Mackinnon, PhD, is an associate professor in the School of Human Movement Studies at the University of Queensland. In 1992, Dr. Mackinnon authored *Advances in Exercise and Immunology,* the first book to explore the intriguing relationship between exercise and the immune system. In 2003 she coauthored *Exercise Management: Concepts and Professional Practice.* She has received grant funding for projects related to overtraining and immune function in athletes.

Dr. Mackinnon is a fellow of the American College of Sports Medicine. She is a former founding board member of the International Society of Exercise and Immunology (ISEI) and the Australian Association for Exercise and Sports Science. In 1997, she served as program chair for the international symposium of ISEI in Germany. She earned her PhD in exercise science at the University of Michigan.

Marcus G. Pandy, PhD, is currently a professor of engineering in the Department of Biomedical Engineering at The University of Texas at Austin. He is also an associate editor of the *Journal of Biomechanical Engineering,* an executive member of the International Society of Biomechanics Computer Simulation Technical Group, and a member of the Scientific Advisory Committee of the Steadman-Hawkins Sports Medicine Foundation.